MATERIALS SCIENCE

[A Textbook for Students of B.E./B. Tech. students of different Technical Universities & for A.M.I.E. Courses]

R.S. KHURMI
R.S. SEDHA

S. CHAND PUBLISHING

S. Chand And Company Limited
(ISO 9001 Certified Company)

S. Chand And Company Limited
(ISO 9001 Certified Company)

Head Office: D-92, Sector–2, Noida – 201301, U.P. (India), Ph. 91-120-4682700
Registered Office: A-27, 2nd Floor, Mohan Co-operative Industrial Estate, New Delhi – 110 044, Phone: 011-49731800
www.schandpublishing.com; e-mail: info@schandpublishing.com

Marketing Offices:

Chennai	:	Ph: 23632120; chennai@schandpublishing.com
Guwahati	:	Ph: 2738811, 2735640; guwahati@schandpublishing.com
Hyderabad	:	Ph: 40186018; hyderabad@schandpublishing.com
Jalandhar	:	Ph: 4645630; jalandhar@schandpublishing.com
Kolkata	:	Ph: 23357458, 23353914; kolkata@schandpublishing.com
Lucknow	:	Ph: 4003633; lucknow@schandpublishing.com
Mumbai	:	Ph: 25000297; mumbai@schandpublishing.com
Patna	:	Ph: 2260011; patna@schandpublishing.com

© S. Chand And Company Limited, 1987

All rights reserved. No part of this publication may be reproduced or copied in any material form (including photocopying or storing it in any medium in form of graphics, electronic or mechanical means and whether or not transient or incidental to some other use of this publication) without written permission of the copyright owner. Any breach of this will entail legal action and prosecution without further notice.

Jurisdiction: All disputes with respect to this publication shall be subject to the jurisdiction of the Courts, Tribunals and Forums of New Delhi, India only.

First Edition 1987
Subsequent Editions and Reprints 1989, 91, 92, 93, 94, 95, 96, 98, 99, 2000, 2001, 2003, 2004, 2005, 2007, 2008 (Twice), 2009, 2010, 2011, 2012, 2013, 2014 (Twice), 2015, 2017, 2018 (Twice), 2020, 2021, 2022, 2023

Reprint 2024

ISBN : 978-81-219-0146-8   **Product Code :** H3MSC62MATS10ENAE0XO

PRINTED IN INDIA

By Vikas Publishing House Private Limited, Plot 20/4, Site-IV, Industrial Area Sahibabad, Ghaziabad – 201 010 and Published by S Chand And Company Limited, A-27, 2nd Floor, Mohan Co-operative Industrial Estate, New Delhi – 110 044.

PREFACE TO THE FIFTH EDITION

No Course module in science or engineering may remain static. Not only does technology advance and scientific understanding increase, the academic framework (syllabus) undergoes changes. Thus, periodic revisions are desirable in an effort to optimize the value of a textbook for students who will be tomorrow's engineers.

Developments in the field of superconductivity has been exciting. Today we have a tunneling electron microscope (TEM) which can be used to gather scientific data of different materials to provide new insights. During the last two decades, the evolving structure of the academic environment has had a more direct impact on introducing material courses within engineering curricula. Whereas, academic environments will continue to have specialists in metals, polymers and ceramics, the trend has been towards departments of materials science and engineering.

The entire text of this book has been rewritten and arranged in a systematic manner. Lot of useful illustrations and photographs have been added to widen the scope and utility of this book. This book is suitable for a first course in engineering materials or materials science. To make the book more useful for A.M.I.E.(I) stream students, the complete solutions of their examination papers up to Winter, 2005 have been included. We hope that the book will continue to earn the appreciation of students for whom it has been written.

Any errors, omissions and suggestions for the improvement of this volume, brought to our notice, will be thankfully acknowledged and incorporated in the next edition.

R.S. Khurmi
R.S. Sedha

PREFACE TO THE FIRST EDITION

We take an opportunity to present 'Materials Science' to the students of A.M.I.E.(I) Diploma Stream in particular, and other engineering students in general. The object of this book is to present the subject matter in a most concise, compact, to the point and lucid manner.

While preparing the book, we have constantly kept in mind the requirements of A.M.I.E.(I) students, regarding the latest trend of their examination. To make it really useful for the A.M.I.E.(I) students, the solutions of their complete examination has been written in an easy style, with full details and illustrations. All along the approach to the subject matter, every care has been taken to arrange the matter from simpler to harder, known to unknown in a self-study style. In short, it is expected that the book will embrace the requirements of the students for which it has been designed.

Although every care has been taken to check mistakes and misprints, yet it is very difficult to claim perfection. Any errors, omissions and suggestions for the improvement of this book brought to our notice, will be thankfully acknowledged and incorporated in the next edition.

R.S. Khurmi
R.S. Sedha

Syllabus
MATERIAL SCIENCE
For A.M.I.E.

Structure of Atoms: Present concept. Rutherford and Bohr's models, Electronic configuration. Bonding — ionic and metallic. Crystal structure–space lattices, ionic and molecular crystals.

Mechanical Properties: Elastic and plastic behaviour of solids. Stress-strain relationships, yield stress, tensile strength, ductility. Ductile and brittle fracture, Variation of properties with temperature. Fatigue and creep: basic definitions and concepts.

Electronic Properties: Conductivity: behaviour of conductors, semi-conductors and insulators. Magnetism: Magnetization, dismagnetism, para-magnetism, ferro-magnetism. Magnetic energy. Dielectric materials. Metallic bindings, cohesive and repulsive forces in metals, electron energies in metals, the zone theory of solids, zones in conductors and insulators, factors affecting the electrical resistance of materials.

Thermal Properties: Characteristics of conductors, inculators, refractories.

Chemical Properties: Polymerization—deformation of plastics. Corrosion—various types, mechanism and control.

Deformation of Metals: Elastic and plastic deformation, slip and twinning dislocation theory, critical resolved shear stress, deformation in polycrystalline material, season cracking, Bauchinger effect and elastic after-effect, work hardening, recovery, re-crystallization and grain growth, preferred orientation, cold and hot working.

Heat Treatment: Annealing, normalizing, critical cooling rate, hardenability, factors affecting hardenability and quench test and determination of hardenability. Martensitic transformation, hardening, carbon and alloy steels: various types of carbon steels, their properties and uses. Effects of alloy additions and characteristics of alloying elements. Heat resisting alloys, tool and die steels, magnetic alloys, non-ferrous alloys, bearing alloys, metals for nuclear energy.

CONTENTS

Chapter	Pages
PART-I : SCIENCE OF METALS	
1. Introduction	1-4
2. Structure of Atoms	5-34
3. Crystal Structure	35-90
4. Bonds in Solids	91-105
5. Electron Theory of Metals	106-133
PART-II : MECHANICAL BEHAVIOUR OF METALS	
6. Mechanical Properties of Metals	134-142
7. Mechanical Tests of Metals	143-169
8. Deformation of Metals	170-196
9. Fracture of Metals	197-207
PART-III : ENGINEERING METALLURGY	
10. Iron-Carbon Alloy System	208-222
11. Heat Treatment	223-234
12. Corrosion of Metals	235-252
PART-IV : ENGINEERING MATERIALS	
13. Ferrous and Non-Ferrous Alloys	253-272
14. Organic Materials	273-293
15. Composite Materials and Ceramics	294-309
16. Semiconductors	310-327
17. Insulating Materials	328-341
18. Magnetic Materials	342-359
Appendix A	360
Appendix B	361
Index	362-367

1
Introduction

1. Definition 2. Beginning and Development of Materials Science 3. Sub-divisions of Materials Science 4. Science of Metals 5. Mechanical Behaviour of Metals 6. Engineering Metallurgy 7. Engineering Materials.

1.1 DEFINITION

The primary job of an engineer, in the present days of complex technology, is the designing, fabrication and smooth functioning of the machines and structures. The ever increasing demand of the available materials, coupled with new applications and requirements, has brought about many changes in the style of their uses. As a matter of fact, in every sphere of modern activities, the materials are used in one form or the other. A systematic and scientific study, selection, as well as use of various materials, is known as *Materials Science.*

1.2 BEGINNING AND DEVELOPMENT OF MATERIALS SCIENCE

In the dim ages of pre-historic time, the primitive man used to live in the Stone Age. About 5,000 years back, he learnt the art of smelting metal in kilns of heat resistant stones. Then, he started extracting and using metals for his stray jobs required for his livelihood. As the time passed, the man, entirely on the basis of his practical experience and common sense, created a few thumb rules to serve as a guide for using timber and metals for his various jobs.

It is only when the human progress has reached its evolutionary height, we call the dawn of civilization, the man actually started to design and fabricate his weapons, utensils, houses etc. according to his requirements. Since then, the man has been curious to know more and more about the various aspects of his requirements and the materials needed. He was equally interested to know the various properties of materials, specially regarding their uses and smooth workability. Experience taught him, and he continued to make progress in this artbit by bit.

As a matter of fact, the Industrial Revolution of Europe gave a new direction to the production of raw materials used in various industries. It gave impetus to the production of iron and steel, which were considered to be the real backbone of industries in those days. In this era, Britain became the leading producer of iron and steel. The scientific history of steel production started with Sir Henry Bessemer of England, who demonstrated that steel could be produced in 15 or 20 minutes, by simply blowing cold air through molten pig iron by means of steam driven turbo-blowers. Since then the production of steel rose steeply. As a result of this, steel and allied metals industries flourished all over the world.

The present day engineers are assigned the job of manufacturing complex and complicated products. In this process, they are always searching for new materials of greater strength, lightness, safety, reliability, cheapness, corrosion and heat resistant, etc. The tremendous increase in available materials, coupled with research in basic sciences, has led to the development of new era in the technology which is popularly known as Materials Science. In fact, the progress in one branch of science enriches other branches of the bordering sciences. Similarly, with the passage of time, physical sciences, mathematics, chemistry, metallurgy, crystallography, X-rays, electrical and electronics contributed a lot in the sphere of Materials Science. And its development gained momentum in the twentieth century.

It has been experienced that majority of modern failures are not due to faulty design or lack of proper supervision during their manufacture or construction. But they are mainly due to the selection or supply of improper materials. It is a problem which needs immediate attention of our technocrats. Our engineers are seldom qualified to carry out this work. Moreover, our materials managers cannot exercise proper control over the quality of materials, as their main job is to arrange sufficient quantity of materials businessmen, are always interested in profits and goodwill rather than supplying absolute perfect quality materials. Most of them even do not have proper research and development departments. A little consideration will show that this gap can only be filled by an engineer who has a considerable knowledge of Materials Science.

There is also a serious defect in the technical education in India, as there is no co-ordination between the educationists of engineering institutions and the requirements of the prospective employers. It has been seen that at the time of revising the syllabus, no effort is made to understand the needs and expectations of the prospective employers for the young engineers. These days, a few universities have included the subject of Materials Science in their curriculum. It is hoped that one day this subject will be taught to all engineers at all levels and also in all institutions.

1.3 SUB-DIVISIONS OF MATERIALS SCIENCE

The subject of Materials Science is very vast and unlimited. But if we see it from the syllabus point of view, we find that the subject matter can be sub-divided into the following four main heads:

1. Science of metals. 2. Mechanical behaviour of metals. 3. Engineering metallurgy.
4. Engineering materials.

Now we shall discuss the above mentioned sub-divisions of the subject in an extremely short form in this chapter to have a birds's eye view. However, we shall discuss these in detail in the appropriate chapters.

1.4 SCIENCE OF METALS

It includes the study of basic science, which is applied to the metals. The knowledge of this science provides a sound base for the detailed study of the various materials used by the engineers these days. It includes the following topics:

1. *Introduction.* This chapter deals with the introduction and importance of the subject to the engineering students. It also gives an idea of the subject to a beginner.
2. *Structure of atom.* This chapter deals with the necessary background of the structure of atom. It also deals with the electron structure of elements and modern periodic table. Though the study of this chapter is purely of scientific interest, yet its few applications are important from the subject point of view.
3. *Crystal structure.* This chapter deals with the crystal and metallic structures of various metals. It also deals with the arrangement of atoms in different types of structures. Though the study of this chapter is also purely of scientific interest, yet its few applications are important from the subject point of view.
4. *Bonds in solids.* This chapter deals with the different types of bonds between the atoms of molecules. As a matter of fact, the type and strength of a bond, in the atoms of the same or different metals has an important effect on the characteristics of the product.

Introduction

The study of bonds in solids is very important for an engineer in selecting suitable metals for his various jobs.

5. *Electron theory of metals.* This chapter deals with the different models and their applications for the study of conductors, insulators and semiconductors. The study of superconductivity and applications of superconductors is very useful for an engineer in selecting suitable metals for modern equipments.

1.5 MECHANICAL BEHAVIOUR OF METALS

It includes the study of various properties and mechanical tests of metals. The knowledge of these properties and tests is very essential for an engineer in selecting suitable materials for his components and structures. It includes the following topics:

1. *Mechanical properties of metals.* This chapter deals with all the mechanical and technological properties of a metal. A sound knowledge of these properties forms the basis of selecting suitable metals for various jobs of an engineer.

2. *Mechanical tests of metals.* This chapter deals with the important mechanical tests, which are carried out to ascertain the various properties of metal at the time of its selection. As a matter of fact, a thorough knowledge of mechanical tests is very important for an engineer at the time of selection of various metals with full confidence.

3. *Deformation of metals.* This chapter deals with the various types of deformations, which take place when a metal specimen is tested in a laboratory or the metal is put in actual use. It also deals with the cold working and hot working of metals along with their advantages. A thorough study of deformation also forms a sound basis for an engineer in selecting suitable metals for his various jobs.

4. *Fracture of metals.* This chapter deals with various types of fractures, which take place when a metal specimen is tested in a laboratory or it is put into actual use. It also deals with the non-destructive tests, which are carried out to ascertain the suitability of a metal. A thorough study of fracture also forms a sound basis for an engineer in selecting suitable metals for his various jobs.

1.6 ENGINEERING METALLURGY

It includes the study of metallurgy, which is of special interest to an engineer. A thorough knowledge of engineering metallurgy is very essential for an engineer at the time of taking final decision regarding the selection of suitable materials for his jobs. It also helps him in deciding the treatment processes and their sequence, which are to be carried out on the finished components and structures. It includes the following topics:

1. *Iron-carbon alloy system.* This chapter deals with the structure of iron and steel as well as iron-carbon equilibrium diagrams. It also deals with the transformation of alloys and steels under various sets of conditions. A thorough study of his topic helps an engineer to decide the suitability of process and the selection of iron alloy for various type of his jobs.

2. *Heat treatment.* This chapter deals with the objectives and processes of heat treatment. A thorough knowledge of heat treatment helps an engineer in deciding the type of process to be undertaken for the smooth and efficient working of the components and structures.

3. *Corrosion of metals.* This chapter deals with the colossal problem of corrosion of metals and its prevention. It also deals with the processes which help the present day engineers in improving the life and outward appearance of the metal components and structures.

1.7 ENGINEERING MATERIALS

It includes the study of basic engineering materials, which are used by the engineers working in various spheres of activities. In fact, a thorough knowledge of engineering materials, coupled with the knowledge of previous sub-divisions, is very essential for all types of engineers to select suitable materials for their day-to-day use. It includes the following topics:

1. *Ferrous and non-ferrous metals.* This chapter deals with properties and uses of different types of steels, cast iron and their alloys used in practice. It also deals with the properties and uses of metals like aluminium, copper, lead, tin, zinc, nickel, magnesium, cadmium, vanadium, antimony and then alloys. A thorough knowledge of this topic helps an engineer for selecting an alloy for a component or structure.

2. *Organic materials.* This chapter deals with the various characteristics of polymers, plastics, rubber, wood etc. It also includes the study of mechanical and electrical behaviour of polymers. A thorough study of this vast sphere of organic material is very essential for an engineer to select the suitable material for different types of his jobs.

3. *Composite materials and ceramics.* This chapter deals with the preliminary study of agglomerated and laminated materials. It also includes preliminary study of reinforced materials. The knowledge of composite materials and ceramics is very essential for an engineer to design the layout of his factories manufacturing such materials.

4. *Semiconductors.* This chapter deals with the study of semiconductors and transistors. It also includes the preliminary study of integrated circuits and semiconductor materials. A thorough study of semiconductors is very essential for an electronic or telecommunication engineer to select suitable materials for his components.

5. *Insulating materials.* This chapter deals with the dielectric properties of insulation materials. It also includes a brief study of important insulating materials. A thorough study of these materials is very essential for the engineers working in electronic and space research activities for selecting suitable materials for their projects.

6. *Magnetic materials.* This chapter deals with the various properties of magnetic materials. It also includes the classification of magnetism and magnetic materials. A thorough study of magnetic materials is also essential for the electrical and electronic engineers in selecting suitable materials for different type of their jobs.

2
Structure of Atoms

1. Introduction. 2. Fundamental Particles. 3. Atomic Number. 4. Atomic Weight. 5. (Isotopes. 6. Isobars. 7. Atomic Models. 8. Thomson's Atomic Model. 9. Rutherford's atomic Model. 10. Difficulties In Rutherford's' Atomic Model. 11. Bohr's Atomic Model. 12. Energy of an Electron in Bohr's Atomic Model. 13. Orbital Frequency of an Electron. 14. Deficiencies in Bohr's Atomic Model. 15. Quantum Numbers. 16. Sommerfeld's Atomic Model. 17. Modern Atomic Model. 18. Maximum Number of Electrons in an Atom. 19. Maximum Number of Electrons in the Main Shells. 20. Maximum Number of Electrons in the Sub-shells. 21. Energy Level Diagram. 22. Energy Level Diagram of a Hydrogen Atom. 23. Energy Level Diagram of a Multielectron Atom. 24. Electronic Configuration. 25. Electron Structure of Elements. 26. Wave Mechanics. 27. Schrodinger Theory. 28. Heisenberg's Uncertainty Principle. 29 Pauli's Exclusion Principle. 30. Periodic Table. 31. Mandeleev's Periodic Table. 32. Modern Periodic Table. 33. Significance of Atomic Number in the Periodic Table.

2.1 INTRODUCTION

In the earlier scientific literature, the atom was considered to be the smallest and chemically indivisible particle of matter. And the smallest quantity of a substance, which can exist freely by itself, in a chemically recognisable form, was known as molecule. It was only in 1803, when John Dalton, in his famous theory, gave the concept of atomicity. In this theory, he suggested that the molecules are composed of atoms of different elements in a fixed proportion. He discovered that the molecules may contain on atom known as monoatomic), two atoms (known as diatomic), three atoms (known as triatomic) or even more atoms (known as polyatomic).

In 1897, J.J. Thomason, while studying the passage of electricity through gases of low pressure, discovered that the gases ionise into positive and negative charges. The discovery of radioactivity showed the emission of small charged particles from radioactive elements. These and several other observations provided sufficient evidence to convince the scientists that the atom, though indivisible, yet consists of smaller particles known as fundamental particles.

2.2 FUNDAMENTAL PARTICLES

We have already discussed in the last article that an atom consists of a few fundamental particles. It has been observed that some of these fundamental particles are stable, whereas others are unstable. The following fundamental particles are important from the subject point of view:

1. *Electrons*. These are negatively charged particles. The mass of an electron is 9.1×10^{-31} kg. It is equal to 1/1836th of the mass of a hydrogen atom. Each electron possesses a unit negative charge of electricity, which is equal to 1.602×10^{-19} coulomb.
2. *Protons*. These are positively charged particles. The mass of a proton is 1.672×10^{-27} kg. Each proton possesses a unit positive charge of electricity, this charge is also equal to 1.602×10^{-19} coulomb.
3. *Neutrons*. These are electrically neutral particles. The mass of each neutron is 1.675×10^{-27} kg. This is approximately equal to the mass of a proton. Each neutron is composed of one proton and one electron.

Apart from the above mentioned three fundamental particles, there are other particles also, which are not of much importance from the subject point of view. It will be interesting to know that all the neutrons and protons together form the nucleus* of the atoms.

2.3 ATOMIC NUMBER

The atomic number of an element is numerically equal to the value of the positive charge on the nucleus of an atom or the number of protons present in the nucleus. Since an atom contains equal of protons and electrons, therefore atomic number of an element may also be defined as the number of electrons present outside the nucleus of its atom. It may be noted that all the atoms of the same element possess the same atomic number which identifies the element. Thus atomic number is a fundamental property of an atom and is generally denoted by the letter 'Z'.

2.4 ATOMIC WEIGHT

It is also called atomic mass or mass number. The atomic weight is numerically equal to the sum of protons and neutrons present in the nucleus.

Mathematically the atomic weight,
$$A = \text{No. of protons} + \text{No. of neutrons}$$

For example, chlorine atom has 17 protons and 18 neutrons in its nucleus. Therefore its atomic weight is 35.

2.5 ISOTOPES

The atoms of the same element, which possess different masses are called isotopes. The atomic number of isotopes of an element is the same, as this is the fundamental characteristic of an element. Since the atomic number of isotopes of an element is the same, therefore they must contain same number of protons and electrons. The difference in their masses is due to the different number of neutrons contained in the nuclei.

Consider a chlorine element. We know that its atomic number is 17. Chlorine has two isotopes with atomic weight (or masses) 35 and 37. Hence two isotopes of chlorine contain 17 protons and 17 electrons each. The chlorine atom with mass number 35 contains 17 protons and 18 neutrons in it. On the other hand, a chlorine atom with mass number 37 contains 17 protons and 20 neutrons in it. Both the atoms have similar electronic configuration and hence similar chemical properties.

Now consider a hydrogen element. We know that its atomic number is 1. Hydrogen exists in 3 isotopic forms as given below.
1. Ordinary hydrogen with atomic weight equal to 1.
2. Deuterium with atomic weight equal to 2.
3. Tritium with atomic weight equal to 3.

All the three isotopes of hydrogen contain 1 proton and 1 electron but different number of neutrons. Thus an ordinary hydrogen atom does not have a neutron, the deuterium atom has 1 neutron in its nucleus, while tritium has 2 neutrons in its nucleus.

2.6 ISOBARS

The atoms possessing different atomic numbers, but same atomic weight (or mass), are called isobars. For example, certain atoms of argon (atomic number = 18) and calcium (atomic number = 20) possess the same atomic weight of 40. Since isobars possess different atomic numbers, they

---

* It is a spherical body, which is positively charged due to the presence of protons. It occupies the central position in an atom. Its position and importance in an atom is somewhat, similar to that of the Sun in our Solar System.

must contain different number of protons and electrons in their atoms. However, their masses being the same, the total number of protons and neutrons in each of their nuclei will be same. Thus argon has 18 protons, 18 electrons and 22 neutrons in its atom. Similarly calcium has 20 protons, 20 electrons and 20 neutrons in its atom.

2.7 ATOMIC MODELS

As a matter of fact, no one had actually seen an atom in the past. But their existence was confirmed as matter, electricity, radiation etc. are all atomic in character. In order to study the physical structure of an atom, from its spectrum characteristics, several theories were put forward in the past. These theories are popularly known as atomic models. It will be interesting to know that these theories have greatly helped the scientists to obtain experimental data. The interpretation of this data has lead our scientists to completely understand the physical structure of an atom. Though there are many atomic models, yet the following are important from the subject point of view :

1. Thomson's atomic model. 2. Rutherford's atomic model. 3. Bohr's atomic model.
4. Sommerfeld's atomic model. 5. Modern atomic model.

Now we shall discuss all the above mentioned atomic models one by one in the following pages.

2.8 THOMSON'S ATOMIC MODEL

It was the first scientific explanation, which was proposed by J.J. Thomson (see picture below) in 1899. According to this model, an atom was supposed to be a sphere filled with positively charged mater of uniform density. In this sphere, just sufficient number of electrons are embedded to balance the positive charge as shown in Fig. 2.1.

The number of electrons in an atom was proportional to the atomic weight of an element. The Thomson's atomic model could not show completely the arrangement of electrons and protons in an atom. But from the observation, that the atom as a whole was stable, he argued that electron must be held together by the positive charges due to electrostatic forces of attraction.

JJ Thomson (father) – discovered electron was a particle (1856-1940) The Nobel Prize in Physics 1906.

In the initial stage, this model was considered to be a reliable scientific explanation. But later on, it was felt that the Thomson's atomic model cannot be accepted due to the following two reasons :

1. It could not explain the hydrogen spectrum completely.
2. No explanation was given for large angle scattering of α–particles.*

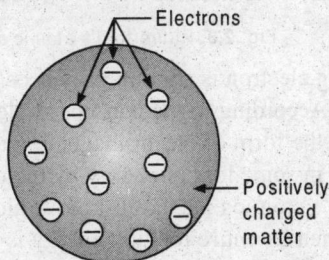

Fig. 2.1. Thomson's atomic model.

---

* An α-particle is a doubly charged helium nucleus. A helium atom like, other atoms, contain one nucleus which has two protons. Therefore, the total charge of one helium nucleus is double the charge of one proton. Thus an α-particle has a charge equal to that of one helium nucleus.

Today, the only importance of Thomson's atomic model in the scientific literature is that Thomson, by proposing this model, drew the attention of other scientists and pioneered the way of modern research in this sphere.

2.9 RUTHERFORD'S ATOMIC MODEL

t is an improved atomic model, which was proposed by Ernest Rutherford in 1911. This model is based on the experiment, which was conducted by Rutherford and his two associates Geiger and Marsden on the scattering of a- particle by thin metal foils.

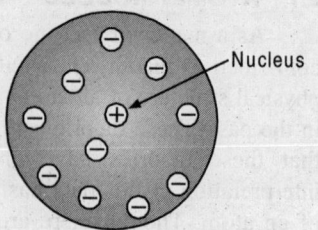

Fig. 2.2. Rutherford's atomic model.

According to this model, an atom is made up of a small positive nucleus with electrons around it in circular orbits as shown in Fig 2.2. The total negative charge of all the electrons is equal to the concentrated positive charge in the nucleus so that the atom, as a whole, is electrically neutral. Such a picture of an atom leaves most of the volume unoccupied.

2.10 DIFFICULTIES IN RUTHERFORD'S ATOMIC MODEL

We have already discussed in the last article that Rutherford suggested that positively charged nucleus is surrounded by negatively charged electrons. Such an arrangement offers some difficulties about the stability of the atom. Assuming that the electrons offers some electrostatic force of attraction, then the protons should make them to fall into the nucleus. Rutherford removed this difficulty by assuming that the electrons were not at rest. But they revolve around the nucleus in various orbits in the same way as the planets revolve around the Sun in our Solar System as shown in Fig. 2.3 (a).

The centrifugal force caused due to the rotation of the electrons is balanced by the electrostatic force of attraction between the electrons and the positive nucleus. This theory appeared to be very reasonable for the stability of the atom. But it introduced another difficulty as described below:

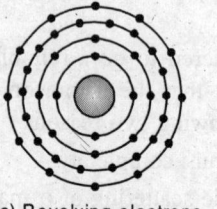 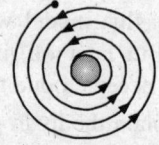

(a) Revolving electrons (b) Collapsing electrons

Fig. 2.3. Rutherford's atomic model.

We know that a revolving electron is continuously accelerated due to the change of direction at every instant of its motion. According to the concept of electromagnetic theory, an accelerating charge must radiate energy in the form of electromagnetic radiations. We know that the emission of these radiations must result in some loss of energy by the orbiting electrons. It is thus obvious, that an electron, in Rutherford's atomic model, must move along a spiral path of decreasing radius till it falls into the nucleus, when its entire rotation energy is spent, on emission, as shown in Fig. 2.3 (b).

It will be interesting to know that on the basis of Classical Mechanics, the planetary system of electrons revolving around the nucleus, as pictured by Rutherford, could last for an extremely short duration (i.e., 10×10^{-10} second). But this is in sharp contrast to the fact that the atoms show no tendency to collapse. Moreover, an electron moving inward along a spiral path, must continuously radiate light of decreasing wavelength (or of increasing frequency). But in actual practice, the spectral lines of fixed wavelength are always observed. As a result of this, we can say the Rutherford's' atomic model led to certain consequences, which were in contradiction to the experimental observations.

2.11 BOHR'S ATOMIC MODEL

It is an improved Rutherford's atomic model, which was proposed by Neils Bohr in 1913. In this atomic model, all the difficulties of the Rutherford's atomic model have been assumed to be non-existent. The modern atomic model is, somewhat similar to the Bohr's atomic model.

In this model, Neils Bohr retained the small positively charged nucleus in the centre. He also agreed that the negatively charged electrons revolve around the nucleus in circular orbits. However, Neils Bohr applied the Planck's Quantum Theory * to the revolving electrons to extend this model. He suggested the following postulates (i.e., simplifying assumptions) to explain the electron motion in an atom:

Niels Bohr David Bohr
(1885–1962)
Nobel Prize in Physics 1922

1. The electron continue to revolve in their respective orbits without loosing energy. According to this postulate, the energy of an electron remains constant, so long as it stays in the same orbit. This concept led to an idea that each orbit is associated with a definite energy. The orbits, therefore, are also known as energy levels or energy shells.

2. The smallest orbit, (i.e., first orbit from the nucleus), has the least energy, and farthest orbit, (i.e., first orbit from the nucleus) has the maximum energy.

3. The energy is emitted by an electron, when it moves from a higher energy level to a lower energy level; or in other words, from a farther orbit to an orbit nearer the nucleus. Similarly, the energy is absorbed by an electron, when it moves from a lower energy level to a higher energy level. The amount of energy emitted or absorbed is derived from the Planck's Quantum Theory and is given by the relation:

$$E_2 - E_1 = h \cdot f$$

Where
E_1 = Energy of initial orbit,
E_2 = Energy of the final orbit,
h = Planck's constant, whose value is 6.626×10^{-34} Joules-second, and
f = Frequency of radiation.

4. The angular momentum of an electron, revolving around the nucleus is given by the relation :

Angular momentum = $m \cdot v \cdot r$

where m = Mass of the revolving electron.
v = Velocity of the revolving electron, and
r = Radius of the orbit.

Bohr postulated that these orbits or energy levels are such that the angular momentum can have only definite or discrete values. It is given by the relation :

Angular momentum = $m \cdot v \cdot r = n \cdot h/2\pi$

---

* According to this theory, the difference of energy between any two orbits is directly proportional to the frequency of radiation.

where n = An integer i.e., 1, 2, 3....etc., representing the orbit, and
h = Planck's constant.

2.12 ENERGY OF AN ELECTRON IN BOHR'S ATOMIC MODEL

Consider an atom in which a single electron is revolving in some orbit around its nucleus as shown in Fig. 2.4.

Let e = Charge of the electron equal to 1.6×10^{-19} C,
M = Mass of the electron equal to 9.1×10^{-31} kg.
Z = No. of protons in the nucleus of the atom,
n = An integer representing the orbit in which the electron is revolving.
r_n = Radius of the orbit in which the electron is revolving,
v_n = Tangential velocity of the electron,
h = Planck's constant equal to 6.626×10^{-34} J-s, and
ε_0 = Absolute permittivity equal to 8.854×10^{-12} F/m.

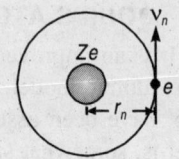

Fig. 2.4. Energy of an electron.

As per Coulomb's Law of Forces, between two charges, the force of attraction between the nucleus and electron,

$$F_e = \frac{Z.e^2}{4\pi\varepsilon_0.r_n^2} \qquad \ldots (i)$$

and the centrifugal force acting on the electron due to its motion along the circular path,

$$F_e = \frac{m.v_n^2}{r_n} \qquad \ldots (ii)$$

Now according to Bohr's postulate, the electrons revolve in the orbits without losing energy. Therefore equating equations (*i*) and (*ii*),

$$\frac{Z.e^2}{4\pi\varepsilon_0.r_n^2} = \frac{m.v_n^2}{r_n}$$

$$\therefore \quad m.v_n^2 = \frac{Z.e^2}{4\pi\varepsilon_0 r_n} \qquad \ldots (iii)$$

We know that the kinetic energy of the revolving electron.

$$K.E. = \frac{1}{2} m.v_n^2$$

Substituting the value of $m.v_n^2$ from equation (*iii*),

$$K.E. = \frac{1}{2} \times \frac{Z.e^2}{4\pi\varepsilon_0.r_n} = \frac{Z.e^2}{8\pi\varepsilon_0.r_n} \qquad \ldots (iv)$$

and potential energy of electron due to its position in the electric field of the nucleus is obtained when it is moved from a radius r_n to the infinity. It is done by integrating the work done against the force of attraction between the nucleus and the electron between the limits r_n and infinity. Therefore potential energy of the electron,

$$P.E. = -\int_{r_n}^{\infty} F_e . dr_n$$

Structure of Atoms

Minus sign indicate that the energy spent in moving the electron from r_n to infinity is against the force of attraction between the nucleus and electron, i.e., F_e. Substituting the value of F_e in the above equation from equation (i),

$$P.E. = -\int_{r_n}^{\infty} \frac{Z.e^2}{4\pi\varepsilon_0 . r_n^2} dr_n = \frac{Z.e^2}{4\pi\varepsilon_0} \int_{r_n}^{\infty} \frac{1}{r_n^2} dr_n$$

$$= -\frac{Z.e^2}{4\pi\varepsilon_0}\left[-\frac{1}{r_n}\right]_{r_n}^{\infty} = -\frac{Z.e^2}{4\pi\varepsilon_0}\left[\frac{1}{r_n}\right] = -\frac{Z.e^2}{4\pi\varepsilon_0 . r_n} \quad \ldots (v)$$

∴ Total energy of the electron,

$$E = K.E. + P.E = \frac{Z.e^2}{8\pi\varepsilon_0 . r_n} - \frac{Z.e^2}{8\pi\varepsilon_0 . r_n}$$

$$= -\frac{Z.e^2}{8\pi\varepsilon_0 . r_n} \quad \ldots (vi)$$

We also know that angular momentum of an electron moving in radius r_n with a tangential velocity v_n, $m.v_n.r_n = \dfrac{nh}{2\pi}$

Where $\qquad h$ = Planck's constant $\qquad \ldots (vii)$

Dividing equation (vii) by equation (iii),

$$\frac{m.v_n.r_n}{m.v_n^2} = \frac{\dfrac{nh}{2\pi}}{\dfrac{Z.e^2}{4\pi\varepsilon_o.r_n}} = \frac{nh.2\varepsilon_0.r_n}{Z.e^2}$$

∴ $$v_n = \frac{Z.e^2}{2\varepsilon_0.n.h} \quad \ldots (viii)$$

Now substituting this value or v_n in equation (vii),

$$m \cdot \frac{Z.e^2}{2\varepsilon_0.nh} \cdot r_n = \frac{nh}{2\pi}$$

∴ $$r_n = \frac{\varepsilon_0.n^2.h^2}{\pi m.Z.e^2}$$

Substituting this value of r_n in equation (vi),

$$E = -\frac{Z.e^2}{8\pi\varepsilon_0 \times \dfrac{\varepsilon_0.n^2.h^2}{\pi m.Z.e^2}} = \frac{m.Z^2.e^4}{8\varepsilon_0^2.n^2h^2}$$

Now substituting the values of constants (*i.e.*, m equal to 9.1×10^{-31} kg, e equal to 1.6×10^{-19}, ε_0 equal to 8.854×10^{-12} and h equal to 6.62×10^{-34}),

$$E = -\frac{9.1 \times 10^{-31} \times Z^2 \times (1.6 \times 10^{-19})^4}{8 \times (8.854 \times 10^{-12})^2 n^2 (6.62 \times 10^{-34})^2} \text{ Joules}$$

$$= -\frac{21.7 \times 10^{-19} Z^2}{n^2} \text{ joules}$$

$$= -\frac{21.7 \times 10^{-19} Z^2}{1.6 \times 10^{-19} n^2} eV \quad \ldots (\because 1eV = 1.6 \times 10^{-19} \text{ Joules})$$

$$= -\frac{13.6\, Z^2}{n^2} eV$$

Now total energy of an electron in the first orbit,

$$E_1 = -\frac{13.6\, Z^2}{(1)^2} - 13.6\, Z^2 \text{ eV}$$

Similarly, total energy of an electron in the second orbit.

$$E_2 = -\frac{13.6\, Z^2}{(2)^2} = -3.4\, Z^2 \text{ eV}$$

and total energy of an electron in the third orbit.

$$E_3 = -\frac{13.6\, Z^2}{(3)^2} = -1.52\, Z^2 \text{ eV}$$

Example 2.1. *Calculate the radius of the first Bohr's orbit in hydrogen atom.*

Solution. Given $n = 1$

We know that radius of the nth Bohr's orbit in hydrogen atom,

$$r_n = \frac{\varepsilon_0 . n^2 . h^2}{\pi . m . Z . e^2}$$

and for first Bohr's orbit, of hydrogen atom, $n = 1$ and $Z = 1$.

$$\therefore \quad r_1 = \frac{\varepsilon_0 . h^2}{\pi . m . e^2} = \frac{8.854 \times 10^{-12} \times (6.626 \times 10^{-34})^2}{\pi \times 9.1 \times 10^{-31} \times (1.602 \times 10^{-19})^2} m$$

$$= 5.3 \times 10^{-11} m \approx 0.53 \text{ Å} \quad \textbf{Ans.}$$

Example 2.2. *The radius of first Bohr orbit of electron in a hydrogen atom is 0.529 Å, calculate the radius of the second Bohr orbit in a singly ionized helium atom.*

(A.M.I.E., Winter, 1993)

Solution. Given the radius of the first Bohr orbit of electron in hydrogen atom., $(r_1)_H = 0.529$ Å.

We know that the radius of an nth orbit of an atom,

$$r_n = \frac{\varepsilon_0 . n^2 . h^2}{\pi . m . Z . e^2} = k \times \frac{n^2}{Z}$$

Where
$$k = \frac{\varepsilon_0 . n^2 . h^2}{\pi . m . e^2} \text{ is a constant}$$

$\therefore$ For the first Bohr orbit of electron in hydrogen atom,

$$(r_1)_H, 0.529 = k \times \frac{(1)^2}{1} = k \text{ or } k = 0.529$$

and radius of the second Bohr orbit ($n = 2$) in helium atom ($Z = 2$)

Structure of Atoms

$$(r_2)_{He} = k \cdot \frac{n^2}{Z} = 0.529 \times \frac{(2)^2}{2} = 1.058 \text{ Å} \textbf{ Ans.}$$

Example 2.3 *Prove that the energy by an electron jumping from orbit number 3 to orbit number 1 and the energy emitted by an electron jumping from orbit number 2 to orbit number 1 are in the ratio 32 : 27.*

(A.M.I.E. Summer, 1993)

Solution. We know that energy of an electron in the 'n'th orbit of an atom,

$$E_n = -\frac{13.6 Z^2}{n^2} \text{ eV}$$

∴ For the first orbit, $\quad E_1 = -\dfrac{13.6 Z^2}{(1)^2}$ eV

and for the second orbit, $\quad E_2 = -\dfrac{13.6 Z^2}{(2)^2}$ eV

Similarly for the third orbit, $E_3 = -\dfrac{13.6 Z^2}{3^2}$ eV

Energy emitted by an electron jumping from orbit number 3 to orbit number 1,

$$E_3 - E_1 = -\frac{13.6 Z^2}{(3)^2} - \left(\frac{-13.6 Z^2}{(1)^2}\right) = -13.6 Z^2 \left(\frac{1}{9} - \frac{1}{1}\right) \quad \ldots (i)$$

and energy emitted by an electron jumping from orbit 2 to orbit umber 1.

$$E_2 - E_1 = \frac{-13.6 Z^2}{(2)^2} - \left(-\frac{13.6 Z^2}{(1)^2}\right) = -13.6 Z^2 \left(\frac{1}{4} - \frac{1}{1}\right) \quad \ldots (ii)$$

Dividing equation (*i*) by equation (*ii*),

$$\frac{E_3 - E_1}{E_2 - E_1} = \frac{-13.6 Z^2 \left(\frac{1}{9} - \frac{1}{1}\right)}{-13.6 Z^2 \left(\frac{1}{4} - \frac{1}{1}\right)} = \frac{\left(\frac{1}{9} - \frac{1}{1}\right)}{\left(\frac{1}{4} - \frac{1}{1}\right)} = \frac{-\frac{8}{9}}{-\frac{3}{4}} = \frac{32}{27} \textbf{ Ans.}$$

Example 2.4. *Calculate the velocity of an electron of hydrogen atom in the Bohr's first orbit. Take* $h = 6.626 \times 10^{-34}$ *J.s,* $\varepsilon_0 = 8.825 \times 10^{-12}$ *F/m and* $e = 1. \times 10^{-19}$ *C.*

Solution. Given $n = 1$ and $Z = 1$ (for hydrogen atom)

We know that the tangential velocity of the electron,

$$v_n = \frac{Z \cdot e^2}{2\varepsilon_0 n h} = \frac{1(1.6 \times 10^{-19})^2}{2 \times (8.825 \times 10^{-12}) \times 1 \times (6.626 \times 10^{-34})} \text{ m/s}$$

$$= 2.189 \times 10^6 \text{ m/s } \textbf{ Ans.}$$

2.13 ORBITAL FREQUENCY OF AN ELECTRON

The term 'orbital frequency of an electron' may be defined as the number of revolutions made by it per second around the nucleus. Consider an electron revolving in some orbit, whose orbital frequency is required to be found out.

Let $\quad \omega$ = Angular velocity of the electron,

r_n = Radius of the orbit, and

f_n = Orbital frequency of the electron.

We know that tangential velocity of the electron,
$$v_n = \omega r_n = 2\pi f_n \cdot r_n \quad \quad \ldots (\because \omega = 2\pi f_n)$$
Substituting the values of r_n and v_n from Art. 2.8,
$$\frac{Z \cdot e^2}{2\varepsilon_0 \cdot n \cdot h} = 2\pi f_n \times \frac{\varepsilon_0 \cdot n^2 \cdot h^2}{\pi m \cdot Z \cdot e^2}$$
$$\therefore \quad f_n = \frac{mZ^2 \cdot e^4}{4\varepsilon_0^2 \cdot n^3 \cdot h^3}$$

Now substituting the values of constants (*i.e.*, m equal to 9.1×10^{-21} kg, e equal to 1.6×10^{-19} C, ε_0 equal to 8.854×10^{-12} F/m, and h equal to 6.626×10^{-34} J-s) in the above equation.

$$f_n = \frac{9.1 \times 10^{-31} \times Z^2 (1.6 \times 10^{-19})^4}{4 \times (8.854 \times 10^{-12})^2 \times n^3 (6.626 \times 10^{-34})^3}$$

$$= 6.56 \times 10^{15} \times \frac{Z^2}{n^3}$$

Example 2.5. *Calculate the orbital frequency of an electron in the first Bohr's orbit in a hydrogen atom.*

Solution. Given $n = 1$ and $Z = 1$ (for hydrogen atom)

We know that the orbital frequency of the electron,
$$f_n = 6.56 \times 10^{15} \times \frac{Z^2}{n^3} = 6.56 \times 10^{15} \times \frac{(1)^2}{(1)^3} \text{ Hz}$$
$$= 6.56 \times 10^{15} \text{ Hz } \textbf{Ans.}$$

Example 2.6. *A hydrogen atom exists with its electron in the $n = 3$ state. The electron undergoes a transition to the $n = 2$ state.*

Calculate (a) the energy of the photon emitted, (b) its frequency and (c) its wavelength.

(A.M.I.E., Summer 2002)

Solution. (*a*) *Energy of the photon emitted.*

We know that energy of the electron / photon
$$E = \frac{-13.6 \, Z^2}{n^2} \text{ eV}$$

For hydrogen, $Z = 1$. Now the energy of an electron, in the third orbit, ($n = 3$),
$$E_3 = \frac{-13.6 \cdot Z^2}{3^2} = -1.51 \text{ eV}$$

and energy of an electron / photon in the second orbit ($n = 2$),
$$E_2 = \frac{-13.6 \cdot Z^2}{2^2} = \frac{-13.6 \times 1^2}{2^2} = -3.4 \text{ eV}$$

$\therefore$ The energy of the photon emitted.
$$\Delta E = E_3 - E_2 = -1.51 - (-3.4) = 1.89 \text{ eV } \textbf{Ans.}$$

(*b*) *Frequency of the photon emitted.*

We know that energy of the photon emitted,
$$= 1.89 \, eV = 1.89 \times (1.6 \times 10^{-19}) \text{ Joules.}$$
$$= 3.024 \times 10^{-19} \text{ Joules}$$

Structure of Atoms

and the frequency of the photon emitted,

$$v = \frac{\text{Energy of the photon emitted (Joules)}}{\text{Mass of an electron (kg)}}$$

$$= \frac{3.024 \times 10^{-19}}{6.626 \times 10^{-34}} = 0.456 \times 10^{14} \text{ Hz or } 4.56 \times 10^{14} \text{ Hz } \textbf{Ans.}$$

(c) *Wavelength of the photon emitted*

We know that wavelength of the photon emitted,

$$\lambda = \frac{C}{v} = \frac{3 \times 10^8}{4.56 \times 10^{14}} = 0.658 \times 10^{-6} m \text{ or } 6.58 \times 10^{-7} \text{ m } \textbf{Ans.}$$

2.14 DEFICIENCIES IN BOHR'S ATOMIC MODEL

The Bohr's atomic model has a considerable degree of accuracy. But this theory cannot be applied to complex atoms. Following are the main deficiencies in the Bohr's atomic model :

1. This theory cannot be applied to complex atoms having two or more planetary electrons.
2. This theory cannot be used to make any calculations about transitions from one orbit to another, such as the rate at which they occur or the selection rules which apply to them.
3. This theory cannot explain as to how the interaction of individual atoms take place, among themselves, resulting in aggregate of matter.

2.15 QUANTUM NUMBERS

Bohr explained the spectrum of hydrogen on the basis of electron jumps from some higher orbits to some lower orbits. But he failed to explain the spectrum of many electron atoms *i.e.*, atoms containing more than one electron. Later on, with the use of sophisticated instruments, the spectral studies revealed that energies of all the electrons, belonging to a given energy level, are not the same but differ from one another. Therefore, it was concluded that it is not possible to fully explain the energy and location of an electron in an atom with the help of only one quantum number (n). Further studies revealed that four quantum numbers are necessary to fully explain the energy and location of an electron in an atom with the help of only one quantum number (n) The studies also revealed that four quantum numbers are necessary to fully explain the energy and location of electrons in an atom. These four quantum numbers are as given below:

1. *Principal quantum number (n).* This quantum number gives us the information about the main energy level to which an electron belongs. This number takes only the integer values *i.e.*, 1, 2, 3, 4......... and so on. Thus for first energy level, $n = 1$, for second energy level $n = 2$, and so on.

2. *Orbital or azimuthal quantum number (l).* This quantum number gives us the information about the shape of the sublevel of the main energy level to which an electron belongs. This number also takes only the integer values. But its value depends upon 'n'. For a particular value of 'n' the different values of 'l' are 0, 1, 2 and 3.

3. *Magnetic quantum number (m_1).* This quantum number gives us the information about the orientation (or arrangement) of sub-level in space to which an electron belongs. Its values are determined by the value of 'l' and range from $-l \rightarrow 0 \rightarrow +l$ *i.e.*, a total of $2l + 1$ values. Thus for $l = 1$, the different values of m_l are -1, 0 and $+1$.

4. *Spin quantum number (m_s).* This quantum number gives us the information about the spin of electrons about their own axis in the orbital *i.e.*, whether the spin is counter clockwise or clockwise. Thus there are only two possible values of m_s *i.e.*, $+1/2$ and $-1/2$.

2.16 SOMMERFELD'S ATOMIC MODEL

It is an improved Bohr's atomic model, which was proposed by Sommerfeld in 1916. He accepted all the postulates of Bohr's theory, except the circular orbits for electron motion. According to the Sommerfeld's model, for a particular value of 'n' there are the same number of possible sub-shells out of which one is a circular and the remaining (*i.e.*, n − 1) are elliptical in shape. This may be best understood from the following examples:

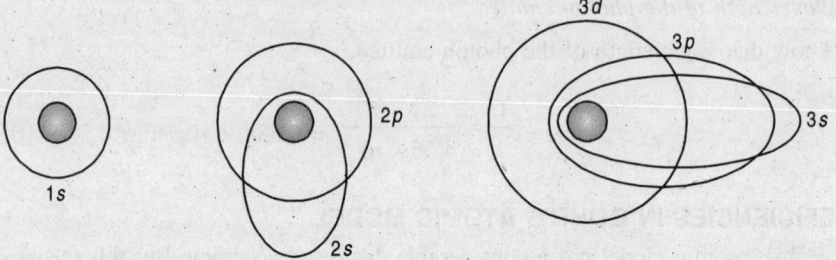

Fig. 2.5. Shells and sub-shells in an atom.

Consider the first energy level ($n - 1$). In this energy level, there is only one orbit or subshell for the electron. This orbit is circular as shown in Fig. 2.5 (a). Similarly, for the second energy level ($n = 2$), there are two permissible sub-shells for the electrons. Out of these two orbits, one is circular and the other is elliptical as shown in Fig. 2.5 (b). And for the third energy level ($n = 3$), there are three permissible sub-shells for the electron. Out of these three orbits, one is circular and the other two are elliptical as shown in Fig. 2.5 (c).

It will be interesting to know that all the sub-shells are designated by the letter 'l'. For a particular value of 'n', the different values of 'l' are 0, 1, 2............(n - 1). The term 'n' is known as Principal Quantum Number and 'l' as Orbital or Azimuthal Quantum Number.

It was shown by Sommerfeld that the energy level 'n' and the sub-shell 'l' are related by the expression:

$$\frac{b}{a} = \frac{l+1}{n}$$

Where a and b are the semi-major and semi-minor axes of an elliptical orbit.

It may be noted that when $n = 1$, then $l = 0$. Or in other words, $a = b$. In this case, we find that the two axes of the figure are equal. As a result of this, the orbit corresponding to $n = 1$ is circular. This sub-shell is designated as s sub-shell. Since this sub-shell belongs to $n = 1$, it is designated as $1s$. Similarly when $n = 2$, l has two values 0 and 1. Substituting these two values of l in the general expression, we find that when

$$l = 0, \quad \frac{b}{a} = \frac{0+1}{2} = \frac{1}{2} \text{ or } b = \frac{a}{2} = 0.5\, a$$

From the above equation, we find that since $b = 0.5\, a$, therefore the sub-shell, corresponding to $l = 0$, is elliptical in shape and is designated as $2s$

Similarly, when $\quad l = 1, \quad \dfrac{b}{a} = \dfrac{1+1}{2} = 1 \text{ or } b = a$

From the above equation, we find that since $b = a$, therefore the sub-shell corresponding to $l = 1$ is circular in shapes, and is designated as $2p$. And when $n = 3$, l has three values 0, 1 and 2. Substituting these values of l in the general expression we find that

when $\quad l = 0 \quad \dfrac{b}{a} = \dfrac{0+1}{1} = \dfrac{1}{3} \text{ or } b = \dfrac{a}{3}$

Structure of Atoms

Similarly when $l = 1, \dfrac{b}{a} = \dfrac{1+1}{3} = \dfrac{2}{3}$ or $b = \dfrac{2a}{3}$

and when $l = 2, = \dfrac{b}{a} = \dfrac{2+1}{3} = 1$ or $b = a$

Note. The sub-shells corresponding to $l = 0, 1$ and 2 are designated as $3s, 3p$ and $3d$ respectively.

2.17 MODERN ATOMIC MODEL

Pictures of Atomic force Microscope and scanning. Tunnelling microscope.

With the recent advances in technology the scientist have studied the structure of an atom using Atomic Force Microscope (A.F.M) and Scanning Tunnelling Microscope. (STM) shown in Fig. 2.6. These equipments allow the scientists to look at the atoms at the surface as well as underneath. The surface of materials. Based upon this studies, the scientists have proposed the modern atomic model.

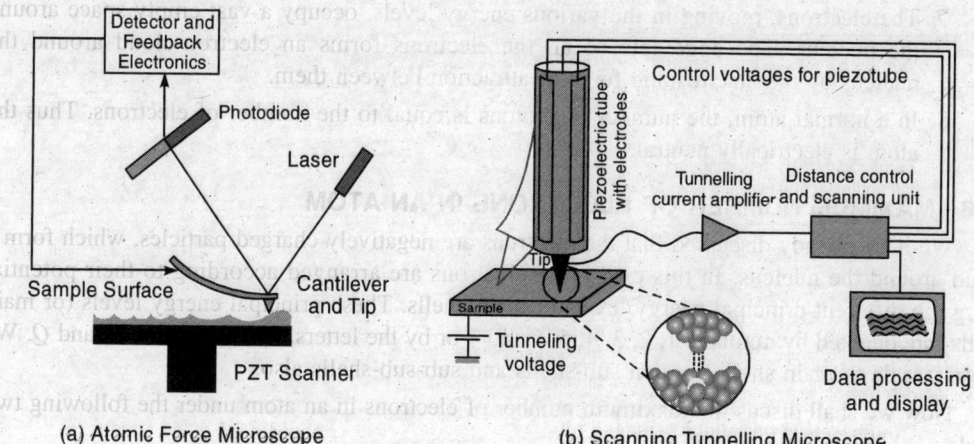

(a) Atomic Force Microscope (b) Scanning Tunnelling Microscope

Fig. 2.6.

According to this model :

1. An atom is made up of three fundamental particles i.e., electrons, protons and neutrons. The electrons have negative charge, protons have positive charge and the neutrons have no charge. Thus they are neutral.

2. The protons and neutrons are located in the small nucleus at the centre of the atom. Due to the presence of protons, the nucleus is positively charged.

3. The electrons revolve around the nucleus in fixed circular as well as elliptical paths known as Principal Energy Levels or main shells. These energy levels are represented by the letters $K, L, M, N, O, P,$ and Q. These energy levels are counted from the nucleus to outwards.

4. Each energy level shell is divided into sub-shells or orbitals. These sub-shells are represented by the letters $s, p, d,$ and f. Fig. 2.7 shows the shape of s-, p- and d-orbital respectively. The orbitals can be seen as "rooms" in which the electrons in an atom "live." The shape of an f-orbital is more complex and is usually not shown in text books. These sub-shells are further sub-divided into sub-sub-shells.

5. Each energy level is associated with a fixed amount of energy. The energy level, which is nearest to the nucleus, has minimum energy. Where as the energy level. which is farthest from the nucleus, has maximum energy.

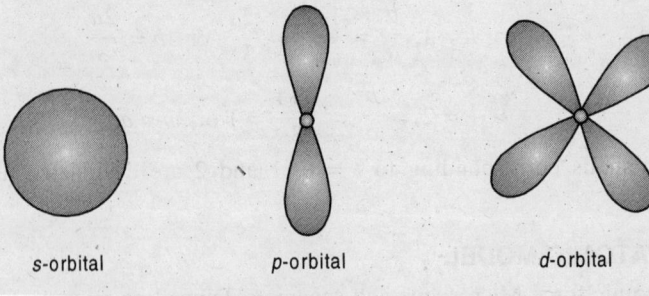

Fig. 2.7. Orbitals.

6. There is no change in energy of electron, so long as it keeps on revolving in the same energy level and the atom remains stable. But when it jumps from a lower energy level to a higher energy level, or when it comes down from the higher energy level to a lower energy level, some change in the energy of an electron takes place.
7. The electrons, moving in the various energy levels, occupy a vast empty space around the nucleus. The aggregate of all the electrons forms an electron-cloud around the nucleus by the electrostatic force of attraction between them.
8. In a normal atom, the number of protons is equal to the number of electrons. Thus the atom is electrically neutral.

2.18 MAXIMUM NUMBER OF ELECTRONS IN AN ATOM

We have already discussed that the electrons are negatively charged particles, which form a cloud around the nucleus. In this cloud, the electrons are arranged according to their potential energy in different principal energy levels or main shells. These principal energy levels (or main shells) are denoted by numbers 1, 2, 3, 4, 5, 6 and 7 or by the letters, K, L, M, N, O, P, and Q. We know that these main shells contain sub-shells and sub-sub-shells also.

Now we shall discus the maximum number of electrons in an atom under the following two heads :

1. Maximum number of electrons in the main shells.
2. Maximum number of electrons in the sub-shells.

2.19 MAXIMUM NUMBER OF ELECTRONS IN THE MAIN SHELLS

The maximum number of electrons in a main shell
$$= 2(n)^2$$
where 'n' is the number of the main shell from the nucleus. Thus for K-shell ($n = 1$). maximum number of electrons $= 2(1)^2 = 2$

Similarly, for L-shell ($n = 2$), maximum number of electrons
$$= 2(2)^2 = 8$$
and for M-shell ($n = 3$), and N-shell ($n = 4$), maximum number of electrons
$$= 2(3)^2 = 18$$
and $= 2(4)^2 = 32$ respectively.

It will be interesting to know that with the present state of knowledge, the above relation, for maximum number of electrons, does not hold good beyond N-shell; The maximum number of electrons in O-shell ($n = 5$) P-shell ($n = 6$), and Q-shell ($n = 7$) are 32 only.

Structure of Atoms

2.20 MAXIMUM NUMBER OF ELECTRONS IN THE SUB-SHELLS

The maximum number of electron in a sub-shell or an orbital

$$= 2(2l + 1)$$

Where 'l' is the number of the sub shell. Thus for the first sub-shell ($l = 0$), maximum number of electrons

$$= 2(2l + 1) = 2(2 \times 0 + 1) = 2$$

This sub-shell is known as s-sub-shell. The symbol is $2s, 3s, 4s,...$ denotes the s-sub shell of the first, second, third and fourth energy levels respectively.

Similarly, in the second sub-shell ($l = 1$) maximum number of electrons

$$= 2(2 \times 1 + 1) = 6$$

This sub-shell is known as p-sub-shell. The symbol $2p, 3p\ 4p$ .... denotes the p-sub-shell of the second, third, and fourth energy levels respectively.

Similarly, in the third sub-shell ($l = 2$) maximum number of electrons

$$= 2(2 \times 2 + 1) = 10$$

This sub-shell is known as d-sub-shell. The symbol $3d, 4d, 5d$ ... denotes the d-sub-shell of the third, fourth and fifth energy levels respectively.

Similarly, in the fourth sub-shell ($l = 3$) maximum number of electrons

$$= 2(2 \times 3 + 1) = 14$$

The sub-shell is known as f-sub-shell. The symbol $4f, 5f$... denotes the f-sub shell of the fourth and fifth energy levels respectively. The following points should be remembered for the number of electrons in the sub-shells:

1. The maximum number of electrons in the first principal energy level is 2. Therefore, it will have only one sub-shell ($1s$) containing two electrons. The first principal energy level cannot have any other sub-shell.
2. The maximum number of electrons in the second principal energy level are 8. Therefore, it will have two sub-shells namely $2s$ (containing two electrons) and $2p$ (containing 6 electrons).
3. The maximum number of electrons in the third principal energy level are 18. Therefore, it will have three sub-shells namely $3s$ (containing two electrons), $3p$ (containing 6 electrons) and $4d$ (containing 10 electrons).
4. The maximum number of electrons in the fourth principal energy level are 32. Therefore, it will have four sub-shells namely $4s$ (containing two electrons) $4p$ (containing 6 electrons), $4d$ (containing 10 electrons) and $4f$ (containing 14 electrons).

2.21 ENERGY LEVEL DIAGRAM

We have already obtained an expression for the total energy of an electron, when it is revolving in any orbit around its nucleus. If we plot the energies E_1, E_2, E_3, ...etc. as calculated in the article 2.8 along the negative side of the vertical axis by horizontal lines, we shall obtain diagram as shown in Fig 2.8. This diagram is popularly known as Energy Level Diagram.

It will be interesting to know that when the electron is present in the first orbit ($n = 1$). It is said to be in its normal or ground state. And if it is present in the higher orbits, it is said to be in the excited state.

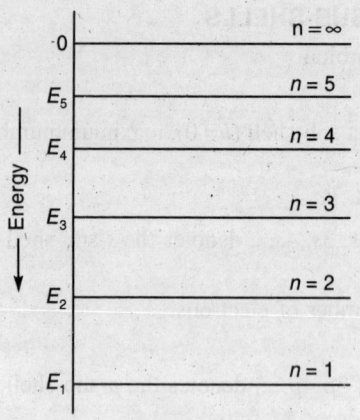
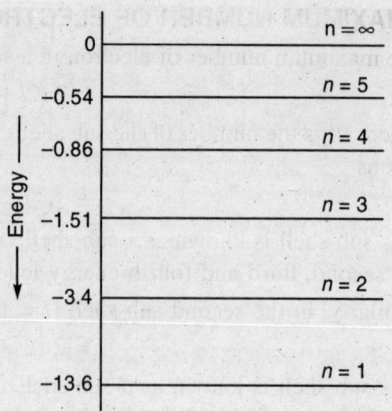

Fig. 2.8. Energy level diagram. Fig. 2.9. Energy level diagram of a hydrogen atom.

2.22 ENERGY LEVEL DIAGRAM OF A HYDROGEN ATOM

We have already discussed in Art. 2.7 that the total energy of an electron, when it is revolving in any orbit around its nucleus is given by the relation:

$$E = -\frac{13.6 Z^2}{n^2} \text{ eV}$$

We know that for hydrogen atom, Z = 1. Therefore, substituting the value of Z in the above equation,

$$E = -\frac{13.6}{n^2} \text{ eV}$$

Now total energy of a hydrogen atom in the first orbit (*i.e.*, n = 1).

$$E_1 = -\frac{13.6}{(1)^2} = -13.6 \text{ eV}$$

Similarly, total energy of the second orbit (*i.e.*, n = 2)

$$E_2 = -\frac{13.6}{(2)^2} = -3.4 \text{ eV}$$

and total energy for the third, fourth and fifth orbit,

$$E_3 = -\frac{13.6}{(3)^2} = -15.1 \text{ eV}$$

$$E_4 = -\frac{13.6}{(4)^2} = -0.86 \text{ eV}$$

$$E_5 = -\frac{13.6}{(5)^2} = -0.54 \text{ eV}$$

Now if we plot the energies E_1, E_2, E_3, ... along the negative side of the vertical axis, we shall obtain a diagram as shown in Fig. 2.8. This diagram is known as energy level diagram of a hydrogen atom. In the same way, we can draw energy level diagram for all the elements.

Structure of Atoms

2.23 ENERGY LEVEL DIAGRAM OF A MULTIELECTRON ATOM

Fig. 2.9 shows the Standard Energy Level Diagram of a multielectron atom (*i.e.*, an atom containing more than one electron). It will be interesting to know that there may be an infinite number of principal energy levels in an atom. But with the present state of knowledge, the distribution of electrons in the elements is represented in seven principal energy levels only (*i.e.*, $n = 1$ to $n = 7$). Or in other words, from bottom to top in the Energy Level Diagram.

The sequence for filling up the electrons in various sub-shells or orbitals is done with the following set of rules :

1. The electrons are added one by one to the orbitals, as we move from one element to the next in the order to increasing atomic number.
2. The orbitals are successively filled in the order to their increasing energy. Their energies increase in the order as shown in Fig. 2.9 *i.e.*, 1s, 2s, 3p, 3s, 4s, 3d, 4p, 5s, 6s, 4f, 5d, 6p, and so on. Here the numbers 1, 2, 3, 4, etc. indicate the principal energy level.
3. Pairing of electrons in any *s, p, d* or *f*-orbitals is not possible, until all the available orbitals of a given set contain one electron each. This is known as Hund's rule of maximum multiplicity.
4. An orbital or a sub-shell in a given principal quantum number can accommodate a maximum of two electrons of opposite spin. In other words, no orbital can accommodate two electrons of the same spin. This is known as Pauli's Exclusion Principle (For details see Art. 2.29).
5. Orbitals in the same sub-level tend to become full or exactly half-full of electrons, because it represents a stabler arrangement of electrons.
6. If two or more orbitals of a given set (*i.e.*, *p, d,* or *f*-orbitals) contain one electron each, they tend to have the same spin. But two electrons in an orbital will always be of opposite spin.

Now we shall discuss in detail the sequence of filling up the electrons in various orbitals in the different energy levels of a multielectron atom. It is done as discussed below :

1. *First principal level (i.e., n = 1).* It has only one orbital represented by blank circle and is known as *ls*-orbital., It can accommodate only two electrons (because each *s*-orbital has only one sub-sub-shell, and it cannot accommodate more than two electrons).
2. *Second principle energy level (i.e., n = 2).* It has two orbitals known as 2s and 2p-orbitals represented by shaded circles in the diagram. The sequence of filling up of these two orbitals is as discussed below:

    2s-orbital. It has only one sub-sub-shell and can accommodate only two electrons (because s-orbital cannot accommodate more than two electrons).

    2p-orbital. It has only three sub-sub-shells and can accommodate at the most six electrons (because each *p*-orbital has three sub-sub-shells and each sub-sub-shell of *p*-orbital cannot accommodate more than two electrons).

    It is thus obvious, that the maximum numbers of electrons, filled in the second energy level is 8. And the total number or electrons, which can be accommodated up to the second principal energy level is 10 (two in the first energy level and 8 in the second energy level).

    It may be noted that the second principal energy level ($n = 2$) will start accommodating the electrons only after the first principal energy level ($n = 1$) is completely filled up.
3. *Third principal energy level (i.e., n = 3).* It has three orbitals known as 3s, 3p and 3d-orbitals represented by black circles in the diagram. It may be noted that the 3d-orbital does not fall within the range of third principal energy level, as the energy of 3d-orbital

is greater than that of 4s orbital (in the fourth principal energy level). It is due to this reason that the 3d-orbital is shown above the 4s-orbital in the diagram. Thus in the third principal energy level, only 3s and 3p-orbitals are shown. The sequence of filling up these two orbitals is as discussed below:

3s-orbital. It has only one sub-sub-shell and can accommodate only two electrons.

3p-orbital. It has only three sub-sub-shells and can accommodate at the most six electrons.

It is thus obvious, that the maximum number of electrons filled up in the third principal energy level is also 8. And the total number of electrons, which can be accommodated up to third principal energy level is 18 (10 up to second principal energy level and 8 in third principal energy level).

It may also be noted that the third principal energy level ($n = 3$) will start accommodating the electrons only after the second principal energy level is completely filled up.

4. *Fourth principal energy level (i.e., $n = 4$)*. It has four orbitals known as 4s, 4p, 4d and 4f orbitals represented by blank circles in the diagram. It may be noted that the energy of 4d and orbitals do not fall within the range of fourth principal energy level. Since the energy of 4d-orbital is greater than that of 5s-orbital in the diagram, Similarly, as the energy of 4f-

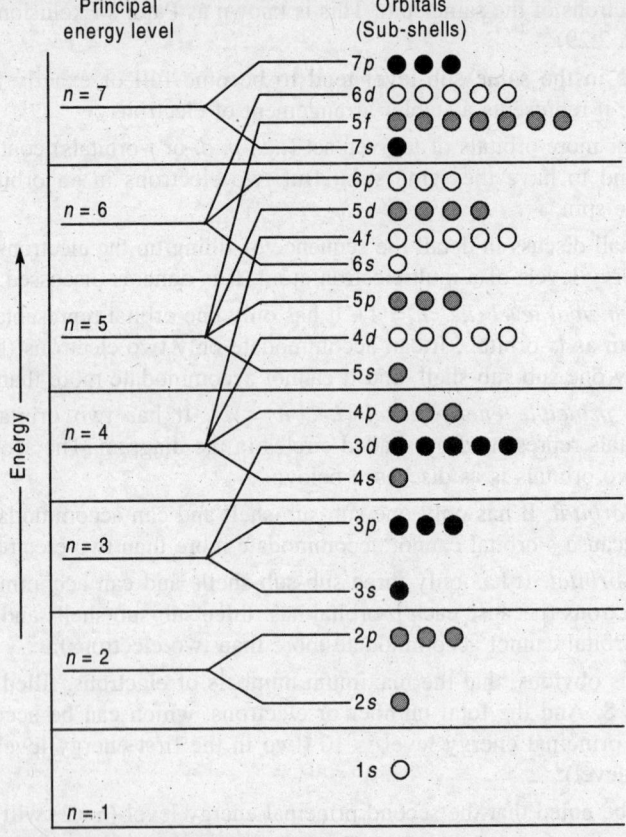

Fig. 2.9. Energy Level Diagram of a multielectron atom.

* The 3d-orbital is filled after filling up of 4s-orbital of the fourth principal energy level.

orbital is greater than 6s-orbital, therefore the *4f-orbital* is shown above the 6s-orbital. Thus in the fourth principal energy level, only 4s, 3d and 4p orbitals are shown. The sequence of filling up these three orbitals is as below:

4s-orbital. It has only one sub-sub-shell and can accommodate only two electrons.

3d-orbital. It has five sub-sub-shells and accommodate at the most 10 electrons (because each *d*-orbital has five sub-sub-shells. And each sub-sub-shells of *d*-orbital cannot accommodate more than 2 electrons).

4p-orbitals. It has three sub-sub-shells and can accommodate at the most 6 electrons.

It is thus obvious that the maximum number of electrons filled up in the fourth principal energy level is 18. And the total number of electrons, which can be accommodated up to the fourth principal energy level is 36.

It may also be noted again that the fourth principal energy level ($n = 4$) will start accommodating the electrons only after the third principal energy level is completely filled up.

Similarly, in the 5th principal energy level the sequence of filling up of the orbitals is 5s, 4d and 5p as shown in diagram. And similarly in the 6th and 7th principal energy levels the sequence of filling up of the orbitals is 6s, 4f, 5d, 6p and 7s, 5f, 6d, 7p respectivel. It may be noted that the maximum number of electrons filled up in the 5th, 6th and 7th principal energy levels are 18, 32 and 32 respectively.

2.24 ELECTRONIC CONFIGURATION

We have already discussed in the last article the distribution of electrons in shells and sub-shells. The scientists have evolved a principle to denote systematically the distribution of electrons in shells and sub-shells. This principle is known as Electronic Configuration. It may be best understood from the following examples:

1. *Hydrogen* (atomic number 1.) We know that the hydrogen atom has only one electron. It is accommodated in the first main shell. Thus the electronic configuration of the hydrogen atom is designated as $1s^1$. It shows that in a hydrogen atom, the prefix '*1*' represents the first main shell '*s*' represents the *s*-sub-shell and the script 1 represents the number of electrons in the sub-shell.

2. *Helium* (atomic number 2.) We know that the Helium atom has two electrons. Both the electrons are accommodated in the first main shell. Thus the electronic configuration of the Helium atom is designated as $1s^2$. It shows that in a Helium atom, the prefix '*1*' represents the first main shell, '*s*' represents the *s*-shell and the script 2 represents the number of electrons in the sub-shell.

3. *Lithium* (atomic number 3.) We know that the Lithium atom has three electrons. Two electrons are accommodated in the first main shell, while the remaining one electron is accommodated in the '*s*' sub-shell of the second main shell. Thus the electronic configuration of the Lithium atom is designated as $1s^2\ 2s^1$.

4. *Beryllium* (atomic number 4.) We know that the Beryllium atom has four electrons. Two electrons are accommodated in the first main shell, while the other two are accommodated in the '*s*' sub-shell of the second main shell. Thus the electronic configuration of the beryllium atom is designated as $1s^2\ 2s^2$.

---

* Each *f*-orbital has 7 sub-sub-shells. And each sub-sub-shell cannot accommodate more than two electrons.

5. *Carbon* (atomic number 6.) We know that the Carbon atom has six electrons. Two electrons are accommodated in the first main shell, the other two are accommodated in the 's' sub-shell of the second main shell while the remaining two electrons are accommodated in the 'p' sub-shell of the second main shell. Thus the electronic configuration of the carbon atom is designated as $1s^2\ 2s^2\ 2p^2$.

6. *Nitrogen* (atomic number 7). We know that the Nitrogen atom has seven electrons. Two electrons are accommodated in the first main shell, the other two are accommodated in the 's' sub-shell of the second main shell, while the remaining three electrons are accommodated in the 'p' sub-shell of the second main shell. Thus the electronic configuration of the Nitrogen is designated as $1s^2\ 2s^2\ 2p^3$.

7. *Oxygen* (atomic number 8.) We know that the oxygen atom has eight electrons. Two electrons are accommodated in the first main shell, the other two are accommodated in the 's' sub-shell of the second main shell, while the remaining four electrons are accommodated in the 'p' sub-shell of the second main shell. Thus the electronic configuration of the Oxygen atom is $1s^2\ 2s^2\ 2p^4$.

8. *Fluorine* (atomic number 9). We know that the Fluorine atom has nine electrons. Two electrons are accommodated in the first main shell, the other two are accommodated in the 's' sub-shell of the second main shell, while the remaining five electrons are accommodated in the 'p' sub-shell of the second main shell. Thus the electronic configuration of the Fluorine atom is $1s^2\ 2s^2\ 2p^5$.

9. *Neon* (atomic number 10). We know that Neon atom has ten electrons. Two electrons are accommodated in the first main shell, the other two are accommodated in the 's' sub-shell of the second main shell, while the remaining six electrons are accommodated in the 'p' sub-shell of the second main shell. Thus the electronic configuration of the neon atom is $1s^2\ 2s^2\ 2p^6$.

10. *Sodium* (atomic number 11). We know that the Sodium atom has eleven electrons. Two electrons are accommodated in the first main shell, the other two are accommodated in the 's' sub-shell of the second main shell, the other six are accommodated in the 'p' sub-shell of second main shell, while the remaining one electron is accommodated in 's' sub-shell of the third main shell. Thus the electronic configuration of the sodium atom is $1s^2\ 2s^2\ 2p^6\ 3s^1$.

11. *Chlorine* (atomic number 17). We know that the Chlorine atom has seventeen electrons. The first ten electrons are accommodated in first and second main shells in the same way as in the Neon and Sodium atoms (*i.e.*, $1s^2\ 2s^2\ 2p^6$). The other two electrons are accommodated in the 's' sub-shell of the third main shell, and the remaining five electrons are accommodated in the 'p' sub-shell of the third main shell. Thus the electronic configuration of the Chlorine atom is $1s^2\ 2s^2\ 2p^6\ 3s^2\ 3p^5$.

12. *Copper* (atomic number 29). We know that the copper atom has twenty nine electrons. The first ten electrons are accommodated in the first and second main shells. The other eighteen electrons are accommodated in the third main shell, while the remaining one electron is accommodated in the 's' sub-shell of the fourth main shell. Thus the electronic configuration of the Copper atom is $1s^2\ 2s^2\ 2p^6\ 3s^2\ 3p^6\ 3d^{10}\ 4s^1$.

13. *Silver* (atomic number 47). We know that the silver atom has forty seven electrons. Its electronic configuration is $1s^2\ 2s^2\ 2p^6\ 3s^2\ 3p^6\ 3d^{10}\ 4s^2\ 4p^6\ 4d^{10}\ 5s^1$.

14. *Mercury* (atomic number 80). Its electronic configuration is $1s^2\ 2s^2\ 2p^6\ 3s^2\ 3p^6\ 3d^{10}\ 4s^2\ 4p^6\ 4d^{10}\ 4f^{14}\ 5s^2\ 5p^6\ 5d^{10}\ 6s^2$.

15. *Uranium* (atomic number 92). Its electronic configuration is $1s^2\ 2s^2\ 2p^6\ 3s^2\ 3p^6\ 3d^{10}\ 4s^2\ 4p^6\ 4d^{10}\ 4f^{14}\ 5s^2\ 5p^6\ 5d^{10}\ 5f^3\ 6s^2\ 6p^6\ 6d^1\ 7s^2$.

2.25 ELECTRON STRUCTURE OF ELEMENTS

We have already discussed the electronic configuration of various elements in the last article. Table 2.1 gives the electronic configuration or electron structure of the 109 known elements in a tabular form (shown on pages 27 to 31)

Example. 2.7. *An element with atomic number 50 has all its inner energy levels filled up except 4f level, which is empty. Determine its expected valence.* (A.M.I.E., Summer, 2004)

Solution. Given, atomic number of the element = 50.

The electronic configuration for an element with atomic number 50 is,

$= 1s^2, 2s^2, 2p^6, 3s^2, 3p^6, 4s^2, 3d^{10}, 4p^6, 5s^2, 4d^{10}, 5p^2.$

As seen from this the 5p-level has a deficiency of 4 electrons. Therefore, the valence of the given element is –4. **Ans.**

2.26 WAVE MECHANICS

In the beginning of 19th century, it was experimentally found that light energy is radiated in discrete bundles of energy called quantas or *photons* of energy. It is called Planck's law. This implies that light radiations have a dual character i.e., a character of *particle* as well as a *wave*. De Broglie made a bold suggestion in 1924 that small particles of matter such as electrons have a corresponding property of behaving like waves. The wave-length of such particles is given by the relation.

$$\lambda = \frac{h}{mv}$$

where h = Planck's constant.

m = Mass of the particle, and

v = Velocity of the particle.

Prince Louis-Victor Pierre Raymond de Broglie (1892–1987)
The Nobel Prize in Physics 1929.

This hypothesis was verified by the scientists in United States and England who demonstrated that electrons are diffracted by crystals similar to X-rays. This discovery put the electrons and photons in the same class *i.e.,* both these particles have a dual character. Since this theory gives the description of the particles in terms of a wave, therefore it commonly known as *wave mechanics*.

2.27 SCHRODINGER THEORY

Following De Broglie's postulate of the wave nature of electrons, Schrodinger showed in 1926, that a wave equation can be used to characterize the behaviour of Bohr's theory. The wave equation in one-dimensional time-independent from may be written as below :

$$\frac{d^2\psi}{dn^2} + \frac{8\pi^2 m}{h^2}(E-V)\psi = 0$$

Where ψ = Wave function,

m = Mass of the electron.

h = Planck's constant,

E = Total energy of an electron, and

V = Potential energy of an electron.

Erwin Schrodinger (1887–1961)

The wave-equation in three-dimensions may be obtained by replacing $d^2\psi/dn^2$ by partial derivatives with respect to x, y and z axes. Thus three-dimensional Schrodinger equation is,

$$= \left(\frac{d^2}{dx^2} + \frac{d^2}{dy^2} + \frac{d^2}{dz^2}\right)\psi + \frac{8\pi m^2(E-V)}{h^2}\psi = 0$$

The wave function in Schrodinger equation can be used to describe the position of an electron. However, it will be interesting to know position (or displacement) of an electron cannot be observed directly. In order to observe an electron, we need either light photons or X-ray photons. These particles interact with electron, changing its position immediately as the signal describing its previous position reaches the observer. Thus the information obtained from the experiment does not correctly describe the State of the electron. Max Born showed in 1926, that the quantity $|\psi^2|$ tells the probability of the presence of an electron in a unit volume of space at the time (t) at which the wave function ψ is being considered.

We shall use the Schrodinger equation to explain the Sommerfeld theory, which will be discussed in chapter on *Free Electron Theory of Metals*.

2.28 HEISENBERG'S UNCERTAINTY PRINCIPLE

It is also called principle of indeterminancy and was pronounced in 1927. According to the Heisenberg's uncertainty principle, it is not possible to determine experimentally, both the position and the velocity of an electron simultaneously. In fact, any attempt to examine the electron disturbs or sets off the electron in some unpredictable motion. Let us suppose, in an experiment, the energy of a particle can be measured to some accuracy ΔE and the time at which the measurement is taken is known to some accuracy Δt, then the product of these uncertainties must be greater than the Planck's constant, *i.e.*,

$$\Delta E \cdot \Delta t \geq h$$

Similarly, it can be shown that the product of the accuracies for a simultaneous measurement of momentum (Δp) and position (Δx) must be greater than Planck's constant. *i.e.*,

$$\Delta p . \Delta x \geq h$$

Suppose, we try to locate the exact position of a particle, thus specifying $\Delta x = 0$. But this would be possible only if all knowledge of its momentum were sacrificed, because Δp to satisfy equation (*i*) must become infinite.

The fundamental in accuracy in the simultaneous measurement of certain pairs of variables is to be expected even if we assume the measuring instruments to be perfect.

2.29 PAULI'S EXCLUSION PRINCIPLE

We know that an electrons' orbit can be described by a set of four quantum numbers namely principal quantum numbers (n), azimuthal quantum number (l), magnetic quantum number (m_l) and spin quantum number (m_s). According to Pauli's Exclusion Principle, no two electrons in any atom have the same four quantum numbers. This principle is the basis on which the electronic structure of any elements is calculated.

Looking at the Pauli's Exclusion Principle from physical point of view, it states that it is not possible to get two electrons in the same orbit with the same spin, since these two electrons will repel each other with great force. It may be noted that an orbit is determined by the three quantum numbers only, *i.e.*, the principal quantum number (n), azimuthal quantum number (l) and the magnetic quantum number (m_l), but not the spin quantum number (m_s).

Table 2.1

Atomic number	Element	Symbol	1 s	2 s p	3 s p d	4 s p d f	5 s p d f	6 s p d f	7 s p
1	Hydrogen	H	1						
2	Helium	He	2						
3	Lithium	Li	2	1					
4	Beryllium	Be	2	2					
5	Boron	B	2	2 1					
6	Carbon	C	2	2 2					
7	Nitrogen	N	2	2 3					
8	Oxygen	O	2	2 4					
9	Fluorine	F	2	2 5					
10	Neon	Ne	2	2 6					
11	Sodium	Na	2	2 6	1				
12	Magnesium	Mg	2	2 6	2				
13	Aluminium	Al	2	2 6	2 1				
14	Silicon	Si	2	2 6	2 2				
15	Phosphorus	P	2	2 6	2 3				
16	Sulphur	S	2	2 6	2 4				
17	Chlorine	Cl	2	2 6	2 5				
18	Argon	Ar	2	2 6	2 6				
19	Potassium	K	2	2 6	2 6	1			
20	Calcium	Ca	2	2 6	2 6	2			
21	Scandium	Sc	2	2 6	2 6 1	2			
22	Titanium	Ti	2	2 6	2 6 2	2			
23	Vanadium	V	2	2 6	2 6 3	2			
24	*Chromium	Cr	2	2 6	2 6 5	1			
25	Manganese	Mn	2	2 6	2 6 5	2			

Atomic number	Element	Symbol	1 s	2 s p	3 s p d	4 s p d f	5 s p d f	6 s p d f	7 s p d f
26	Iron	Fe	2	2 6	2 6 6	2			
27	Cobalt	Co	2	2 6	2 6 7	2			
28	Nickel	Ni	2	2 6	2 6 8	2			
29	*Copper	Cu	2	2 6	2 6 10	1			
30	Zinc	Zn	2	2 6	2 6 10	2			
31	Gallium	Ga	2	2 6	2 6 10	2 1			
32	Germanium	Ga	2	2 6	2 6 10	2 2			
33	Arsenic	As	2	2 6	2 6 10	2 3			
34	Selenium	Se	2	2 6	2 6 10	2 4			
35	Bromine	Br	2	2 6	2 6 10	2 5			
36	Krypton	Kr	2	2 6	2 6 10	2 6			
37	Rubidium	Rb	2	2 6	2 6 10	2 6	1		
38	Strontium	Sr	2	2 6	2 6 10	2 6	2		
39	Yttrium	Y	2	2 6	2 6 10	2 6 1	2		
40	Zirconium	Zr	2	2 6	2 6 10	2 6 2	2		
41	*Niobium	Nb	2	2 6	2 6 10	2 6 4	1		
42	*Molybdenum	Mo	2	2 6	2 6 10	2 6 5	1		
43	Technetium	Tc	2	2 6	2 6 10	2 6 5	2		
44	*Ruthenium	Ru	2	2 6	2 6 10	2 6 7	1		
45	*Rhodium	Rh	2	2 6	2 6 10	2 6 8	1		
46	*Palladium	Pd	2	2 6	2 6 10	2 6 10			
47	*Silver	Ag	2	2 6	2 6 10	2 6 10	1		
48	Cadmium	Cd	2	2 6	2 6 10	2 6 10	2		

Structure of Atoms

Atomic number	Element	Symbol	1 s	2 s p	3 s p d	4 s p d f	5 s p d f	6 s p d f	7 s p d f
49	Indium	In	2	2 6	2 6 10	2 6 10	2 1		
50	Tin	Sn	2	2 6	2 6 10	2 6 10	2 2		
51	Antimony	Sb	2	2 6	2 6 10	2 6 10	2 3		
52	Tellurium	Te	2	2 6	2 6 10	2 6 10	2 4		
53	Iodine	I	2	2 6	2 6 10	2 6 10	2 5		
54	Xenon	Xe	2	2 6	2 6 10	2 6 10	2 6		
55	Caesium	Cs	2	2 6	2 6 10	2 6 10	2 6	1	
56	Barium	Ba	2	2 6	2 6 10	2 6 10	2 6	2	
57	Lanthanum	La	2	2 6	2 6 10	2 6 10	2 6 1	2	
58	Cerium	Ce	2	2 6	2 6 10	2 6 10 1	2 6 1	2	
59	Praseodymium	Pr	2	2 6	2 6 10	2 6 10 3	2 6	2	
60	Neodymium	Nd	2	2 6	2 6 10	2 6 10 4	2 6	2	
61	Promethium	Pm	2	2 6	2 6 10	2 6 10 5	2 6	2	
62	Samarium	Sm	2	2 6	2 6 10	2 6 10 6	2 6	2	
63	Europium	Eu	2	2 6	2 6 10	2 6 10 7	2 6	2	
64	Gadolinium	Gd	2	2 6	2 6 10	2 6 10 7	2 6 1	2	
65	Terbium	Tb	2	2 6	2 6 10	2 6 10 9	2 6	2	
66	Dysprosium	Dy	2	2 6	2 6 10	2 6 10 10	2 6	2	
67	Holmium	Ho	2	2 6	2 6 10	2 6 10 11	2 6	2	
68	Erbium	Er	2	2 6	2 6 10	2 6 10 12	2 6	2	
69	Thullium	Tm	2	2 6	2 6 10	2 6 10 13	2 6	2	

Atomic number	Element	Symbol	$\dfrac{1}{s}$	$\dfrac{2}{s\ p}$	$\dfrac{3}{s\ p\ d}$	$\dfrac{4}{s\ p\ d\ f}$	$\dfrac{5}{s\ p\ d\ f}$	$\dfrac{6}{s\ p\ d\ f}$	$\dfrac{7}{s\ p\ d\ f}$
70	Ytterbion	Yb	2	2 6	2 6 10	2 6 10 14	2 6	2	
71	Lutetium	Lu	2	2 6	2 6 10	2 6 10 14	2 6 1	2	
72	Hafnium	Hf	2	2 6	2 6 10	2 6 10 14	2 6 2	2	
73	Tantalum	Ta	2	2 6	2 6 10	2 6 10 14	2 6 3	2	
74	Tungsten	W	2	2 6	2 6 10	2 6 10 14	2 6 4	2	
75	Rhenium	Re	2	2 6	2 6 10	2 6 10 14	2 6 5	2	
76	Osmium	Os	2	2 6	2 6 10	2 6 10 14	2 6 6	2	
77	Iridium	Ir	2	2 6	2 6 10	2 6 10 14	2 6 7	2	
78	Platinum	Pt	2	2 6	2 6 10	2 6 10 14	2 6 9	1	
79	*Gold	Au	2	2 6	2 6 10	2 6 10 14	2 6 10	1	
80	Mercury	Hg	2	2 6	2 6 10	2 6 10 14	2 6 10	2	
81	Thallium	Ti	2	2 6	2 6 10	2 6 10 14	2 6 10	2 1	
82	Lead	Pb	2	2 6	2 6 10	2 6 10 14	2 6 10	2 2	
83	Bismuth	Bi	2	2 6	2 6 10	2 6 10 14	2 6 10	2 3	
84	Polonium	Po	2	2 6	2 6 10	2 6 10 14	2 6 10	2 4	
85	Astatine	At	2	2 6	2 6 10	2 6 10 14	2 6 10	2 5	
86	Radon	Rn	2	2 6	2 6 10	2 6 10 14	2 6 10	2 6	
87	Francium	Fr	2	2 6	2 6 10	2 6 10 14	2 6 10	2 6	1
88	Radium	Ra	2	2 6	2 6 10	2 6 10 14	2 6 10	2 6	2
89	Actinum	Ac	2	2 6	2 6 10	2 6 10 14	2 6 10	2 6 1	2
90	Thorium	Th	2	2 6	2 6 10	2 6 10 14	2 6 10	2 6 2	2
91	Protactinum	Pa	2	2 6	2 6 10	2 6 10 14	2 6 10 2	2 6 1	2
92	Uranium	U	2	2 6	2 6 10	2 6 10 14	2 6 10 3	2 6 1	2
93	Neptunium	Np	2	2 6	2 6 10	2 6 10 14	2 6 10 4	2 6 1	2
94	Plutonium	Pu	2	2 6	2 6 10	2 6 10 14	2 6 10 6	2 6	2

Structure of Atoms

Atomic number	Element	Symbol	1 s	2 s p	3 s p d	4 s p d f	5 s p d f	6 s p d f	7 s p d f
95	Amerilium	Am	2	2 6	2 6 10	2 6 10 14	2 6 10 7	2 6	2
96	Curium	Cm	2	2 6	2 6 10	2 6 10 14	2 6 10 7	2 6 1	2
97	Berkelium	Bk	2	2 6	2 6 10	2 6 10 14	2 6 10 9	2 6	2
98	Californium	Cf	2	2 6	2 6 10	2 6 10 14	2 6 10 10	2 6	2
99	Einsteinium	Es	2	2 6	2 6 10	2 6 10 14	2 6 10 11	2 6	2
100	Fermium	Fm	2	2 6	2 6 10	2 6 10 14	2 6 10 12	2 6	2
101	Mendelevium	Mv	2	2 6	2 6 10	2 6 10 14	2 6 10 13	2 6	2
102	Nobelium	No	2	2 2	2 6 10	2 6 10 14	2 6 10 14	2 6	2
103	Lawrencium	Lw	2	2 6	2 6 10	2 6 10 14	2 6 10 14	2 6 1	2
104	Uniquadium	Unq	2	2 6	2 6 10	2 6 10 14	2 6 10 14	2 6 2	2
105	Unipentium	Unp	2	2 6	2 6 10	2 6 10 14	2 6 10 14	2 6 3	2
106	Unilhexium	Unh	2	2 6	2 6 10	2 6 10 14	2 6 10 14	2 6 4	2
107	Uniseptium	Uns	2	2 6	2 6 10	2 6 10 14	2 6 10 14	2 6 5	2
108	Uniloctium	Uno	2	2 6	2 6 10	2 6 10 14	2 6 10 14	2 6 6	2
109	Unilenium	Une	2	2 6	2 6 10	2 6 1014	2 6 10 14	2 6 7	2

* It may be noted that the electric configurations of the elements such as Chromium, Copper, Molybdenum, Silver, Gold, etc. differs from the normal sequence of filling up the electrons in the energy levels of their atoms. It is due to the inter electronic repulsions. As a result of this, the electrons differing slightly in their energies have a preference to enter into those sub-shells which either get completely filled or just half-filled.

Hence two electrons, with opposite spins, can exist in the same orbit (*i.e.*, same n, l and m_l). However, two electrons with same spin will never take up the same orbit. Hence in a stable atomic system, each electron must have either a different orbit or different spin, *i.e.*, for any two electrons, the four quantum numbers can never be same.

2.30. PERIODIC TABLE

There are 109 elements, which have been discovered so far. It is difficult to study the various properties of all these elements separately. In order to reduce their study, the elements having similar properties are divided into a few groups. All these groups (along with their elements) are then arranged in a tabular form which is known as periodic table. A periodic table consists of horizontal rows of elements known as periods and vertical columns as groups. Following two periodic tables are important from the subject point of view:

2.31. MENDELEEV'S PERIODIC TABLE

It was proposed by Mendeleev, a Russian scientist in 1869, when he discovered that if the elements are arranged in the order of their increasing atomic weights, the elements with similar properties occur at regular intervals. Based on this observation, he gave a law which states that the properties of the elements are periodic functions of their atomic weights.

Mendeleev arranged all the elements (known at that time) in the order of their increasing atomic weights in horizontal rows in such a way that the elements having similar properties come directly under one another in the same vertical columns or groups. In this table, there were seven horizontal rows (or periods) and eight vertical columns (or groups). In order to make sure that the elements having similar properties fall in the same vertical column or group, Mendeleev left some gaps in his periodic table. These gaps were left for the elements, which were not discovered at that time.

Though the Mendeleev's Periodic Table was of great significance in the study of elements, yet it could not explain the position of isotopes. The isotopes were not given a separate place in the periodic table. In certain cases, it was found that the elements with higher atomic weight came first and the elements with lower atomic weight came later. As a result of this, it was felt that the atomic weight cannot be taken as the basis for the classification of elements.

2.32. MODERN PERIODIC TABLE

In 1913, Mosley discovered that the properties of elements are periodic functions of their atomic numbers and not atomic weights. We know that the atomic number of an elements is fixed. And no two elements can have the same atomic number. It is thus obvious that the atomic number is more a fundamental property than the atomic weight for the classification of elements. After this discovery, Mosey proposed a periodic table, in which the elements were arranged in the order of their increasing atomic numbers.

Later on, the periodic table proposed by Mosley was modified by Bohr, which is now popularly known as Modern Periodic Table. In this table, the elements are also arranged in the order of their increasing atomic number as shown in Fig. 2.10

The Modern Periodic Table, is divided into the following four blocks:

1. *s-block*. The elements, whose valence electrons lie in the *s*-sub shell, are known as *s*-block elements. These are shown in the left hand side of the periodic table. This block consists of 2 groups namely IA and IIA. The group IA consists of elements of atomic numbers 1, 3, 11, 19, 37, 55 and 87. All these elements have one valence electron and are said to have similar electronic configuration. The elements in this group are known as Alkali metals. Similarly, the group *IIA* consists of elements of atomic numbers 4, 12,

Structure of Atoms

20, 38, 56, and 88. All these elements have two valence electrons and are also said to have similar electronic configuration.

2. *p-block.* The elements, whose valence electrons lie in the *p*-sub shell, are known as *p*-block elements. These are shown in the shell, are known as *p-block* elements. These are shown in the right hand side of the periodic table. This block consists of 6 groups namely *IIIA, IVA, VA, VIA, VIIA and zero.* This group number indicates the number of valence electrons. The zero group has no valence electron as its outermost shell is completely filled with electrons. The elements in this group are known as inert gases or noble gases.

3. *d-block.* The elements, whose valence electrons lie in the *d-sub-shell* are known as *d*-block elements. These are shown in the centre of the periodic table. This block consists of 8 groups namely *IIIB, IVB, VB, VIB, VIIB, IB and IIB.* The group *VIII* consists of three sub-groups.

4. *f*-block. The elements, whose valence electrons lie in the *f*-sub shell are known as *f*-block elements. These are shown in the bottom of the periodic table.

It will also be interesting to know that the periodic table shown above, is also divided into seven horizontal rows or periods:

1. *First period.* It contains only two elements with atomic number 1 and 2. The element with atomic number 1 lies in group IA (of *s*-block) and that with atomic number 2 in group zero (of *p*-block). It is known as a short period.

2. *Second period.* It contains eight elements with atomic number 3 to 10. The elements with atomic number 3, 4 lie in group IA and IIA (of *s*-block) and those with atomic number 5 to 10 in groups *IIIA* to zero (in *p*-block). It is also known as a short period.

3. *Third period.* It also contains eight elements with atomic number 11 to 18. The elements with atomic number 11, 12 lie in groups IA and IIA (of *s*-block) and those with atomic number 13 to 18. In groups IIIA to zero (in *p*-block). It is also known as a short period.

4. *Fourth period.* It contains 18 elements with atomic number 19 to 36. The elements with atomic number 19, 20 lie in *s*-block, 21 to 30 in *d-block* and 31 to 36 in *p*-block. It is known as a long period.

5. *Fifth period.* It also contains 18 elements with atomic number 37 to 54. The elements with atomic number 37, 38 lie in *s*-block, 39 to 48 in *d*-block and 49 to 54 in *p*-block. It is also known as a long period.

6. *Sixth period.* It contains 32 elements with atomic number 55 to 86. The elements with atomic number 55, 56 lie in *s*-block 57 and 72 to 80 in *d*-block, 58 to 71 in *f*-block and 81 to 86 in *p*-block. It is known as very long period.

7. *Seventh period.* It contains the remaining elements with atomic number 87 to 105. The elements with atomic number 87, 88 lie in *s*-block 89, 104 to 105 in *d*-block and 90 to 103 in *f*-block. This period is still incomplete.

All the elements in *d*-block are called transition elements. And the elements with atomic number 58 to 71 in *f*-block are called lanthanides and those with atomic number 90 to 103 are actinides.

These two series of elements have similar properties. Thus they have been placed at the bottom of the periodic table.

2.33. SIGNIFICANCE OF ATOMIC NUMBER IN THE PERIODIC TABLE

We have already discussed in the last article that the properties of an element depend upon the number of valence electrons in its atoms. The elements having same number of valence electrons, show similar properties. In other words, the elements having similar electronic configuration show similar properties.

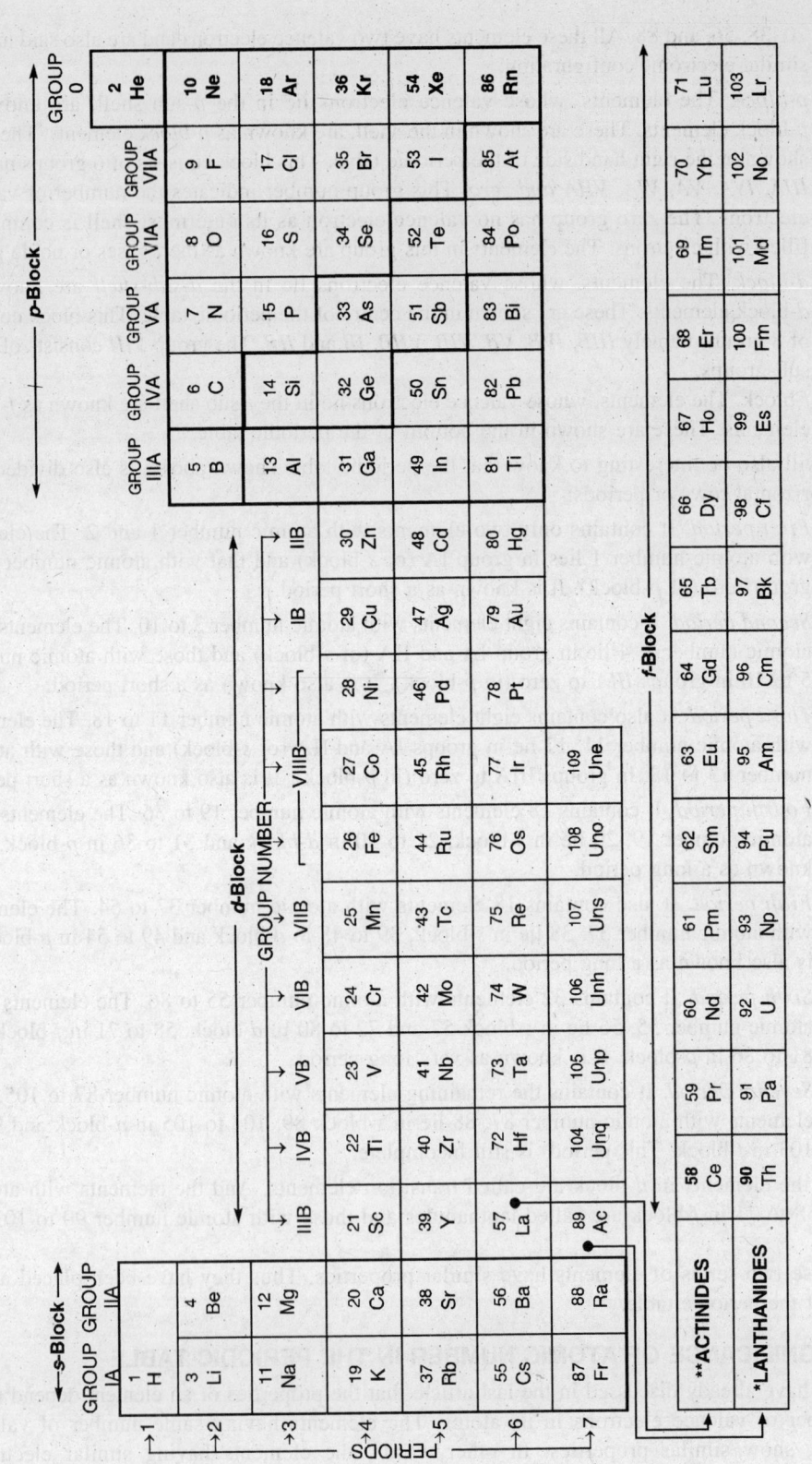

Fig. 2.10. Modern Periodic Table.

Structure of Atoms

The atomic number of an element gives us the number of electrons present in an atom. Moreover, with the given atomic number, we can write the electronic configuration of any element. All the elements with a similar electronic configuration, can be placed in the same group. Thus the significance of atomic number lies in the classification of elements in arranging them in the periodic table.

A.M.I.E. (I) EXAMINATION QUESTIONS

1. State and explain Bohr's model of an electron. *(Summer, 1993)*
2. Describe Bohr's theory of hydrogen atom. *(Winter, 1993)*
3. Write electronic configuration of iron. *(Winter, 1993)*
4. Write electronic configuration of silicon and germanium. *(Summer, 1995)*
5. Briefly discuss the various quantum numbers. *(Winter, 1996)*
6. (a) Enumerate Bohr's postulates of atomic structure.
   (b) What is understood by the electronic configuration ? *(Summer, 1997)*
7. What is the maximum number of electrons in a sub-shell ? *(Summer, 1997)*
8. Describe briefly Bohr's atomic model, quantum numbers and Pauli's exclusion principle. *(Summer, 1998)*
9. Explain briefly the electronic structure of an atom. *(Winter, 1999)*
10. Briefly explain the model proposed by Rutherford. *(Summer, 2001)*
11. What were the major deficiencies in Rutherford's atomic model ? *(Winter 2001; 2003)*

3

Crystal Structure

1. Introduction. 2. Crystal. 3. Single Crystal, 4. Whiskers. 5. Crystal Lattice or Space Lattice. 6. Unit Cell. 7. Lattice Parameters of a Unit Cell. 8. Primitive Cell. 9. Types of Crystal Systems. 10. Bravis Lattices. 11. Metallic Structure. 12. Body Centred Cubic (BCC) Structures. 13. Face Central Cubic (FCC) Structures. 14. Hexagonal Close Packed (HCP) Structures. 15. Miller Indices. 16. Procedure for Finding Miller Indices of Crystal Planes. 17. Important Features of Miller Indices of Crystal Planes.18. Representation of Crystal Planes in a Cubic Unit Cell. 19. Procedure for Sketching the Plane from the Given Miller Indices. 20. Common Planes in Simple Cubic Structure. 21. Crystal Directions. 22. Procedures for Finding Miller indices of Crystal Directions. 23. Representation of Crystal Direction in a Cubic Unit Cell. 24.Procedure for Sketching Directions from the Given Miller Indices 25. Representation of Crystal planes and Directions in HCP unit cell. 26. Atomic Radius in Cubic and Hexagonal systems. 27. Number of Atoms in a Cubic Structure. 28. Number of Atoms per Square Millimeter in Cubic Crystal Planes. 29. Number of Atoms per Square Millimeter in Simple Cubic Structure. 30. Number of Atoms per Square Millimeter in a Body Centered Cubic (BCC) Structure. 31. Number of Atoms per Square Millimetre in a Face Centered Cubic (FCC) Structure. 32. Atomic Packing Factor. 33. Close Packed Structures (FCC and HCP). 34. Perpendicular Distance between a Plane and Origin of the Cube. 35. Angle between Two Planes or Directions. 36. Dimensions of a Unit Cell. 37. Calculation of Number of Atoms per Cubic Centimeter. 38. Mean Distance between Atoms. 39. Co-ordination Number 40. Crystalline and Amorphous Solids. 41. Polymorphism. . 42. Crystal Defects or Imperfections. 43. Point Defects. 44. Line Defects. 45. Surface Defects 46. Determination of Crystal Structure. 47. Bragg's Law.

3.1 INTRODUCTION

If we examine a few metallic solid materials, with which we daily come across, we shall find that most of them do not have any characteristic difference in their outer appearance. But if we see them under a scanning tunnelling microscope, we shall find that some materials have a crystalline structure. The examples of such materials are iron, copper, aluminium etc. There are some other materials, which do not have crystalline structure. Such materials. are called non-crystalline or amorphous materials. The examples of such materials are wood, plastics, paper glass etc. In this chapter, we shall discuss only the crystalline materials.

A crystalline material may be either in the form of single crystal or an aggregate of many crystals (known as polycrystalline). The polycrystalline material is always separated by well-defined boundaries.

In a single crystal solid, the arrangement of atoms is perfectly periodic and repeated throughout the whole specimen without interruption. All the unit cells (covered in Art. 3.6) inter-lock in the same way and have the same orientation. Although single crystals do exist in nature yet they may also be produced artificially. They are ordinarily difficult to grow because the environment must be carefully controlled. Within the past few decades, single crystals have become extremely important in many of our modern technologies. This is especially the case in particular electronic microcircuits (integrated circuits).

In a polycrystalline material, the solid is composed of many small crystals or grains. Fig. 3.1 shows the diagram of a polycrystalline material. Notice that this type of material there exists some mismatch within the region where two grains meet. This area is called grain boundary.

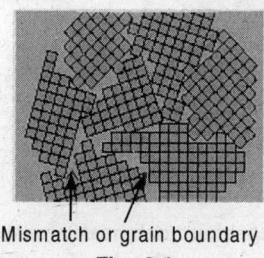

Mismatch or grain boundary
Fig. 3.1

3.2 CRYSTAL

The term 'crystal of a material may be defined as a small body having a regular polyhedral form, bounded by smooth surfaces, which are acquired under the action of its intermolecular forces. The crystals are also known as grains. The boundary, separating the two adjacent grains, is called grain boundary. It has been observed that a crystal is symmetrical about its certain elements like points, lines or planes. If a crystal is rotated about these elements, it is not possible to distinguish its new position form the original position.

This symmetry of a crystal is an important characteristic, which is based on its internal structure. This characteristic helps us in the classification of crystals and describing their behaviour.

3.3 SINGLE CRYSTAL

We have already discussed in Art. 3.1 that most of the materials exist in polycrystalline form. But there are some materials, which exist in the form of single crystals. The common examples of such materials are sugar, common salt (sodium chloride) diamond etc.

The single crystals of materials are produced artificially from their vapour or liquid state. These crystals represent a material in its ideal condition. They also help us in studying their behaviour and defects.

3.4 WHISKERS

The term 'whiskers' may be defined as single crystals, which are produced under some special conditions so that they are without any structural defect. It has been observed that the presence of crystal defects, in a crystal, decrease its bond strength. But the whiskers, being defect-free, have their bond strength equal to their ideal strength.

The whiskers are produced in the form of fine filaments *i.e.*, thread-like structure. Their diameter may be as small as 10^{-5} millimeter. It has been observed that the increase in diameter of the whiskers decreases its strength and increases its ductility.

3.5 CRYSTAL LATTICE OR SPACE LATTICE

We have already discussed in Art. 3.1 that a metal consists of a number of crystals. And each crystal, in turn, consists of a large number of atoms.

To discuss the crystalline structure, we will consider atoms as being hard spheres with well defined radii. In this hard sphere model, the shortest distance between two like atoms is one diameter.

Let us now consider the three-dimensional crystalline structure in which the atoms are arranged in a regular pattern as shown in Fig. 3.2 (a).

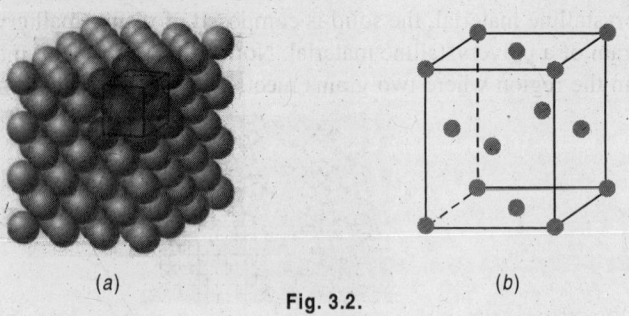

(a)   (b)

Fig. 3.2.

It may be noted from the figure that each atom, present in the crystal has its surrounding identical to that of every other atom. Now if we replace all the atoms by small circle or points (corresponding to the centre of spheres or atoms), we shall get a group of small circles or points as shown in Fig. 3.2(b). In this group, each point has its surrounding identical to that of every other point. These points are known as lattice points. This group of lattice points is known as crystal lattice or space lattice.

Note: The usual assumption, while discussing the crystal lattice of metal crystals, is that its structure continues to extend to infinity in all directions.

3.6 UNIT CELL

We have already discussed in the last article that in a crystal system, the atoms or molecules are arranged in a particular fashion. Moreover, these atoms or molecules are assumed to continue to extend to infinity in all the three axes in the same fashion as the bricks are laid in a massive masonry structure. In the study of geometry of the arrangement of these atoms, the general method adopted is that we study its small part, which has the same arrangement of its atoms as that of the crystal. This small part is called unit cell.

Thus to summarize, unit cell is the smallest structural unit or building block that can describe the crystals structure. Repetition of the unit cell generates the entire crystal.

3.7 LATTICE PARAMETERS OF A UNIT CELL

Fig. 3.3 shows a unit cell of a three-dimensional crystal lattice. It is formed by primitives or intercepts a, b and c along the three axes respectively. The three angles (α, β and γ) are called interfacial angles. A little consideration will show, that both the intercepts and interfacial angles constitute the lattice parameters of the unit cell. It is thus obvious, that if the values of intercepts and interfacial angles are known, we can easily determine the form and actual size of the unit cell.

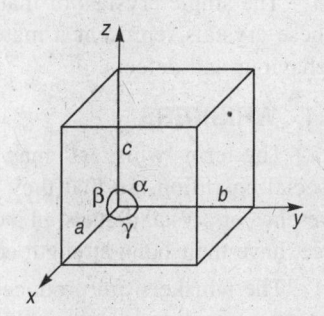

Fig. 3.3. Lattice parameters of a unit cell.

3.8 PRIMITIVE CELL

The term 'primitive cell' may be defined as a unit cell, which possesses lattice points at its corners only. In other words, a primitive cell is the simplest type of unit cell which contains one lattice point per unit cell. A simple cubic cell is an example of a primitive cell. The unit cells, which contain more than one lattice point, are called non-primitive cells. For example, the unit cells of * body-centred and face-centred cubic structures contain more than one lattice point per unit cell. Thus they are non-primitive cells.

---

* For details please refer to Art. 3.12.

It will be interesting to know that if the number of lattice points per unit cell are two, the unit is called doubly primitive cell. Similarly, if the number of lattice points per unit cell are three or four, the unit is called triply primitive cell or quadruply primitive cell respectively.

3.9 TYPES OF CRYSTAL SYSTEMS

There are 32 classes of crystal systems based on the geometrical considerations (*i.e.,* symmetry and internal structure). But it is a common practice to divide all the crystal systems into seven groups or basic systems. These seven basic crystal systems are distinguished from one another by the angle between the three axes and the intercepts of the faces along them. These seven basic crystal systems are:

1. Cubic. 2. Tetragonal. 3. Hexagonal. 4. Orthorhombic. 5. Rhombohedral. 6. Monoclinic. 7. Triclinic.

In the following pages, we shall discuss all the above mentioned seven types of basic crystal systems one by one.

1. *Cubic crystal system*

This system includes all those crystals, which have three equal axes and at right angles to each other as shown in Fig. 3.4 (*a*)

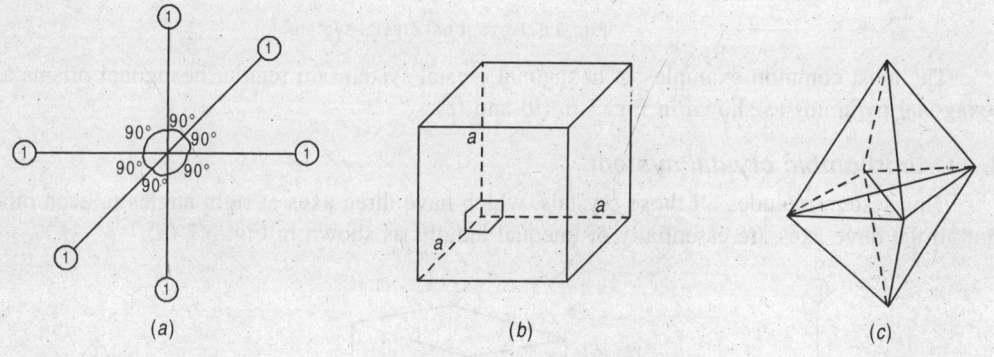

Fig. 3.4. Cubic crystal system.

The most common examples of cubic crystal system are cube and octahedron as shown in 3.4 (*b*) and (*c*).

2. *Tetragonal crystal system*

This system includes all those crystals, which have three axes at right angles to each other. Two of these axes (say horizontal) are equal, while the third (say vertical) is differnt (*i.e.* either longer or shorter than the other two) as shown in Fig. 3.5(*a*).

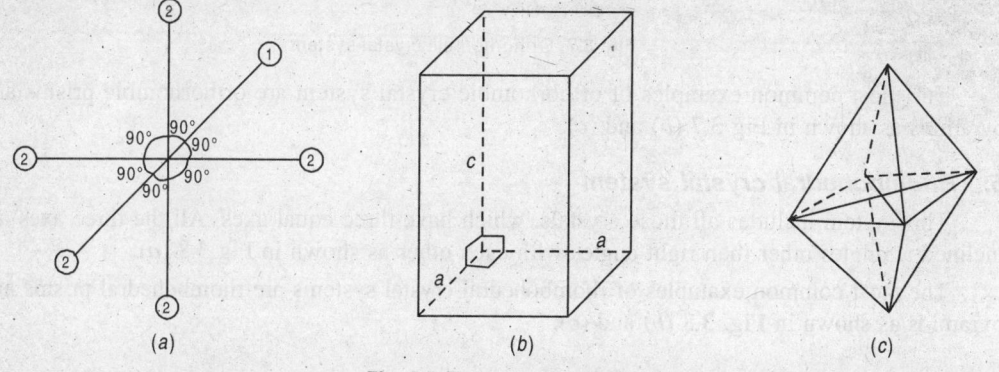

Fig. 3.5. Tetragonal crystal system.

The most common examples of tetragonal crystal system are regular tetragonal prisms and pyramids as shown in Fig. 3.5 (b) and (c).

3. Hexagonal crystal system

This system includes all those crystals, which have four axes. Three of these axes (say horizontal) are equal and meet each other at an angle of 60°. The fourth axis (say vertical) is different (*i.e.*, either longer or shorter than the other three axes) as shown in Fig. 3.6 (a).

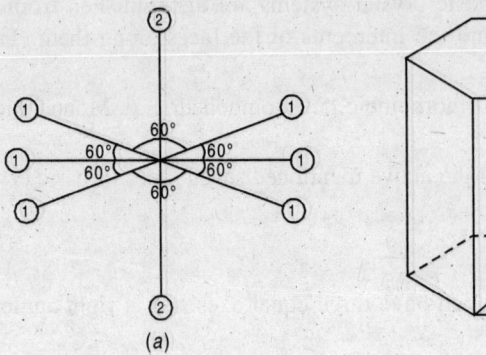

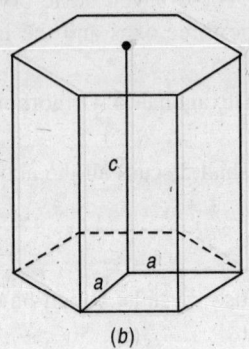

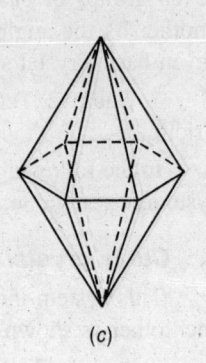

Fig. 3.6. Hexagonal crystal system.

The most common examples of hexagonal crystal system are regular hexagonal prisms and hexagonal pyramids as shown in Fig. 3.6. (b) and (c).

4. Orthorhombic crystal system

This system includes all those crystals, which have three axes at right angles to each other. But all the three axes are essentially of unequal lengths as shown in Fig. 3.7 (a).

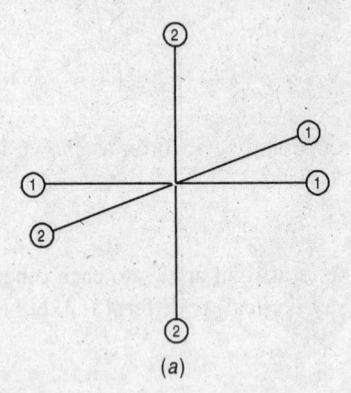

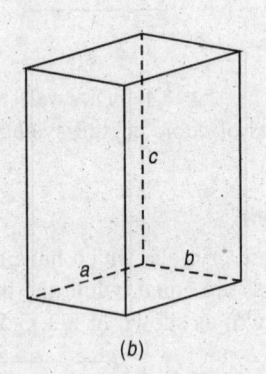

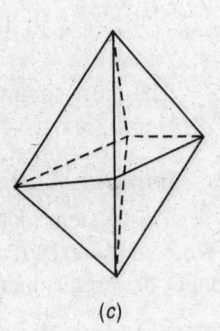

Fig. 3.7. Orthorhombic crystal system.

The most common examples of orthorhombic crystal system are orthorhombic prisms and pyramids as shown in Fig 3.7 (b) and (c).

5. Rhombohedral crystal system

This system includes all those crystals, which have three equal axes. All the three axes are inclined at angles other than right angles with each other as shown in Fig 3.8 (a).

The most common examples of rhombohedral crystal systems are rhombohedral prisms and pyramids as shown in Fig. 3.8 (b) and (c).

Crystal Structure

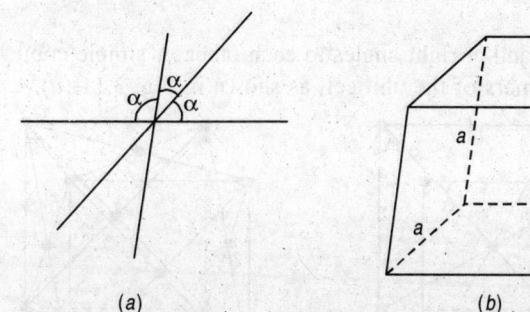

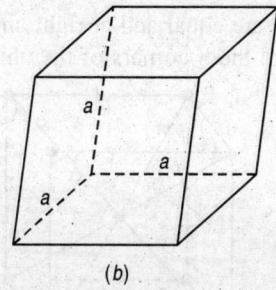

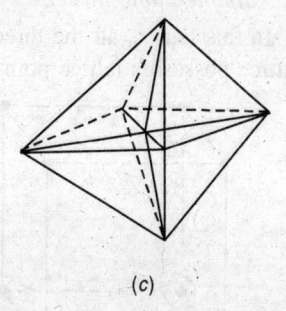

Fig. 3.8. Rhombohedral crystal system.

6. Monoclinic crystal system

This system includes all those crystals, which have three essentially unequal axes. One of these axes is at right angles with the other two. But the other two axes are not at right angles to each other as shown in Fig 3.9.

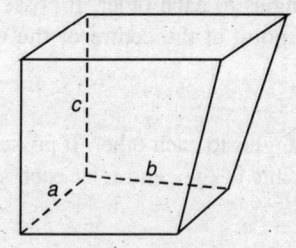

Fig. 3.9. Monoclinic crystal system.

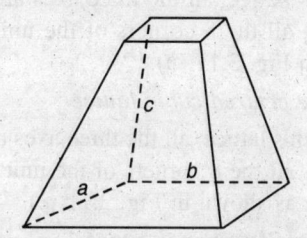

Fig. 3.10. Triclinic crystal system.

7. Triclinic crystal system

This system includes all those crystals, which have three unequal axes and none of them is at right angles to the other two axes as shown in Fig. 3.10. A little consideration will show that all the irregular crystals belong to this crystal system.

Table 3.1 shows the seven basic crystal systems along with their characteristics:

Table 3.1

S.No.	Crystal system	Axial lengths (a, b, and c)	Interaxial angles (α, β, γ)
1.	Cubic	$a = b = c$	$\alpha = \beta = \gamma\ 90°$
2.	Tetragonal	$a = b \neq c$	$\alpha = \beta = \gamma = 90°$
3.	Hexagonal	$a = b \neq c$	$\alpha = \beta = 90°; \gamma = 120°$
4.	Orthorhombic	$a \neq b \neq c$	$\alpha = \beta = \gamma = 90°$
5.	Rhombohedral	$a = b = c$	$\alpha = \beta = \gamma \neq 90°$
6.	Monoclinic	$a \neq b \neq c$	$\alpha = \beta = 90°; \gamma \neq 90°$
7.	Triclinic	$a \neq b \neq c$	$\alpha \neq \beta \neq \gamma \neq 90°$

3.10 BRAVAIS LATTICES

We have already discussed in the last article that there are seven crystal systems. But according to Bravais, there are 14 possible types of space lattices in these 7 crystal systems. Now we shall discuss all the fourteen Bravais lattices one by one.

1. *Simple cubic lattice*

In this lattice, all the three axes are equal and at right angles to each other. A simple cubic lattice possesses lattice points at all the 8 corners of the unit cell as shown in Fig. 3.11 (*a*).

(a) Simple cubic   (b) Body centred   (c) Face centred

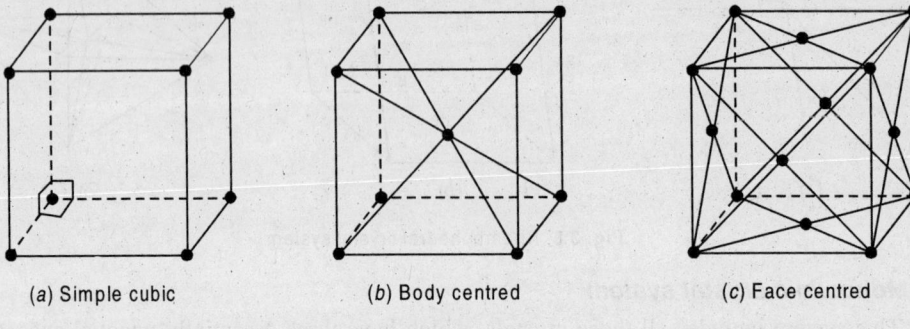

Fig. 3.11

2. *Body centred cubic lattice*

In this lattice, all the three axes are equal and at right angles to each other. It possesses lattice points at all the 8 corners of the unit cell and one lattice point at the centre of the unit cell as shown in Fig. 3.11 (*b*).

3. *Face centred cubic lattice*

In this lattice, all the three axes are equal and at right angles to each other. It possesses lattice points at all the 8 corners of the unit cell and six lattice points at the centres of each six faces of the cube as shown in Fig. 3.11 (*c*).

4. *Simple tetragonal lattice*

(a) Simple tetragonal   (b) Body centred tetragonal   (c) Simple hexagonal

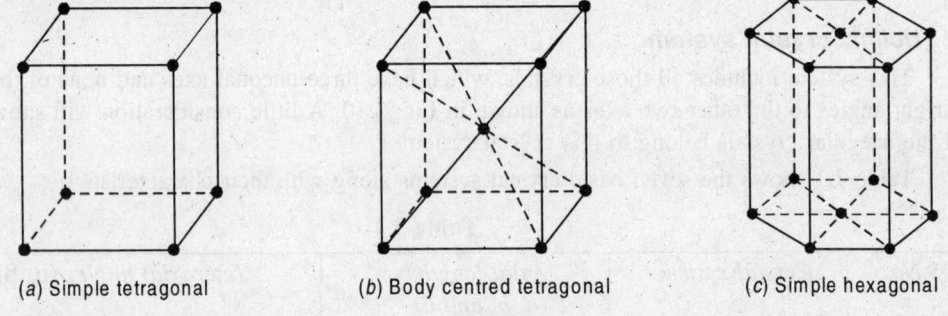

Fig. 3.12

In this lattice, all the three axes are at right angles to each other. Two of these axes (say horizontal) are equal, while the third one (say vertical) is different (*i.e.*, either longer or shorter than the other two). It possesses lattice points at all the 8 corners of the unit cell as shown in Fig 3.12 (*a*).

5 *Body centred tetragonal lattice*

In this lattice, all the three axes are at right angles to each other. Two of these (say horizontal) are equal, while the third one (say vertical) is different (*i.e.,* either longer or shorter than the other). It possesses lattice points at all the 8 corners of the unit cell and also one lattice point at the centre of the body as shown in Fig. 3.12(b).

6. *Simple hexagonal lattice*

In this lattice, there are 4 axes. Three of these axes (say horizontal) are equal and meet each other at an angle of 60°. The fourth axis (say vertical) is different (*i.e.*, either longer or shorter than the other three axes). It possesses lattice points at all the 12 corners of the hexagonal prism

Crystal Structure

and also two lattice points each at the centre of the two hexagonal faces of the unit cell as shown in Fig. 3.12 (c).

7. Simple orthorhombic lattice

In this lattice, all the three axes are at right angles to each other. But all the three axes are essentially of unequal lengths. It possesses lattice points at all the 8 corners of the unit cell as shown in Fig. 3.13 (*a*).

(a) Simple orthorhombic   (b) End centred   (c) Body centred   (d) Face centred

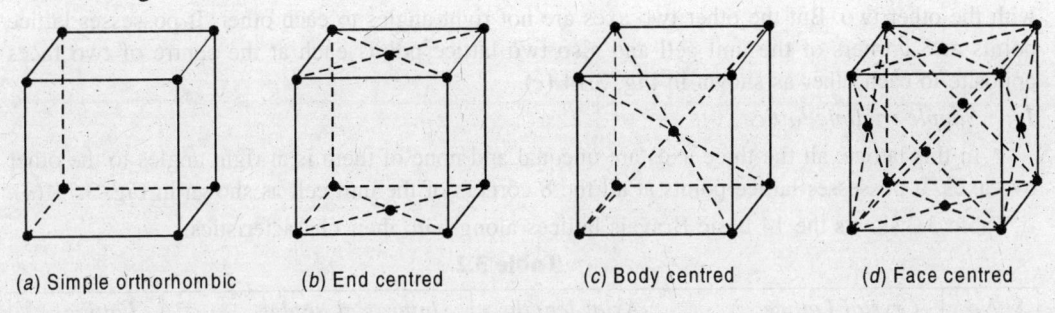

Fig. 3.13

8. End centred orthorhombic lattice

In this lattice, all the three axes are at right angles to each other. But all the three axes are essentially of unequal lengths. It possesses lattice points at all the 8 corners of the unit cell and also two lattice points each at the centre of two faces opposite to each other. If the lattice points are at the top and bottom faces of the lattice, it is known as base-centred orthorhombic lattice as shown in Fig 3.13 (*b*).

9. Body centred orthorhombic lattice

In this lattice, all the three axes are at right angles to each other and all the three axes are essentially of unequal lengths. It possesses lattice points at all the 8 corners of the unit cell and also one lattice at the centre of the body as shown in Fig. 3.13 (*c*).

10. Face centred orthorhombic lattice

In this lattice, all the three axes are at right angles to each other, and all the three axes are essentially of unequal lengths. It possesses lattice points at all the 8 corners of the unit cell and also six lattices at the centre points of each six faces of the unit cell as shown in Fig. 3.13 (*d*).

11. Simple rhombohedral lattice

In this lattice, all the three axes are equal and inclined at equal angles, but not at right angles with each other. It possesses lattice points at all the 8 corners of the unit cell as shown in Fig 3.14(*a*).

(a) Simple rhombohedral   (a) Simple monoclinic   (a) End centred monoclinic   (a) Simple triclinic

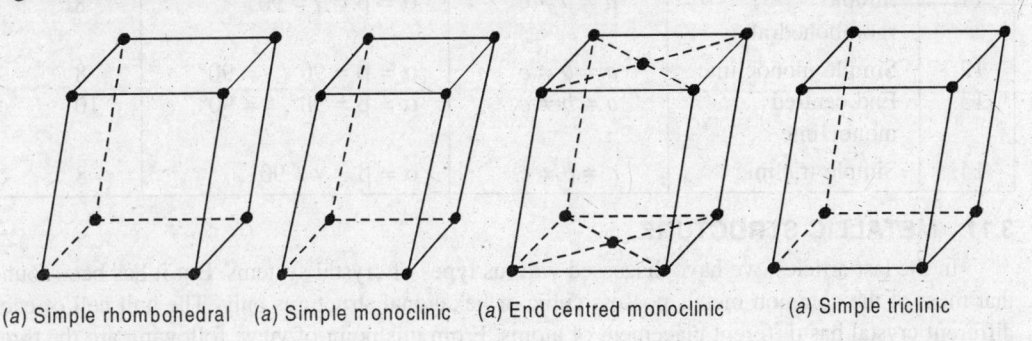

Fig. 3.14

12. Simple monoclinic lattice

In this lattice, all the three axes are essentially unequal. One of these axes is at right angles with the other two. But the other two axes are not at right angles to each other. It possesses lattice points at all the 8 corners of the unit cell as shown in Fig. 3.14 (b).

13. End centred monoclinic lattice

In this lattice, all the three axes are essentially unequal. One of these axes is at right angles with the other two. But the other two axes are not right angles to each other. It possesses lattice points at 8 corners of the unit cell and also two lattice points each at the centre of two faces opposite to each other as shown in Fig. 3.14 (c).

14. Simple triclinic lattice

In this lattice, all the three axes are unequal and none of them is at right angles to the other two axes. It possesses lattice points at all the 8 corners of the unit cell as shown in Fig. 3.14 (d).

Table 3.2 shows the 14 basic Bravais lattices along with their characteristics:

Table 3.2

S. No.	Crystal Lattice	Axial lengths (a,b and c)	Interaxial angles	Lattice point
1.	Simple cubic	$a = b = c$	$\alpha = \beta = \gamma = 90°$	8
2.	Body-centred cubic	$a = b = c$	$\alpha = \beta = \gamma = 90°$	9
3.	Face-centred cubic	$a = b = c$	$\alpha = \beta = \gamma = 90°$	14
4.	Simple tetragonal	$a = b \neq c$	$\alpha = \beta = \gamma = 90°$	8
5.	Body-centred tetragonal	$a = b \neq c$	$\alpha = \beta = \gamma = 90°$	9
6.	Simple hexagonal	$a = b \neq c$	$\alpha = \beta = 90°, \gamma = 120°$	14
7.	Simple orthorhombic	$a \neq b \neq c$	$\alpha = \beta = \gamma = 90°$	8
8.	End centred orthorhombic	$a \neq b \neq c$	$\alpha = \beta = \gamma = 90°$	10
9.	Body-centred orthorhombic	$a \neq b \neq c$	$\alpha = \beta = \gamma = 90°$	9
10.	Face centred orthorhombic	$a \neq b \neq c$	$\alpha = \beta = \gamma = 90°$	14
11.	Simple rhombohedral	$a \neq b \neq c$	$\alpha = \beta = \gamma = 90°$	8
12.	Simple monoclinic	$a \neq b \neq c$	$\alpha = \beta = 90°, \gamma \neq 90°$	8
13.	End centred monoclinic	$a \neq b \neq c$	$\alpha = \beta = 90°, \gamma \neq 90°$	10
14.	Simple triclinic	$a \neq b \neq c$	$\alpha \neq \beta \neq \gamma \neq 90°$	8

3.11 METALLIC STRUCTURE

In the last articles, we have discussed various types of crystal systems. But it has been found that most of the common metals possess cubic or hexagonal structures only. The unit cell of each different crystal has different placement of atoms. From this point of view, following are the three types of metallic structures:

Crystal Structure

1. Cubic structures
   (a) Body centred cubic (*BCC*) structure
   (b) Face centred cubic (*FCC*) structure
2. Hexagonal close packed (*HCP*) structures

In the following pages, we shall discuss all the above mentioned types of metallic structures one by one.

3.12 BODY CENTRED CUBIC (BCC) STRUCTURES

In this type of structure, the unit cell (in the shape of a cube) contains one atom at each of its 8 corners and another atom at the body centre as shown in Fig. 3.15. It is thus obvious, that each unit cell shares 8 atoms one on each of its corners in addition to one atom at the body centre.

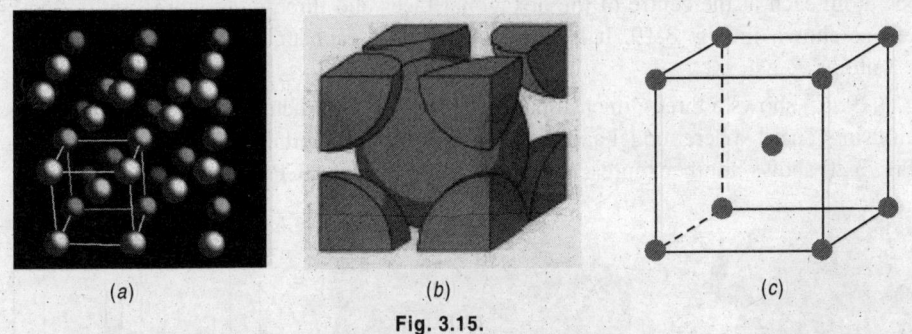

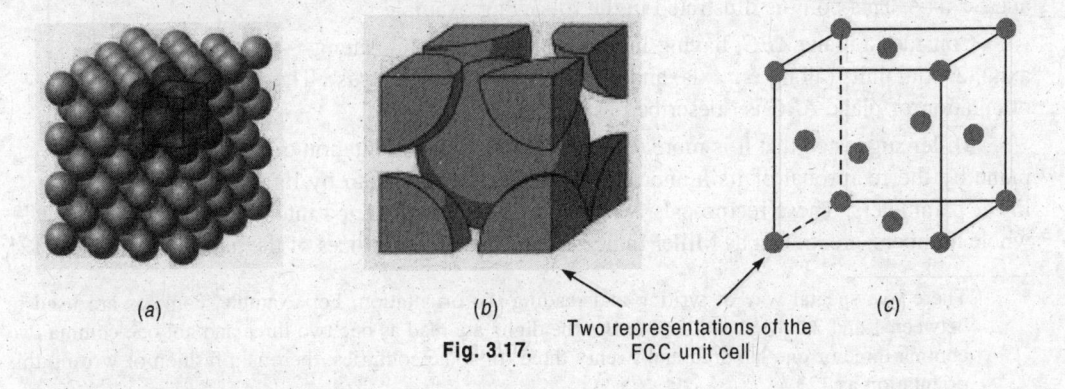

(a)   (b)   (c)

Fig. 3.15.

Fig. 3.15 (*a*) shows a three-dimensional model of a BCC structure while Fig. 3.15 (*b*) shows unit cell using a hard sphere model and Fig. 3.15 (*c*) shows a unit cell indicating only the lattice points. Fig. 3.16 shows another representation for the unit cell of the BCC structure.

The common examples of this type of structure are α-iron (below 910 °C), δ-iron (1400 °C to 1839 °C), tungsten, vanadium, molybdenum, chromium and alkali metals (*i.e.*, sodium and caesium etc).

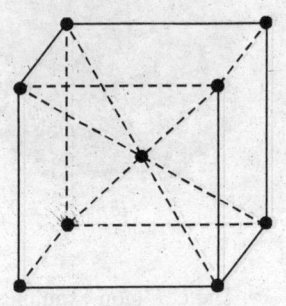

Fig. 3.16. BCC structure.

3.13 FACE CENTRED CUBIC (FCC) STRUCTURES

In this type of structure, the unit cell (in the shape of a cube) contains one atom at the centre of its each face, in addition to one atom at each of its 8 corners as shown in Fig. 3.17. It may be noted that this type of structure does not contain any atom at the centre of the unit cell. It is thus obvious that each unit cell shares 14 atoms.

(a)   (b)   (c)

Fig. 3.17.

Two representations of the FCC unit cell

Fig. 3.17 (a) shows a three-dimensional model of a FCC structure, while Fig. 3.17 (b) shows a unit cell using hard sphere model and Fig. 3.17 (c) shows a unit cell indicating only the lattice points. Fig. 3.18 shows another representation for the unit cell of FCC structure.

The common examples of this type of structure are Υ-iron (910°C to 1400°C), copper, silver, gold, aluminium, nickel lead and platinum etc. The face centred cubic structures are also found in ceramic crystals.

Fig. 3.18. FCC structure.

3.14 HEXAGONAL CLOSE PACKED (HCP) STRUCTURES

In this type of structure, the unit cell contains one atom at each corner of the hexagonal prism, one atom each at the centre of the hexagonal faces and three more atoms within the body of the cell as shown in Fig. 3.19. It is thus obvious that each unit cell shares 14 atoms, and contains 3 atoms.

Fig. 3.19. (a) shows a three-dimensional model of a HCP structure while Fig. 3.19 (b) shows a unit cell using a hard sphere model and Fig. 3.19. (c) shows a unit cell indicating only the lattice points. Fig. 3.20 shows another representation of a unit cell of HCP structure.

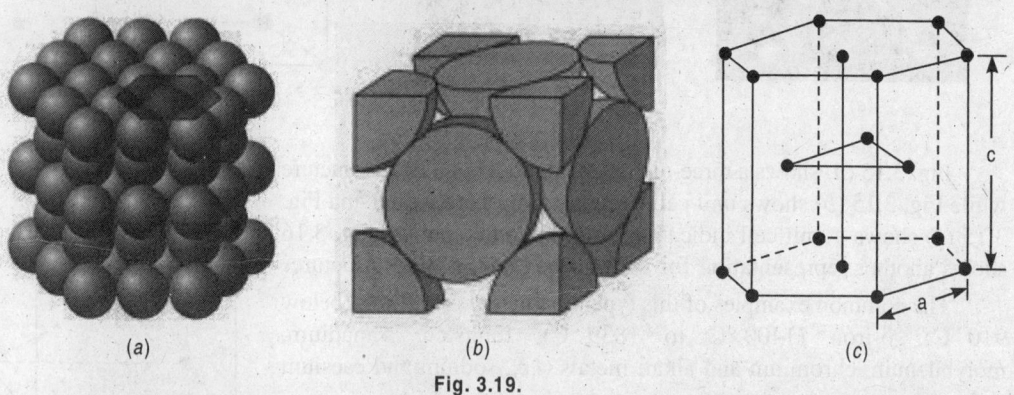

(a)   (b)   (c)

Fig. 3.19.

The common examples of this type of structure are magnesium, zinc, titanium, zirconium, beryllium, and cadmium etc.

3.15 MILLER INDICES

The orientation of a plane, in every crystal system, is described in terms of coordinates through which they pass along x-x axis, y-y axis, and z-z axis. This point is illustrated in the following example:

Consider a plane *ABC*, having intercepts of 1 axial unit (along x-x axis), 2 axial units (along y-y axis) and 3 axial units (along z-z axis). The orientation of plane *ABC* is *described as 1, 2, 3.

Miller suggested that it is more useful to describe the orientation of plane by the reciprocal of its numerical parameters, rather than by its linear parameters. These reciprocals, when appropriately converted into whole numbers, are known as Miller Indices. Thus the Miller indices of the plane *ABC* in Fig. 3.21

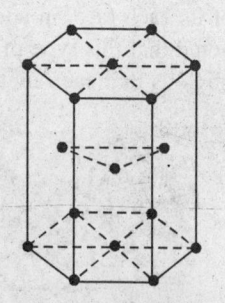

Fig. 3.20. HCP structure.

---

* There is a special way of writing and reading the orientation. For example, commas are used in between 1 and 2 as well as 2 and 3. But the digits are read as one two three and not one comma two comma three or one hundred and twenty three. In some countries, there is a fashion of writing this orientation as 123.

Crystal Structure

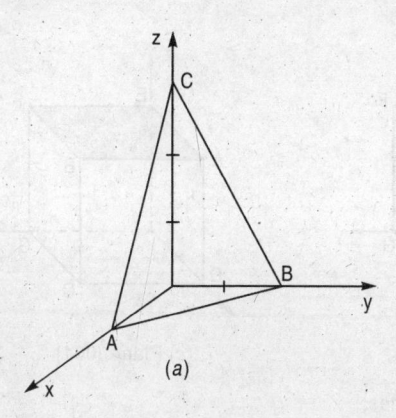

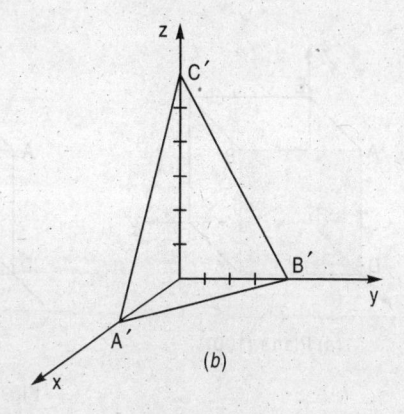

Fig. 3.21. Miller indices.

(a) are (1 : 1/2: 1/3) or (6 : 3 : 2). It may be noted that for getting the whole numbers, all the three reciprocals have been multiplied by 6. The ratio sign (:) is also omitted. It is due to the fact that multiplying all intercepts or reciprocal by the same number does not change the orientation of the plane. The numbers for the planes are enclosed in small bracket. Thus the Miller indices for plane ABC in Fig. 3. 21 (a) may be written as (632). The digits in the bracket are read as six three two and not six hundred and thirty two.

Similarly, the numerical parameters of the plane A' B' C' in Fig. 3.21 (b) are 2, 4 and 6 (*i.e.*, 2 axial units along *x-x* axis), 4 axial units along *y-y* axis and 6 axial units along *z-z* axis. We know that the Miller indices of this plane are (1/2 : 1/4 : 1/6) or (6 : 3 : 2) or (632). It may be noted that Miller Indices of the planes ABC and A' B' C' are the same (*i.e.*, 632). It is thus obvious, that both the planes are parallel to each other and belong to the same family of parallel planes.

3.16 PROCEDURE FOR FINDING MILLER INDICES OF CRYSTAL PLANES

The following procedure is adopted for finding the Miller indices of any plane along the three axes:

1. Find the intercepts *x, y* and *z* of the plane along the three reference axes.
2. Express the intercepts in terms of axial units (*i.e.*, $x=pa$; $y=qb$ and $x=rc$).
3. Now find the ratio of their reciprocals (*i.e.*, $1/p : 1/q : 1/r$).
4. Convert these reciprocals into whole numbers by multiplying each one of them with their L.C.M.
5. Enclose these numbers in small bracket. This represents the indices of the given plane.

3.17 IMPORTANT FEATURES OF MILLER INDICES OF CRYSTAL PLANES

Following are the important features of the Miller indices:

1. All the parallel planes have the same Miller indices. Thus the Miller indices define a set of parallel planes.
2. A plane parallel to one of the co-ordinate axes has an intercept of infinity.
3. If the Miller indices of two planes have the same ratio (*i.e.*, 844 and 422 or 211) then the planes are parallel to each other.

3.18 REPRESENTATION OF CRYSTAL PLANES IN A CUBIC UNIT CELL

We have already discussed in Art. 3.15 that the crystal planes are represented by the smallest set of integers enclosed in round brackets (or parenthesis) *e.g.*, (100), (010) and (001) represent the Miller indices of the cubic planes ABCD, BFGC and AEFB respectively as shown in Fig. 3.22 (*a*), (*b*) and (*c*).

(a) Plane (100) (b) Plane (010) (c) Plane (001)

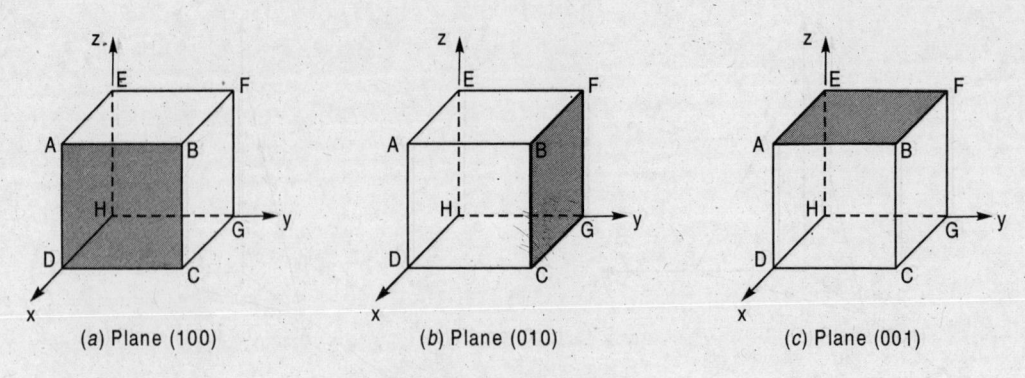

Fig. 3.22. Crystal planes.

In other words, the above mentioned three planes represent the three faces of the cubic unit cell. The other three faces of the cube may be represented by shifting the origin of the coordinated system to another corner of the unit cell e.g., the plane EFGH may be represented by shifting the origin from point 'H' to point 'D' as shown in Fig. 3.23. (a).

Now the values of the intercepts are minus one along x-x axis, infinite along y-y and z-z axes. The indices of this plane are represented by taking the reciprocals of the intercepts and placing a negative sign over the intercepts having negative values. Thus the Miller indices are ($\bar{1}$00) as shown in Fig. 3.23. (a).

Similarly, the planes AEHD and DHGC may be represented by the Miller indices (0$\bar{1}$0) and (00$\bar{1}$) by shifting the origin to the points G and E respectively as shown in Fig. 3.23 (b) and (c).

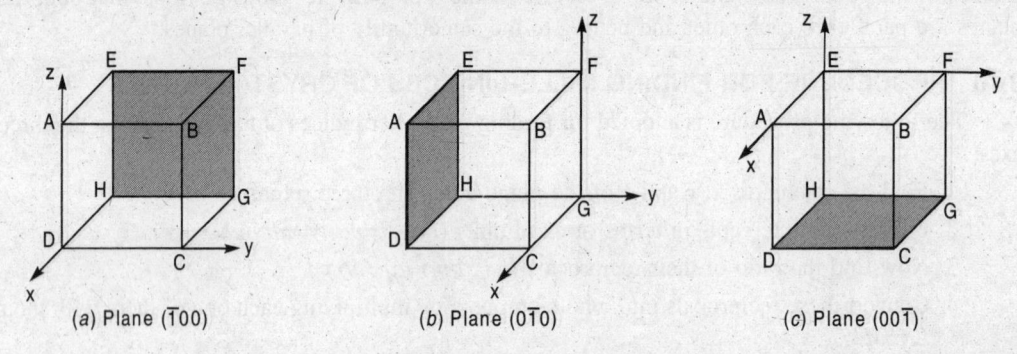

(a) Plane ($\bar{1}$00) (b) Plane (0$\bar{1}$0) (c) Plane (00$\bar{1}$)

Fig. 3.23. Crystal planes.

We know that all the planes or six faces of the cubic unit cell are of the same form (i.e., have same geometry). Therefore, the Miller indices of such planes are represented by the indices of one of the planes enclosed in curly brackets or braces { }. Thus for a unit cubic cell, {100}, represents a set of six planes (100), (010), (001), ($\bar{1}$00), (0$\bar{1}$0) and (00$\bar{1}$). Similarly {110} represents a set of six planes (110), (101), (011), ($\bar{1}$10), (10$\bar{1}$), and (01$\bar{1}$).

It may be noted that these planes represent diagonal planes. Each diagonal plane is formed by joining two opposite edges of the unit cubic cell. Now we know that there are 12 edges of a cubic unit cell. Thus there are 6 diagonal faces represented by {110} as shown in Fig. 3.24. (a) to (f).

Crystal Structure

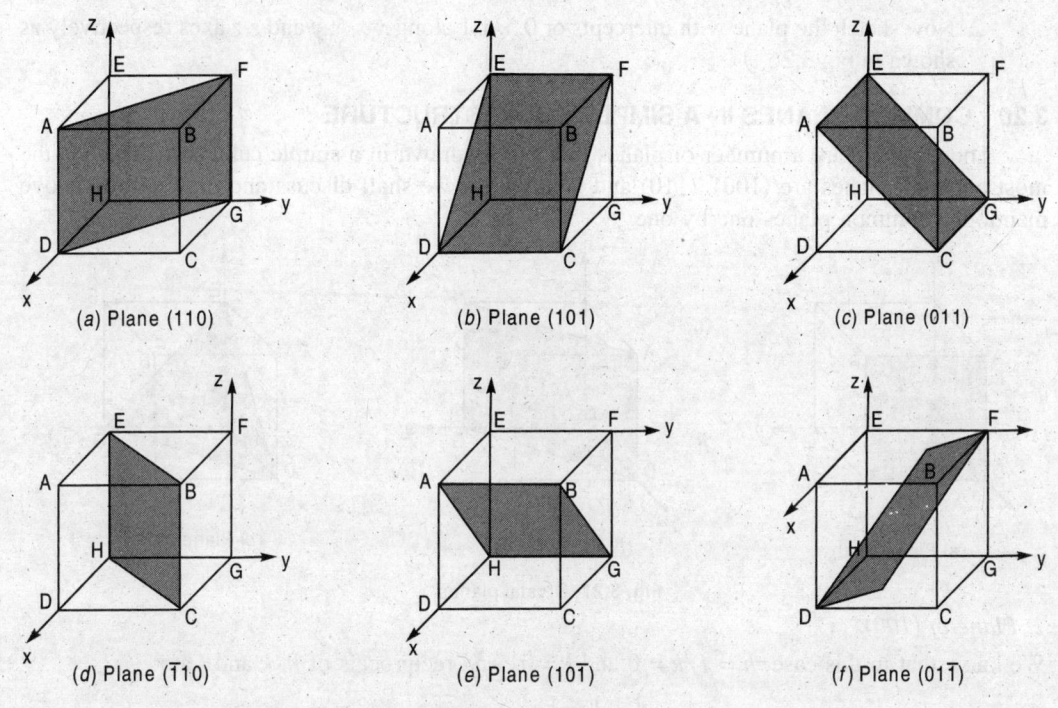

(a) Plane (110) (b) Plane (101) (c) Plane (011)

(d) Plane ($\bar{1}$10) (e) Plane (10$\bar{1}$) (f) Plane (01$\bar{1}$)

Fig. 3.24. Crystal planes.

Similarly, {111} represents a set of four planes (111), ($\bar{1}\bar{1}$1), ($\bar{1}\bar{1}\bar{1}$) and (1$\bar{1}\bar{1}$). These planes represent triangular planes of the unit cubic cell. Thus there are 4 triangular planes represented by {111} as shown in Fig. 3.25 (a) to (d).

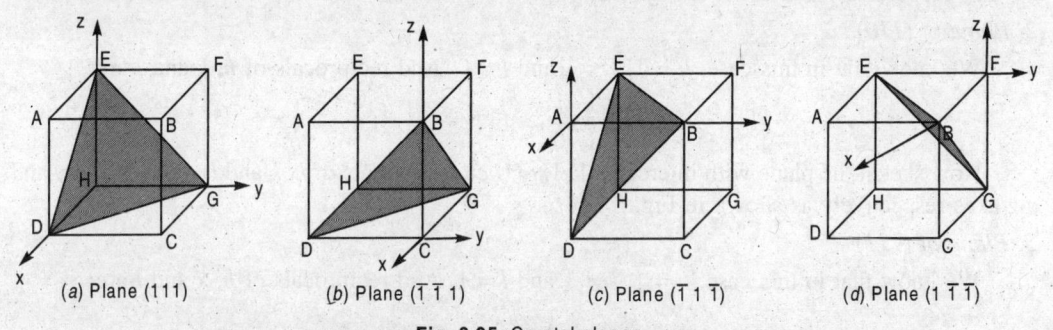

(a) Plane (111) (b) Plane ($\bar{1}\bar{1}$1) (c) Plane ($\bar{1}$1$\bar{1}$) (d) Plane (1$\bar{1}\bar{1}$)

Fig. 3.25. Crystal planes.

3.19 PROCEDURE FOR SKETCHING THE PLANE FROM THE GIVEN MILLER INDICES

Sometimes, we are required to sketch the planes, when its Miller indices are given. In such a case, the following procedure is adopted for sketching any plane.

1. First of all, take the reciprocals of the given Miller indices. These reciprocals represent the intercepts in terms of the axial units along x-x, y-y and z-z axes respectively, e.g., if the Miller indices be (211), its reciprocals or intercepts will be 1/2, 1/1 and 1/1 or 0.5, 1,1 respectively.

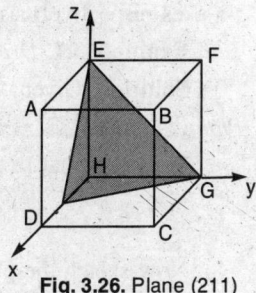

Fig. 3.26. Plane (211)

2. Now sketch the plane with intercepts or 0.5, 1,1 along x-x, y-y and z-z axes respectively as shown in Fig 3.26.

3.20 COMMON PLANES IN A SIMPLE CUBIC STRUCTURE

Though there are a number of planes that can be drawn in a simple cubic structure, yet the most common planes are (100), (110) and (111). Now we shall discuss and draw all the above mentioned common planes one by one.

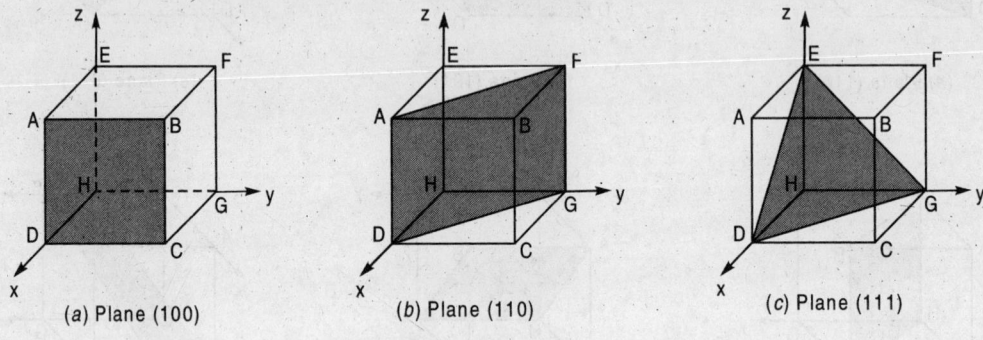

(a) Plane (100) (b) Plane (110) (c) Plane (111)

Fig. 3.27. Crystal planes.

1. Plane of (100)

We know that in this case, $h = 1$, $k = 0$ and $l = 0$. And reciprocals of h, k and l are:

$$= \frac{1}{1}, \frac{1}{0}, \frac{1}{0} = 1, \infty, \infty$$

Now sketch the plane with intercepts 1, ∞, ∞ (*i.e.*, reciprocals of h, k and l along x-x, y-y and z-z axis respectively as shown in Fig. 3.27 (a).

2. Plane of (110)

We know that in this case, $h = 1$, $k = 1$ and $l = 0$. And reciprocals of h, k and l are:

$$= \frac{1}{1}, \frac{1}{1}, \frac{1}{0} = 1, 1, \infty$$

Now sketch the plane with intercepts 1, 1, ∞ (*i.e.*, reciprocals of h, k and l) along x-x, y-y and z-z axes respectively as shown in Fig. 3.27 (b).

3. Plane of (111)

We know that in this case, $h = 1$, $k = 1$ and $l = 1$. And reciprocals of h, k and l are:

$$= \frac{1}{1}, \frac{1}{1}, \frac{1}{1} = 1, 1, 1$$

Now sketch the plane with intercepts 1, 1, 1, (*i.e.*, reciprocals of h, k and l) along x-x, y-y and z-z axes respectively as shown in Fig. 3.27 (c).

Example 3.1. *Draw the planes in the following a F.C.C. Structures: (112), (001) and (101).*

Solution. Given: F.C.C. structure (112), (001) and (101).

Plane of (112)

We know that in this case, $h = 1$ $k = 1$ and $l = 2$. And reciprocals of h, k and l are:

$$= \frac{1}{1}, \frac{1}{1}, \frac{1}{2} = 1, 1, 0.5$$

Now sketch the plane with intercepts 1, 1, 0.5 (*i.e.*, reciprocals of h, k and l) along x-x, y-y and z-z axes respectively as shown in Fig. 3.28 (a).

Crystal Structure

Plane of (001)

We know that in this case, $h = 0$, $k = 0$ and $l = 1$. And reciprocals of h, k and l are;

$$= \frac{1}{0}, \frac{1}{0}, \frac{1}{1} = \infty, \infty, 1$$

Now sketch the plane with intercepts $\infty, \infty, 1$ (*i.e.*, reciprocals of h, k and l) along *x-x, y-y* and *z-z* axes respectively as shown in Fig 3.28 (*b*).

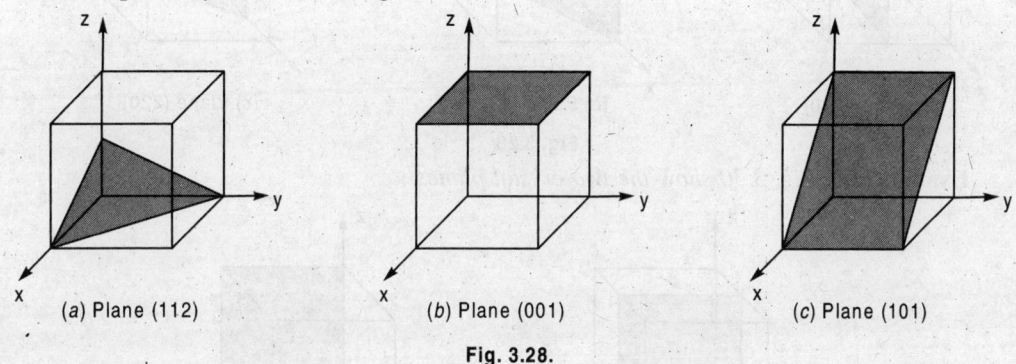

(a) Plane (112)   (b) Plane (001)   (c) Plane (101)

Fig. 3.28.

Plane of (101)

We know that in this case, $h = 1$, $k = 0$ and $l = 1$. And reciprocals of h, k and l are:

$$= \frac{1}{1}, \frac{1}{0}, \frac{1}{1} = 1, \infty, 1$$

Now sketch the plane with intercepts $1, \infty, 1$ (*i.e.*, reciprocals of h, k and l) along *x-x, y-y*, and *z-z* axes respectively as shown in Fig 3.28 (*c*).

Example 3.2. *Draw the planes (020), (120) and (220) in a face centred cubic structure.*

Solution. Given: Face-centred cubic structures (020), (120) and (220).

Plane of (020)

We know that in this case $h = 0$, $k = 2$ and $l = 0$. And reciprocals of h, k and l are:

$$= \frac{1}{0}, \frac{1}{2}, \frac{1}{0} = \infty, 0.5 \infty,$$

Now sketch the plane with intercepts $\infty, 0.5, \infty$ (i.e., reciprocals of h, k and l) along *x-x, y-y* and *z-z* axes respectively as shown in Fig. 3.29 (*a*).

Plane of (120)

We know that in the case, $h = 1$, $k = 2$ and $l = 0$. And reciprocals of h, k and l are:

$$= \frac{1}{1}, \frac{1}{2}, \frac{1}{0} = 1, 0.25 \infty,$$

Now sketch the plane with intercepts $1, 0.5, \infty$ (*i.e.*, reciprocals of h, k and l) along *x-x, y-y* and *z-z* axes respectively as shown in Fig: 3.29 (*b*).

Plane of (220)

We know that in this case, $h = 2$, $k = 2$ and $l = 0$. And reciprocals of h, k and l are:

$$= \frac{1}{2}, \frac{1}{2}, \frac{1}{0} = 0.5, 0.5, \infty,$$

Now sketch the plane with intercepts $0.5, 0.5$ and ∞ (*i.e.*, reciprocals of h, k and l) along *x-x, y-y* and *z-z* axes respectively as shown in Fig. 3.29 (*c*).

Materials Science

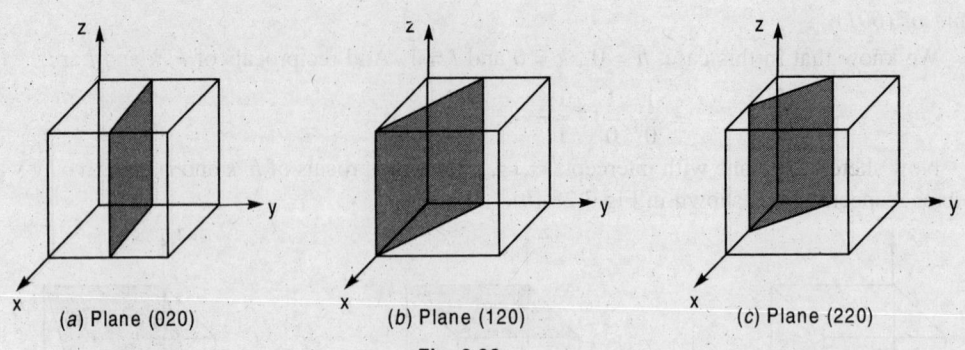

(a) Plane (020)   (b) Plane (120)   (c) Plane (220)

Fig. 3.29

Example 3.3. *Fig. 3.30 show the two crystal planes.*

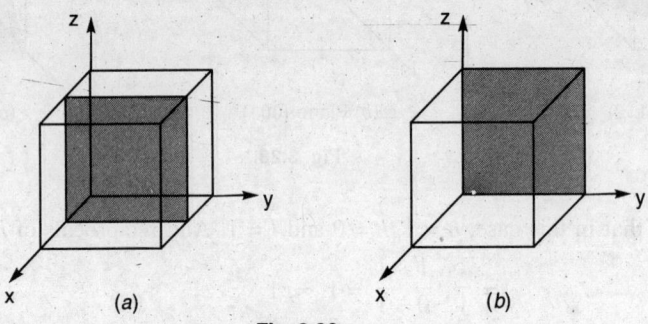

(a)   (b)

Fig. 3.30

Compute the Miller indices of the above two planes.

Solution. Given: Two crystal planes.

First case

We see in Fig. 3.30 (a) that the given plane is parallel to 'y-y' and 'z-z' axes. Thus its numerical intercepts on these two axes are infinity. Moreover, the numerical intercept on x-x axis is 1/2. Thus the numerical intercepts of the plane

$$= \frac{1}{2}, \infty, \infty = 0.5, \infty, \infty$$

∴ Miller indices of the plane

$$= \frac{1}{0.5}, \frac{1}{\infty}, \frac{1}{\infty} = (200) \quad \textbf{Ans.}$$

Second case

We see in Fig. 3.30 (b) that the given plane is parallel to 'y-y' and 'z-z' axes. Thus its numerical intercepts of these two axes are infinity. Moreover, the numerical intercept on x-x axis is –1. Thus the numerical intercepts of the plane

$$= -1, \infty, \infty$$

∴ Miller indices of the plane

$$= \frac{1}{-1}, \frac{1}{\infty}, \frac{1}{\infty} = (-100) = (\bar{1}00) \quad \textbf{Ans.}$$

Example. 3.4. *Topaz, an orthorhombic semi-precious stone has a ratio of a:b:c of 0.529:1:0.477. Find the Miller indices of the face whose intercept is as follows: 0.264:1:0.238.*

(A.M.I.E., Summer, 2004)

Solution. We know that the Miller indices of the given plane,

$$= \frac{0.529}{0.264} : \frac{1}{1} : \frac{0.477}{0.238}$$
$$= 2 : 1 : 2$$

∴ Miller indices of the given plane = (212) **Ans.**

Example. 3.5. *Draw the planes (110) and (111) in a simple cubic lattice.*

(*AMIE., Summer, 2002*)

Solution. Given : Simple cubic structure.

Plane (110)

We know that in this case, $h = 1$, $k = 1$ and $l = 0$. And reciprocals of h, k and l are :

$$= \frac{1}{1}, \frac{1}{1}, \frac{1}{0} = 1, 1, \infty$$

Now sketch the plane with intercepts 1, 1, ∞ along $x-$, $y-$ and $z-$axis, respectively as shown in Fig. 3.31(*a*).

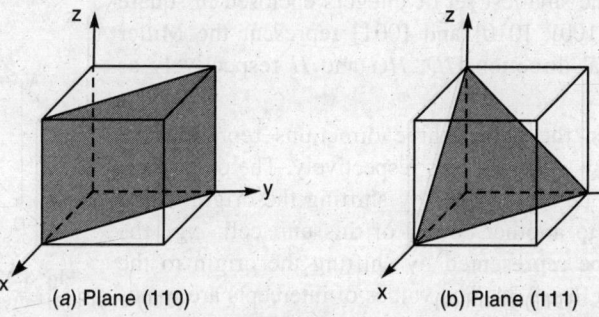

(a) Plane (110) (b) Plane (111)

Fig. 3.31.

Plane (111)

We know that in this case, $h = 1$, $k = 1$, $l = 1$. And reciprocals of h, k and l are :

$$= \frac{1}{1}, \frac{1}{1}, \frac{1}{1} = 1, 1, 1$$

Now sketch the plane with intercepts 1, 1, 1 along $x-$, $y-$ and $z-$axis respectively as shown in Fig. 3.31(*b*).

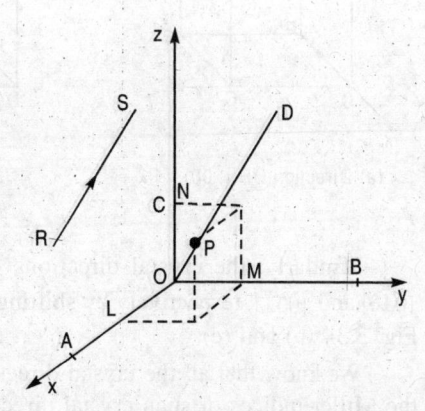

3.21 CRYSTAL DIRECTIONS

A direction, in general, may be represented in terms of three axes with references to the origin. In a crystal system, the line joining the origin and a lattice point represents the direction of the lattice point.

Fig. 3.32. shows the unit cell of a cubic crystal, in which the point '*O*' represents the origin and *A*, *B* and *C* represent the lattice points along *x-x*, axis, *y-y* axis and *z-z* axis respectively. Now the line joining *O* and *A* (*i.e.*, *OA*) represents the crystal direction of the lattice point *A*. Similarly, *O B* and *OC* represent the crystal directions of the lattice points *B* and *C* respectively.

3.22 PROCEDURE FOR FINDING MILLER INDICES OF CRYSTAL DIRECTIONS

Fig. 3.32. Crystal directions.

The following procedure is adopted for finding the Miller indices of a crystal direction:

1. Draw a straight line *OD* passing through the origin *O* and parallel to the crystal direction *RS*, whose Miller indices are required to be determined as shown in Fig. 3.32.

2. Now take any point P on the line OD, and draw perpendiculars PL, PM and PN on x-x axis, y-y axis and z-z axis respectively.
3. Find the intercepts of OL, OM and ON in terms of axial units. Let these intercepts be 0.5, 0.25 and 1.0 as shown in Fig. 3.32.
4. Reduce these intercepts in the smallest integers. This is done by dividing the values of three intercepts with the least value of the three intercepts. For example, the smallest integers of the intercepts 0.5, 0.25 and 1.0 are 0.5/ 0.25, 0.25/ 0.25 and 1.0/0.25 *i.e.*, 2, 1, 4.
5. Enclose the smallest integers 2, 1, 4 in square brackets, i.e., [214] which represents the Miller indices of the line RS.

3.23 REPRESENTATION OF CRYSTAL DIRECTIONS IN A CUBIC UNIT CELL

We have already discussed in Art 3.22 that crystal directions are represented by the smallest set of integers enclosed in square brackets [ ]. e.g., [100], [010] and [001] represent the Miller indices of the crystal direction HD, HG and H respectively as shown in Fig. 3.33.

In other words, the above three directions represent the directions along x-x, y-y and z-z axes respectively. The other three directions may now be represented by shifting the origin of the coordinated system to another corner of this unit cell, *e.g.*, the direction DH may be represented by shifting the origin to the point D, as shown in Fig. 3.34. The values of intercepts are minus one along x-x axis, and zero along y-y and z-z axes. The indices of this direction are represented by placing a negative sign on the intercepts having negative value. Thus its Miller indices are [100] as shown in Fig. 3.34 (*a*).

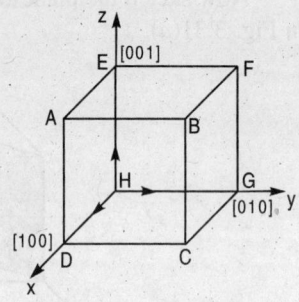

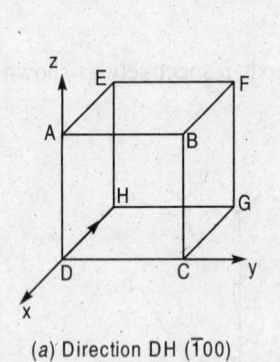

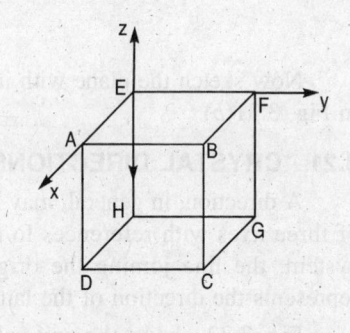

Fig. 3.33. Representation of crystal directions.

(a) Direction DH ($\bar{1}$00)

(b) Direction GH (0$\bar{1}$0)

(c) Direction EH (00$\bar{1}$)

Fig. 3.34

Similarly, the crystal directions GH and EH may be represented by the Miller indices [0$\bar{1}$0] and [00$\bar{1}$] respectively by shifting the origin to the points G and E respectively as shown in Fig. 3.34 (b) and (c).

We know that all the crystal directions, in a cubic unit cell are of the same form. Therefore, the Miller indices of such crystal directions are represented by the indices of one of the crystal directions enclosed in a carat < >. Thus for a cubic unit cell < 100 > represents a set of six directions. *i.e.*, [100], [010], [001], [$\bar{1}$00], [0$\bar{1}$0] and [00$\bar{1}$].

Crystal Structure

It may be noted from the figures that the crystal direction is perpendicular to the plane *ABCD*, as they have the same Miller indices, *i.e.*, 100. Similarly the crystal directions *HG* and *HE* are also perpendicular to the planes *BFGC* and *AEFB* respectively. Thus in a cubic system, all the directions and planes with identical indices are perpendicular.

Fig. 3.35 shows some of the important crystal directions in a cubic unit cell. Thus *HA, HC, HF* and *HB* represent [101], [110], [011], [111] crystal directions respectively. It may be noted that the crystal directions, having negative values along *x-x*, *y-y* and *z-z* axes respectively, are represented in the same manner as that of the crystal planes.

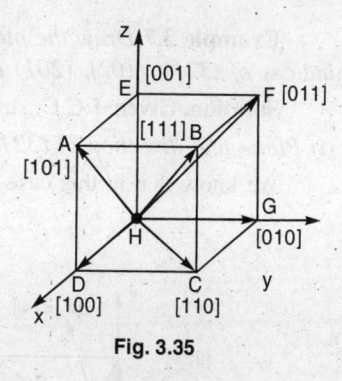

Fig. 3.35

3.24 PROCEDURE FOR SKETCHING DIRECTIONS FROM THE GIVEN MILLER INDICES

Sometimes, we are required to sketch the direction, when its Miller indices are given. In such a case, the following procedure is adopted for sketching any direction:

1. First of all, sketch the plane with the given Miller indices.
2. Now through the origin, draw a line normal to the sketched plane, which will give the required direction.

Examples 3.6. *Draw the following planes and directions in F.C.C. structure with the following Miller indices (112), (001) and (101);*

Solution. *Given: F.C.C. structures with Miller indices of (112), (001) and (101) Direction for (112).*

First of all, sketch a plane with intercepts equal to 1, 1 and 0.5 (*i.e.*, reciprocals of 1, 1, 2) along *x-x*, *y-y* and *z-z* axes respectively. Now through the origin, draw a line normal to the sketched plane, which gives the required direction as shown in Fig. 3.36 (*a*).

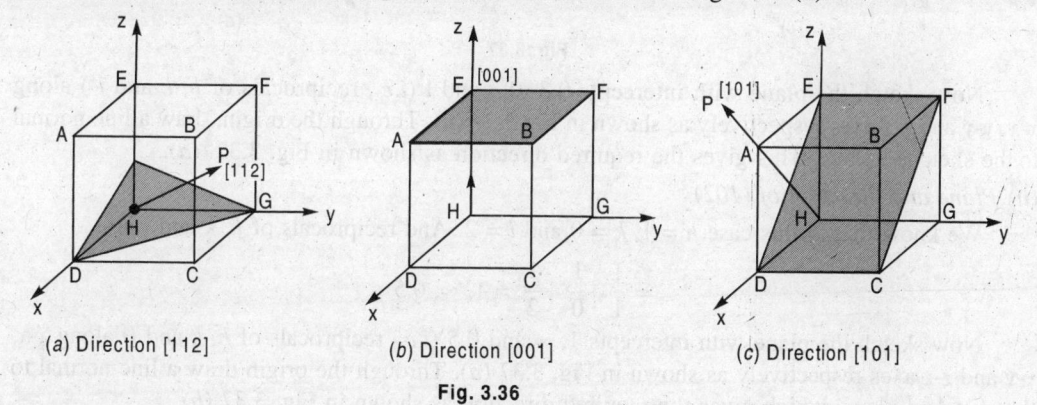

(a) Direction [112] (b) Direction [001] (c) Direction [101]

Fig. 3.36

Direction of (001)

First of all, sketch a plane with intercepts equal to ∞, ∞ and 1 (*i.e.*, reciprocals of 0,0,1) along *x-x*, *y-y* and *z-z* axes respectively. Now through the origin, draw a line normal to the sketched plane, which gives the required directions as shown in Fig. 3.36 (b).

Direction of (101)

First of all, sketch a plane with intercepts equal to 1,∞ and 1 (*i.e.*, reciprocals of 1, 0, 1) along *x-x*, *y-y* and *z-z* axes respectively. Now through the origin, draw a line normal to the sketched plane, which gives the required direction as shown in Fig. 3.37 (*c*).

Example 3.7 *Draw the planes and directions of the following F.C.C. structures with Miller indices of (321), (102), (201) and (111).*

Solution. Given: F.C.C: structures with Miller indices of (321), (102), (201) and (111).

(*i*) *Plane and direction of (321)*

We know that in this case $h = 3$, $k = 2$ and $l = 1$. And reciprocals of h, k and l are:

$$= \frac{1}{3}, \frac{1}{2}, \frac{1}{1} = 0.3, 0.5, 1$$

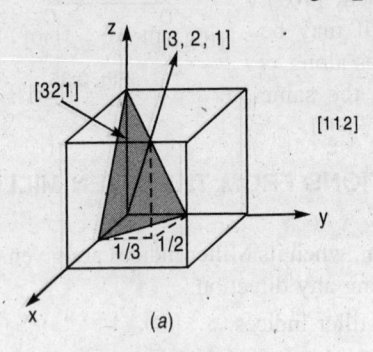

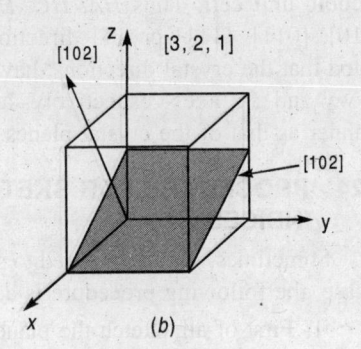

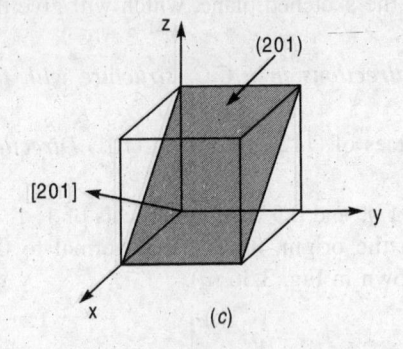

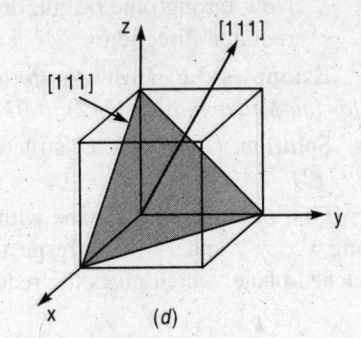

Fig. 3.37

Now sketch the plane with intercepts 0.3, 0.5 and 1 (*i.e.*, reciprocals of h, k and l) along x-x, y-y and z-z axes respectively as shown in Fig 3.37 (*a*). Through the origin, draw a line normal to the sketched plane. This gives the required direction as shown in Fig. 3.37. (*a*).

(*ii*) *Plane and direction of (102)*

We know that in this case $h = 1$, $k = 0$ and $l = 2$. And reciprocals of h, k and l are:

$$= \frac{1}{1}, \frac{1}{0}, \frac{1}{2} = 1, \infty, 0.5$$

Now sketch the plane with intercepts 1, ∞ and 0.5 (*i.e.*, reciprocals of h, k and l) along x-x, y-y and z-z axes respectively as shown in Fig. 3.37 (*a*). Through the origin draw a line normal to the sketched plane, which gives the required direction as shown in Fig. 3.37 (*b*).

(*iii*) *Plane and direction of (201)*

We know that in this case $h = 2$, $k = 0$ and $l = 1$. And intercepts of h, k and l are:

$$= \frac{1}{2}, \frac{1}{0}, \frac{1}{1} = 0.5, \infty, 1$$

Now sketch the plane with intercepts 0.5, ∞ and 1 (*i.e.*, reciprocals of h, k and l) along x-x, y-y and z-z axes respectively as shown in Fig. 3.37 (*c*). Though the origin draw a line normal to the sketched plane, which gives the required direction as shown in Fig. 3.37. (*c*).

Crystal Structure

(iv) Plane and direction of (111)

We know that in this case, $h = 1$, $k = 1$ and $l = 1$. And reciprocals of h, k, and l are:

$$= \frac{1}{1}, \frac{1}{1}, \frac{1}{1} = 1, 1, 1$$

Now sketch the plane with intercepts 1, 1 and 1 (*i.e.*, reciprocals of h, k and l) along x-x, y-y and z-z axes respectively as shown in Fig. 3.37 (d) Through the origin draw a line normal to the sketched plane, which gives the required direction as shown in Fig. 3.37. (d).

Example 3.8. *Draw the planes and directions in a F.C.C. structure (010), (111), (011) and (001).*

Solution. Given: F.C.C. structure (010), (111), (011) and (001).

(i) Plane and direction of (010)

We know that in this case, $h = 0$, $k = 1$ and $l = 0$. And reciprocals of h, k and l are:

$$= \frac{1}{0}, \frac{1}{1}, \frac{1}{0} = \infty, 1, \infty$$

Now sketch the plane with intercepts ∞, 1 and ∞ (*i.e.*, reciprocals of h, k and l) along x-x, y-y and z-z axes respectively as shown in Fig. 3.38 (a). Through the origin draw a line normal to the sketched plane, which gives the required direction as shown in Fig. 3.38 (a).

(ii) Plane and direction of (111)

We know that in this case, $h = 1$, $k = 1$ and $l = 1$. And reciprocals of h, k and l are:

$$= \frac{1}{1}, \frac{1}{1}, \frac{1}{1} = 1, 1, 1$$

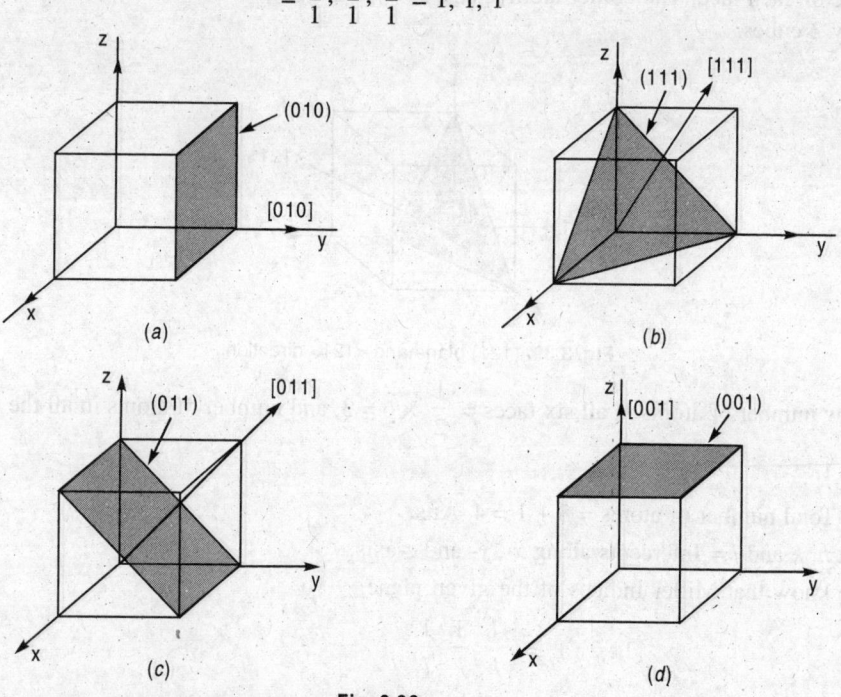

Fig. 3.38

Now sketch the plane with intercepts, 1, 1 and 1 (*i.e.*, reciprocals of h, k and l) along x-x, y-y and z-z axes respectively as shown in Fig. 3.38 (b). Through the origin draw a line normal to the sketched plane, which gives the required direction as shown in Fig. 3.38 (b).

(iii) Plane and direction of (011)

We know that in this case, $h = 0$, $k = 1$ and $l = 1$. And reciprocals of h, k and l are:

$$= \frac{1}{0}, \frac{1}{1}, \frac{1}{1} = \infty, 1, 1$$

Now sketch the plane with intercepts ∞, 1 and 1 (*i.e.*, reciprocal of h, k and l) along x-x, y-y and z-z axes respectively as shown in Fig. 3.38 (c). Through the origin draw a line normal to the sketched plane, which gives the required direction as shown in Fig. 3.38 (c).

(iv) Plane and direction of (001)

We know that in this case $h = 0$, $k = 0$ and $l = 1$. And reciprocals of h, k and l are:

$$= \frac{1}{0}, \frac{1}{0}, \frac{1}{1} = \infty, \infty, 1$$

Now sketch the plane with intercepts ∞, ∞ and 1 (*i.e.* reciprocals of h, k and l) along x-x, y-y and z-z axes respectively as shown in Fig. 3.33 (d). Through the origin draw a line normal to the sketched plane, which gives the required direction as shown in Fig. 3.38 (d).

Example 3.9. *Calculate the atoms per unit cell of metallic Zinc. Draw (121) plane and <121> direction in a cubic lattice.* (A.M.I.E, Winter 2000)

Solution. We know that Zinc has FCC structure. The total number of atoms per unit cell may be obtained as follows:

The FCC structure has 8 corners and 6 faces. It has an atom at each corner and another at the centre of each face. The corner atom is shared by 8 cubes, and each atom at the face centre is shared by 2 cubes.

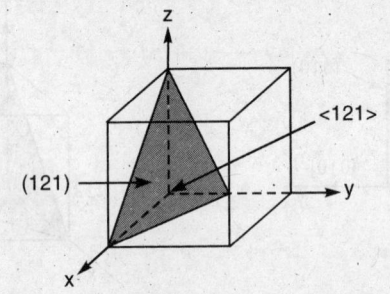

Fig. 3.39. (121) plane and <121> direction.

Now number of atoms in all six faces = $\frac{1}{2} \times 6 = 3$, and number of atoms in all the corners = $\frac{1}{8} \times 8 = 1$.

∴ Total number of atoms = 3 + 1 = 4. **Ans.**

Let h, k and l = Intercepts along x-, y- and z-axis.

We know that Miller Indices of the given plane,

$$1 : 2 : 1 = \frac{1}{h} : \frac{1}{k} : \frac{1}{l}$$

or $$h : k : l = \frac{1}{1} : \frac{1}{2} : \frac{1}{1} = 1 : \frac{1}{2} : 1$$

Therefore, the intercept along x-axis is one unit, one-half along y-axis and one unit along z-axis. Knowing this, the (121) plane can be sketched as shown by the shaded plane as shown in Fig. 3.39. The <121> direction can be sketched by drawing a line perpendicular to the plane (121) as shown in Fig. 3.39.

3.25 REPRESENTATION OF CRYSTAL PLANES AND DIRECTIONS IN HCP UNIT CELL

We have already discussed the representation of crystal planes in a cubic unit cell in the last article. As discussed, the crystal planes in a cubic unit cell can be represented by three principal axes. However, this system is not valid for representing crystal planes in hexagonal close packed (HCP) unit cells.

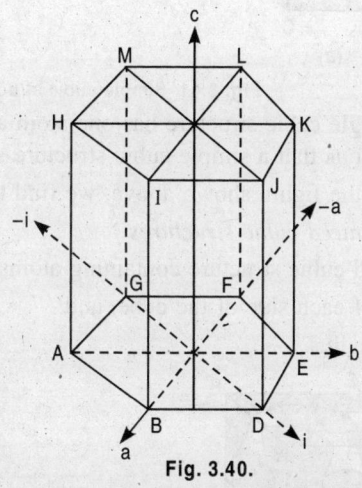

Fig. 3.40.

Fig. 3.40. shows a way to represent crystal planes in a HCP crystal structure. As seen from this diagram, the system has four axes a, b, i and c. The three axes (a, b, and i) are at an angle of $120°$ to each other. The fourth axis (c) is perpendicular to the base plane.

To find out the Miller indices of a given plane in a HCP unit cell, we follow the steps given below.

1. Find the intercepts of the given crystal plane with the axis a, b, i and c.
2. Express the intercepts in terms of axial units.
3. Find the reciprocals of the intercepts.
4. Convert into whole numbers by multiplying each one of them with their least common multiplier (L.C.M.).
5. Enclose these four numbers in small brakets.

For example, it can be shown easily that plane IJBD have the Indices (1010). Similarly, the plane ABIH has the Miller indices of (1$\bar{1}$00) and the plane DEJK, (0110), the plane HIJKLM, (0001). The procedure for sketching the planes and directions is similar to what we have discussed for the cubic unit cell.

3.26 ATOMIC RADIUS IN A CUBIC AND HEXAGONAL SYSTEM

In a cubic system, the atoms are assumed to be placed in such a way that any two adjacent atoms touch each other. The radius of such atoms is always found out in terms of length of the side of the cube, which is usually taken as 'a'. This radius of atoms is known as atomic radius. It may be noted that the atomic radius is numerically equal to half of the distance between the centres of two adjacent atoms, placed systematically in the cube. Now we shall discuss atomic radius for three types of cubic structures and the hexagonal structure one by one.

1. Atomic radius of simple cubic structures

Consider a simple cubic structure containing atoms as shown in Fig. 3.41 (*a*) and (*b*).
Let a = Length of each side of the cube, and
r = Atomic radius.

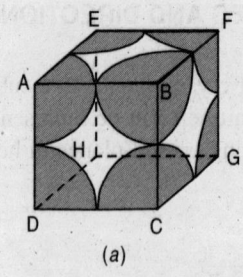

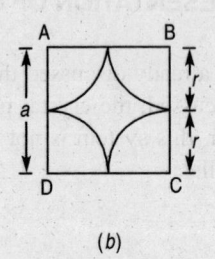

Fig. 3.41. Simple cubic structure.

We know that each single cubic structure has one atom at each of its 8 corners as shown in the Fig. 3.41. It is thus obvious that a simple cubic structure shares 8 atoms.

From the geometry of the figure shown above, we find that the atomic radius, $r = a/2$.

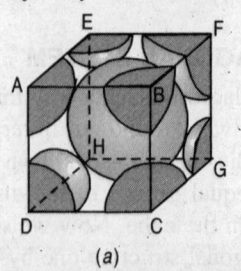

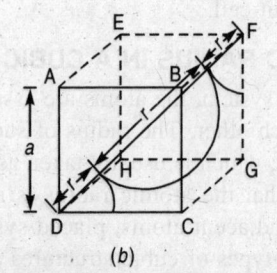

2. Atomic radius of face centred cubic structures

Consider a face centred cubic structure containing atoms as shown in Fig. 3.42 (a) and (b).

Let a = Length of each side of the cube, and
  r = Atomic radius.

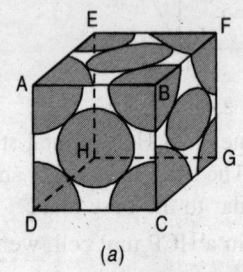

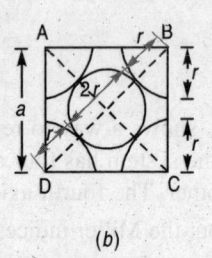

Fig. 3.42. Face centred cubic structure.

We know that each face centred cubic structure has one atom at each of its 8 corners. In addition to this, it has one atom at the centre of its 6 faces also as shown in the figure. It is thus obvious that a face centred cubic structure shares 14 atoms.

From the geometry of the figure, shown above, we find that the diagonal

$$BD = 4r = \sqrt{a^2 + a^2} = \sqrt{2a^2} = a\sqrt{2}$$

$$\therefore \quad r = \frac{a\sqrt{2}}{4}$$

3. Atomic radius of body centred cubic structures

Fig. 3.43. Body centred cubic structure.

Consider a body centred cubic structure containing atoms as shown in Fig. 3.43 (a) and (b).

Let a = Length of each side of the cube, and
  r = Atomic radius.

Crystal Structure

We know that each body centred cubic structure contains one atom at the body centre and shares one atom at each of its 8 corners as shown in the figure. It is thus obvious that a body centred structure contains one full atom and shares 8 atoms.

From the geometry of the figure, we find that the diagonal

$$DF = 4r = \sqrt{DC^2 + CG^2 + GF^2} = \sqrt{a^2 + a^2 + a^2}$$
$$= \sqrt{3a^2} = a\sqrt{3}$$
$$\therefore \quad r = \frac{a\sqrt{3}}{4}$$

4. Atomic radius of hexagonal close packed structures

Consider a hexagonal close packed (HCP) structure containing atoms as shown in Fig. 3.44 (a) and (b).

Let
- a = Length of each side of the base of hexagonal prism
- r = atomic radius
- c = height of HCL structure.

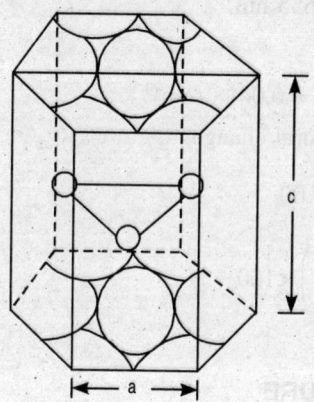

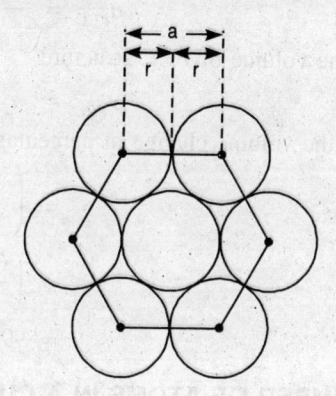

Fig. 3.44. Hexagonal close packed structure.

As seen from Fig. 3.44 (b), each hexagonal plane in HCP consists of one atom at the centre and six atoms at all the six corners. The centre atom is shared by two unit cells and each atom at the corner is shared by six atoms. From the geometry of the hexagonal plane, we find that

$$a = 2r$$

or

$$r = \frac{a}{2}.$$

Example 3.10. *Calculate the largest diameter of an atom which could fit interstially in a copper crystal without distorting it. The edge length of the FCC unit cell of copper is 3.61Å..*

(A.M.I.E., Summer, 1999)

Solution. Given, the edge length of the FCC unit cell of copper, $a = 3.61$Å.

Let r = radius of an atom

We know that for a FCC unit cell,

$$r = \frac{a\sqrt{2}}{4} = \frac{3.61 \times \sqrt{2}}{4} = 1.276 \text{ Å}$$

and diameter of one atom,

$$d = 2r = 2 \times 1.276 = 2.552 \text{ Å. \textbf{Ans.}}$$

Example 3.11. *Metallic iron changes from BCC to FCC at 910°C. At this temperature, the atomic radii of the iron atom in the two structures are 0.1258 nm and 0.1292 nm respectively. Calculate the volume change in percentage during this structural change.*

(Anna Univ., April 2002.)

Solution. Given: Atomic radii in BCC structure = 0.1258 nm = 0.1258 × 10^{-9} m; atomic radii in FCC structure = 0.1292 nm = 0.1292 × 10^{-9} m.

We know that in BCC structure, the atomic radii (r_{BCC})

$$0.1258 = \frac{a\sqrt{3}}{4}, \quad \therefore a_{BCC} = \frac{0.1258 \times 4}{\sqrt{3}} = 0.2905 \text{ nm}$$

and the volume of BCC structure,

$$v_{BCC} = a_{BCC}^3 = (0.2905)^3 = 0.0245 \text{ (nm)}^3$$

We also know that in FCC structure, the atomic radii (r_{FCC}),

$$0.1292 = \frac{a\sqrt{2}}{4}$$

$$\therefore a_{FCC} = \frac{0.1292 \times 4}{\sqrt{2}} = 0.3655 \text{ nm}$$

and the volume of *FCC* structure,

$$v_{FCC} = a_{FCC}^3 = (0.3655)^3 = 0.0488 \text{ (nm)}^3$$

Now the volume change in percentage during structural change,

$$= \left(\frac{v_{FCC} - v_{BCC}}{v_{BCC}}\right) \times 100$$

$$= \left(\frac{0.0488 - 0.0245}{0.0245}\right) \times 100$$

$$= 99.2\% \text{ Ans.}$$

3.27 NUMBER OF ATOMS IN A CUBIC STRUCTURE

In the last article, we have discussed the atomic radius of various types of cubic structures (or unit cells). We have also discussed that a cubic structure shares a number of atoms in each of its 8 corners. A little consideration will show, that each corner atom is shared by eight surrounding cubes. Thus the share of each cubic structure comes to 1/8 of an atom. As there are 8 corner atoms in all, therefore each cubic structure has 8 × 1/8 = 1 atom. Now we shall discuss the number of atoms contained in three types of cubic structures.

1. *Simple cubic structure.* We know that a simple cubic structure shares 8 atoms, one in each of its 8 corners. Therefore total number of atoms in a simple cubic structure is 1.
2. *Face centred cubic structure.* We know that a face centred cubic structure shares 8 atoms one in each of its 8 corners, in addition to one atom in the centre of its *each face. There are 6 atoms at the face centres of the cube. Since every face of the cube is shared by two such cubes, therefore the number of face atoms per cube is 3. Thus total number of atoms in a face centred cubic structure is 3 + 1 = 4.
3. *Body centred cubic structure.* We know that a body centred cubic structure shares 8 atoms, one in each of its 8 corners, in addition to one atom at the body centre. Therefore total number of atoms in a body centred cubic structure is 1+1 = 2.

---

* Each face contains half of the atom. Since there are six face in a cubic structure, therefore number of atoms in the six faces comes out to be 6 × 1/2 = 3 atoms.

Crystal Structure

3.28 NUMBER OF ATOMS PER SQUARE MILLIMETRE IN CUBIC CRYSTAL PLANES

We have already discussed that a cubic crystal has three different structures, namely simple cubic (*SC*), body centred cubic (*BCC*) and face centred cubic (*FCC*). In order to determine the number of atoms per square millimetre in a plane, we shall consider the most common planes, *i.e.*, (100), (110) and (111) in each cubic structure one by one.

Note. In all these cases, we shall consider the lattice constant (a) in terms of millimeters.

3.29 NUMBER OF ATOMS PER SQUARE MILLIMETER IN SIMPLE CUBIC STRUCTURE

1. First of all, let us consider the plane (100).

Fig. 3.45 (*a*) and (*b*) shows actual view and plane (100) of a simple cubic structure containing one atom each on all the 8 corners of the cube. We know that the plane *ABCD* in Fig. 3.45. (*b*) contains one-fourth atom in each of its four corners. Therefore total number of atoms contained in the plane *ABCD*.

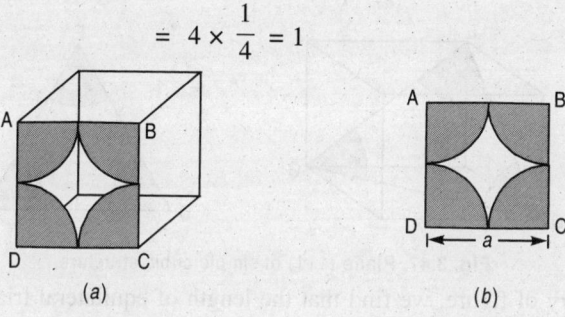

$$= 4 \times \frac{1}{4} = 1$$

Fig. 3.45. Plane (100) of simple cubic structure.

From the geometry of the figure, we find that the lengths *AB* and *AD* of the plane *ABCD*

$$= a$$

∴ Area of the plane *ABCD*

$$= AB \times AD = a \times a = a^2$$

and number of atoms per square mm

$$= \frac{\text{No. of atoms}}{\text{Area of plane}} = \frac{1}{a^2}$$

2. Now consider the plane (110)

Fig. 3.46 (*a*) and (*b*) shows actual view and plane (110) of the simple cubic structure containing one atom each on all the 8 corners of the cube. We know that the plane *ABCD* in Fig 3.46 (*b*) contains one-fourth atom in each of its four corners. Therefore total number of atoms contained in the plane *ABCD*

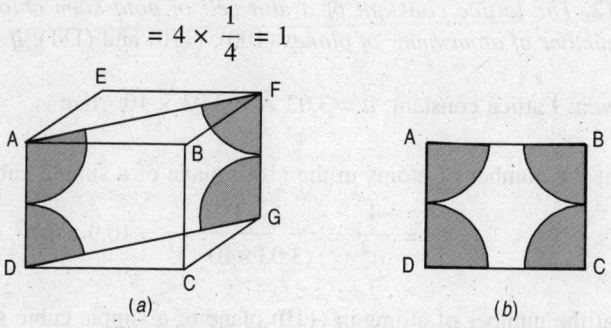

$$= 4 \times \frac{1}{4} = 1$$

Fig. 3.46. Plane (110) of simple cubic structure.

From the geometry of the figures, we find that the length $AD = a$ and the length of $AF = a\sqrt{2}$

∴ Area of the plane $AFGD$

$$= AB \times AF = a \times a\sqrt{2} = a^2\sqrt{2}$$

and number of atoms per square mm

$$= \frac{\text{No. of atoms}}{\text{Area of plane}} = \frac{1}{a^2\sqrt{2}} = \frac{0.707}{a^2}$$

3. *Now consider the plane (111)*

Fig. 3.47 (*a*) and (*b*) shows the actual view and plane (111) of the single cubic structure containing one atom each on all the 8 corners of the cube. We know the plane *EDG* in Fig. 3.47 (*b*) contains one-sixth atom in each of its three corners. Therefore total number of atoms contained in the plane *EDG*

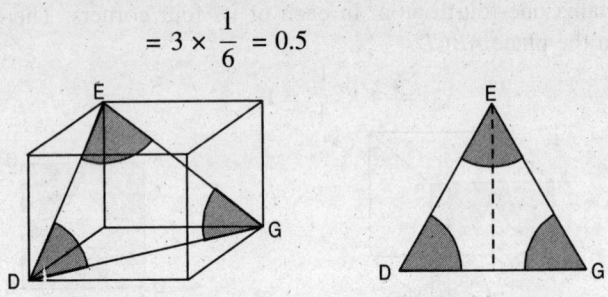

$$= 3 \times \frac{1}{6} = 0.5$$

Fig. 3.47. Plane (111) of simple cubic structure.

From the geometry of figure, we find that the length of equilateral triangle *EDG*

$$= a\sqrt{2}$$

and height of perpendicular *EK*

$$= a\sqrt{2} \sin 60° = a\sqrt{2} \times 0.866$$

∴ Area of the plane *EDG*

$$= \frac{1}{2} \times DG \times EK = \frac{1}{2} \times a\sqrt{2} \times a\sqrt{2} \times 0.866$$

$$= 0.866 \, a^2$$

and number of atoms per square mm

$$= \frac{\text{No. of atoms}}{\text{Area of plane}} = \frac{0.5}{0.866 \, a^2} = \frac{0.58}{a^2}$$

Example 3.12. *The lattice constant of a unit cell of potassium chloride (KCl) crystal is 3.03 Å. Find the number of atoms/mm² of planes* (100), (110) *and* (111), *if KCl has simple cubic structure.*

Solution. Given: Lattice constant, $a = 3.03$ Å $= 3.03 \times 10^{-7}$ mm

(100) plane

We know that the number of atoms in the (100) plane of a simple cubic structure

$$= \frac{1}{a^2} = \frac{1}{(3.03 \times 10^{-7})^2} = 10.9 \times 10^{12} \textbf{ Ans.}$$

(110) plane

We know that the number of atoms in (110) plane of a simple cubic structure

Crystal Structure

$$= \frac{0.707}{a^2} = \frac{0.707}{(3.03 \times 10^{-7})^2} = 7.7 \times 10^{12}. \textbf{ Ans.}$$

(111) plane

We also know that the number of atoms in (111) plane of a simple cubic structure

$$= \frac{0.58}{a^2} = \frac{0.58}{(3.03 \times 10^{-7})^2} = 6.3 \times 10^{12} \textbf{ Ans.}$$

3.30 NUMBER OF ATOMS PER SQUARE MILLIMETER IN BODY CENTRED CUBIC (BCC) STRUCTURE

1. First of all consider the plane (100)

Fig. 3.48. (a) and (b) shows actual view and plane (100) of a body centred cubic structure containing one atom on each of the 8 corners of the cube and one atom in the centre of the body. We know that the plane ABCD in Fig 3.48 (b) contains one-fourth atom in each of its four corners and the body centred atom will not appear on the plane. Therefore total number of atoms contained in the plane ABCD.

$$= 4 \times \frac{1}{4} = 1$$

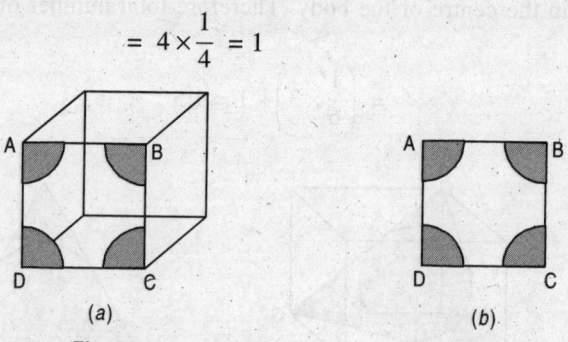

Fig. 3.48. Plane (100) of simple cubic structure.

From the geometry of the figure, we find that the lengths AB and AD of the plane ABCD

$$= a \text{ mm}$$

∴ Area of the plane ABCD

$$= AB \times AD = a \times a = a^2$$

and number of atoms per square mm

$$= \frac{\text{No. of atoms}}{\text{Area of Plane}} = \frac{1}{a^2}$$

2. Now consider the plane (110)

Fig. 3.49 (a) and (b) shows actual view and plane (110) of the body centred cubic structure containing one atom on each of the 8 corners of the cube and one atom in the centre of the body. We know that the plane ABCD in Fig 3.41 (b) contains one-fourth atom in each of its four corners and one full atom in the centre of the body. Therefore total number of atoms contained in the plane AFGD

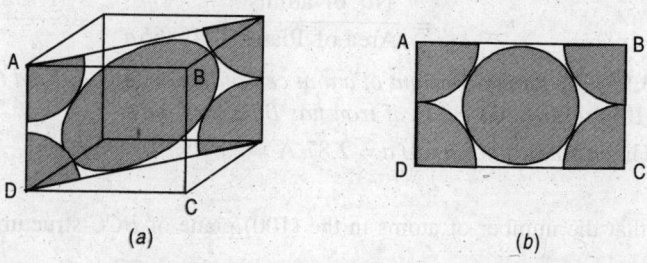

Fig. 3.49. Plane (110) of body centred cubic structure.

$$= \left(4 \times \frac{1}{4}\right) + 1 = 2$$

From the geometry of the figure, we find that the length AD
$$= a \text{ mm}$$
and length AE
$$= a\sqrt{2} \text{ mm}$$
∴ Area of the plane AFGD
$$= AB \times AD = a \times a\sqrt{2} = a^2 \sqrt{2} \text{ mm}^2$$
and number of atoms per square mm
$$= \frac{\text{No. of atoms}}{\text{Area of Plane}} = \frac{2}{a^2 \sqrt{2}} = \frac{1.414}{a^2}$$

3. *Now consider plane (111)*

Fig. 3.50 (*a*) and (*b*) shows actual view and plane (111) of the body centred cubic structure containing one atom of each of the 8 corners of the cube and one atom in the centre of the body. We know that the plane EDG in Fig. 3.50 (*b*) contains one-sixth atom in each of its three corners and one full atom in the centre of the body. Therefore total number of atoms contained in the plane EDG

$$= \left(\frac{1}{6} \times 3\right) + 1 = 1.5$$

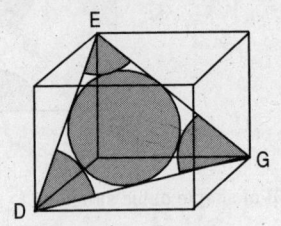

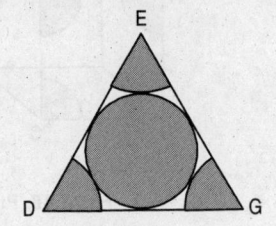

Fig. 3.50. Plane (111) of simple cubic structure.

From the geometry of the figure, we find that the length DG of equilateral triangle EDG
$$= a\sqrt{2}$$
and height of perpendicular EK
$$= a\sqrt{2} \sin 60° = a\sqrt{2} \times 0.866$$
∴ Area of the plane EDG
$$= \frac{1}{2} \times DG \times EK = \frac{1}{2} \times a\sqrt{2} \times a\sqrt{2} \times 0.866$$
$$= 0.866 \, a^2$$
and total number of atoms per square mm
$$= \frac{\text{No. of atoms}}{\text{Area of Plane}} = \frac{1.5}{0.866 \, a^2} = \frac{1.732}{a^2}$$

Example 3.13. *The lattice constant of a unit cell of iron is 2.87 Å. Find the number of atoms/ mm^2 of planes* (100), (110) *and* (111), *if iron has BCC structure.*

Solution. Given Lattice constant, $a = 2.87 \text{ Å} = 2.87 \times 10^{-7} \text{ mm}$

(100) plane

We know that the number of atoms in the (100) plane of BCC structure

$$= \frac{1}{a^2} = \frac{1}{(2.87 \times 10^{-7})^2} = 1.21 \times 10^{13} \text{ Ans.}$$

(110) plane

We know that the number of atoms in the (110) plane of BCC structure

$$= \frac{1.414}{a^2} = \frac{1.414}{(2.87 \times 10^{-7})^2} = 1.72 \times 10^{13} \text{ Ans.}$$

(111) plane

We also know that the number of atoms in the (111) plane of BCC structure.

$$= \frac{1.732}{a^2} = \frac{1.732}{(2.87 \times 10^{-7})^2} = 2.1 \times 10^{13} \text{ Ans.}$$

3.31 NUMBER OF ATOMS PER SQUARE MILLIMETRE IN FACE CENTRED CUBIC (FCC) STRUCTURE

1. *First of all consider plane (100).*

Fig. 3.51 (*a*) and (*b*) shows actual view and plane (100) of a face centred cubic structure containing one atom each on all the 8 corners and one atom each on all the 7 faces of the cube. We know that the plane (100) in Fig. 3.51 (*b*) contains one-fourth atom in each of its four corners and one atom in the centre of the face. Therefore total number of atoms contained in the plane.

$$= \left(4 \times \frac{1}{4}\right) + 1 = 2$$

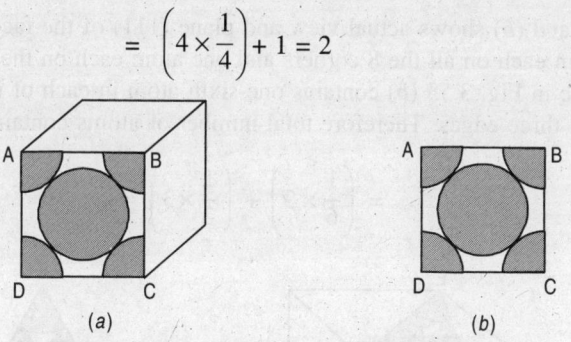

Fig. 3.51. Plane (100) of face centred cubic structure.

From the geometry of the figure, we find that the lengths *AB* and *AD* of the plane *ABCD*

$$= a \text{ mm}$$

∴ Area of the plane *ABCD*

$$= AB \times AD = a \times a = a^2$$

and number of atoms per square mm

$$= \frac{\text{No. of atoms}}{\text{Area of Plane}} = \frac{2}{a^2}$$

2. *Now consider the plane (110).*

Fig. 3.52 (*a*) and (*b*) shows actual view and plane (110) of the face-centred cubic structure containing one atom each on all the 8 corners and one atom each on all the 6 faces of the cube. We know that the plane (110) in Fig. 3.52 (*b*) contains one-fourth atom in each of its four corners and half atom in each of its two long edges *AF* and *DG*. Therefore total number of atoms contained in the plane

$$= \left(4 \times \frac{1}{4}\right) + \left(2 \times \frac{1}{2}\right) = 2$$

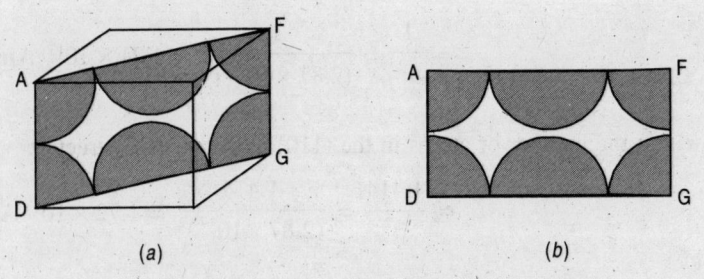

Fig. 3.52. Plane (110) of face centred cubic structure.

From the geometry of the figure, we find that the length AD

$$= a \text{ mm}$$

and length AF

$$= a\sqrt{2} \text{ mm}$$

∴ Area of the plane

$$= AD \times AF = a \times a\sqrt{2} = a^2\sqrt{2} \text{ mm}^2$$

and number of atoms per square mm

$$= \frac{\text{No. of atoms}}{\text{Area of Plane}} = \frac{2}{a^2\sqrt{2}} = \frac{1.484}{a^2}$$

3. Now Consider plane (111).

Fig. 3.53 (a) and (b) shows actual view and plane (111) of the face centred cubic structure containing one atom each on all the 8 corners and one atom each on the 6 faces of the cube. We know that the plane in Fig. 3.53 (b) contains one-sixth atom in each of its three corners and half atom in each of its three edges. Therefore total number of atoms contained in the plane

$$= \left(\frac{1}{6} \times 3\right) + \left(\frac{1}{2} \times 3\right) = 2$$

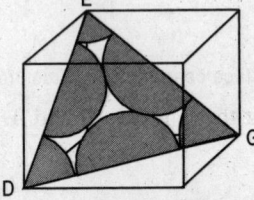

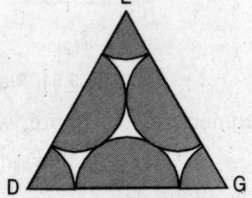

Fig. 3.53. Plane (111) of face centred cubic structure.

From the geometry of the figure, we find that the length of one side of the equilateral triangle

$$= a\sqrt{2}$$

and height of perpendicular

$$= a\sqrt{2} \sin 60° = a\sqrt{2} \times 0.866$$

∴ Area of the plane

$$= \frac{1}{2} \times DG \times EK = \frac{1}{2} \times a\sqrt{2} \times a\sqrt{2} \times 0.866 = 0.866 \, a^2$$

and total number of atoms per square nm.

$$= \frac{\text{No. of atoms}}{\text{Area of Plane}} = \frac{2}{0.866 \, a^2} = \frac{2.31}{a^2}$$

Crystal Structure

Example 3.14. *How many atoms per square millimetre surface area are there in (100) plane, (110) plane and (111) plane for lead, which has FCC structure and lattice constant, $a = 4.93$ Å ?*

Solution. Given: Lattice constant, $a = 4.93$ Å $= 4.93 \times 10^{-7}$ mm

(100) Plane

We know that number of atoms in (100) plane of F.C.C. Structure

$$= \frac{2}{a^2} = \frac{2}{(4.93 \times 10^{-7})^2} = 8.2 \times 10^{12} \text{ Ans.}$$

(110) Plane

We know that the number of atoms in (110) plane of F.C.C. structure

$$= \frac{1.414}{a^2} = \frac{1.414}{(4.93 \times 10^{-7})^2} = 5.8 \times 10^{12} \text{ Ans.}$$

(111) Plane

We know that the number of atoms in (110) plane of F.C.C. structure

$$= \frac{2.31}{a^2} = \frac{2.31}{(4.93 \times 10^{-7})^2} = 9.5 \times 10^{12} \text{ Ans.}$$

Example 3.15. *Calculate the planar density of atoms in (111) plane of aluminium. Atomic radius of aluminium = 0.143 nm.* (A.M.I.E., Winter, 2000)

Solution. Given, plane (111) and atomic radius of aluminium $= 0.143$ nm $= 0.143 \times 10^{-6}$ mm.

We know that aluminium has an FCC structure. And the number of atoms in (111) plane of FCC structure,

$$= \frac{2.31}{a^2} = \frac{2.31}{(0.143 \times 10^{-6})} = 12.22 \times 10^{12} \text{ atoms/mm}^2. \text{ Ans.}$$

3.32 ATOMIC PACKING FACTOR

We have already discussed in the last articles the number of atoms in a cubic structure of unit cell and their atomic radii in various types of structures. The ratio of actual volume of atoms per unit cell, to volume of the unit cell, is known as atomic packing factor, density of packing or packing efficiency. Now we shall discuss the atomic packing factor for three types of cubic structures one by one.

1. *Simple Cubic structure:* We know that in a simple cubic structure, the total number of atoms per unit cell is one. And the atomic radius, $r = a/2$. Therefore volume of one atom with radius 'r',

$$v = \frac{4}{3}\pi r^3 = \frac{4\pi}{3}\left(\frac{a}{2}\right)^3 = \frac{\pi a^3}{6}$$

and volume of one unit cell with side 'a'

$$V = a^3$$

∴ Atomic packing factor

$$= \frac{v}{V} = \frac{\frac{\pi a^3}{6}}{a^3} = \frac{\pi}{6} = 0.52$$

2. *Body centred cubic structure:* We know that in a body centred cubic structure, the total number of atoms per unit cell are two and the atomic radius, $r = a\sqrt{3}/4$. Therefore volumes of two atoms with radius 'r',

$$v = 2 \times \frac{4}{3} \pi r^3 = \frac{8\pi}{3} \left(\frac{a\sqrt{3}}{4}\right)^3 = \frac{\pi a^3 \sqrt{3}}{8}$$

and volume of one unit cell with side '*a*'
$$V = a^3$$

∴ Atomic packing factor

$$= \frac{v}{V} = \frac{\dfrac{\pi a^3 \sqrt{3}}{8}}{a^3} = \frac{\pi \sqrt{3}}{8} = 0.68$$

3. *Face centred cubic structure:* We know that in a face centred cubic structure, total the number of atoms per unit cell are four and the atomic radius, $r = a\sqrt{2}/4$. Therefore volume of four atoms with radius '*r*'.

$$v = 4 \times \frac{4}{3} \pi r^3 = \frac{16\pi}{3}\left(\frac{a\sqrt{2}}{4}\right)^3 = \frac{\pi a^3 \sqrt{2}}{6}$$

and volume of one unit cell with side '*a*'
$$V = a^3$$

∴ Atomic packing factor

$$= \frac{v}{V} = \frac{\dfrac{\pi a^3 \sqrt{2}}{6}}{a^3} = \frac{\pi \sqrt{2}}{6} = 0.74.$$

Fig. 3.54 shows a face-centred cubic (FCC) structure by a stack of close packed planes *i.e.*, planes with highest density of atoms.

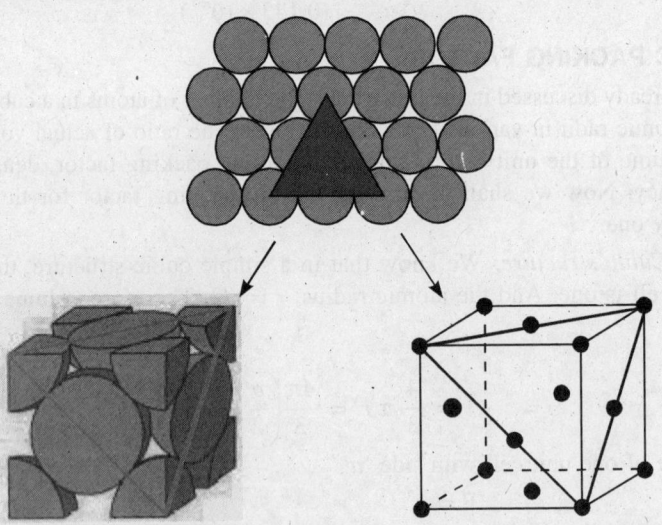

Fig. 3.54.

4. *Hexagonal close packed structure:* We know that in hexagonal close packed (HCP) structure, the effective number of atoms per unit cell is six and the atomic radius, $r = a/2$. Therefore volume of six atoms with radius '*r*'.

$$V = 6 \times \frac{4}{3}\pi r^3 = 8\pi \left(\frac{a}{2}\right)^3 = \pi a^3$$

Crystal Structure

We also know that volume of a HCP unit cell,
$$V = \text{(area of base)} \times \text{(height of unit cell)}$$

As seen from Fig. 3.55, the area of the base,

$$= 6 \times \text{Area } \triangle OAB$$

$$= 6 \times \left(\frac{1}{2} \times OG \times AB\right)$$

$$= 6 \times \frac{1}{2} \times \left(\frac{\sqrt{3}}{2}\right) a \times a$$

$$= \frac{3\sqrt{3}}{2} a^2$$

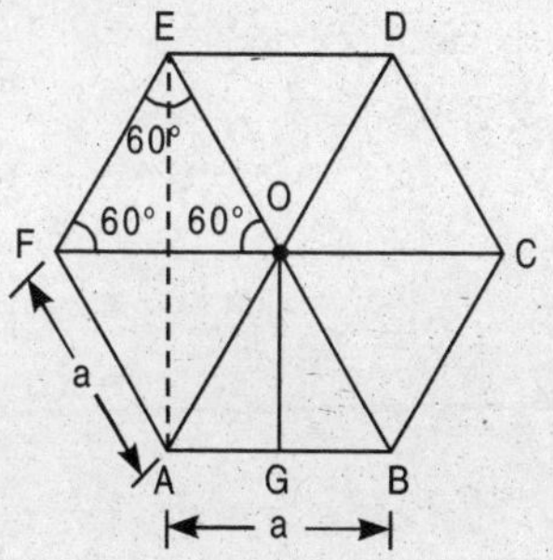

This volume of a HCP unit cell,

$$V = \left(\frac{3\sqrt{3}}{2} a^2\right) \times c$$

$$= \frac{3\sqrt{3}}{2} a^2 c$$

Fig. 3.55.

Now atomic packing factor,

$$= \frac{v}{V} = \frac{\pi a^3}{3\frac{\sqrt{3}}{2} a^2 c} = \frac{2\pi a}{3\sqrt{3}\,c}$$

In a hexagonal close packed structure, the ratio of c/a is 1.633 ideally. Substituting this value for (c/a) in the above equation, we get,

The atomic packing factor,

$$= \frac{2\pi}{3\sqrt{3}} \cdot \frac{1}{(c/a)} = \frac{2\pi}{3\sqrt{3} \times 1.633} = 0.74.$$

Example 3.16. *Calculate the volume of zinc crystal unit cell by using the following data: Zinc has HCP crystal structure with a = 0.2665 nm and c = 0.4947 mm.*

(A.M.I.E., Summer, 2002)

Solution. Given: Zinc has *HCP* crystal structure, $a = 0.2665$ nm and $c = 0.4947$ mm.

We know that volume of *HCP* unit cell,

$$V = \frac{3\sqrt{3}}{2} a^2 C$$

$$= \frac{3\sqrt{3}}{2} \times (0.2665)^2 \times (0.4947)$$

$$= 0.0913 \text{ mm}^3 \text{ Ans.}$$

3.33 CLOSE-PACKED STRUCTURES (FCC AND HCP)

As discussed earlier, both *FCC* (face-centred cubic) and HCP (hexagonal close-packed) structures have atomic packing factor of 0.74. The value 0.74 is considered to be a maximum possible value. Both the FCC and HCP crystal structures may be generated by the stacking of close-packed planes. The difference between the two structures is in the stacking sequence.

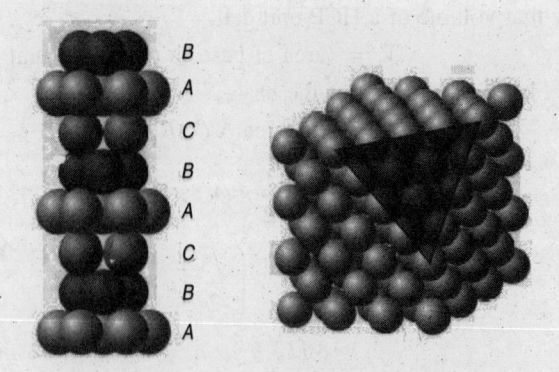

Fig. 3.56. Stacking sequence of FCC crystal structure.

Fig. 3.56 shows the stacking sequence of FCC crystal structure. Notice that the stacking sequence in a FCC crystal structure is : ABC ABC ABC ... . The third plane is placed above the "holes" of the first plane not covered by the second plane.

Fig. 3.57, shows the stacking sequence of HCP crystal structure. Notice that the stacking sequence in a HCP crystal structure is: AB AB AB ... . The third plane is placed directly above the first plane of atoms.

Fig. 3.57. Stacking sequence of HCP crystal structure.

Example 3.17. *Find the packing efficiency of Ge (DC) crystal. If the radius of Ge atom is 1.22 Å, find its lattice parameter.*

Solution. Given: Radius, $r = 1.22$ Å

Packing efficiency

We know that Ge crystal have a body cubic centred structure. Therefore packing efficiency (or atomic packing factor)) of Ge crystal,

$$= 0.68 \text{ Ans.}$$

Lattice parameter

We also know that lattice parameter of the Ge crystal,

$$a = \frac{4r}{\sqrt{3}} = \frac{4 \times 1.22}{\sqrt{3}} = 2.82 \text{ Å} \quad \textbf{Ans.}$$

3.34 PERPENDICULAR DISTANCE BETWEEN A PLANE AND ORIGIN OF THE CUBE

Consider a cube and a plane *PQR* as shown in Fig. 3.58. Let the Miller indices of the plane be h, k and l. Now draw *HL* perpendicular from the origin (*H*) of the cube to the plane *PQR*.

Crystal Structure

Let a = Cubic edge or the lattice constant of the cube.

d = Perpendicular distance between the origin (H) and the plane (*i.e.*, HL),

α, β and γ = Angles which the perpendicular makes with x-x, y-y and z-z axes respectively, and

HP, QH and HR = Intercepts of the plane along x-x, y-y, and z-z axes respectively.

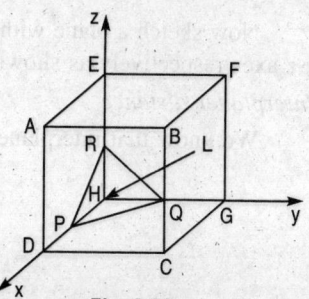

Fig. 3.58.

We know that the Miller indices of a plane are the smallest integers of the reciprocals of its intercepts. Therefore, the intercepts may also be expressed as reciprocals of Miller indices. Or in other words,

$$HP : HQ : HR = \frac{1}{h} : \frac{1}{k} : \frac{1}{l} = \frac{a}{h} : \frac{a}{k} : \frac{a}{l}$$

$$\therefore \quad HP = \frac{a}{h} ; HQ = \frac{a}{k} \text{ and } HR = \frac{a}{l}$$

From the geometry of the right angle triangles HPL, HQL and HRL we know that

$$\cos \alpha = \frac{HL}{HP} = d \times \frac{h}{a} = \frac{dh}{a}$$

$$\cos \beta = \frac{HL}{HQ} = d \times \frac{k}{a} = \frac{dk}{a}$$

and $$\cos \gamma = \frac{HL}{HR} = d \times \frac{l}{a} = \frac{dl}{a}$$

Since $\cos^2 \alpha + \cos^2 \beta + \cos^2 \gamma = 1$, therefore

$$\left(\frac{dh}{a}\right)^2 + \left(\frac{dk}{a}\right)^2 + \left(\frac{dl}{a}\right)^2 = 1$$

$$\frac{d^2}{a^2}(h^2 + k^2 + l^2) = 1$$

$$\therefore \quad d^2 = \frac{a^2}{(h^2 + k^2 + l^2)} \quad \text{or} \quad d = \frac{a^2}{(h^2 + k^2 + l^2)}$$

Note. If we draw a plane through the cube edge (H) and parallel to the plane PQR, then this relation may be used for obtaining the distance between these two planes. In this case, the distance HL is known as interplanar distance.

Example 3.18. *Show in a diagram the* (111) *plane of a cubic lattice. Calculate their interplanar distance.*

Solution. Given, Miller indices, $h = 1$, $k = 1$ and $l = 1$

(111) plane

We know that the reciprocals of $h = 1$, $k = 1$ and $l = 1$ are

$$= \frac{1}{1}, \frac{1}{1}, \frac{1}{1} = 1, 1, 1$$

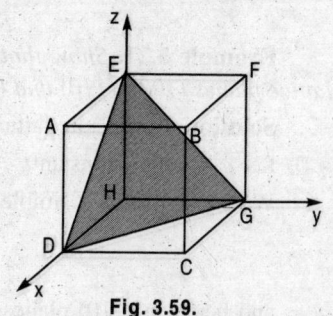

Fig. 3.59.

Now sketch a plane with intercepts 1, 1, 1 (*i.e.*, reciprocals of *h, k* and *l*) along *x-x, y-y* and *z-z* axes respectively as shown in Fig. 3.59.

Interplaner distance

We know that interplaner distance,

$$d = \frac{a}{\sqrt{h^2 + k^2 + l^2}} = \frac{a}{\sqrt{(1)^2 + (1)^2 + (1)^2}} \text{ Å}$$

$$= \frac{a}{\sqrt{3}} \text{ Å Ans.}$$

Example 3.19. *A F.C.C crystal has an atomic radius of* 1.246 Å. *What are the* d_{200}, d_{220}, *and* d_{111} *spacing ?*

Solution, Given: Atomic radius, *r* = 1.246 Å

We know that for a *F.C.C.* structure, the lattice constant,

$$a = \frac{4r}{\sqrt{2}} = \frac{4 \times 1.246}{\sqrt{2}} = 3.52 \text{ Å}$$

and spacing with Miller indices (*h, k, l*)

$$d = \frac{a}{\sqrt{h^2 + k^2 + l^2}}$$

$$\therefore \quad d_{200} = \frac{3.52}{\sqrt{(2)^2 + (0)^2 + (0)^2}} = \frac{3.52}{2} = 1.76 \text{ Å Ans.}$$

Similarly, $\quad d_{220} = \frac{3.52}{\sqrt{(2)^2 + (2)^2 + (0)^2}} = \frac{3.52}{2.828} = 1.24 \text{ Å Ans.}$

and $\quad d_{111} = \frac{3.52}{\sqrt{(1)^2 + (1)^2 + (1)^2}} = \frac{3.52}{1.732} = 2.03 \text{ Å Ans.}$

Example 3.20. *Copper has an FCC crystal structure and a unit cell with a lattice constant of 0.316 nm,. What is the interplanar spacing* d_{220}. (A.M.I.E., Summer, 2002)

Solution. Given: Copper has FCC structure and *a* = 0.316 nm. We know that the interplanar spacing,

$$d = \frac{a}{\sqrt{h^2 + k^2 + l^2}} = \frac{0.316}{\sqrt{2^2 + 2^2 + 0^2}}$$

$$= \frac{0.316}{\sqrt{8}} = 0.117 \text{ nm Ans.}$$

Example 3.21. *Show that in a simple cubic lattice the separation between the successive lattice planes (100), (110) and (111) are in the ratio of 1 : 0.71 : 0.58.* (Anna Univ. April, 2002)

Solution. Given: the lattice planes = (100), (110) and (111).

Let *a* = lattice constant.

We know that in a simple cubic lattice, the interplanar distance between (100) planes,

$$d_{100} = \frac{a}{\sqrt{h^2 + k^2 + l^2}} = \frac{a}{\sqrt{1^2 + 0^2 + 0^2}} = a$$

and between (110) planes,

$$d_{110} = \frac{a}{\sqrt{h^2 + k^2 + l^2}} = \frac{a}{\sqrt{1^2 + 1^2 + 0}} = \frac{a}{\sqrt{2}}$$

Similarly the interplanar distance between (111) planes,

$$d_{111} = \frac{a}{\sqrt{h^2 + k^2 + l^2}} = \frac{a}{\sqrt{1^2 + 1^2 + 1^2}} = \frac{a}{\sqrt{3}}$$

The ratio of interplanar distances,

$$d_{100} : d_{110} : d_{111} :: a : \frac{a}{\sqrt{2}} : \frac{a}{\sqrt{3}}$$

$$= 1 : \frac{1}{\sqrt{2}} : \frac{1}{\sqrt{3}}$$

$$= 1 : 0.71 : 0.58 \ \textbf{Ans.}$$

3.35 ANGLE BETWEEN TWO PLANES OR DIRECTIONS

Consider a cube having two planes *ABCD* and *EFCD* inclined at an angle (θ) with each other as shown in Fig. 3.60.

Let h_1, k_1, and l_1 = Miller indices of plane *ABCD*

and h_2, k_2 and l_2 = Miller indices of plane *EFCD*.

The angle between these two planes is given by the relation:,

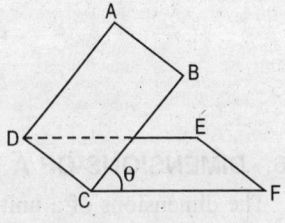

Fig. 3.60. Angle between two planes.

$$\cos \theta = \frac{h_1 h_2 + k_1 k_2 + l_1 l_2}{\sqrt{h_1^2 + k_1^2 + l_1^2} \times \sqrt{h_2^2 + k_2^2 + l_2^2}}$$

Similarly, the angle (ϕ) between the two directions having Miller indices (h_1, k_1 and l_1) and (h_2, k_2 and l_2) respectively is given by the relation:

$$\cos \phi = \frac{h_1 h_2 + k_1 k_2 + l_1 l_2}{\sqrt{h_1^2 + k_1^2 + l_1^2} \times \sqrt{h_2^2 + k_2^2 + l_2^2}}$$

Note. The derivation of this relation is beyond the scope of this book.

Examples 3.22. *Find the perpendicular distance between the two planes indicated by the Miller indices (111) and (222) in a unit cell of a cubic lattice with a lattice constant parameter 'a'.* (A.M.I.E., Summer. 1993)

Solution. Given: the two planes (111) and (222).

We know that the perpendicular distance between the origin and the plane (111),

$$d_1 = \frac{a}{\sqrt{h_1^2 + k_1^2 + l_1^2}} = \frac{a}{\sqrt{(1)^2 + (1)^2 + (1)^2}} = \frac{a}{\sqrt{3}}$$

and the perpendicular distance between the origin and the plane (222),

$$d_2 = \frac{a}{\sqrt{h_2^2 + k_2^2 + l_2^2}} = \frac{a}{\sqrt{(2)^2 + (2)^2 + (2)^2}} = \frac{a}{2\sqrt{3}}$$

∴ The perpendicular distance between the planes (111) and (222) are,

$$d_1 - d_2 = \frac{a}{\sqrt{3}} - \frac{a}{2\sqrt{3}} = \frac{2a - a}{2\sqrt{3}} = \frac{a}{2\sqrt{3}} \quad \textbf{Ans.}$$

Example 3.23. *Find the angle between the directions (122) and (111) in a cubic crystal.*
Solution. Given. Directions (122) and (111).

We know that the general expression for the angle (ϕ) between two directions with Miller Indices $(h_1, k_1$ and $l_1)$ and $(h_2, k_2$ and $l_2)$.

$$\cos\phi = \frac{h_1 h_2 + k_1 k_2 + l_1 l_2}{\sqrt{h_1^2 + k_1^2 + l_1^2} \times \sqrt{h_2^2 + k_2^2 + l_2^2}}$$

Now substituting the values of h_1, k_1 and $l_1 = (122)$ and h_2, k_2 and $l_2 = (111)$.

$$\cos\phi = \frac{1 \times 1 + 2 \times 1 + 2 \times 1}{\sqrt{(1)^2 + (2)^2 + (2)^2} \times \sqrt{(1)^2 + (1)^2 + (1)^2}} = \frac{5}{3\sqrt{3}}$$

$$= 0.962 \text{ or } \phi = 15°\,48' \quad \textbf{Ans.}$$

3.36 DIMENSIONS OF A UNIT CELL

The dimensions of a unit cell (or interatomic distance in a crystal lattice) may be obtained mathematically as discussed below:

Let
a = Lattice constant of a unit cell,
n = Number of atoms per unit cell,
A = Atomic weight of the crystalline substance or compound, and
N = Avogadro's number (6.023×10^{23})

We know that as per the Avogadro's hypothesis, one g-mol of crystalline compound contains 6.023×10^{23} molecules. Therefore mass of N molecules

$$= A \text{ g}$$

and mass of n molecules

$$= \frac{An}{N} \text{ g}$$

∴ Density of a unit cell,

$$\rho = \frac{\text{Mass of a unit cell}}{\text{Volume of unit cell}} = \frac{An/N}{a^3} = \frac{An}{a^3 N}$$

or
$$a^3 = \frac{An}{N\rho}$$

Note. In case of pure metal, the lattice constant (a) is equal to the distance between any two atoms at the corners of the cube. But in case of crystals like Sodium Chloride, Potassium Chloride etc, it is twice the distance between adjacent atoms. Therefore in such a case, $d = a/2$.

Crystal Structure

Example 3.24. *Iron has a density of 7.87 g/cc and atomic weight 55.85. What is the concentration of iron atoms per cc?* (A.M.I.E., Winter, 1999)

Solution. Given: density of iron (d) = 7.87 g/cc; atomic weight of iron = 55.85

We know that mass of an iron atom,

$$= \frac{\text{Atomic weight}}{\text{Avogadro's number}} = \frac{55.85}{6.023 \times 10^{23}} = 9.273 \times 10^{-23}$$

and the number of atoms per unit volume,

$$= \frac{\text{Mass of unit volume of iron}}{\text{Mass of an iron atom}} = \frac{\text{Density of iron}}{\text{Mass of an Iron atom}}$$

$$= \frac{7.87}{9.273 \times 10^{-23}} = 0.849 \times 10^{23} \text{ atoms/cm}^3 \quad \textbf{Ans.}$$

Example 3.25. *The density of α-iron is 7.87×10^3 kg/m^3 and its atomic weight is 55.8 if α-iron crystallizes in B.C.C. space lattice, find the value of lattice constant. (Given Avogadro number 6.02×10^{26} per kg mole).* (A.M.I.E., Winter 1999)

Solution. Given: Density of α-iron ρ = 7.87×10^3 kg/m^3: Atomic weight, A = 55.8; Avogadro's number, $N = 6.02 \times 10^{26}$ per kg mole

We know that lattice constant,

$$a^3 = \frac{An}{N\rho} = \frac{55.8 \times 2}{6.02 \times 10^{26} \times 7.87 \times 10^3} = 2.355 \times 10^{-29}$$

($\because$) number of atoms per unit cell in a B.C.C. lattice = 2)

$\therefore \quad a = (2.355 \times 10^{-29})^{1/3} = 2.866 \times 10^{-10}$ m = 2.866 Å **Ans.**

Note: In the expression for determining lattice constant, the value of ρ is in g/cc., if $N = 6.02 \times 10^{23}$ per kg mole. But in the present case, ρ is in kg/m^3 and hence $N = 6.02 \times 10^{26}$ per kg mole.

Example. 3.26. *Copper has a F.C.C. structure. Its atomic radius is 1.278 Å. Calculate its density (atomic weight of copper is 63.5. Avogadro's No. is 6.02×10^{23}).*

Solution. Given : Atomic Radius of copper atom, r = 1.278 Å = 1.278×10^{-8} cm. Atomic weight of copper, A 63.5 Avogadro's number, $N = 6.02 \times 10^{23}$

We know that atomic radius of a F.C.C. lattice (r),

$$1.278 \times 10^{-8} = \frac{a\sqrt{2}}{4}$$

or $\quad a = \dfrac{1.278 \times 10^{-8} \times 4}{\sqrt{2}} = 3.61 \times 10^{-8}$ cm

and density of copper,

$$\rho = \frac{An}{a^3 N} = \frac{63.5 \times 4}{(3.61 \times 10^{-8})^3 \times 6.02 \times 10^{23}} \text{ g/cc}$$

($\because n$ = 4 in a F.C.C. lattice)

$= 8.97$ g/cc **Ans.**

Example 3.27. *Calculate the number of atoms per unit cell of a metal having a lattice parameter of 2.9 Å and density of 7.87 g/cc. Atomic weight of the metal is 55.85 and Avogadro's constant is 6.023×10^{23}.*

Solution. Given: Lattice constant a = 2.9 Å = 2.9×10^{-8} cm; Density of metal, ρ = 7.87 g/cc; Atomic weight of the metal A = 55.85; Avogadro's constant, $N = 6.023 \times 10^{23}$.

Let n = Number of atoms per unit cell.

We know that in a unit cell

$$a^3 = \frac{An}{N\rho}$$

$$(2.9 \times 10^{-3})^3 = \frac{55.85 \times n}{6.023 \times 10^{23} \times 7.87} = 1.18 \times 10^{-23} n$$

$$n = \frac{(2.9 \times 10^{-8})^3}{1.18 \times 10^{-23}} = 2 \text{ Ans.}$$

Example 3.28. *A substance with face centred cubic lattice has density 6250 kg/m³ and molecular weight 60.2. Calculate the lattice constant 'a'. Given Avogadro No. = 6.02 × 10²⁶ kg. mole⁻¹.* *(Chartrapati Shiva Ji Maharaj Univ., 2004)*

Solution. Given: Density $(d) = 6250$ kg/m³, Molecular weight $(m) = 60.2/$ g/mol $= 60.2 \times 10^{-3}$ kg/mol. Avogadro number $= 6.02 \times 10^{26}$ atoms/mole.

We know that density (d),

$$6250 = \frac{\text{Mass/unit cell}}{\text{Volume/unit cell}} = \frac{M}{V}$$

Now Mass, $$M = \frac{(\text{No. of atoms/unit cell}) \times \text{Atomic mass}}{\text{Avogadro's number}}$$

$$= \frac{4 \times (60.2 \times 10^{-3})}{6.02 \times 10^{23}} = 40 \times 10^{-26} \text{ kg.}$$

and density, d,

$$6250 = \frac{40 \times 10^{-26}}{V}; \quad \therefore V = \frac{40 \times 10^{-26}}{6250} m^3 = 6.4 \times 10^{-29} m^3$$

We also know that volume of a unit cell, (V),

$$6.4 \times 10^{-29} = a^3$$

$$\therefore \quad a = \sqrt[3]{6.4 \times 10^{-29}} = 0.1857 \times 10^{-9} \text{ m} = 0.1857 \text{ nm. Ans.}$$

Example 3.29. *Calculate the number of atoms per unit cell of a metal having the lattice constant 2.9 Å and density 7.87 g/cm³. Atomic weight of the metal is 55.85 and Avogadro's constant is 6.02 × 10²³.*

Solution. Given: Lattice constant $(a) = 2.9$ Å $= 2.9 \times 10^{-8}$ cm; density $(d) = 7.87$ g/cm³; atomic weight of the metal = 55.85; Avogadro's constant $(N) = 6.023 \times 10^{23}$.

Let n = number of atoms per unit cell.

We know that density of a unit cell (ρ),

$$7.87 = \frac{\text{Mass of a unit cell}}{\text{Volume of a unit cell}} = \frac{A.n/N}{a^3} = \frac{A.n}{a^3 N}$$

$$= \frac{55.85 \times n}{(2.9 \times 10^{-8})^3 \times (6.023 \times 10^{23})} = 3.8 n$$

or $$n = 7.87 / 3.8 \approx 2. \text{ Ans.}$$

Note. Since the number of atoms per unit cell is 2. Therefore, it has a body-centred cubic *(BCC)* structure. Further you can relate to your priory knowledge that iron has a density of 55.85 and it has a BCC structure at room temperature.

Crystal Structure

3.37 CALCULATION OF NUMBER OF ATOMS PER CUBIC CENTIMETER

Let d = density of the material (g/cm^3)

N = Avogadro's number (6.023 × 10^{23} atoms / mol)

M = atomic mass or weight (g/mol)

Then the number of atoms per cm^3.

$$n = \frac{N \times d}{M} \text{ atoms/cm}^3$$

For example, graphite (carbon) : d = 2.3 g/cm^3, M = 12 g/mol

$$\therefore \quad n = \frac{6.023 \times 10^{23} \times 2.3}{12} = 11.5 \times 10^{22} \text{ atoms/cm}^3$$

For diamond (carbon) : d = 3.5 g/cm^3 ; M = 12 g/mol

$$\therefore \quad n = \frac{6.023 \times 10^{23} \times 3.5}{12} = 17.5 \times 10^{22} \text{ atoms/cm}^3$$

For water (H$_2$O), d = 1 g/cm^3, M = 18 g/mol

$$\therefore \quad n = \frac{6.023 \times 10^{23} \times 1}{18} = 3.3 \times 10^{22} \text{ atoms/cm}^3.$$

3.38 MEAN DISTANCE BETWEEN ATOMS

The mean distance between the atoms is given by,

$$L = \left(\frac{1}{n}\right)^{1/3}$$

For example, a material with n = 6 × 10^{22} atoms/cm^3, the mean distance between the atoms,

$$L = \left(\frac{1}{n}\right)^{1/3} = \left(\frac{1}{6 \times 10^{22}}\right)^{1/3} = 3.04 \times 10^{-8} \text{ cm} = 0.304 \text{ nm}.$$

3.39 CO-ORDINATION NUMBER

The term 'co-ordination number' may be defined as the number of atoms directly surrounding a given atom. We know that in a simple cubic structure, each corner atom is directly surrounded by six other atoms of the adjacent unit cells as shown in Fig. 3.61.

It may be noted that two atoms (1 and 2) surround the corner atom of a unit cell along *x-x* axis. Similarly, two atoms (3 and 4) surround the corner atom along *y-y* axis and two atoms (5 and 6) along *z-z* axis respectively. Thus the co-ordination number of a simple cubic structure is six.

We know that in a body centred cubic structure, each corner atom is surrounded by eight other body centred unit cells. Moreover, their body centred atoms are more near the corner atoms than those in the other corners. Therefore, the nearest adjacent neighbours are the body centred atoms of the eight surrounding unit cells. Thus the co-ordination number of a body centred cubic structure is eight. Similarly, in a face centred cubic structure, each corner atom is surrounded by eight other face centred unit cells. It may be noted that the twelve face centred atoms are more near the corner atom than those in the other corners. Thus the co-ordination number of a face centred cubic structure is twelve.

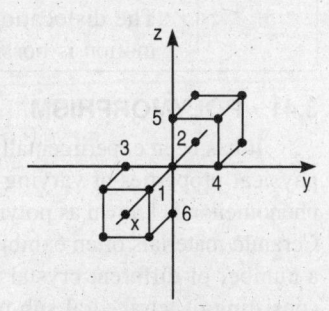

Fig. 3.61. Co-ordination number.

3.40 CRYSTALLINE AND AMORPHOUS SOLIDS

The crystalline solids are those in which the atoms or molecules are arranged in a very regular and orderly fashion in all the three dimensional pattern. Each atom or molecule is fixed at a definite point in the space at a definite distance and in a definite angular orientation, with all other atoms or molecules surrounding it. Metals and alloys are the examples of crystalline solids. The amorphous solids are those in which the atoms or molecules are arranged in an irregular pattern. These solids are noncrystalline in nature. Glass, wood, plastics and rubbers are the examples of amorphous solids.

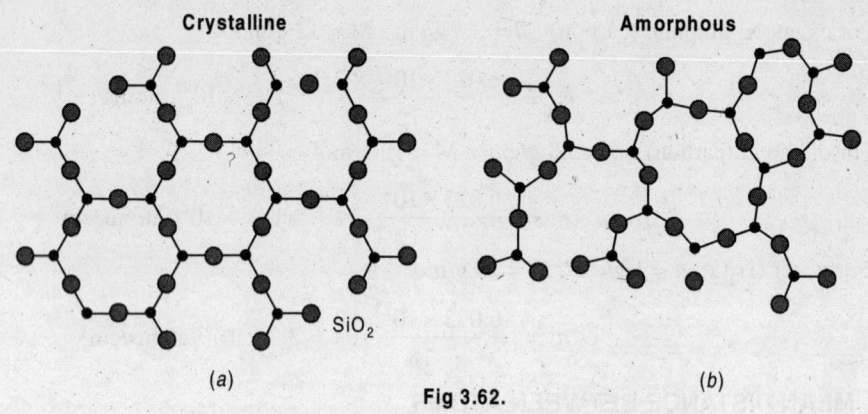

Fig 3.62.

Fig. 3.62 (a) and (b) shows the arrangement of SiO_2 atoms in a crystalline and amorphous form. Notice the systematic or periodic arrangement of atoms in a crystalline form. However, the amorphous form lacks systematic atomic arrangement.

Table 3.3 gives the difference between the crystalline and amorphous solids:

Table 3.3

S. No	Crystalline	Amorphous
1.	All the atoms or molecules are arranged in a regular and orderly fashion in all the three dimensions.	All the atoms or molecules are arranged in an irregular fashion in all the three dimensions.
2.	The solids fracture in a ductile manner.	The solids fracture in a brittle manner.
3.	The solids behave elastically up to their yield points.	The solids do not behave elastically.
4.	The tensile strength is high.	The tensile strength is low.
5.	The dislocation defects in crystals motion is possible.	The dislocation defects in crystals motion is not possible.

3.41 POLYMORPHISM

It has been experimentally found that certain materials exhibit different crystal structures and physical properties at varying temperatures while possessing the same chemical behaviour. This phenomenon is known as polymorphism. If the material is an elemental solid, it is called **allotropy**. Ceramic materials often exhibit polymorphism. For example, pure crystalline silica (SiO_2) assumes a number of different crystal structures by the formation of ordered three-dimensional network consisting of tetrahedral sub-units which share corners. Different crystalline forms of compound are called polymorphs.

Crystal Structure

An example of allotropy is carbon, which can exist as diamond, graphite and amorphous carbon. As shown in Fig. 3.62 pure solid carbon occurs in three crystalline forms, namely: diamond, graphite and large hollow fullerenes. Two kinds of fullerenes are shown here: buck minster fullerene (bucky ball) and carbon nanotube.

Another example is the simplest polymorph of SiO_2, called crystobalite. It is stable at high temperature. Two other polymorphs of silica are tridymite and quartz. Their stability and transformation temperatures are as follows:

$$\text{Quartz} \xrightarrow{867°C} \text{Tridymite} \xrightarrow{1470°C} \text{Crystobalite} \xrightarrow{1726°} \text{Melt}$$

(specific gravity = 2.65)   (specific gravity = 2.26)   (specific gravity = 2.32)

All the three polymorphs of SiO_2 namely quartz, tridymite and crystobalite have their SiO_2 tetrahedrals linked together in different ways. Transformation from one polymorph to another involves breaking of Si–O–Si bond and rejoining of tetrahedra according to new schemes. However, this process is very slow and because of this, the three polymorphs are capable of existing in metastable condition for long periods. Each of these forms can also exist in low temperature form (represented by α) and a high temperature form (represented by β.)

3.42 CRYSTAL DEFECTS OR IMPERFECTIONS

It has been observed that the crystals are rarely found to be perfect. The atoms do not have their full quota of electrons in the lowest energy level. But the atoms vibrate due to thermal effect, and the electrons also change their positions. There are many other types of defects found in the structure of the crystals. It will be interesting to know that the type and magnitude of any defect, in the structure of the metal crystals, plays an important role at the time of the selection as well as actual use. Though there are many types of defects in the crystals, yet we shall discuss only those defects which are found in the arrangement of atoms. Following three types of defects are important from the subject point of view:

1. Point defects.   2. Line defects.   3. Surface defects.

3.43 POINT DEFECTS

The defects, which take place due to imperfect packing of atoms during crystallisation, are known as point defects. The point defects also take place due to vibrations of atoms at high temperatures. Following types of point defects are important from the subject point of view:

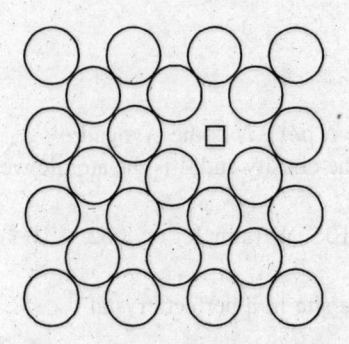

(a) Vacancy defect 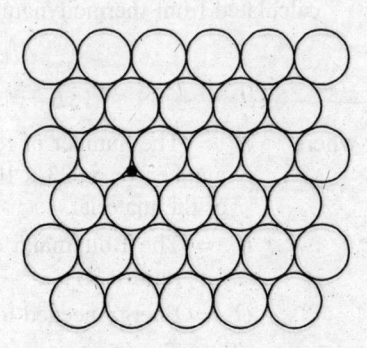   (b) Interstitial defect

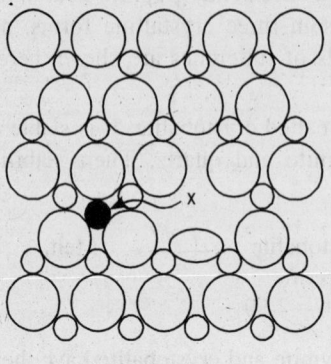

(c) Frenkel defect

(d) Substitutional defect

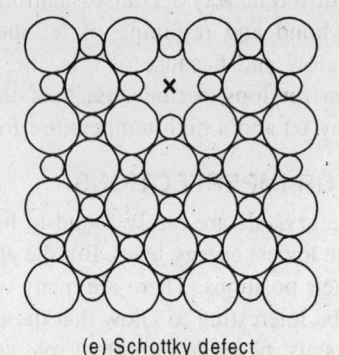

(e) Schottky defect

Fig. 3.63. Crystal defects.

1. *Vacancies.* Whenever one or more atoms are missing from a normally occupied position, as shown in Fig. 3.63 (a), the defect caused is known as vacancy. Such defects can be a result of imperfect packing during the formation of crystals. They also arise from thermal vibration of the atoms at high temperatures. It may be noted that there may be a single vacancy (if one atom is missing), di-vacancies (if two atoms are missing), tri-vacancies (if three atoms are missing) and so on.

    The equilibrium number of vacancies formed as a result of thermal vibrations may be calculated from thermodynamics as follows,

    $$N_v = N_s . e^{\left(\frac{Q_v}{K_B T}\right)}$$

    where N_s = The number of regular lattice sites (= $N\rho/A$). N is the Avogadro's number (= 6.023×10^{23} atoms / mol, ρ is the density and A is the atomic weight of the material.

    K_B = The Boltzmann constant (= 1.38×10^{-23} J/ (atm K) or 8.62×10^{-5} eV/(atm – K)

    Q_v = Energy needed to form a vacant lattice site in a perfect crystal

    T = Temperature in Kelvin.

2. *Interstitial defects.* Whenever an extra atom occupies interstitial position (*i.e.*, voids) in the crystal system, without dislodging the parent atom as shown in Fig. 3.63 (b), the defect caused is known as interstitial defect. It may be noted that the atom, which occupies the interstitial position is generally smaller than the parent atoms. In close

Crystal Structure

packed structures (such as *FCC* and *HCP*), the largest size of the atom that can fit in the voids have radius about 22.5% of the radii of parent atoms. However, if an atom, with larger size than that mentioned above fits in the voids, they produce distortion in atoms.

3. *Frenkel defect.* Whenever a missing atom (responsible for vacancy) occupies interstitial position (responsible for interstitial defect) as shown in Fig. 3.63 (c), the defect caused is known as Frenkel defect. It may be noted that a Frenkel defect is a combination of vacancy and interstitial defects. This type of defect is more common in ionic crystals, because the positive ions, being smaller in size, get lodged easily in the interstitial positions.

4. *Substitutional defect.* Whenever a foreign atom (*i.e.*, other than the parent atoms) occupies a position, which was initially meant for a parent (atom or in other words replaces a parent atom), as shown in Fig. 3.63 (d), the defect caused is known as substitutional defect. It may be noted that in this type of defect, the atom which replaces the parent atom may be of the same size (or slightly smaller or larger) than that of the parent atom.

5. *Schottky defect.* Whenever a pair of positive and negative ions is missing from a crystal, as shown in Fig. 3.63 (e), the defect caused is known as schottky defect. It may be noted that in this type of defect, the crystal is electrically neutral.

6. *Phonon.* Whenever a group of atoms is displaced from its ideal location, the defect caused is known as phonon. It may be noted that such a defect is caused by thermal vibrations. It happens as the atoms interact with one another, they tend to virbrate in synchronism in the same way as waves on the ocean surface. It will be interesting to know that this defect effects electrical and magnetic properties to a great extent.

Example 3.30. *Calculate the number of vacancies in copper at room temperature. Assume Boltzmann constant = 1.38×10^{-23} J/(atom – K) (or 8.62×10^{-5} eV/(atom – K)); $Q_v = 0.9$ eV/atom, room temperature, $t = 27°C$. $\rho_{cu} = 8.4$ g/cm^3, $A_{cv} = 63.5$ g/mol.*

Solution. Given: Boltzmann constant, $k_B = 1.38 \times 10^{-23}$ J/(atom – K) or 8.62×10^{-5} eV/(atom –K); $Q_v = 0.9$ ev/atom; $T = 27°C + 273 = 300$ K.

We know that the number of regular lattice sites,

$$N_S = \frac{N_v P}{A_{cu}} = \frac{(6.023 \times 10^{23}) \times 8.4}{63.5} = 8 \times 10^{22} \text{ atoms/cm}^3.$$

and the number of vacancies in copper,

$$N_v = N_s \cdot e^{-\left(\frac{Q_v}{K_B T}\right)}$$

$$= (8 \times 10^{22}) \times e^{-\left(\frac{0.9}{(8.62 \times 10^{-5}) \times 300}\right)} = (8 \times 10^{22}) \times e^{-\left(\frac{0.9}{0.026}\right)}$$

$$= 7.4 \times 10^7 \text{ vacancies / cm}^3 \textbf{ Ans.}$$

3.44 LINE DEFECTS

The defects, which take place due to dislocation or distortion of atoms along a line, in some direction, are called as line defects. The line defects also take place when a central portion of a crystal lattice slips without effecting the outer portion. Following two types of line defects are important from the subject point of view:

1. *Edge dislocation.* Whenever a half plane of atom is inserted between the planes of atoms in a perfect crystal, the defect so produced is known as edge dislocation.

Fig. 3.64 (*a*) shows a cross-section of a crystal where dots represent atoms arranged in an orderly manner. Fig 3.64 (*b*) shows the displacement of atoms when an extra half plane is inserted from the top. It may be noted from this figure that top and bottom of the crystal above and below the line *XY* appears perfect. If the extra half plane is inserted from top, the defect so produced, is represented by ⊥ (inverted Tee) as shown in figure. And if the extra half plane is inserted from the bottom, the defect so produced is represented by T (Tee).

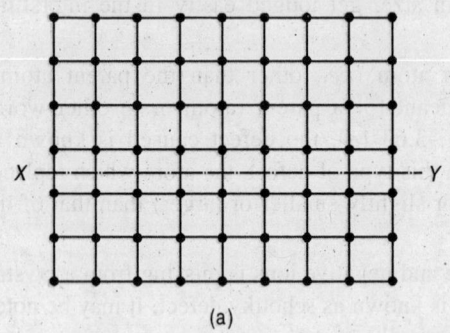

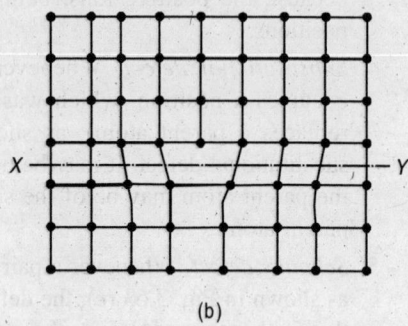

Fig. 3.64. Edge dislocation.

2. *Screw dislocation.* Whenever the atoms are displaced in two separate planes perpendicular to each other, the defect so produced is known as screw dislocation. Fig. 3.65 (a) represents an isometric view of a perfect crystal. Fig. 3.65 (b) shows the displacement of atoms in the region ABC. In screw dislocation, the arrangement of atoms appear like that of a screw or a helical surface.

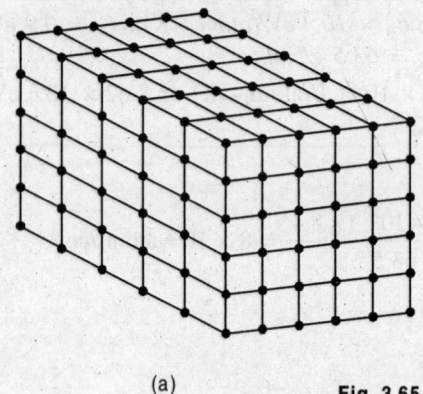

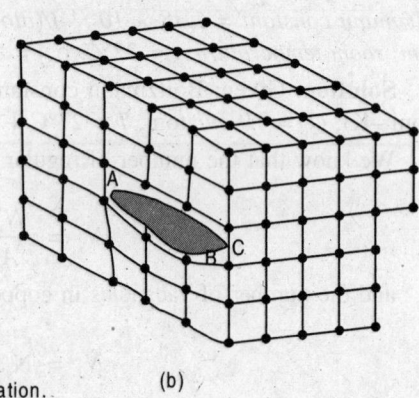

Fig. 3.65. Screw dislocation.

3.45 SURFACE DEFECTS

The defects, which takes place on the surface of a material, are known as surface defects or plane defects. It may be noted that surface defects take place either due to imperfect packing of atoms during crystallisation or defective orientation of the surface. Following types of surface defects are important from the subject point of view:

1. *Grain boundary.* Whenever grains of different orientation separate the general pattern of atoms and exhibits a boundary, as shown in Fig. 3.66, the defect caused is known as grain boundary. The type of defect generally takes place during the solidification of the liquid metal.

Fig. 3.66. Grain boundary.

Crystal Structure

2. *Twin boundary.* When the boundaries in which the atomic arrangement on one side of the boundary is a somewhat mirror image of the arrangement of atoms of the other side, as shown in Fig. 3.67, the defect caused is known as twin boundary. The region in which a twin boundary defect occurs is between the twinning planes as shown in the figure.

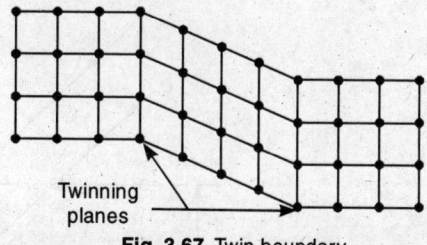

Fig. 3.67. Twin boundary.

3. *Stacking fault.* Whenever the stacking of atoms is not in proper sequence throughout the crystal, the fault caused is known as stacking fault.

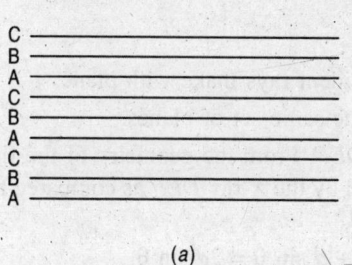

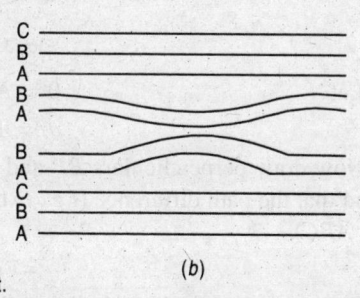

(a)   (b)

Fir. 3.68. Stacking fault.

Fig. 3.68 (a) shows the proper sequence of atomic planes if we read from bottom to top is A-B-C-A-B-C-A-B-C. Fig. 3.68 (b) shows the sequence of atomic planes as A-B-C-A-B-A-B-A-B-C. The region in which the stacking fault occurs (A-B-A-B) forms a thin region of hexagonal close packing in a F.C.C. crystal.

3.46 DETERMINATION OF CRYSTAL STRUCTURE

The determination of crystal structure is an important matter in the field of science and engineering. Following are the important methods of determining crystal structure as given below:

1. Using Bragg's law
2. X-ray diffraction method
3. Electron method
4. Neutron diffraction method

In this chapter, we shall discuss the method using Bragg's law only.

3.47 BRAGG'S LAW

Bragg's law is based on the concept that *X-rays are reflected by a set of parallel planes. As a result of this, the X-ray spectrum is produced which consists of series of maxima and minima intensity. It has been found that maxima are abtained for the reflected. X-rays, which are in phase with each other. And the minima are obtained for the reflected rays, which are out of phase with each other. The condition for the X-rays, to be in phase with each other, may be obtained as follows:

Consider a set of planes consisting of plane-1, plane-2 and plane-3 which are rich in atoms. Let the incident rays AB and DE be reflected at B and E, from the atoms, along BC and EF respectively as shown in Fig. 3.69.

---

* X-rays are the electromagnetic radiations with wavelength around 1Å (= 10 nanometre). The X-rays are produced whenever a fast moving electron beam strickes a metal having high atomic weight like tungsten kept in vacuum.

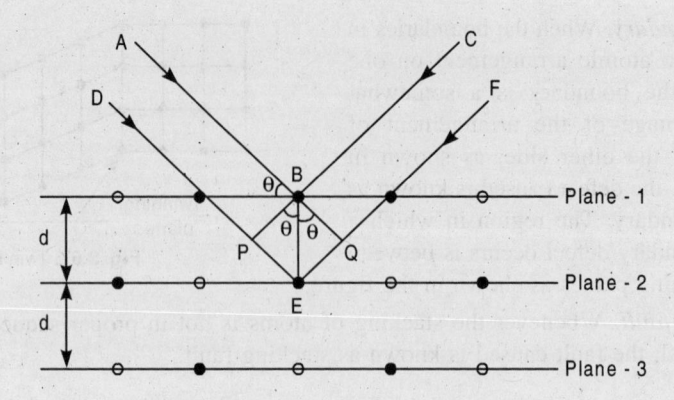

Fig. 3.69. Bragg's Law.

Let θ = Angle which the incident rays make with plane,

d = Distance between the same set of planes.

Now draw perpendiculars *BP* and *BQ* on the X-ray *DEF*. From the geometry of the figure, we find that the path difference (*i.e.*, extra distance travelled by the X-ray *DEF* as compared to the X-ray *ABC*)

$$= PE + EQ = d \sin \theta + d \sin \theta = 2d \sin \theta$$

We know that if this path difference is equal to an integral multiple of wavelength of X-rays (λ), the two rays are in phase with each other. And if it is equal to an odd multiple of $\lambda/2$, the two rays are out of phase with each other. Thus the condition for maxima may be written as:

$$2d \sin \theta = n \lambda$$

Where '*n*' is equal to 1, 2, 3, etc. for the first order, second order, third order .... respectively.

Example 3.31. *Determine the interplanar spacing when a beam of X-ray of wavelength 1.54 Å is directed towards the crystal at angle 20.3° to the atomic plane.* (A.M.I.E. Summer, 1993)

Solution. Given, $\lambda = 1.54$ Å and $\theta = 20.3°$

assuming the first order reflection, *i.e.*, $n = 1$,

We know that as per Bragg's law,

$$2d \sin \theta = n \lambda$$
$$2d \sin 20.3° = 1 \times 1.54$$

$$d = \frac{1.54}{2 \sin 20.3°} = \frac{1.54}{2 \times 0.3469} = 2.22 \text{ Å } \textbf{Ans.}$$

Example 3.32. *X-rays of wavelength 1.5418 Å are diffracted by (111) planes in a crystal at an angle 30° in the first order. Calculate the interatomic spacing.*

Solution. Given: Wavelength of X-rays = 1.5418 Å; $\theta = 30°$, Order of reflection, $n = 1$;

Let d = distance between the same set of planes (*i.e.*, interatomic spacing),

We know that as per Bragg's law,

$$2 d \sin \theta = n\lambda$$
$$2 d \sin 30° = 1 \times 1.5418 = 1.5418$$

$$\therefore \qquad d = \frac{1.5418}{2 \sin 30°} = \frac{1.5418}{2 \times 0.5} = 1.5418 \text{ Å } \textbf{Ans.}$$

Crystal Structure

Example 3.33. *For a certain BCC crystal, the (110) plane has a separations of 1.181 Å. These planes are indicated with X-rays of wavelength 1.540 Å. How many orders of Bragg's reflections can be observed in this case ?*

Solution. Given : $d = 1.181$ A; $\lambda = 1.540$ Å

We know that maximum order of the Bragg's reflection can be observed with the maximum value of reflection angle (θ) equal to $90°$. Therefore number of orders as per Bragg's law,

$$n = \frac{2d \sin \theta}{\lambda} = \frac{2 \times 1.181 \sin 90°}{1.540} = 1.53$$

Since the value of 'n' can be integer only, therefore the highest permissible value of 'n' in this case is 1. **Ans.**

Example 3.34. *X-rays with a wavelength of 0.58 Å are used for calculating d_{200} in nickel. The reflection angle is $9.5°$. What is the size of unit cell ?* (A..M.I.E. Summer 1982)

Solution. Given $\lambda = 0.58$ Å and $\theta = 9.5°$

Let $a =$ Size of unit cell or interatomic distance.

We know that for a cubic crystal system (of nickel), the interplanar distance,

$$d = \frac{a}{\sqrt{h^2 + k^2 + l^2}}$$

$$\therefore \quad d_{200} = \frac{a}{\sqrt{2^2 + 0^2 + 0^2}} = \frac{a}{2} = 0.5\, a$$

and according to Bragg's law,

$$2d \sin = n\, \lambda$$

$$\therefore \quad 2d_{200} \sin 9.5° = 1 \times 0.58$$

$$2 \times 0.5\, a \times 0.165 = 0.58$$

or $\quad a = \dfrac{0.58}{1.165} = 0.498$ Å **Ans.**

Example 3.35. *Calculate the Bragg angle if (111) planes of a cube ($a = 3.57$ Å) crystal are exposed to X-rays (wavelength is 1.54 Å)* (A.M.I.E. Summer, 1991)

Solution. Given: Miller indices of the (111) planes, $h = 1$ $k = 1$ and $l = 1$; $a = 3.57$ Å and $\lambda = 1.54$ Å.

Let $\theta =$ Bragg's angle for first order reflection.

We know that for a cubic crystal system, the interplanar distance,

$$d = \frac{a}{\sqrt{h^2 + k^2 + l^2}}$$

$$\therefore \quad d_{111} = \frac{3.57}{\sqrt{(1)^2 + (1)^2 + (1)^2}} = 2.06 \text{ Å}$$

and according to Bragg's law,

$$2d_{111} \sin \theta = n\, \lambda$$

$$2 \times 2.06 \times \sin \theta = 1 \times 0.54$$

$$\sin \theta = \frac{1 \times 0.54}{2 \times 2.06} = 0.131$$

$$\theta = 7°\ 32'\ \textbf{Ans.}$$

Example 3.36. *A diffraction pattern of a cubic crystal of lattice parameter a = 3.16 Å. is obtained with a monochromatic X-rays beam of wavelength 1.54 Å. The first line on this pattern was observed to have θ = 20.3 degrees. Determine the interplanar spacing and the Miller indices of the reflecting plane.*

Solution. Given: $\quad a = 3.16$ Å; $\lambda = 1.54$ Å; $n = 1$; $\theta = 20.3°$

We know that as per Bragg's law.
$$2d \sin \theta = n\lambda$$
$$2d \sin 20.3° = 1 \times 1.54 = 1.54$$

$\therefore \quad d = \dfrac{1.54}{2 \sin 20.3°} = \dfrac{1.54}{2 \times 0.3469} = 2.22$ Å

We also know that for a cubic crystal system, the interplanar distance,

$$d = \dfrac{a}{\sqrt{h^2 + k^2 + l^2}}$$

$$2.22 = \dfrac{3.16}{\sqrt{h^2 + k^2 + l^2}}$$

$$\sqrt{h^2 + k^2 + l^2} = \dfrac{3.16}{2.22} = 1.423$$

This value is very close to $\sqrt{2}$ (*i.e.*, 1.414)

$\therefore \quad \sqrt{h^2 + k^2 + l^2} = \sqrt{2} \quad$ or $\quad h^2 + k^2 + l^2 = 2$

The above equation may be solved by assuming any one of the indices (say l) equal to zero.
$\therefore \quad h^2 + k^2 = 2$

This is only possible if $h = 1$, $k = 1$, Similarly, if we assume $h = 0$ or $k = 0$, we get the values as $k = 1$, $l = 1$ and $h = 1$, $l = 1$ respectively. Thus the Miller indices of the reflecting planes are (110), (011) or (101). **Ans.**

Example 3.37. *For BCC iron, compute (i) the interplanar spacing and (ii) the diffraction angle for the (211) set of planes. The lattice parameter for iron is 0.2866 nm (2.866 Å). Also assume the monochromatic radiation having a wavelength of 0.1542 nm (1.542 Å) is used and order of reflection is 1.* (A.M.I.E., Summer 2003)

Solution. Given: The lattice parameter for iron = 0.2866 nm. Wavelength of monochromatic radiation, $\lambda = 0.1542$ nm. Order of reflection $n = 1$.

We know that Miller indices of the (211) plane,
$$h = 2 \; ; k = 1 \; ; l = 1$$

(i) Interplanar spacing

We know that interplanar spacing,
$$d_{211} = \dfrac{a}{\sqrt{h^2 + k^2 + l^2}} = \dfrac{0.2866}{\sqrt{2^2 + 1^2 + 1^2}} = 0.1433 \text{ nm}$$

(ii) Different angle

Let $\quad \theta = $ Bragg's angle of first order diffraction.

We know that the condition for the maxima.
$$2d \sin \theta = n\lambda$$

Crystal Structure

For the present case, $d = d_{211}$ and $n = 1$.

$\therefore \quad 2 d_{211} \sin \theta = 1.\lambda = 1 \times 0.1542 = 0.1542$

$\therefore \quad \sin \theta = \dfrac{0.1542}{2 d_{211}} = \dfrac{0.1542}{2 \times 0.1433} = 0.5380$

or $\quad \theta = \sin^{-1}(0.5380) = 32°33'$ **Ans.**

A.M.I.E. (I) EXAMINATION QUESTIONS

1. Write a short note on space lattice. *(Winter, 1992)*
2. Draw atomic arrangement in a planar surface imperfection with a boundary. *(Summer, 1993)*
3. State and explain the Bragg's law of x-ray diffraction. What are its uses? *(Summer, 1994)*
4. What are Miller indices? How are they determined? *(Winter, 1994)*
5. Discuss Bragg's law of x-ray diffraction. Describe the powder method for the determination of crystal structure. *(Summer, 1996)*
6. Draw neat sketches of unit cells of simple cubic, *BCC* and *FCC* crystal structures. Calculate the number of atoms in each case. *(Winter, 1996)*
7. Using neat sketches show (*a*) vacancy defect and (*b*) interstitial defect in crystal. *(Winter, 1996)*
8. What is the number of atoms per unit cell in the *FCC* structure? *(Summer, 1997)*
9. Show that the atomic packing factor of FCC crystal is 0.74. *(Summer, 1998)*
10. What is the difference between atomic structure and crystal structure? *(Summer, 1998)*
11. What are Miller Indices? How are they determined? Give suitable examples. *(Winter, 1998)*
12. Give a neat sketch of a unit cell of Copper. Indicate Miller indices for its slip plane and slip direction. Why Copper is so ductile. *(Summer, 1999)*
13. Explain the terms crystal : unit cell, coordination number and packing factor. *(Winter 1999)*
14. (*a*) Differentiate between a grain and a crystal? Illustrate with sketches.
    (*b*) Give a neat sketch of unit cell of Copper, indicate Miller indices for its slip palne and slip direction. Why Copper is so ductile. *(Winter 1999)*
15. Calculate the volume of a FCC unit cell in terms of the atomic radius. Calculate its packing factor also. *(Summer, 2000)*
16. Calculate the atoms per unit cell of metallic Zinc. Draw (121) plane and < 121 > direction in a cubic lattice. *(Winter, 2000)*
17. Differentiate between grain and a crystal. *(Summer 2001; 1999)*
18. What is polymorphism? Give at least two examples of polymorphism in materials. *(Summer 2001)*
19. Show that the atomic packing factor of BCC crystal is 0.68. *(Winter 2001)*
20. Name the crystal structures for the following metals:
    Mg, α-Fe, Copper and aluminium *(Winter 2001)*
21. Draw the following lattices: BCC, FCC and HCP. Determine for each lattice
    (*i*) effective number of atoms,
    (*ii*) packing factor and
    (*iii*) coordination number *(Winter, 2003)*

MULTIPLE CHOICE QUESTIONS

1. In a unit cell of aluminium, the aluminium atoms occupy
    (*a*) 80% volume of the cube
    (*b*) 90% volume of the cube
    (*c*) 74% volume of the cube
    (*d*) 84% volume of the cube
    (AMIE, Summer 2000)

2. In a cubic unit cell whose lattice constant is 'a', the distance between two {hkl} planes is,

(a) $\dfrac{a}{\sqrt{h^2+k^2+l^2}}$

(b) $\dfrac{a^2}{\sqrt{h^2+k^2+l^2}}$

(c) $a\sqrt{h^2+k^2+l^2}$

(d) $\dfrac{1}{\sqrt{h^2+k^2+l^2}}$

(AMIE, Summer 2000)

3. The Miller indices of a set of parallel planes which makes intercepts in the ratio $3a : 4b$ on the X and Y axes and are parallel to the Z axis (a, b and c are the primitive vectors of the lattice) are:

(a) 0, 4, 3   (b) 4, 3, 0   (c) 3, 3, 0   (d) 3, 4, 0

(Chatrapati Shivaji Maharaj Univ. 2004)

4. The number of nearest neighbour in FCC lattice is

(a) 6   (b) 8   (c) 12   (d) 10

(Chatrapati Shivaji Maharaj Univ. 2004)

ANSWERS

1. (c)   2. (a)   3. (b)   4. (c)

4
Bonds in Solids

1. Introduction. 2. Inertness of Noble Gases. 3. Cause of Bonding. 4. Classification of Bonds. 5. Primary Bonds. 6. Secondary Bonds. 7. Types of Primary Bonds. 8. Ionic Bonds. 9. Properties of Ionic Solids. 10. Covalent Bonds. 11. Properties of Covalent Solids. 12. Metallic Bonds. 13. Properties of Metallic Solids. 14. Comparison between Ionic, Covalent and Metallic Bonds. 15. Types of Secondary Bonds. 16. Dispersions Bonds. 17. Dipole Bonds. 18. Hydrogen Bonds. 19. Bond Energy. 20. Bond Length.

4.1 INTRODUCTION

It is a well known fact that an atom, by itself, rarely exists independently. Most of the elements are found in the form of clusters or aggregates of atoms. Any such cluster, in which the atoms are held together and is electrically neutral, is known as a molecule.

It has been observed that when the atoms of the elements combine to form molecules, a force of attraction is developed between the atoms, which hold them together. This force is known as bond. For example, in an oxygen molecule (O_2) the two oxygen atoms are held together by a bond. The formation of a bond is accompanied by a decrease in energy of the reacting atoms. It will be interesting to know that the resulting molecule has less energy, that is why it is more stable than the individual atoms. For example, an oxygen molecule (O_2) has less energy than its two oxygen atoms. Thus an oxygen molecule is more stable than the two separate atoms. In order to understand the formation of bonds in solids, its very essential to know the electronic arrangement of inert gases or noble gases (*i.e.*, gases whose outermost shell is completely filled up with electrons).

4.2 INERTNESS OF NOBLE GASES

We have already discussed in Art. 2.31 the Modern Periodic Table. The elements in the zero group of the periodic table, which do not combine with other elements, are known as inert gases or noble gases. These elements are helium, neon, argon, krypton, xenon and radon. As a matter of fact, only the outermost electrons of an atom take part in any chemical reaction. Since the noble gases have their outermost shell completely filled up with electrons, therefore they are unreactive. Their atoms do not allow the outermost electrons to take part in any chemical reaction.

The Table 4.1 gives the electronic configuration of the inert or noble gases. A careful study of the table will give the reason for their inertness.

Table 4.1

S. No.	Noble gas	Symbol	Atomic number	Electronic configuration	No. of valence electrons
1.	Helium	He	2	$1s^2$	2
2.	Neon	Ne	10	$1s^2\ 2s^2\ 2p^6$	8
3.	Argon	Ar	18	$1s^2\ 2s^2\ 2p^6\ 3s^2\ 3p^6$	8
4.	Krypton	Kr	36	$1s^2\ 2s^2\ 2p^6\ 3s^2\ 3p^6\ 3d^{10}\ 4s^2\ 4p^6$	8
5.	Xenon	Xe	54	$1s^2\ 2s^2\ 2p^6\ 3s^2\ 3p^6\ 3d^{10}$ $4s^2\ 4p^6\ 4d^{10}\ 5s^2\ 5p^6$	8
6.	Radon	Rn	86	$1s^2\ 2s^2\ 2p^6\ 3s^2\ 3p^6\ 3d^{10}$ $4s^2\ 4p^6\ 4d^{10}\ 4f^{14}\ 5s^2$ $5p^6\ 5d^{10}\ 6s^2\ 6p^6$	8

It may be noted from the above table, that the number of electrons in the outermost shell of the noble gases are 8 (except the helium, which has only 2 electrons). We know that the atoms of all the noble gases are very stable and have 8 electrons in their outermost shells. As a result of this, we may state that an atom, having 8 electrons in its outermost shell, is considered to have the most stable arrangement of the electrons. Thus they cannot take part in chemical reactions. In other words, they cannot form a bond with any other atom.

4.3 CAUSE OF BONDING

As a matter of fact, every molecule or atom of every material has a tendency to become more stable. In case of atoms, the stability means to have its electronic configuration of noble gases. It is thus obvious, that the atoms of most of the elements, form bonds with one another, in order to achieve such an electronic configuration. It has been observed that the atoms can achieve stable configuration in the following three ways:

1. By loosing one or more electrons to another atom.
2. By gaining one or more electrons from another atom.
3. By sharing one or more electrons with other atoms.

4.4 CLASSIFICATION OF BONDS

The bonds may be broadly classified into the following two types :
    1. Primary bonds.   2. Secondary bonds.

4.5 PRIMARY BONDS

A primary bond is an interatomic bond, in which electrostatic force holds the atoms together. It may be noted that a primary bond is strong and more stable.

4.6 SECONDARY BONDS

A secondary bond is an intermolecular bond, in which the weak forces hold the molecules together. The weak forces are also known as van der Waal's forces. It may be noted that a secondary bond is weak and less stable than the primary bond.

4.7 TYPES OF PRIMARY BONDS

The primary bonds are the following three types:
    1. Ionic bonds. 2. Covalent bonds. 3. Metallic bonds.

Bonds in Solids

4.8 IONIC BONDS

An ionic bond is formed when one or more electrons, from the outermost shell of one atom, are transferred to the outermost shell of another atom.

In this process, both the atoms acquire electronic configuration of a noble gas. We know that the atom, which loses the electron, acquires a positive charge and becomes a positive ion. And the atom which gains the electron acquires a negative charge and becomes a negative ion. In this way, both the atoms become oppositely charged ions. The electrostatic attraction, between the oppositely charged ions, forms the ionic bond. Following two condition are necessary for the bond formation:

1. The atom, which loses electron, should have low ionization energy. It means that a small amount of energy is required to remove the electron from its outermost energy level.
2. The atom, which gains electron should have high electron affinity. It means that the atom should be able to readily accept and electron in its outermost energy level.

It has been observed that alkali metals like Lithium, Sodium, Potassium and Rubidium etc. have low ionization energy. Whereas Halogens like Fluorine, Chlorine, Bromine, Iodine etc. have high electrons affinity. It is thus obvious, that an ionic bond is readily formed between atoms of alkali metals and halogen atoms. The ionic bond is also formed between the atoms of metals like magnesium, aluminium, calcium etc. and the atoms of non-metals like oxygen, chlorine, bromine etc. The formation of an ionic bond in the case of sodium chloride is illustrated below:

We know that the atomic number of sodium is 11 and its electronic configuration is $1s^2\, 2s^2\, 2p^6\, 3s^1$. Similarly, the atoms number of chlorine is 17 and its electronic configuration is $1s^2\, 2s^2\, 2p^6\, 3s^2\, 3p^5$. The atomic structures of both the sodium and chlorine atoms are shown in Fig. 4.1 (a) Thus we see that the sodium atom has one electron in its outermost energy level ($3s^1$). and the chlorine atom has seven electrons in its outermost energy level ($3s^2\, 3p^5$). we also know that the sodium atom has a tendency to lose its electron present in its outermost energy level to have a stable configuration as $1s^2\, 2s^2\, 2p^6$. Moreover, the chlorine atom has a tendency to accept one electron (lost by sodium atom) in its outermost energy level to have a stable configuration of argon with electronic configuration as $1s^2\, 2s^2\, 2p^6\, 3s^2\, 3p^6$. The transfer of this electron is shown in Fig. 4.1 (b).

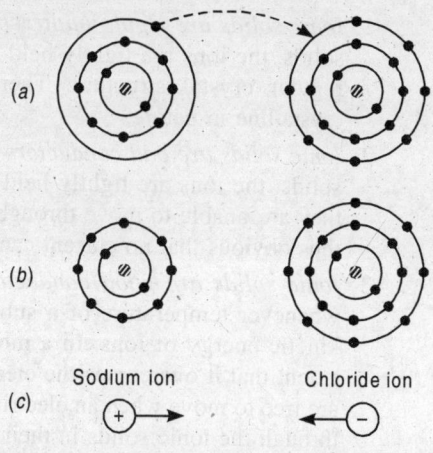

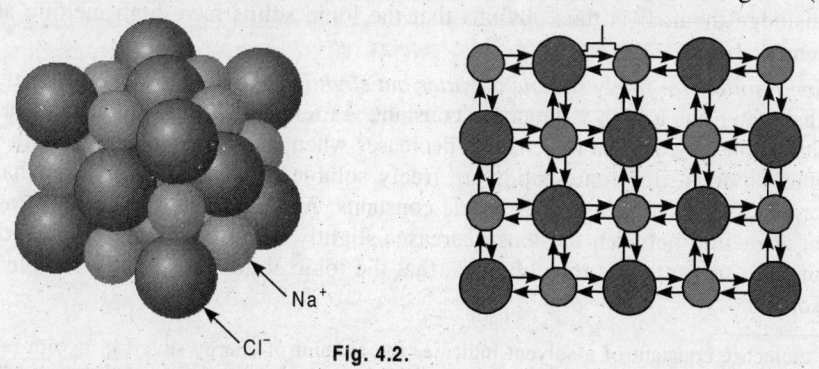

Fig. 4.1. Formation of ionic bond.

Fig. 4.2.

It will be interesting to know that after this transfer, both the sodium and chlorine atoms have stable configuration. This happens as the sodium atom (by losing an electron) becomes a positive ion. And the chlorine atom (by accepting that electron) becomes a negative ion. Thus a bond is now formed due to the electrostatic force of attraction between the two oppositely charged ions as shown in Fig. 4.1 (a). It is thus obvious, that in this process sodium atom loses one electron while the chlorine atom accepts it. Now a new solid is formed which is known as sodium chloride (Refer to Fig. 4.2). Similarly, the formation of an ionic bond in the case of magnesium oxide may be explained as follows:

The atomic number of magnesium is 12 and its electronic configuration is $1s^2\ 2s^2\ 2p^6\ 3s^2$. Similarly, the atomic number of oxygen is 8 and its electronic configuration is $1s^2\ 2s^2\ 2p^4$. From the above two electronic configurations, we find that the magnesium atom has 2 electrons in its outermost energy level. Moreover, it has a tendency to become stable by losing its 2 electrons from the outermost energy level. Similarly, the oxygen atom has a tendency to become stable by accepting these two electrons (lost by the magnesium atom). In this process, both the atoms acquire stable configuration of neon whose electronic configuration is $1s^2\ 2s^2\ 2p^6$. The magnesium atom becomes positive ion and the oxygen atom a negative ion. Due to the electrostatic force of attraction between two ions, a new solid is formed which is known as magnesium oxide.

4.9 PROPERTIES OF IONIC SOLIDS

We know that ionic solids consist of oppositely charged ions. As a result of this, the ions are tightly bound to one another by the electrostatic force of attraction. Though there are a number of properties of ionic solids, yet the following are important from the subject point of view:

1. *Ionic solids are rigid, unidirectional and crystalline in nature.* We know that in ionic solids, the ions are tightly held together by electrostatic force of attraction and form a regular crystal structure. Therefore the ionic solids are rigid, unidirectional and crystalline in nature.

2. *Ionic solids are bad conductors of electricity in their solid state.* We know that in ionic solids, the ions are tightly held together by electrostatic force of attraction. Therefore they are unable to move through a large distance, when an electric field is applied. It is thus obvious that no current can flow through the ionic solids in their solid state.

3. *Ionic solids are good conductors of electricity in their molten state.* We know that whenever temperature of a substance is increased, there is always an increase in the kinetic energy of ions. In a molten state, the kinetic energy increases to such a large extent that it overcomes the electrostatic force of attraction. As a result of this, the ions are free to move when an electric field is applied. It is thus obvious, that current can flow through the ionic solids in their molten state.

4. *Ionic solids have high melting and boiling temperatures.* We know that in all the ionic solids, the ions are tightly held together. Therefore a considerable energy is required to dislodge them. It is thus obvious that the ionic solids have high melting and boiling temperatures.

5. *Ionic solids are freely soluble in water but slightly soluble in organic solvents.* We know that water has a high * dielectric constant. As a result of this, the electrostatic force of attraction between the ionic solids decreases when they come in contact with water. It is thus obvious, that ionic solids are freely soluble in water. On the other hand, all the organic solvents have low dielectric constants. As a result of this, the electrostatic force of attraction between the ions decreases slightly when they come in contact with any organic solvent. It is thus obvious, that the ionic solids are slightly soluble in organic solvents.

---

* The dielectric constant of a solvent indicates the amount of energy stored in it with respect to the energy stored in air or vacuum. It is this energy, which is responsible for dissolving any material. It may be noted that the electrostatic force of attraction varies inversely to the dielectric constant of the solvent.

Bonds in Solids

4.10 COVALENT BONDS

A covalent bond is formed, when two or more electrons of an atom, in its outermost energy level, are shared by electrons of other atoms. It happens, when the sharing of electrons takes place, in such a way, that each atom, in the resulting molecule, sets a stable configuration. This may be best understood from the formation of covalent bond of chlorine molecule, which is discussed below:

We know that the atomic number of chlorine atom is 17 and its electronic configuration is $1s^2\ 2s^2\ 2p^6\ 3s^2\ 3p^5$. Thus we see that each chlorine atoms has 7 electrons in its outermost energy level. Moreover, it needs 1 more electron to achieve stable configuration of argon. It will be interesting to know, that one atom of chlorine gets this electron, by sharing with another chlorine atom as shown in Fig. 4.3. The two shared electrons are now counted with both the chlorine atoms for the purpose of determining the stable configuration. Thus we say that a chlorine molecule has two chlorine atoms. Now both the chlorine atoms have 8 electrons in their outermost energy level. A little consideration will show that the chlorine molecule is more stable than its two separate chlorine atoms. Also refer to the Fig. 4.4 showing formation of chlorine molecule.

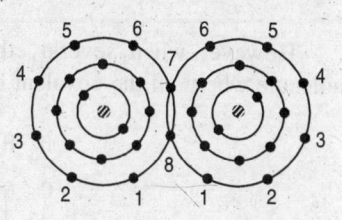

Fig. 4.3. Formation of covalent bond in chlorine molecule.

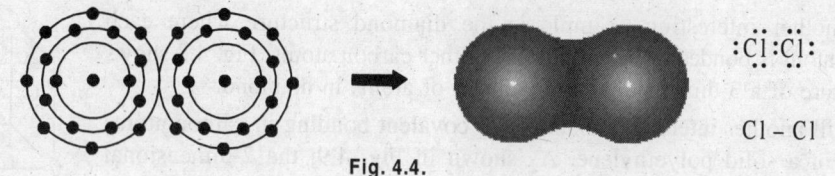

Fig. 4.4.

Sometimes, a covalent bond is also formed when two atoms of different non-metal share one or more pairs of electrons in their outermost energy level. It will be interesting to know that when the atoms share a single pair of electrons, the bond formed is known as a single covalent bond. Similarly, when the atoms share two pairs of electrons, the bond formed is known as double bond. And when the atoms share three pairs of electrons, the bond formed is known as triple bond.

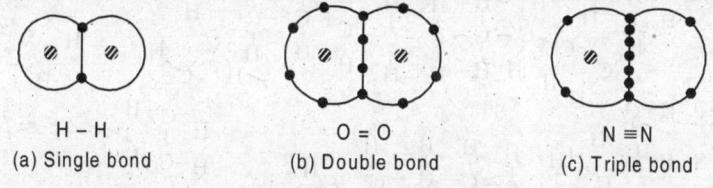

(a) Single bond H – H
(b) Double bond O = O
(c) Triple bond N ≡ N

Fig. 4.5. Covalent bonds.

The formation of single, double and triple bonds in hydrogen, oxygen and nitrogen respectively are shown in Fig. 4.5. It may be noted that the atomic number of hydrogen atom is 1 and its electronic configuration is $1s^1$. Therefore it needs one electron to achieve a stable configuration of helium. Now the hydrogen atom gets one electron by sharing it with one electron of another hydrogen atom. Thus two hydrogen atoms share one electron each to form hydrogen molecule (H_2). Similarly, two oxygen atoms share two electrons each to form an oxygen molecule (O_2). And two nitrogen atoms share three electrons each to form a nitrogen molecule (N_2).

The formation of a covalent bond in the case of carbon atoms is very important, because this bond is very common among most of the organic compounds. The atomic number of carbon is 6 and its electronic configuration is $1s^2\ 2s^2\ 2p^2$. It needs four electrons to achieve stable electronic configuration of neon. It gets these electrons by sharing one electron in each of its four neighbours. For example, in a methane (CH_4) molecule, the carbon atom shares its 4 electrons with 4 hydrogen atoms.

Refer to Fig. 4.6 (a) below. However in the ethylene molecule; the two carbon atoms are bonded together through a double covalent bond whereas the same carbon atoms are also bonded to the hydrogen atoms through a single covalent bond. Refer to Fig. 4.6 (b).

```
        H                           H   H
        |                           |   |
    H - C - H                       C = C
        |                           |   |
        H                           H   H

 (a) Methane (CH₄) molecule    (b) Ethylene (C₂H₆) molecule
```
Fig. 4.6.

However when several ethylene molecules come together. They form a polyethylene molecule where all the covalent bonds are the single covalent bonds as shown in Fig. 4.7.

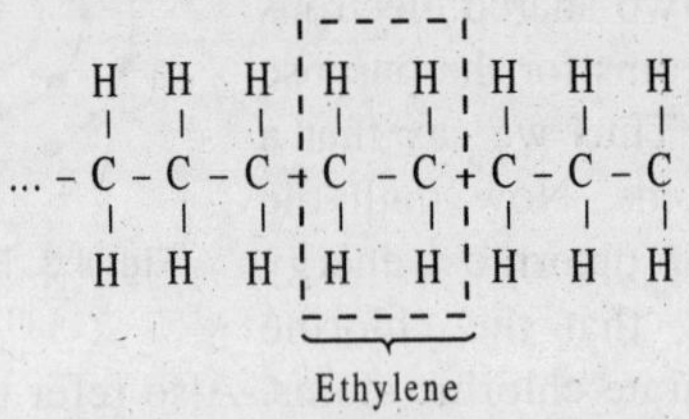

```
          H  H  H │H  H│ H  H  H
          |  |  | |  |  |  |  |  |
    ...-  C- C- C+C - C+C - C- C
          |  |  | |  |  |  |  |  |
          H  H  H │H  H│ H  H  H
                  └────┘
                 Ethylene
```
Fig. 4.7. Polyethylene molecule.

Another interesting example is the diamond structure where each carbon atom is bonded covalently to four other carbon atoms. Fig. 4.8 shows the picture of a 3-dimensional arrangement of atoms in diamond.

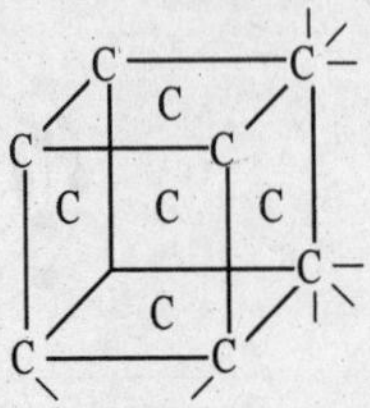

Still another interesting example of covalent bonding in carbon atoms is that of a solid polyethylene. As shown in Fig. 4.9, the 2-dimensional schematic of the solid polyethylene structure looks like "spaghethi" or (Noodles).

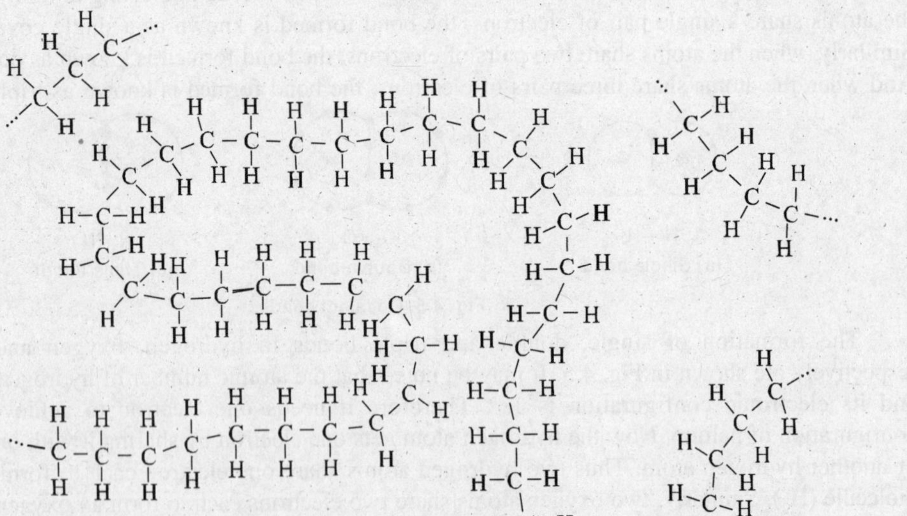

Fig. 4.8.

Fig. 4.9.

Bonds in Solids

Using the concepts of quantum and wave mechanics, it is possible to explain why a covalent bond tends to be the strongest in the directions when ψ^2 is a maximum. The explanation is given below.

The Schrodinger's equation in three dimensions is given by the relation,

$$\left(\frac{\partial^2}{\partial x^2} + \frac{\partial^2}{\partial y^2} + \frac{\partial^2}{\partial z^2}\right)\psi + \frac{8\pi m^2 (E-V)\psi}{h^2} = 0$$

where, ψ = The wave function related to the amplitude of the wave.
m = Mass of an electron
h = Planck's constant
E = Total energy of an electron, and
V = Potential energy of an electron.

Now ψ^2 represents the probability of locating an electron. Its value is used to show the extent of bonding between two orbitals. The value or ψ^2, would be maximum when electron density between two overlapping orbitals is maximum. Thus we can say that the greater value of ψ^2, the greater would be the overlapping of the orbitals, this in turn implies that greater the overlapping of two orbitals, greater would be the directional strength of the covalent bond. Thus depending upon direction, in which the value of ψ^2 is maximum, the covalent bond would also be stronger in that direction.

4.11 PROPERTIES OF COVALENT SOLIDS

The covalent compounds, under normal pressure and temperature, exist in the form of solids, liquid or even gases. It will be interesting to know that the covalent compounds, which have high molecular weights, exist as solids, For example, chlorine (having molecular weight as 71) is gas bromine (having molecular weight as 160) is liquid whereas iodine (having molecular weight as 254) is a solid. It is due to this reason, we use the term as covalent compound and not covalent solids (as in the case of ionic solids). However, depending upon the state of the compound, we use the terms covalent solid, covalent liquid and covalent gas. Though there are a number of properties of covalent compounds, yet the following are important from the subject point of view:

1. *Covalent compounds are bad conductors of electricity.* We know that the covalent compounds are formed by the mutual sharing of electrons. And there is no transfer of electrons from one atom to another. As a result of this, there is no ion formation. Since the compounds exist as molecules and not ions, therefore they are bad conductors of electricity in fused or dissolved state.

2. *Covalent compounds have low melting and boiling temperatures.* We know that in all the covalent compounds, the electrostatic force of attraction is less than the ionic solids. It is thus obvious, that the covalent solids have low melting temperature than the ionic solids. Similarly, the covalent liquids also have low boiling temperatures.

3. *Covalent compounds are not soluble in water, but are soluble in organic solvents such as benzene, toluene etc.* We know that the covalent compounds do not form ions in water and hence cannot dissolve into it. However, some of the covalent compounds like hydrogen chloride, ammonia, sugar, urea etc. are soluble in water.

4.12 METALLIC BONDS

It has been observed that in all the metal atoms, the electrons in their outermost energy levels are loosely held by their nucleii. As a result of this, all the electrons, in their outermost energy level, require very small amount of energy to detach themselves from their nucleii. It will be interesting to know that at room temperature, all the metal atoms lose electrons from their outermost energy levels, which form an electron cloud or common pool of electrons. These atoms, after leaving their outermost electrons, acquire positive charges and become positive ions. Thus a metal may be considered as a cluster of positive ions surrounded by a large number of free electrons, forming electron gas or electron cloud as shown in Fig. 4.10 (*a*) and (*b*).

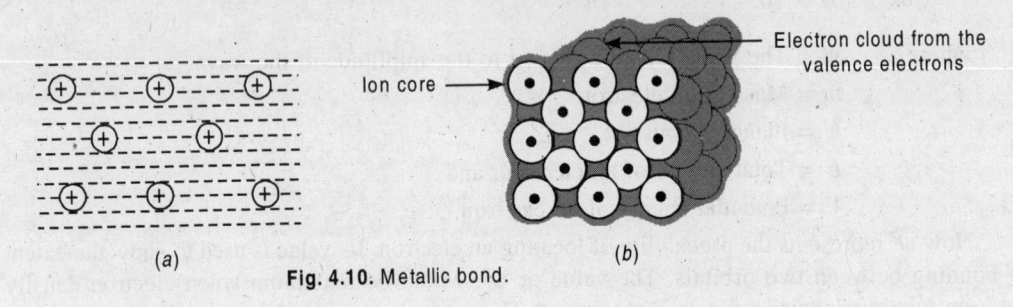

Fig. 4.10. Metallic bond.

The electrostatic force of attraction between the electron cloud and positive ions forms a bond, which is known as metallic bond (as it takes place in metals only).

4.13 PROPERTIES OF METALLIC SOLIDS

We know that free electrons are bonded to different atoms at different times, and that too for an extremely small period. It is due to this reason, that a metallic bond is weaker than the ionic and covalent bonds. Though there are a number of properties of metallic solids, yet the following are important from the subject point of view.

1. *Metallic solids have high electrical and thermal conductivities*. We know that in a metallic bond, the ions are immersed freely in the electronic cloud. As a result of this, there exists a variable electrostatic force of attraction, and the electrons are free to move through large distance when an electric field is applied. Thus the metallic solids have high electrical and thermal conductivity.

2. *Metallic solids have bright lustre*. We know that the electrons can absorb light energy. Thus whenever light falls on a metallic surface, its electrons absorb light energy and get excited. As a result of this, the electron start oscillating and emit light radiations. And the solids appear to have bright lustre.

3. *Metallic solids are malleable and ductile*. We know that the metallic bond is weaker and the planes of atoms can slip over each other but a little force. Moreover, the surrounding of each ion remains the same even after slip, as the electrons are available everywhere within the metal. Thus the metallic solids are malleable and ductile.

4. *Metallic solids have low melting and boiling temperatures*. We know that the electrostatic force of attraction between the positive ions and free electrons is less than ionic bonds. Therefore less energy is required to melt or boil them. Thus the metallic solids have low melting and boiling temperatures.

4.14 COMPARISON BETWEEN IONIC, COVALENT AND METALLIC BONDS

The following table gives the comparison between ionic, covalent and metallic bonds:

S. No.	Ionic bond	Covalent bond	Metallic bond
1.	It exists due to electrostatic force of attraction between positive and negative ions of different elements.	It exists due to the electrostatic force of attraction between atoms which share the electron pairs to form a covalent bond.	It exists due to electrostatic force of attraction between electron cloud and positive ions of same or different metals.
2.	It is formed between two different elements. One of the atom loses its valence electron and the other accepts it. The ions so formed attract each other to form an ionic bond.	It is formed due to the sharing of electron pairs between the atoms of same or different elements.	It is formed when the valence electrons detach themselves from their parent atoms and form a common pool. The force which binds the electron cloud and positive ions of the metal forms the metallic bond.
3.	The ionic solids have very low electrical and thermal conductivities.	The covalent solids have low electrical and thermal conductivities.	The metallic solids have high electrical and thermal conductivities.
4.	The ionic solids have high hardness due to their crystalline structure.	The covalent solids have low hardness except diamond, silicon, carbide etc.	The metallic solids have crystalline structure. But are soft in nature.
5.	The ionic solids are not malleable and ductile.	The covalent solids are not malleable and ductile.	The metallic solids are malleable and ductile.
6.	The ionic solids have high melting and boiling temperatures.	The covalent solids have lower melting and boiling temperatures than the ionic solids.	The metallic solids have slightly lower melting and boiling temperatures than ionic solids.
7.	The ionic compounds exist in the form of solids only.	The covalent compounds exist in the form of solids, liquids or gases.	The metallic compounds exist in the form of solids only.
8.	The ionic solids are soluble in water.	The covalent solids are soluble in benzene, toluene etc.	The metallic solids are neither soluble in water nor in benzene etc.

4.15 TYPES OF SECONDARY BONDS

The secondary bonds are of the following three types:

1. Dispersion bonds.
2. Dipole bonds.
3. Hydrogen bonds.

4.16 DISPERSION BONDS

We know that in a symmetrical molecule, the electrons are uniformly distributed in the space around the nucleus and they are constantly in motion. Therefore, in a symmetrical molecule, the centres of positively charges and negative charges coincide with each other as shown in Fig. 4.11(a). But it has been observed that at certain times the distribution of electrons in the molecule is not symmetrical around its nucleus. This results in the displacement of the centres of positive and negative charges as shown in Fig. 4.11 (b). The electronic imbalance of the charge is known as polarisation. This polarisation is of the fluctuating nature and is known as dispersion effect. Due to this effect, there exists a weak force of attraction between the two molecules of the same element. And a bond is formed between them, which is known as dispersion bond. The molecules of noble gases which consist of single atoms are held together by dispersion bonds, when they are solidified at very low temperatures.

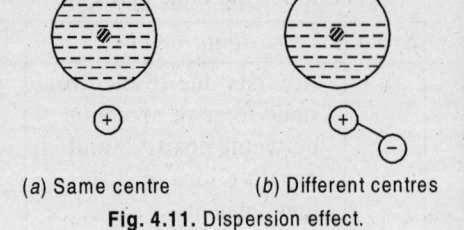

(a) Same centre   (b) Different centres

Fig. 4.11. Dispersion effect.

4.17 DIPOLE BONDS

We have already discussed in Art. 4.9 that a covalent bond is formed when the outermost electrons of an atom are shared by two or more atoms. But sometimes there is an unequal sharing of electrons between two atoms. It has been observed that the unequal sharing of electrons takes place only in those substances in which one of the atoms, in a molecule, has a high affinity to attract electrons than the other. The effect of this unequal sharing of electrons is to create opposite charges on the parent atoms. As a result of this, permanent dipoles are produced. It has also been observed that such dipoles attract each other and a bond is formed which is known as dipole bond.

The formation of dipole bond in the case of hydrogen fluoride molecule is illustrated below :

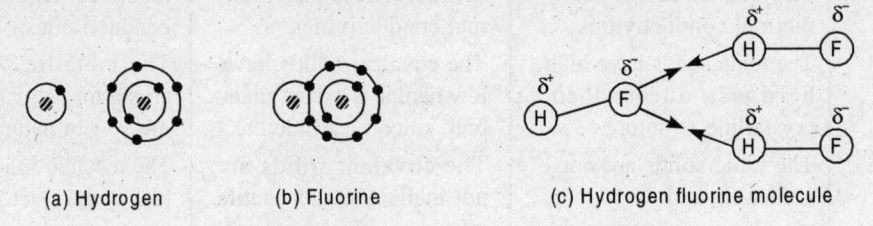

(a) Hydrogen   (b) Fluorine   (c) Hydrogen fluorine molecule

Fig. 4.12. Dipole bond in a hydrogen fluoride molecule.

We know that the atomic number of hydrogen atom is 1, and its electronic configuration is $1s^2$. Similarly, the atomic number of fluorine atom is 9 and its electronic configuration is $1s^2\ 2s^2\ 2p^5$. Thus we see that the hydrogen atom has only one electron in its outermost energy level. Similarly, the fluorine atom has electrons in its outermost energy level ($2s^2\ 2p^6$) as shown in Fig. 4.12 (a). It has been observed that the hydrogen atom requires one more electron to acquire stable configuration. Therefore both the hydrogen and fluorine atoms share a pair of electrons to acquire stable configuration as shown in Fig. 4.12 (b). This leads to the formation of covalent bond, and the molecule so formed is known hydrogen fluoride.

It well be interesting to know that in this covalent bond, the fluorine atom has a high affinity than the hydrogen atom. Thus the shared electron pair shifts towards the fluorine atom. This shifting of electron pair produces a dipole as shown in Fig. 4.12 (c). It may be noted that a similar dipoles are formed in other molecules as well. All such dipoles attract each other to form a dipole bond.

4.18 HYDROGEN BONDS

We have already discussed in the last article the formation of dipole bond. Strictly speaking, a hydrogen bond is a particular type of dipole bond, in which one of the atom is a hydrogen atom. The other atom has a high affinity to attract electron from the hydrogen atom.

Bonds in Solids

The formation of hydrogen bond in the case of water molecule, is illustrated below:

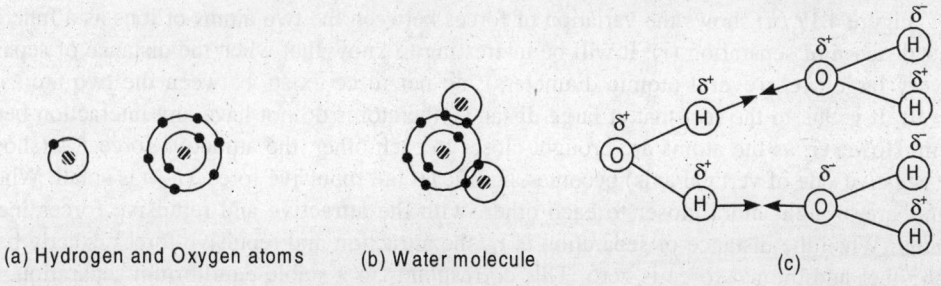

(a) Hydrogen and Oxygen atoms (b) Water molecule (c)

Fig. 4.13. Hydrogen bond.

We know that the atomic number of hydrogen atom is 1, and its electronic configuration is $1s^1$ refer to Fig. 4.13 (a). Similarly, the atomic number of oxygen is 8 and its electronic configuration is $1s^2\,2s^2\,2p^4$ refer to Fig. 4.13 (a). Thus we see that the hydrogen atom has only one electron in its outermost energy level. Similarly, the oxygen atom has 6 electrons in its outermost energy level ($2s^2\,2p^4$). It has been observed that the hydrogen atom requires one more electron to acquire stable configuration. Similarly, the oxygen atom requires two more electrons to acquire stable configuration. Therefore each oxygen atom shares one electron from two hydrogen atoms to acquire stable configuration. This leads to the formation of covalent bond, and the molecule so formed is known as water molecule (refer to Fig. 4.13 (b).

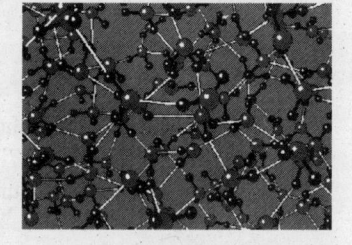

Fig. 4.14.

It will be interesting to know that in this covalent bond, the oxygen atom has higher affinity for the shared electron pair than the hydrogen atom. Thus the shared electron pair shifts towards the oxygen atom. This shifting produces a dipole. Similar dipoles are formed in the other molecules as well. All such dipoles attract each other to form a liquid water as shown in Fig. 4.13 (c). Also refer to Fig. 4.14 which shows a better view of the liquid water. In this picture, the bigger circles represent oxygen atoms and small circles, hydrogen atoms. A pair of dark solid lines indicate covalently bonded atoms while a pair of broken lines represent hydrogen bonded atoms.

Fig. 4.15 represents the crystal structure of ice. Notice the arrangement of oxygen and hydrogen atoms in the ice.

Fig. 4.16 shows the arrangement of water molecules in ice snowflakes. Notice the hexagonal symmetry of the molecules.

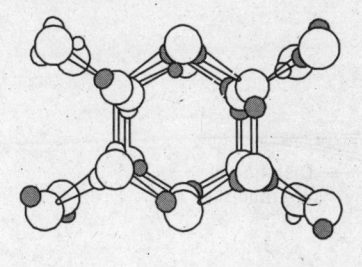

Fig. 4.15.    **Fig. 4.16.**

The hydrogen bonds are directional in character. They exist is various tissues, organs, bloods, skin and bonds in animal life. The hydrogen bonds of the type $N...H....O$, which exist in proteins, play an important role in determining their structures.

4.19 BOND ENERGY

Figure 4.17 (a) shows the variation of forces between the two atoms or ions as a function of their distance of separation (r). It will be interesting to know that when the distance of separation is very large (i.e., several atomic diameters), no net force exists between the two atoms (i.e., F = 0). It is due to the fact that at large distance, the atoms do not have any interaction between them. However, as the atoms are brought closer to each other, the attractive force F_A (shown as the negative side of vertical axis) becomes significant but repulsive force, (F_R) is small. When the atoms are brought more closer to each other, with the attractive and repulsive forces increase rapidly. When the distance of separation is r_0, the attraction and repulsive forces exactly balance each other and the net force is zero. This corresponds to a stable equilibrium separation. If the distance of separation is reduced further between the atoms, the repulsive forces increase sharply. This indicated that the assembly of two atoms becomes unstable.

The sum of attractive and repulsive forces (i.e., net force) provides us a basis for bonding energy. Since the product of force and distance is energy, therefore

$$E = \int_{\infty}^{r_0} (F_A + F_R)\, dr = \int_{\infty}^{r_0} F\, dr$$

The above integration show that the potential energy between the two atoms can be derived from the net force. Fig. 4.17 (b) shows the variation of attractive (F_A) repulsive (F_R) and net potential energy (E). This figure shows that the net potential energy between the two atoms is zero

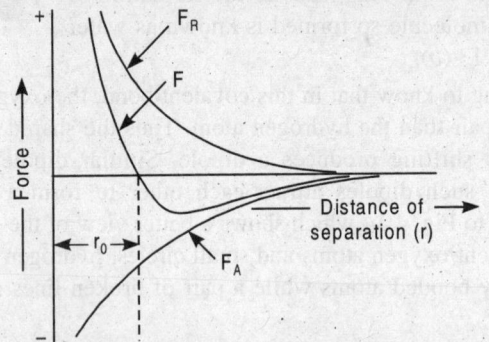

(a) Variation of interatomic forces between two atoms as function of distance of separation (r).

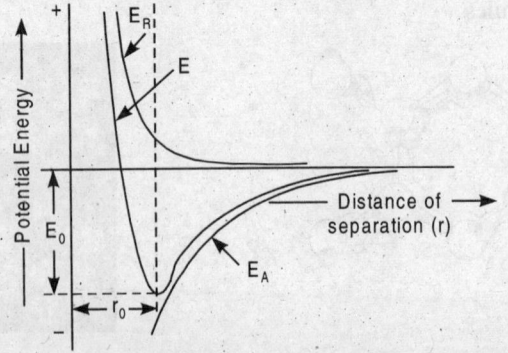

(b) Variation of potential energy between the two atoms as a function of distance of separation (r).

Fig. 4.17.

Bonds in Solids

(*i.e.*, $E = 0$) when both the atoms are in their ground state and infinitely far apart, so that they do not interact with each other. As the atoms are brought closer to each other, the net or potential energy becomes negative as shown in the figure. The potential energy reaches to a maximum value at a r_0 i.e., the point where the net force is zero. The potential energy corresponding to the minimum values (designated by E_0) is called *bond energy*. As the atoms are further brought closer to each other, potential energy increases which indicate that the assembly of two atoms is unstable.

The bond energy may also be defined as the amount of energy required to break one mole (6.023×10^{23}) of bonds. The bond energy is expressed in kilojoules per mole (kJ/mole), electron volt per bond (eV/ bond) or kilocalorie per mole (kcal/ mole). It is useful to mention that 1 eV/ bond $\approx$ 100 kJ/mole and 1 kcal $\approx$ 4.18 kJ. Table 4.2 shows the bond energy for some of the covalent bonds.

Table 4.2. Bond energies for some of the covalent bonds

Bond	Bond Energy (kJ/mole)	Bond	Bond Energy (kJ/mole)
C–C	370	C–Cl	340
C=C	680	O–O	220
C≡C	890	N–H	430
C–H	435	H–H	435

It is evident from the above table that bond energy for C≡C is 890 kJ/mole which is greater than either C=C or C–C.

Note: The bond energy is considered as negative during the formation of a bond because the energy is released to form a bond. On the other hand, the bond energy is considered as positive when the bonds are broken because the energy required to break a bond. For example, if a C=C bond is broken, the bond energy is +680 kJ/ mole and when a C=C is formed, the bond energy is – 680 kJ/ mole.

4.20 BOND LENGTH

The bond length may be defined as the equilibrium separation (r_0) between the centre-to-centre of two bonding atoms. Greater the force of attraction between the two bonding atoms, smaller will be the equilibrium separation and hence smaller the bond length. We know that the primary bonds are stronger than secondary bonds. Therefore the bond lengths of primary bonds are smaller than those of secondary bonds. The bond lengths of primary bonds are in the range of 1–2 Å and those of secondary bonds, 2–5 Å.

It will be interesting to know that bond lengths can be used to determine the diameters of atoms or ions. When the bonding between two neighbouring atoms is of the same kind, the atomic diameter is simply equal to the bond length as shown in Fig. (4.18 (*a*)). For example, the

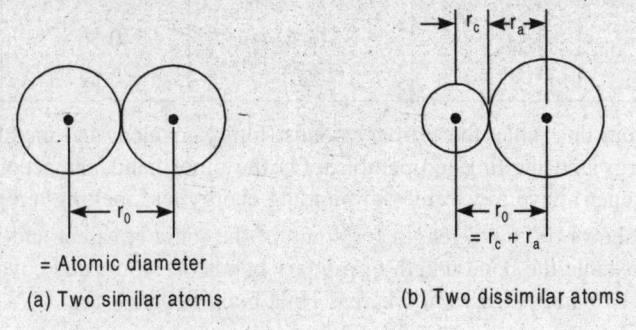

Fig. 4.18. Bond length.

(a) Two similar atoms — r_0 = Atomic diameter

(b) Two dissimilar atoms — $r_0 = r_c + r_a$

equilibrium separation between two nearest bonding nickel atoms is 2.5 Å. Therefore the atomic diameter of nickel atom is equal to 2.5 Å. However, when two bonding atoms are of different types as in ionic bonds, as shown Fig. 4.18 (b), the bond length is equal to the sum of the atomic radii i.e., $r_0 = r_c + r_a$ where r_c is the radii of a cation and r_a is the radii of an anion.

SUMMARY

To summarise the discussion of bonding in solids, refer to the diagram shown in Fig. 4.19. As seen from this diagram, all the metals and alloys have metallic bonding. The semiconductors like silicon, germanium, gallium arsenide etc are covalently bonded. Whereas the polymers (organic solids discussed in later part of the book) are bonded through covalent as well as secondary bonds. Finally the ceramics and glass are bonded through ionic plus covalent bonds.

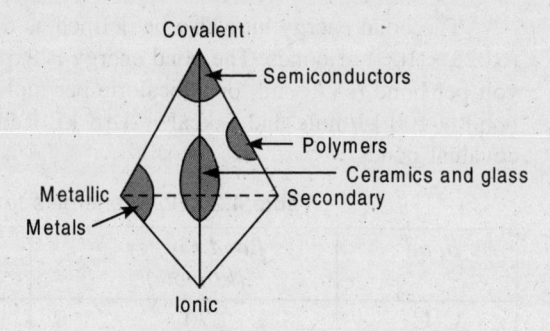

Fig. 4.19.

In order to understand a comparison among the different types of bonds refer to table 4.3. This table shows the bonding energy and melting temperature for various substances.

Table 4.3.

Bonding type	Substance	Bonding energy kJ/mol (kcal/mol)	Energy eV/atom ion, molecule	Melting Temperature (°c)
Ionic	NaCl	640 (153)	3.3	801
	MgO	1000 (239)	5.2	2800
Covalent	Si	450 (108)	4.7	1410
	C(diamond)	713 (170)	7.4	73550
Metallic	Hg	68 (16)	0.7	–39
	Al	324 (77)	3.4	660
	Fe	406 (97)	4.2	1538
	W	849 (203)	8.8	3410
van der Waals	Ar	7.7 (1.8)	0.08	–189
	Cl$_2$	31 (7.4)	0.32	–101
Hydrogen	NH$_3$	35 (8.4)	0.36	–78
	H$_2$O	51 (12.2)	0.52	0

As seen from this table, the primary bonds (ionic, covalent and metallic) have high values of bonding energy and melting temperature. On the other hand, the secondary bonds (van der Waals and hydrogen) have low values of bonding energy and melting temperature.

Table 4.4 shows the bond length for some of the ionic covalent and hydrogen bonds. It is evident from this table the bond length of primary bonds (i.e. ionic and covalent) like C – C bond is 0.154 nm; C–Cl bond is 0.18 nm whereas H–H bond is 0.074 nm.

Table 4.4

Bond	Bond length (nm)	Bond	Bond length (nm)
C–C	0.154	C–Cl	0.18
C=C	—	O–O	0.15
C≡C	—	N–H	0.10
C–H	0.11	H–H	0.074

A.M.I.E. (I) EXAMINATION QUESTIONS

1. Explain electron cloud. What is the role of electron cloud in metallic bond? *(Winter, 1993)*
2. Describe the binding of atoms in metals. *(Summer, 1994)*
3. Distinguish between covalent and ionic bonds in solids. Illustrate with examples. *(Winter, 1994)*
4. (a) Explain the origin of metallic bonding. How does it differ from ionic bonding? Illustrate your answer with suitable examples.
   (b) Using the concepts of quantum and wave mechanics, discuss why a covalent bond tends to be the strongest in the directions when ψ^2 is a maximum.
   (c) Explain why carbon atoms in diamond bond covalently, while lead atoms bond metallically, even though carbon and lead have four valence electrons each. *(Summer, 1995)*
5. Compare and contrast metallic, covalent and ionic bonds. Give examples. *(Winter, 1996)*
6. Explain with the help of suitable sketches the various types of bonding in crystals. *(Summer, 1997)*
7. State the basic differences between metallic bond and ionic bond. *(Winter, 2000)*
8. Cite the difference between atomic mass and atomic weight. Also cite the differences between ionic, covalent and metallic bonding. *(Winter 1998, Summer 1998)*
9. How a single molecule of hydrogen is formed? What are the typical intermolecular bonding mechanisms? *(Summer, 2001)*
10. Describe primary and secondary bonds with examples. *(Summer, 2002)*
11. What do you understand by metallic bonding? Why is it non-directional nature? *(Winter, 2002)*
12. Covalent bonded solids are poor electrical conductors. Explain why? *(Winter, 2002)*
13. Ionically bonded solids show high melting points. Why? *(Winter, 2002)*
14. How is the equilibrium distance of separation between two like atoms is fixed? What is meant by bond energy? *(Winter, 2002)*
15. Why repulsive forces between two like atoms is of shorter range in nature? *(Summer, 2003)*

5

Electron Theory of Metals

1. Introduction. 2. Classification of Electron Theory of Metals. 3. Drude-Lorentz Theory. 4. Applications of Drude-Lorentz Theory. 5. Sommerfeld Theory. 6. Fermi-Dirac Distribution. 7. Brillouin Zone Theory. 8. Brillouin Zone. 9. Relation between Energy and Wave Number. 10. Density of Energy Levels in Sommerfeld's Model. 11. Density of Energy Levels in Brillouin Zone Model. 12. Classification of Solids based on Zone Theory. 13. Conductors. 14. Insulators. 15. Semiconductors. 16. Energy Bands. 17. Classification of Solids based on Band Theory. 18. Conductors. 19. Insulators. 20. Semiconductors. 21. Factors Affecting Resistivity of Metal. 22. Calculation of Resistance and Resistivity of a Wire. 23.Superconductivity. 24. Applications of Superconductors.. 25. Equation of Motion of an Electron 26. Resistivity and Conductivity. 27. Expression for Current Density in a Metal. 28. Mobility. 29. Mean Free Path. 30. Thermoelectricity. 31. Origin of Thermo E.M.F. 32. Magnitude and Direction of Thermo E.M.F. 33. Measurement of Temperature with Thermocouple.

5.1 INTRODUCTION

We have already discussed in Chapter 2 that an atom consists of a nucleus and electrons revolve around it, in various shells and sub-shells. The nucleus of an atom contains protons and neutrons which virtually account for the whole mass of the atom. As a matter of fact, the arrangement of electrons, around the nucleus, determines the manner in which the atoms interact with each other to form a solid. It has been observed that whenever two or more atoms are brought together, the outermost orbits of the different atoms overlap and interact to form a solids.

It has also been observed that in a solid, each overlapping orbit or energy level (of two or more atoms) becomes a band, i.e., group of sub-levels whereas the inner orbits or levels remain unchanged. The study of the valence electrons present in a band, which controls the various properties of metals like electrical as well as thermal conductivity and magnetic properties etc., is known as Electron Theory of Metals.

5.2 CLASSIFICATION OF ELECTRON THEORY OF METALS

The Electron theory of Metals may be classified into the following three types:

1. Drude-Lorentz Theory.
2. Sommerfeld's Theory.
3. Brillouin Zone Theory.

Now we shall discuss all the above mentioned theories one by one in the following pages.

5.3 DRUDE-LORENTZ THEORY

This theory is also known as classical free-electron theory of metals. This theory was initially proposed by Drude in 1900. He suggested that the metals are composed of positive ions whose valence electrons are free to move in the ionic array. The only restriction is that the electrons are confined to remain within the boundaries of a metal crystal. The metal ions are bonded to the electrons by an electrostatic force of attraction. The valence electrons are merely dissociated from their parent atoms. But they are assumed to be free to move within the metal in the same way as the atoms or molecules of a perfect gas. In general, the motion of electrons is random. But in an electric field, the negatively charged electrons move in the positively charged electrons in the positive-field direction and produce current within the metal. In order to prevent the electrons from accelerating indefinitely, it was assumed in this model that the electrons collide elastically with the metal ions. This leads to a steady state current, which is proportional to the applied voltage and explains the origin of Ohm's Law.

Lorentz in 1909, suggested that free electrons in a metal could be treated as a perfect gas. He applied Maxwell-Boltzmann Statistics to the electron gas with the following two assumptions:

1. The mutual repulsion between the negatively charged electrons is neglected.
2. The potential field due to positive ions within the crystal is assumed to be constant everywhere.

Drude-Lorentz Theory explained a number of properties of metal such as electrical conductivity, thermal conductivity, lustre and opacity. However, it has the following two notable failures:

1. Although the Drude-Lorentz theory correctly predicted the magnitudes of electrical resistivity (or conductivity) of most metals, yet the predicted resistivity dependence on temperature is proportional to $\sqrt{T}$ instead of observed linear dependence.
2. It yielded incorrect magnitudes for the specific heat and the paramagnetic susceptibility of metals.

These failures were removed by Sommerfeld in 1928. He applied Fermi-Dirac distribution theory instead of Maxwell-Boltzmann Statistics.

5.4 APPLICATIONS OF DRUDE-LORENTZ THEORY

We have already discussed the Drude-Lorentz Theory in last article. Now we shall discuss how this theory can explain the properties of metals:

1. *High electrical and thermal conductivity*. We know that metal contains free electrons, which can move under the influence of an applied field. This explains high electrical and thermal conductivity. Since the Drude-Lorentz Theory does not depend upon the crystal structure, it indicates that the ratio of electrical to thermal conductivity should be constant for all metals at the same temperature. This relation is called the Wiedemann-Franz Law and has been observed, experimentally, to be true.

2. *High lustre and complete opacity*. When a beam of light falls on a metal crystal, the free electrons start oscillating about their mean positions. The frequency of oscillation is same as that of incident light. Thus all the incident energy is absorbed by the free electrons and the metal appears to be opaque. When an electron oscillates with a greater frequency of light, it jumps into higher energy levels. When such an electron returns to its initial state, it emits a photon of energy which is equal to the absorbed energy. The light emitted by an electrons spreads equally in all directions. But only those rays, which are directed to words the metal surface, can get through. Thus the metal appears to reflect, virtually, all the light which falls on it. This gives the characteristic metallic lustre.

3. *Specific heat and paramagnetism*. These properties can be understood on the basis of interaction between the free electrons and the external energy source, either thermal or magnetic in nature. Drude-Lorentz Theory gives incorrect magnitudes of specific heat and paramagnetic susceptibility of metals. This difficulty arises because Maxwell

Boltzmann Statistics permits all the free electrons to gain energy, which leads to much larger predicted quantities than actually observed.

5.5 SOMMERFELD THEORY

We have already discussed in the last article that Drude-Lorentz Theory could not explain properly, temperature dependence of resistivity, specific heat and paramagnetic susceptibility of metals. These failures were removed by Sommerfeld in 1928. Sommerfeld applied the Schrondinger equation (discussed in Chapter 2) to obtain the expression for electron energies. Following are some of the important assumptions used in Sommerfeld Theory:

1. The potential energy of an electron is uniform or constant within the crystal.
2. The electrons are free to move within the crystal, but are prevented from leaving the crystal by very high energy barriers at its surfaces.
3. The allowed energy levels of an electron bound to a single atom, are quantized.

Consider an electron limited to remain within a one-dimensional crystal of length L as shown in Fig. 5.1. The potential energy everywhere within this crystal is assumed to be constant and equal to zero. The electron is prevented from leaving the crystal by very high potential energy barrier Vs ($\approx \infty$), Now we know that one-dimension Schrodinger equation is:

$$\frac{d^2\psi}{dx^2} + \frac{8\pi^2 m}{h^2}(E-V) = 0$$

Since the potential energy (V) inside the crystal is zero, therefore the Schrodinger equation becomes,

$$\frac{d^2\psi}{dx^2} + \frac{8\pi^2 mE}{h^2} = 0$$

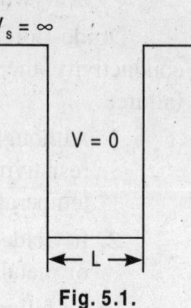

Fig. 5.1.

The solution of the above equation gives the following expression for discrete values of electron energies:

$$E_n = \frac{h^2}{8mL^2} \times n^2 \qquad \ldots(i)$$

where $n = 1, 2, 3\ldots$

The above expression for energies is one-dimension only. In order to obtain this expression for the three dimensional case, we consider the crystal as a cube of edge L, inside which the potential is zero. The corresponding form of the energy expression is:

$$E_{n_x, n_y, n_z} = \frac{h^2}{8mL^2}(n_x^2 + n_y^2 + n_z^2)$$

The integers n_x, n_y and n_z are the first three quantum numbers of an electron. The equation (ii) gives the energies of free electrons in a metal. It may be noted that the equations (i) and (ii) are of the same form, except for the number of integers. Therefore it is possible to obtain a qualitative picture of metallic state by considering a one dimensional case. It may also be noted that for various combination of three integers n_x, n_y and n_z (e.g., 211, 121 and 112) gives the same energy value or level. However, each combination of integers represents a different wave function, having the same energy. Such an energy level is said to be three-fold degenerate.

If we plot a graph with n along the horizontal axis and E along the vertical axis, we shall obtain a curve as shown in Fig. 5.2. It will be interesting to know that although the energy variation is drawn as a continuous curve, it actually consists of discrete points corresponding to

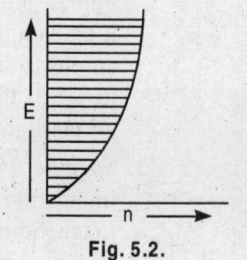

Fig. 5.2.

Electron Theory of Metals

the values of E_n. Since adjacent energy levels differ by less than 10^{-18} eV, therefore it is not possible to show actual breaks in this curve.

Note: If we substitute $\lambda = 2L/n$ and $K = 2\pi/\lambda$ (called wave number) then equation (*i*) may be written as follows:

$$E = \frac{h^2 . k^2}{8\pi^2 . m}$$

5.6 FERMI-DIRAC DISTRIBUTION

The Sommerfeld Theory makes use of Fermi-Dirac distribution of energy levels. According to Fermi-Dirac distribution, the probability of a particular quantum state having an energy E is given by the relation:

$$f(E) = \frac{1}{1 + e^{(E-E_F)/k.T}} \qquad \ldots (i)$$

where
$f(E)$ = Fermi-Dirac function,
E_F = Fermi energy of metal, (eV)
k = Boltzmann's constant (8.62×10^{-5}) eV/K, and
T = Absolute temperature (K)

When $T = 0$ K, the following two conditions arise: (1) If $E > E_F$, the exponential term in equation (*i*) becomes infinite and $f(E) = 0$. This means that there is no probability of finding an occupied quantum state of energy greater than E_F at absolute zero temperature. (2) If $E < E_F$, the exponential term becomes zero and $f(E) = 1$. This means that all quantum states with energies less than E_F are completely occupied at absolute zero temperature.

Fig. 5.3 (*a*) shows a graph of $f(E)$ versus E. It may be noted from this graph, that there are no electrons at 0 K which have energies in excess of E_F, *i.e.*, Fermi energy is the maximum energy at an electron may have at 0 K. The filling up of energy levels are shown in fig. 5.3 (*b*). It is evident that some of the energy states lying above are virtually empty. The importance of applying Fermi-Dirac distribution to Sommerfeld Theory is that it is possible to determine how many electrons can gain energy from the external source. The amount of energy that an electron can gain from either a thermal source of an electric or magnetic field is of the order of kT. At room temperature (*i.e.*, 300 K), $kT = 26$ mV. Only those electrons whose energy is close to the Fermi level can actually gain more energy. It is because of the fact that electrons, occupying lower energy levels, would have to jump to already occupied levels, which is clear violation of Pauli's Exclusion Principle. The shaded area in Fig. 5.3 (*a*) shows the fraction of energy levels that have been filled by the electrons after obtaining an energy equal to kT.

(a) Fermi-Dirac distribution   (b) Filling up of Energy level

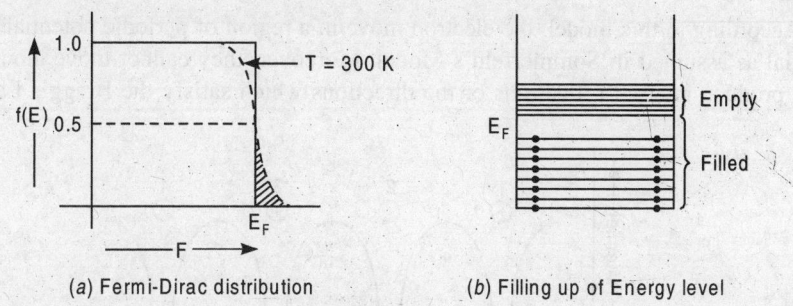

Fig. 5.3.

It is evident from the above discussion that all the free electrons do not gain energy from the external source as permitted by Drude-Lorentz Theory. In fact, only those electrons, whose energy is closer to Fermi level, can actually gain more energy. This is an important difference between the Sommerfeld's and Drude-Lorentz Theories.

Example 5.1. *What is the probability of an electron being thermally permitted to conduction bond in (i) diamond (E_G = 5.6 eV); (ii) Silicon (E_G = 1.07 eV) at room temperature, 25°C. Given k = 86.2 × 10^{-6} eV K^{-1}.* *(A.M.I.E., Summer 2002)*

Solution. Given: for diamond E_G = 5.6 eV; for silicon, E_G = 1.07 eV; k = 86.2 × 10^{-6} eV K^{-1} and room temperature, T = 273 + 25 = 298 K.

(i) For Diamond

We know that bottom of the conduction band in diamond corresponds to,

$$E - E_F = \frac{5.6}{2} = 2.8 \text{ eV}$$

and the probability of an electron being thermally promoted to conduction band,

$$f(E) = \frac{1}{1 + e^{[E - E_F]/k.T}}$$

$$= \frac{1}{1 + e^{2.8/[(86.2 \times 10^{-6})] \times 298]}}$$

$$= 4.58 \times 10^{-48} \textbf{ Ans.}$$

(ii) For Silicon

We know that bottom of the conduction band in silicon corresponds to,

$$E - E_F = \frac{1.07}{2} = 0.5535 \text{ eV}$$

and the probability of an electron being thermally promoted to conduction band,

$$f(E) = \frac{1}{1 + e^{(E - E_F)/k.T}}$$

$$= \frac{1}{1 + e^{0.5535/[(86.2 \times 10^{-5}) \times 298]}}$$

$$= 4.39 \times 10^{-10} \textbf{ Ans.}$$

5.7 BRILLOUIN ZONE THEORY

The Sommerfeld's Free Electron Theory could explain various properties of metals such as electron emission properties, thermo-electricity, etc. But at the same time, it could not distinguish a metal from a non-metal. This drawback was removed by Brillouin Zone Theory of Metals.

According to this model, the electron move in a region of periodic potential and not uniform potential as assumed in Sommerfeld's Model. Moreover, they cannot move along the directions, which produce internal reflections or the directions which satisfy the Bragg's Law.

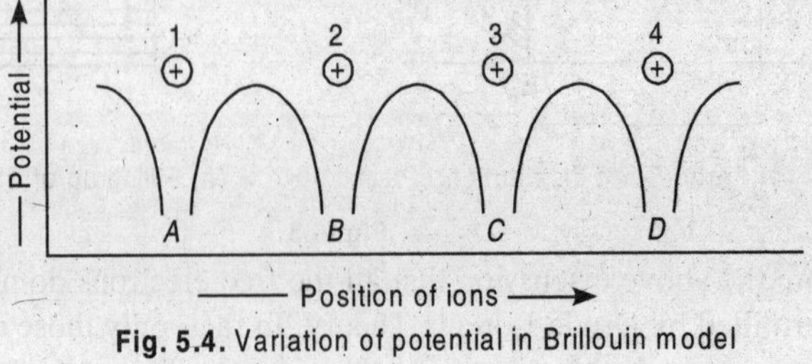

Fig. 5.4. Variation of potential in Brillouin model

Fig. 5.4 shows the variation of potential, when an electron moves in perfectly periodic lattice in which the circles with positive sign represent the positively charged metal ions. It may be noted from the graph that whenever an electron moves closer to a positively charged metal ion, its potential energy falls off sharply as shown by the point A. It may also be noted that when the electron moves away from the metal ion (1), the potential energy sharply increases, and again falls off when it approaches the next metal ion (2). Similarly, for the other metal ions, the potential energy varies in the same manner as shown in the figure.

5.8 BRILLOUIN ZONE

We have already discussed in the last article that in Brillouin Zone Model, the electrons cannot move along the directions which satisfy the Bragg's Law,

$$n\lambda = 2d \sin \theta$$

or

$$\frac{n\pi}{d} = \frac{2\pi}{\lambda} \sin \theta = K \sin \theta$$

where K is equal to $2\pi/\lambda$. It is a vector quantity and is known as a wave number. Its magnitude is directly proportional to the momentum of the electron, while its direction is same in which the electron moves. The term ($K \sin \theta$) is called the normal component of the wave number.

It may be noted from the above equation that the values of normal component of the wave number ($K \sin \theta$) is inversely proportional to the lattice spacing (d). Therefore the crystal planes, which are widely spaced, have greater affect on the motion of electrons.

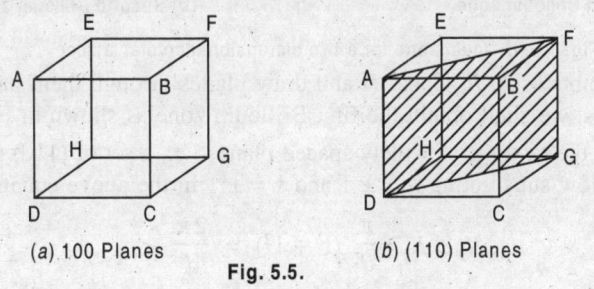

(a) 100 Planes (b) (110) Planes

Fig. 5.5.

In a simple cubic crystal lattice, the most widely spaced planes are a set of 100 planes (*i.e.*, all the six planes of the cube (*ABCDEFGH*) as shown in Fig. 5.5 (*a*). Similarly, the next widely spaced planes are the set of (110) planes. The volume enclosed by the set of these planes is known as Brillouin Zones as shown in Fig. 5.5 (*b*).

The zones obtained from a set of (100) planes are generally termed as first Brillouin zones, whereas the zones obtained from a set of (110) planes are termed as second Brillouin zones and so on.

A Brillouin zone can be obtained from plotting the values of K for different sets of planes. In order to plot the various values of K, the following equation is used:

$$h.K_x + k.K_y + l.K_z = \frac{\pi}{d}(h^2 + k^2 + l^2)$$

where K_x, K_y, K_z = Components of the K-vector along x, y and z axes respectively,

h, k, l = Miller indices of the reflecting planes, and

d = Interplanar spacing.

The above equation may be simplified by ignoring the z component of both the reflecting planes and the vector K. Thus for a two-dimensional Brillouin zone, the above equation reduces to:

$$h.K_x + k.K_y = \frac{\pi}{d}(h^2 + k^2)$$

Now for the first set of widely spaced planes *i.e.*, set of (*100*) planes either $h = \pm 1$, $k = 0$ or $h = 0$, $k = \pm 1$. Now substituting $h = \pm 1$ and $k = 0$ in the above equation, we get

$$K_x = \pm \frac{\pi}{d}$$

Similarly, substituting $K = \pm 1$ and $h = 0$ in the above equation, we get

$$K_y = \pm \frac{\pi}{d}$$

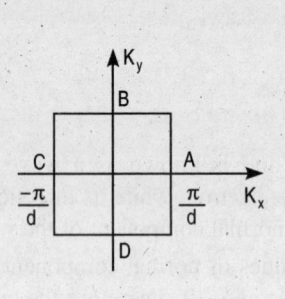

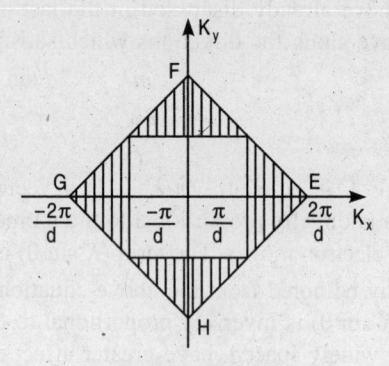

(a) First Brillouin zone (b) Second Brillouin zone

Fig. 5.6. Brillouin zone for a two dimensional crystal lattice.

If we plot these points as *A, B, C* and *D* and draw planes through them parallel to X – X and Y – Y axes, we shall obtain the first Brillouin zone as shown in Fig. 5.6. (*a*).

Similarly, for the next set of widely spaced planes i.e., a set of (110) planes both '*h*' and '*k*' are equal to ± 1. Now substituting $h = \pm 1$ and $k = \pm 1$ in the above equation, we get

$$K_x \pm K_y = \frac{\pi}{d}(1^2 + 1^2) = \frac{2\pi}{d}$$

or

$$K_x + K_y = \pm \frac{2\pi}{d}$$

If we plot these point, as *E, F, G* and *H* and draw planes through them at 45° with X – X axis and Y – Y axis, we shall obtain the second Brillouin zone (over the first Brillouin zone) as shown in Fig. 5.6(b). It will be interesting to know that the different zones for a given crystal have the same area. Thus the area enclosed by the first Brillouin zone is equal to the area enclosed by the second Brillouin zone.

5.9 RELATION BETWEEN ENERGY AND WAVE NUMBER

According to Sommerfeld Model, the expression for the energy of an electron moving in a region of uniform potential is given by:

$$E = \frac{h^2.K^2}{8\pi^2 m} \qquad \ldots (i)$$

where h = Planck's constant,

K = Wave number of the electron (equal to $\frac{2\pi}{\lambda}$), and

m = Mass of the electron.

Consider the movement of electrons parallel to the *x-x* axis. The energy of an electron will be given by the expression:

Electron Theory of Metals

$$E = h^2 \cdot \frac{K_x^2}{8\pi^2 m} \qquad \ldots (ii)$$

We know that as h, π and m are constants, therefore

$$E \propto K_x^2$$

Now if we plot a graph with energy of the electron along vertical axis and wave number K_x along the horizontal axis, we shall obtain a curve as shown in Fig. 5.7 (a). This curve is known as energy-wave number curve or simply $E - K$ curve.

Now consider the movement of the electron in a periodic potential instead of uniform potential. It will be interesting to know that the equation (i) still holds good for relationship between the energy and wave number, over major part of the spectrum. But we have discussed in Art. 5.5 that the normal motion of an electron is disrupted due to the reflection from the crystal planes. This introduces discontinuities in the relationship between energy and wave number. Now for a set of (100) planes (i.e., x-x) direction we know that

$$K_x \sin\theta = \frac{n\pi}{d} \text{ or } K_x = \frac{n\pi}{d.\sin\theta}$$

Now if θ equal $90°$, then $\sin\theta$ is equal to 1. Therefore

$$K_x = \frac{n\pi}{d}$$

A little consideration will show, that as the wave number (K) approaches the values, which are multiples of π/d, discontinuities corresponding to these values of K_x will be noticed. As a result of this, the E-K curve will deviate from its normal parabolic path as shown in Fig. 5.7(b).

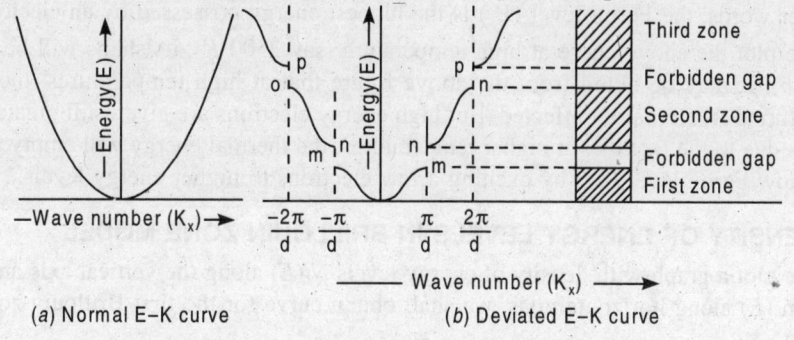

(a) Normal E–K curve   (b) Deviated E–K curve

Fig. 5.7.

The first discontinuity, indicated by m-n and m'-n' on the E-K curve, will occur when $K_x = \pm \pi/d$. This discontinuity is due to the reflection of the electron wave from the set of (110) planes. The second discontinuity, indicated by o-p and o'-p' on the curve, will occur when $K_x = \pm 2\pi/d$. This discontinuity is due to the reflection of the electron wave from the set of (110) planes. Similarly, the third discontinuity on the curve is due to the reflection from the set of (111) planes.

It will be interesting to know that the energy corresponding to the values of $K_x = n\pi/d$ has two values. The energy between these two values is shown as a forbidden gap, because the electrons cannot have energy values lying within these energy gaps. The actual position of this energy gap depends upon the direction of motion of electrons, entering the crystal lattice. Thus it depends upon the point where it crosses the zone boundary.

5.10 DENSITY OF ENERGY LEVELS IN SOMMERFELD'S MODELS

The density of energy levels, occupied by the free electrons, is given by the relation:

$$N(E) = \frac{C\sqrt{E}}{1 + e^{[E-E_F]/k.T}}$$

where C = A constant,
 E = Energy of the electron,
 E_F = Fermi level,
 k = Boltzmann's constant, and
 T = Absolute temperature (K).

If we plot a graph with density of energy levels $N(E)$ along the vertical axis and the energy of the electron (E) along horizontal axis at absolute zero temperature, we shall obtain a curve as shown in Fig. 5.8 (a).

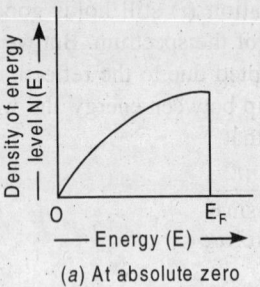

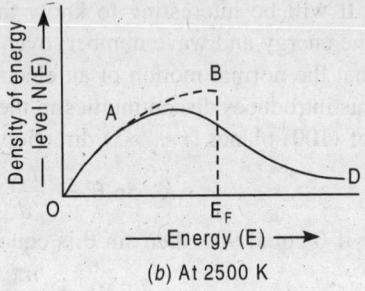

(a) At absolute zero (b) At 2500 K
Fig. 5.8.

It may be noted from the above graph that when $E < E_F$, the value of $N(E)$ increases in a parabolic manner. But when $E > E_F$, the value of $N(E)$ reduces to zero, which indicates that all the energy levels above the Fermi level (E_F) are completely empty at absolute zero temperature. Or in other words, the Fermi level (E_F) is the highest energy possessed by an electron.

If we plot the above curve at high temperature (say 2500 K), its shape will be as shown in Fig. 5.8 (b). It may be noted from the above figure that at high temperatures, the low energy electrons (up to A) are hardly affected. But high energy electrons are given still greater energy (up to D). It is due to the fact that at higher temperatures, the thermal energy will empty a few energy levels, below Fermi level (E_F) by exciting a few electrons to higher energy levels.

5.11 DENSITY OF ENERGY LEVELS IN BRILLOUIN ZONE MODEL

If we plot a graph with density of energy levels $N(E)$ along the vertical axis and the energy of electron (E) along horizontal axis, we shall obtain curve for the first Brillouin zone as shown in Fig. 5.9.

It may be noted from the graph that the relationship between the density of energy levels and energy up to A is the same as for the Sommerfeld's Model. As the energy increases beyond this point, the electrons travelling normal to the set (100) planes, suffer reflections from those planes. As a result of this, the density of energy level increases and the curve rises sharply from the point A and B as shown in the figure. As the energy is further increased beyond the point B, the density of energy level decreases and finally reduces to zero. It is due to

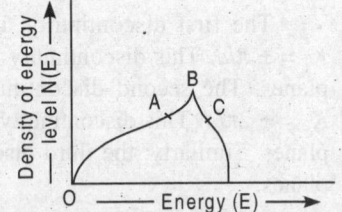

Fig. 5.9. Density of energy level with one electron.

the fact that number of energy levels have a finite value in a given zone. It has been observed that in the body centred and face centred cubic metals, the number of energy levels in a zone is exactly equal to the number of atoms in the metal.

5.12 CLASSIFICATION OF SOLIDS BASED ON ZONE THEORY

The solids may be classified on the basis of zone theory into the following three categories:

1. Conductors.   2. Insulators.   3. Semiconductors.

Electron Theory of Metals

5.13 CONDUCTORS

First of all, let us consider the case of monovalent metals such as sodium, potassium, etc. We know that in such metals there is only one electrons in their valency shell of each atom, which can move freely. It has been found that the number of energy levels in each zone is equal to the number of atoms in a crystal. If there are 'N' atoms in a crystal, then $N/2$ energy levels will be filled up in the first zone. This will happen as all the 'N' electrons will move into $N/2$ energy levels since each energy level cannot accommodate more than two electrons. This may be represented on the density of energy level curve as shown in Fig. 5.10.

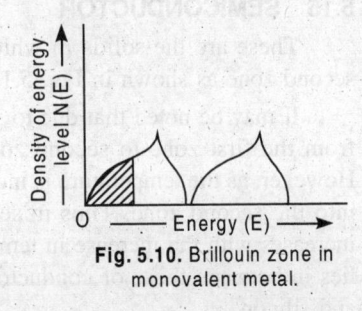

Fig. 5.10. Brillouin zone in monovalent metal.

A little consideration will show, that the application of an electric field to these metals can move electrons to higher unoccupied energy levels in the first zone. This results in the conduction process.

Now consider the case of divalent metals such as magnesium, beryllium etc. We know that in such metals, there are two electrons in their valence shell of each atom, which can move freely. Thus if there are 'N' atoms in a crystal then all the 'N' energy levels will be filled up in the first Brillouin zone. This may be represented on the density of energy level curve. But due to overlapping of the first and second zone (which is empty) the higher energy electrons of the first zone can be made to move to the lower energy levels of the second zone by the application of an electric field, resulting in the conduction process.

It may be noted that the conductors are the solids in which the first zone is completely filled with electrons and the second zone overlaps the first zone as shown in Fig. 5.11(a).

It may also be noted that due to overlapping of the first zone with the second zone (which is empty), the electrons require a slight amount of energy to jump into the second zone, and thus the solids are known as conductors. The examples of such solids are silver, copper, aluminium etc.

5.14 INSULATORS

These are the solids in which there exists a large energy gap between the first zone and the second zone as shown in Fig. 5.11 (b).

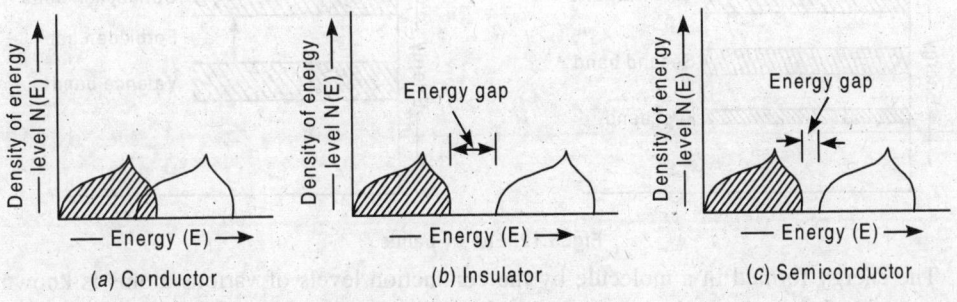

(a) Conductor   (b) Insulator   (c) Semiconductor

Fig. 5.11. Brillouin zone in solids.

It may be noted that due to large energy gap, between the two zones, the electrons cannot jump from the first zone to the second zone (which is empty) at ordinary temperature. However, they require a large amount of energy to jump. Therefore the electrical conduction in insulators cannot take place. It has been observed that if the temperature of an insulator is increased, some electrons do jump into the second zone and therefore a small electrical conduction may take place. The examples of which solids are rubber, bakelite, mica, etc.

5.15 SEMICONDUCTOR

These are the solids in which there exists a small energy gap between the first zone and second zone as shown in Fig. 5.11 (c).

It may be noted that due to small energy gap, between the two zones, no electron can jump from the first zone to second zone (which is empty) at absolute zero temperature (i.e., 0 K). However, as the temperature is increased (say up to room temperature) some of the electrons jump into the second zone. Thus in semiconductors, the electrical conductivity (or flow of current) increases with the increase in temperature. Moreover, the electrical conductivity in such material lies in between those of conductors and insulators. The examples of such solids are germanium and silicon.

5.16 ENERGY BANDS

We have already discussed in chapter 'Structure of Atom' that a matter is composed of small particles known as atoms. When the atoms combine to form a molecule, the electron shell is influenced not only by its now nucleus but also by the nuclei and electrons of surrounding atoms. It has been observed that nearest atoms have the greatest influence over the energy level (electron shell) of an atom. Thus the energy level of each electron in a given atom is controlled, to some extent, by all the neighbouring atoms in the molecule.

It will be interesting to know that the energy levels of all electrons, in a molecule, are different. But the electrons readjust themselves in such a way that their energy levels come closer to each other. Or in other word, all electrons, in the molecule, have their first energy levels in the vicinity of each other, which form a cluster or band known as energy band. Similarly, all the electrons in the second shell form the second energy band and so on. It may be noted that the energy band, corresponding to the outermost shell, is called the valence band as shown in Fig. 5.12 (a).

It has been observed that when the valence electrons are removed from their respective atoms, they are called free or condition electrons. The total energy of a free electron is slightly higher than that in the valence orbit. It is due to this fact that a valence electron absorbs energy at the time of leaving its atom. Therefore the energy of a free electron is very close to the valence level.

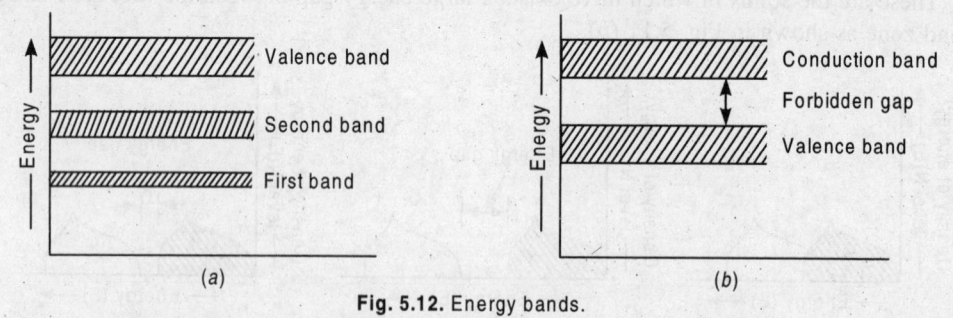

Fig. 5.12. Energy bands.

The energy formed in a molecule by the conduction levels of various atoms is known as conduction band. It is represented in the energy band diagram above the valence band. The energy gap between the conduction band and the valence band is known as forbidden gap as shown in Fig. 5.12 (b). It may be noted that there is no energy shell in this forbidden gap, which the electron can occupy.

5.17 CLASSIFICATION OF SOLIDS BASED ON BAND THEORY

The solids may also be classified on the basis of band theory into the following three categories:

1. Conductors.   2. Insulators.   3. Semiconductors.

Electron Theory of Metals

5.18 CONDUCTORS

These are the solids in which the valence band overlaps the conduction band as shown in Fig. 5.13. (a).

It may be noted that due to overlapping of the two bands, the electrons require a slight amount of energy to jump from the valence band to the conduction band, and thus the solids are known as conductors. The examples of such solids are silver, copper, aluminium etc.

5.19 INSULATORS

These are the solids in which there exists a large energy gap (*i.e.*, forbidden gap) in between the valence band and conduction band as shown in Fig. 5.13 (b).

It may be noted that due to the large energy gap between the two bands, the electrons cannot jump from the valence band to the conduction band (which is empty) at ordinary temperatures. However, they require a large amount of energy to jump. Therefore the electrical conduction in insulators cannot take place. It has been observed that if the temperature of an insulator is increased, some electrons do jump to the conduction band. Therefore a small electrical conduction may take place. The examples of such solids are rubber, bakelite, mica etc.

5.20 SEMICONDUCTORS

These are solids in which there exists a small energy gap (*i.e.*, forbidden gap) in between the valence band and conduction band as shown in Fig. 5.13 (c).

(a) Conductors (b) Insulators (c) Semi-conductors

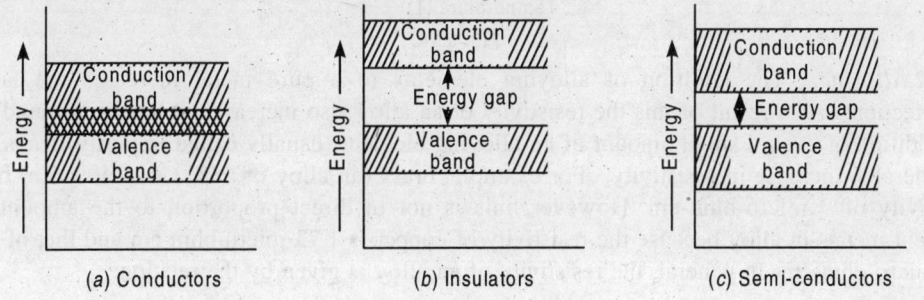

Fig. 5.13.

It may be noted that due to the small energy gap between the two bands, no electron can jump from the valence band to the conduction band (which is empty) at absolute zero temperature (*i.e.*, 0 K). However, as the temperature is increased say up to room temperature), some of the electrons jump into the conduction band. Thus in a semiconductor the conductivity (or flow of current) increases with the increase in temperature. Moreover, the electrical conductivity in such materials lies in between those of conductors and insulators. The examples of such solids are germanium and silicon.

5.21 FACTORS AFFECTING RESISTIVITY OF METALS

We have already discussed that the electric current flows due to the movement of electrons. But sometimes, the electrons experience opposition to their movement through the conductors. This happens when the electrons come across valencies, impurity atoms or other lattice defects. This property of the conductor, due to which it opposes the flow of current, is known as resistivity of a metal. Following are the main factors which affect resistivity of metals:

1. *Temperature*. The resistivity of a metal increases linearly with the temperature according to the relation:

where $\rho_t = \rho_{20}[1 + \alpha(t - 20)]$

ρ_t = Resistivity of the metal at $t°C$,

ρ_{20} = Resistivity of the metal at 20°C (*i.e.*, standard room temperature), and α = Coefficient of temperature resistance.

It may be noted that the metals have positive value of coefficient of temperature resistance. At high temperature, the thermal vibrations of the atoms increase. This reduces the motion of electrons through the conductor. Or in other words, the resistivity of a conductor increases with an increase in temperature. But the resistivity of insulators, semiconductors and electrolytes decreases with rise in temperature.

Fig. 5.14 shows the variation of resistivity with temperature for a pure metal. As seen from this diagram, the resistivity at low temperature exists due to defects in the micro structure of a pure metal. In general, the resistivity of a metal increases with the increase in temperature. The increase in resistivity happens mainly due to increase in thermal vibrations.

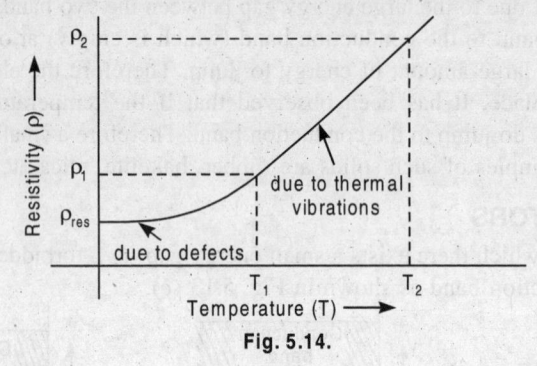

Fig. 5.14.

2. *Alloying.* The addition of alloying elements to a pure metal increases the lattice imperfections. As a result of this the resistivity of an alloy also increases. It has been found that the addition of even a small amount of an alloying element (usually called impurity) leads to a considerable increase in resistivity. For example, brass (an alloy of 63% Cu —40% Zn) has a resistivity of 9 micro-ohm-cm. However, this is not in direct proportion to the amounts of different metals in alloy because the resistivity of copper is 1.73 micro-ohm-cm and that of zinc is 6 micro-ohm-cm. In general, the resistivity of an alloy is given by the relation:

$$\rho_{alloy} = \rho_{metal} + x\, \rho_I$$

where

ρ_{metal} = Resistivity of the parent metal,

x = Amount of impurity added in the metal. This is expressed as atomic per cent.

ρ_I = resistivity of the impurity metal.

Thus if nickel is added to copper, the resistivity of copper goes up by 1.3 micro-ohm-cm for each atomic % addition of nickel. This may be expressed by following relation:

$$\rho_{Ni\text{-}Cu} = \rho_{Cu} + x \times 1.3 \text{ (micro-ohm-cm)}$$

3. *Mechanical process.* Any mechanical process, which increases the number of dislocations, results in the increase of electrical resistivity of a metal. For example, strain hardening result in higher resistivity than *annealed samples of the same metal.

4. *Age hardening.* The electrical resistivity of a metal or alloy increases due to age hardening. During this process, the crystal lattice undergoes some distortion due to which the movement of electrons is reduced.

As a result of this, the resistivity of the metal is increased.

---

* For details, please refer to the chapter on Heat Treatment.

Electron Theory of Metals

5.22 CALCULATION OF RESISTANCE AND RESISTIVITY OF A WIRE

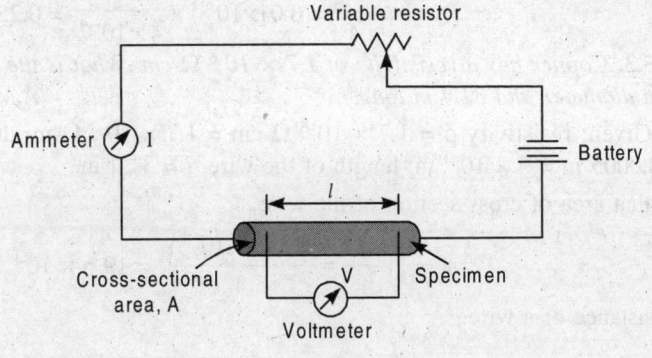

Fig. 5.15.

Fig. 4.15 shows a circuit indicating a wire connected to a battery through an ammeter and a variable resistor. A voltmeter (V) is connected across the two ends of the specimen wire to measure the voltage drop. The variable resistor is to limit the current in the circuit. This circuit can be used to measure the resistance of a wire.

Let V = Voltage measured across the two ends of the specimen wire (volts)

I = Current measured through the specimen (amperes), Then resistance of the wire.

$$R = \frac{V}{I} \ \Omega \quad \ldots (i)$$

The resistance is expressed in ohms (Ω). The resistance of a wire can also be calculated from the dimensions of the wire provided the resistivity of a material (of which the wire is made of) is known. This is shown below

Let
ρ = resistivity of a wire
l = length of a wire (m)
d = diameter of the wire (m)

Then area of cross-section of a wire,

$$A = \frac{\pi d^2}{4} \ m^2$$

and resistance of a wire,

$$R = \rho \frac{l}{A} \ (\Omega) \quad \ldots (ii)$$

The equation (ii) can also be used to find resistivity of a wire as follows. Let us suppose we know the dimensions of the wire, and we have measured the voltage drop and current through the wire, so we can calculate the resistance of a wire using equation (i) then resistivity can be obtained using equation (ii). Thus,

$$\rho = R \cdot \frac{A}{l}$$

Notice that resistivity is expressed in (Ω-m). Thus if resistance is in ohms, A in metre^2, l in metres, then resistivity is in Ω-m (or ohms-metre).

Example 5.2. *Calculate the resistance of a wire whose length $l = 100$ cm, area of cross-section, $A = 4$ cm^2, resistivity, $\rho = 0.01$ Ω-cm.* (A.M.I.E., Winter, 2003)

Solution. Given: $l = 100$ cm = 1 m; $A = 4$ cm^2 = 4×10^{-4} m^2; and $\rho = 0.01$ Ω-cm = 0.01×10^{-2} Ω-m.

We know that resistance of a wire,

$$R = \rho \cdot \frac{l}{A} = (0.01 \times 10^{-2}) \times \frac{1}{4 \times 10^{-4}} = 0.25 \; \Omega \; \textbf{Ans.}$$

Example 5.3. *Copper has a resistivity of 1.7×10^{-6} Ω-cm. What is the resistance of a wire which is 0.5 mm diameter and 31.4 m long.* (A.M.I.E., Winter, 2001)

Solution. Given: resistivity $\rho = 1.7 \times 10^{-6}$ Ω-cm $= 1.7 \times 10^{-8}$ Ω-m; diameter of the wire, $d = 0.5$ mm $= 0.0005$ m $= 5 \times 10^{-4}$ m; length of the wire, $l = 31.4$ m

We know that area of cross-section of the wire,

$$A = \frac{\pi d^2}{4} = \frac{\pi \times (5 \times 10^{-4})}{4} = 19.6 \times 10^{-8} \; m^2$$

and the resistance of a wire.

$$R = \rho \frac{l}{A} = (1.7 \times 10^{-8}) \times \frac{31.4}{19.6 \times 10^{-8}} = 2.72 \; \Omega \; \textbf{Ans.}$$

Example 5.4. *A wire sample (1 mm in diameter by 1 m in length) of an aluminium alloy (containing 1.2% Mn) is placed in an electrical circuit. A voltage drop of 0.432 V is measured across the length of the wire as it carries 10A current. Calculate the conductivity of this wire.*

Solution. Given: Voltage drop across the wire, $V = 0.432$ V; current through the wire $I = 10$A

We know that the resistance of a wire,

$$R = \frac{V}{I} = \frac{0.432}{10} = 0.0432 \; \Omega \; \text{or} \; 43.2 \; m\Omega.$$

and the resistivity,

$$\rho = \frac{RA}{l} = \frac{0.0432 \times (\pi d^2/4)}{l}$$

$$= \frac{0.0432}{1} \times \frac{\pi \times (1 \times 10^{-3})^2}{4} = 33.9 \times 10^{-9} \; \Omega.m$$

$\therefore$ The conductivity of the wire,

$$\sigma = \frac{1}{\rho} = \frac{1}{33.9 \times 10^{-9}} = 29.5 \times 10^{6} \; \Omega^{-1}.m^{-1} \; \textbf{Ans.}$$

5.23 SUPERCONDUCTIVITY

The term 'superconductivity' may be defined as a state of material in which it has zero resistivity. We have already discussed in the last article that the resistivity of all materials increases with the temperature. Or in other words, if the material is cooled below the room-temperature, the resistivity decreases with the decrease in temperature. Prof. Heike discovered in 1911 that mercury is a good conductor of electricity at room temperature. Its resistivity decreases slightly with the decrease in temperature up to 4.27 K. Beyond this temperature, its resistivity decreases sharply and finally becomes zero at 4.22 K. This transition occurs over a very narrow range of temperature of the order of 0.05 K as shown in Fig. 5.16.

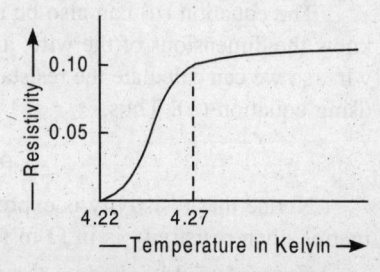

Fig. 5.16. Superconductivity.

As a matter of fact, the resistivity of mercury is a function of temperature. The temperature, at which there is a sudden transition temperature.

Electron Theory of Metals 121

It has been observed that only a few metals such as tin, lead, tantalum, bismuth, antimony, tellurium etc. show superconductivity. It will be interesting to know, that all the above mentioned metals are poor conductors of electricity at room temperature. On the other hand, the metals such as copper, silver and gold which are good conductors of electricity at room temperature do not show superconductivity. The transition temperature of superconducting metal may range from 0.01 K to 9.15 K. It has also been observed that the superconductivity of the materials may be destroyed by the application of sufficiently strong magnetic field. The critical value of magnetic field, for the destruction of superconductivity, is a function of temperature. It varies according to the following relation:

$$H_c = H_0 \left[1 - \left(\frac{T}{T_c} \right)^2 \right]$$

Where H_c = Critical magnetic field at any temperature,
H_0 = Critical magnetic field at absolute zero temperature,
T_c = Transition temperature of the material, and
T = Any temperature below T_c

5.24 APPLICATIONS OF SUPERCONDUCTORS

Though there are many application of superconductors, yet the following are important from the subject point of view:

1. In research and development field.
2. In electrical machines, transformers and cables.
3. In superconducting solenoids for low temperatures.
4. In electrical switching elements.

5.25 EQUATION OF MOTION OF AN ELECTRON

According to Drude-Lorentz Theory (as discussed in Art 5.3) the motion of electrons is random when no electric field is applied. It means that the number one electrons in a metal moving from left to right at any time is the same as that from right to left. As a result of this, no net current flows through the metal. However, if an electric field is applied across a metal, the electrons move in the positive direction of the field and produce current.

Consider an electron within a metal moving in any direction and at any time under the applied electric field 'E'.

Let m = Mass of an electron,
v = Velocity of the electron, and
e = Charge of an electron.

Then force experienced by an electron due to the applied electric field

$$= e.E$$

Because of this force, the electron moves with an average acceleration of d^2x/dt^2 (or dv/dt). We know that the force with which the electrons moves

Mass × Acceleration $= m \dfrac{dv}{dt}$

According to the laws of kinetic theory,

$$m \frac{dv}{dt} = e.E \text{ or } dv = \frac{eE}{m}.dt \qquad \ldots(i)$$

Integrating the above equation,

$$\int dv = \frac{eE}{m}\int dt$$

$$v = \frac{e.E}{m} \times t + K \qquad \ldots (ii)$$

Where K= Constant of integration, which represents the random velocity of the electrons. The average value of random velocity must be zero. Otherwise there will be a flow of current even in the absence of electric field.

Substituting K = 0 in equation (ii)

$$v = \frac{e.E}{m} \times t \qquad \ldots (iii)$$

It may be noted from equation (iii) that the velocity of electron is directly proportional to the time 't'. This implies that the velocity of an electron continues to increase with the time till the collision does not occur. If 't' is the collision time i.e., average time between the two successive collisions, then the average velocity of the electron,

$$v = \frac{e.E.t}{m} \quad \text{or} \quad \frac{V}{E} = \frac{e.t}{m}$$

The above equation is called an equation of motion of an electron under the applied electric field. The average velocity is also referred to as drift velocity, because the drift in electrons is due to applied field (E).

5.26 RESISTIVITY AND CONDUCTIVITY

We have already defined in Art 5.21 that the property (or ability) of a metal due to which it opposes the flow of current through it is called resistivity. It is designated by the Greek letter ρ and is expressed in ohm-m (Ω-m). The reciprocal of resistivity is called conductivity. It is designated by the letter σ and is expressed in per ohm-per metre written as ohm^{-1}-m^{-1} or mhos per metre (mhos/m) such that

$$\sigma = \frac{1}{\rho}$$

Table 5.1

Metal	Electrical Conductivity (Ω-m)$^{-1}$
Silver	6.8×10^7
Copper	6.0×10^7
Gold	4.3×10^7
Aluminium	3.8×10^7
Iron	1.0×10^7
Brass (70% Cu-30% Zn)	1.6×10^7
Platinum	0.94×10^7
Plain Carbon Steel	0.6×10^7
Stainless Steel	0.2×10^7

Table 5.1 shows the electrical conductivity of few metals. Notice that silver has the highest value of conductivity, followed by copper, gold and aluminium. The conductivity of

Electron Theory of Metals

semiconductors is quite low as compared to metals. For example, for silicon, the conductivity is 5 × 10⁻⁴ per ohm per metre (Ω^{-1} m⁻¹) whereas for germanium, the value is 2 × 10⁻⁴ Ω^{-1} m⁻¹. On the other hand for carbon (diamond), the conductivity is extremely low (< 10⁻¹⁶ Ω^{-1} m⁻¹).

5.27 EXPRESSION FOR CURRENT DENSITY IN METAL

Consider N number of free electrons distributed uniformly throughout a metal or conductor of length L and cross sectional area A as shown in Fig. 5.15. When an electric field (E) is applied to such a conductor, the electrons travel a distance L metres in T seconds. This makes the average velocity of electrons equal to L/T.

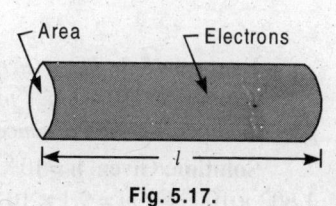

Fig. 5.17.

The number of electrons passing through any area per second,

$$= \frac{N}{T}$$

∴ Total charge passing through any area per second (called current)
= Charge of an electron × Number of electrons per second crossing any area

$$= e \times \frac{N}{T} = \frac{eN}{T}$$

Multiplying and dividing the R.H.S. of the above equation, by the total charge passing through any area per second.

$$= \frac{e.N}{h} \times \frac{L}{T} = \frac{e.N.v}{h} \qquad \ldots (i)$$

Thus the current in a conductor,

$$I = \frac{e.N.v}{h} \qquad \ldots (ii)$$

And the current per unit area (called current density),

$$J = \frac{I}{A} = \frac{e.N.v}{L.A} \qquad \ldots (iii)$$

Since $L.A$ is the volume of the conductor containing N electrons, therefore concentration of electrons per unit volume.

$$n = \frac{N}{LA}$$

And $\qquad J = e.n.v \qquad \ldots (iv)$

Substituting the value of v from Art. 5.24 in the above equation,

$$J = e.n \left(\frac{e.E.t}{m} \right) = \frac{e^2.n.E.t}{m} \qquad \ldots (v)$$

It may be noted that equation e, n, t, and m are all constants for any conductor. Therefore, the term $e^2 n.t/m$ is a constant and is equal to conductivity of a metal, i.e.,

$$\sigma = \frac{e^2.n.t}{m} \qquad \ldots (vi)$$

Substituting this value σ in equation (v),

$$J = \sigma.E$$

It is evident from the above relation that the current density is a conductor is directly proportional to the applied electric field (*E*). Now substituting $\frac{v}{E} = \frac{e.t}{E}$ from Art. 5.24 in equation (*vi*),

$$\sigma = \frac{n.e.v.}{E}$$

Example 5.5. *There are 10^{19} electrons /m^3 to serve as carriers in a material, that has a conductivity of 0.01 ohm $^{-1} m^{-1}$. What is the drift velocity of these carriers, when 0.17 volts is placed across 0.27 mm distance within the material? Given $e = 1.602 \times 10^{-9}$ and $m = 9.1 \times 10^{-31}$ kg.*

Solution. Given: $n = 10^{19} m^3$; $\sigma = 0.01$ ohm $^{-1} m^{-1}$; $V = 0.17$ Volt; $d = 0.27$ mm $= 0.27 \times 10^{-2}$ m; $e = 1.602 \times 10^{-19}$ C ; $m = 9.1 \times 10^{-31}$ kg

Let $v = $ Drift velocity of the electrons.

We know that electric field.

$$E = \frac{V}{d} = \frac{0.17}{0.27 \times 10^{-3}} = 630 \text{ Volt/m}$$

and conductivity (σ),

$$0.01 = \frac{n.e.v.}{V} = \frac{10^{19} \times 1.602 \times 10^{-19} \times v}{630} = 2.54 \times 10^{-3} v$$

$\therefore \qquad v = 0.01/ (2.54 \times 10^{-3}) = 3.93$ m/sec **Ans.**

Example 5.6. *The collision time for electron scattering in copper at 300 K is 2×10^{-14} sec. Calculate the conductivity of copper at 300 K, Given that density of copper = 8960 kg/m^3; Atomic weight of copper = 63.54 a.m.u. and Mass of an electron = 9.1×10^{-31} kg.*

(A.M.I.E., Winter, 1992)

Solution. Given: Temperature = 300 K: Time t = 2×10^{-14} sec; Density of copper = 8960 kg/m³; Atomic weight of copper = 63.54 amu and Mass of an electron = 9.1×10^{-31} kg.

We know that 63.54 grams of copper contains 6.023×10^{23} free electrons (*i.e.*, Avogadro's number) since one atom contributes one electron. The volume of 63.54 grams of copper is 8.9 cubic centimetre (*c.c*). Hence number of electrons per unit volume (*c.c*),

$$n = \frac{6.023 \times 10^{23}}{63.54/8.9} = 8.5 \times 10^{22}$$

and the number of electrons per m³,

$$n = 8.5 \times 10^{22} \times 10^6 = 8.5 \times 10^{28}$$

We also know that conductivity.

$$\sigma = \frac{e^2.n.t}{m} = \frac{(1.6 \times 10^{-19})^2 \times 8.5 \times 10^{28} \times 2 \times 10^{-14}}{9.1 \times 10^{-31}}$$

$$= 4.782 \times 10^7 \text{ mho/m } \textbf{Ans.}$$

5.28 MOBILITY

It has been found that average velocity of the electrons in a conductor is directly proportional to the applied electric field, *i.e.*,

$$v \propto E$$
$$= \mu.E.$$

The constant of proportionality μ is called mobility of the electrons. It is expressed in m²/volt-sec or cm²/volt-sec. Its magnitude is given by the relation:

Electron Theory of Metals

$$\mu = \frac{e.t}{m}$$

where e = Charge of an electron,
m = Mass of an electron, and
t = Collision time.

Substituting the value of μ (equal to v/e) in equation for conductivity in Art. 5.26.

$$\sigma = \frac{n.e.v.}{E} = n.e.\mu$$

Note: Sometimes it is required to determine the concentration of free electrons per unit volume (n). We know that the number of free electrons per unit volume,

$$n = \frac{\text{Avogadro's number (N)}}{\text{Molar volume}}$$

Where molar volume is given by the relation:

$$= \frac{N \times \text{Density}}{\text{Atomic weight}}$$

Thus

$$n = \frac{N \times \text{Density}}{\text{Atomic weight}} = \frac{N \times d}{\text{Atomic weight}}$$

Example 5.7. *What would be the mobility of electrons, when the mean free time between the collisions is 10^{-14} sec?*

Take $e = 1.602 \times 10^{-19}$ C and $m = 9.1 \times 10^{-31}$ kg.

Solution. Given: $t = 10^{-14}$ sec; $e = 1.602 \times 10^{-19}$ C; $m = 9.1 \times 10^{-31}$ kg

We know that the mobility of electrons,

$$\mu = \frac{e.t}{m} = \frac{(1.602 \times 10^{-19}) \times 10^{-14}}{9.1 \times 10^{-31}}$$

$$= 1.76 \times 10^{-3} \text{ m}^2/\text{volt.sec} \textbf{ Ans.}$$

Example 5.8. *The density of silver is 10.5 gm/c.c. and its atomic weight is 107.9. Assuming that each silver atom provides one conduction electron, calculate the number of free electrons per c.c. Take conductivity of silver as 6.8×10^7 mhos/m. Calculate the mobility of the electrons (given, $e = 1.6 \times 10^{-19}$ Coulomb.)*

Solution. Given: $d = 10.5$ gm/c.c.; Atomic weight = 107.9; $\sigma = 6.8 \times 10^7$ mhos/m = 6.8×10^5 mhos/cm and $e = 1.6 \times 10^{-19}$ C

Let μ = Mobility of free electrons

We know that number of free electrons.

$$n = \frac{N \times d}{\text{Atomic weight}} = \frac{6.023 \times 10^{23} \times 10.5}{107.9} = 5.86 \times 10^{22}$$

and conductivity (σ)

$$6.8 \times 10^5 = n.e.\mu = 5.86 \times 10^{22} \times 1.602 \times 10^{-19} \times \mu = 9.39 \times 10^3 \mu$$

$$\therefore \quad \mu = \frac{6.8 \times 10^5}{9.39 \times 10^3} = 72.42 \text{ m}^2/\text{volt-sec} \textbf{ Ans.}$$

Example 5.9. *The conductivity of silver is 6.5×10^7 ohm^{-1}-m^{-1} and the number of conduction electrons/m^3 is 6×10^{28}. Calculate the mobility of conduction electrons and the drift velocity in an electric field of 1 V/m. Given, $e = 1.602 \times 10^{-19}$ C and $m = 9.1 \times 10^{-31}$ kg.*

Solution. Given: $\sigma = 6.5 \times 10^7$ ohm^{-1} –m^{-1}; $n = 6 \times 10^{28}$/m^3; $E = 1$ V/m; $e = 1.602 \times 10^{-19}$ C and $m = 9.1 \times 10^{-31}$ kg.

Mobility of conduction electrons

Let μ = Mobility of conduction electrons.

We know that conductivity σ, $6.5 \times 10^7 = n.e.\mu = 6 \times 10^{23} \times 1.602 \times 10^{-19} \times \mu = 9.612 \times 10^4 \mu$

$$\therefore \quad \mu = \frac{6.5 \times 10^7}{9.612 \times 10^4} = 6.76 \times 10^3 \text{ m}^2/\text{volt.sec. } \textbf{Ans.}$$

Drift velocity

We also know that drift velocity,

$$v = \mu E = 6.76 \times 10^{-3} \times 1 \text{ m/sec}$$
$$= 6.76 \times 10^{-2} \text{ m/sec. } \textbf{Ans.}$$

Example 5.10. *Assuming that the conductivity for copper ($= 58 \times 10^6$ Ω^{-1} m^{-1}) is entirely due to free electrons, calculate the density of free electrons in copper at room temperature. Given mobility of free electrons $= 3.5 \times 10^{-3}$ m^2/(V.S). Also calculate the drift velocity for an electric field strength of 0.5 V/m.*

Solution. Given: $\sigma = 58 \times 10^6$ Ω^{-1} m^{-1} ; $\mu_n = 3.5 \times 10^{-3}$ m^2/(V-s); $E = 0.5$ V/m

We know that conductivity, (σ)

$$58 \times 10^6 = n.e.\mu$$
$$= n \times (1.602 \times 10^{-19}) \times (3.5 \times 10^{-3})$$

$$\therefore \quad n = \frac{58 \times 10^6}{(1.602 \times 10^{-19}) \times (3.5 \times 10^{-3})} = 104 \times 10^{27} \text{ m}^{-3}. \textbf{ Ans.}$$

We also know that the drift velocity,

$$v = \mu E = (3.5 \times 10^{-3}) \times 0.5 = 1.75 \times 10^{-3} \text{ m/s } \textbf{Ans.}$$

5.29 MEAN FREE PATH

A metal may be defined as a substance, which consists of a lattice of positive ion cores held together by means of loosely bound valence electrons (also called gas of electrons or delocalized electrons,). These electrons have a wave characteristics as they move throughout the metal. Waves proceed with a minimum of interruption, when travelling through a periodic structure, *i.e.*, a structure which has uniform repetition. Any irregularity in repetitive structure, through which the wave travels, will deflect the wave. Thus if an electron has been travelling towards the positive electrode, a displaced atom or a foreign atom could cause it to be reflected towards the negative electrode.

It may be noted that when the electrons move towards the positive electrode, they continuously acquire additional momentum and hence more velocity. And when they move towards the negative electrode, the continuously lose momentum and hence velocity. Thus the distance between reflections and deflections determines the net or drift velocity. This representative distance in a metal is called mean free path. Thus mean free path is an average distance which an electron can travel in its wave like pattern without any reflection or deflection. Mathematically, the mean free path is given by the relation,

$$\lambda = v.t$$

where v = Velocity of an electron and

t = Collision time or mean free time.

Electron Theory of Metals

For metal, the velocity of an electron corresponds to that of Fermi energy. It is given by the relation

$$v_F = \sqrt{\frac{2W_F}{m}}$$

W_F = Fermi energy in joules. Its value is equal to 1.602×10^{-19} times the fermi energy in electron volts, (*i.e.*, $1.602 \times 10^{-19}\ E_F$) and

m = Mass of an electron (equal to 9.1×10^{-31} kg)

Example 5.11. *The Fermi level for potassium is 2.1 eV. Calculate the velocity of electrons at the fermi level. Given electronic charge = 1.6×10^{-19} C and mass of the electron = 9.109×10^{-31} kg.*

Solution Given: $E_F = 2.1$ eV; $e = 1.602 \times 10^{-19}$ C and $m = 9.109 \times 10^{-31}$ kg and $t = 10^{14}$ sec

We know that fermi energy,

$$W_F = 1.662 \times 10^{-19}\ E_F = 1.602 \times 10^{-19} \times 2.1 \text{ J}$$
$$= 3.364 \times 10^{-19} \text{ J}$$

and the velocity electrons,

$$v_F = \sqrt{\frac{2W_F}{m}} = \sqrt{\frac{2 \times 3.364 \times 10^{-19}}{9.109 \times 10^{-31}}} \text{ m/s}$$

$$= 0.86 \times 10^6 \text{ m/s } \textbf{Ans.}$$

Example 5.12. *(a) What is the maximum velocity of an electron in a metal in which the fermi energy has a value of 3.75 eV?*

Given: $e = 1.602 \times 10^{-19}$ C and $m = 9.1 \times 10^{-31}$ kg.

(b) What would be the mobility of electrons when the mean free time between the collisions is 10^{-14} sec?

Solution. Given: $E_F = 3.75$ eV; $e = 1.602 \times 10^{-19}$; $m = 9.1 \times 10^{-31}$ kg. and $t = 10^{14}$ sec

(a) Maximum velocity

We know that fermi energy

$$W_F = 1.602 \times 10^{-19}\ E_F = 1.602 \times 10^{-19} \times 3.75$$
$$= 6 \times 10^{-19} \text{ J}$$

And the velocity of electrons,

$$v_F = \sqrt{\frac{2W_F}{m}} = \sqrt{\frac{2 \times 6 \times 10^{-19}}{9.1 \times 10^{-31}}} \text{ m/s}$$

$$= 1.76 \times 10^6 \text{ m/s } \textbf{Ans.}$$

(b) Mobility of electrons

We know that mobility of electrons,

$$\mu = \frac{e.t}{m} = \frac{1.602 \times 10^{-19} \times 10^{-14}}{9.1 \times 10^{-31}} \text{ m}^2/V\text{-sec}$$

$$= 1.76 \times 10^{-3} \text{ m}^2\ /V\text{-sec } \textbf{Ans.}$$

Example 5.13. *Estimate the mean free path of free electrons in pure copper at 4K. The collision time for photon scattering at 4K is 10^{-9} sec. The Fermi energy level for copper is 7 eV. Given electronic charge = 1.602×10^{-19} C and electronic mass = 9.109×10^{-31} kg.*

Solution. Given: $t = 10^{-9}$ sec; $E_F = 7$ eV; $e = 1.602 \times 10^{-19}$ C and electronic mass $m = 9.109 \times 10^{-31}$ kg.

Let λ = Mean free path of free electrons.

We know that fermi energy,
$$W_F = 1.602 \times 10^{-19} E_F = 1.602 \times 10^{-19} \times 7$$
$$= 11.2 \times 10^{-19} \text{ J}$$

and the velocity of electrons,
$$v_F = \sqrt{\frac{2W_F}{m}} = \sqrt{\frac{2 \times 11.2 \times 10^{-19}}{9.109 \times 10^{-31}}} \text{ m/s}$$
$$= 1.57 \times 10^6 \text{ m/s}$$

∴ That mean free path,
$$\lambda = v_F \cdot t = 1.57 \times 10^6 \times 10^{-9} = 1.57 \times 10^{-3} \text{ m}$$
$$= 1.57 \text{ mm. } \textbf{Ans.}$$

Example 5.14. *The following data are known for copper.*

Density = 8.92×10^3 kg/m³

Resistivity of copper = 1.73×10^{-8} Ω-m

Atomic weight = 63.5

Calculate the mobility and the average time of collision of the electrons in copper obeying classical laws.

Solution. Given: Density $(d) = 8.92 \times 10^3$ kg/m³ ; Resistivity $(\rho) = 1.73 \times 10^{-8}$ Ω-m; and Atomic weight = 63.5

Let μ = mobility,
τ = Average time of collision

We know that number of free electrons
$$n = \frac{\text{Avogadro's Number} \times \text{Density}}{\text{Atomic Weight}} = \frac{N \times d}{\text{Atomic Weight}}$$
$$= \frac{(6.023 \times 10^{22}) \times (8.9 \times 10^3)}{63.5}$$
$$= 0.844 \times 10^{25}$$

We also know that conductivity,
$$\sigma = \frac{1}{\rho} = \frac{1}{1.73 \times 10^{-8}} = 0.578 \times 10^8 \text{ (Ω-m)}^{-1}$$

and conductivity (σ)
$$0.578 \times 10^8 = n.e.\mu$$
$$= (0.844 \times 10^{25})_\mu \times (1.602 \times 10^{-19}) \times \mu$$
$$= 1.352 \times 10^6 \mu$$

∴
$$\mu = \frac{0.578 \times 10^8}{1.352 \times 10^6} = 42.75 \text{ m}^2/\text{V-s.}$$

We also know that mobility (μ)
$$42.75 = \frac{e \cdot t}{m}$$

Electron Theory of Metals

$$= \frac{(1.602 \times 10^{-19}) \times t}{9.1 \times 10^{-31}} = (176 \times 10^9)t$$

$$\therefore \quad t = \frac{42.75}{176 \times 10^9} = 2.428 \times 10^{-10} \text{ s or } 0.243 \text{ ns } \textbf{Ans.}$$

Example 5.15. *Sodium metal with BCC structure has two atoms per unit cell. The radius of sodium atom is 1.86 Å. Calculate is electrical resistivity at 0 deg. celsius if the classical value of mean free time at this temperature is 3×10^{-14} s. Avogadro's number = 6.023×10^{23} atoms/mol.*

Solution. Given: Radius of sodium atom = 1.86 Å = 1.86×10^{-10} m and mean free time, $t = 3 \times 10^{-14}$ s; Atomic mass of sodium = 23 g/mol = 23×10^{-3} kg/m.

Let d = density of sodium metal.

We know that density of sodium,

$$d = \frac{\text{Mass / unit cell}}{\text{Volume / unit cell}}$$

Let us first determine the mass/unit cell and then the volume/unit cell.

We know that mass/unit cell,

$$m = \frac{(\text{No. of atoms / unit cell}) \times (\text{Atomic Mass})}{\text{Avogadro's number}}$$

$$= \frac{2 \times (23 \times 10^{-3})}{6.023 \times 10^{23}} = 7.637 \times 10^{-26} \text{ kg}$$

and the volume/unit cell,

$$V = a^3$$

where a is the lattice constant. Now for a BCC structure, we know that the lattice constant $a = (4/\sqrt{3})r$. Therefore

$$V = \left[(4/\sqrt{3}) \times (1.86 \times 10^{-10})\right]^3 = 0.0797 \times 10^{-27} \text{ m}^3$$

$\therefore$ Density of sodium,

$$d = \frac{7.637 \times 10^{-26}}{0.0797 \times 10^{-27}} = 958.2 \text{ kg/m}^3$$

We know that mobility

$$\mu = \frac{e \cdot t}{m} = \frac{(1.602 \times 10^{-9}) \times (3 \times 10^{-14})}{(9.1 \times 10^{-31})} = 0.528 \times 10^8 \text{ m}^2/\text{V-s}.$$

We also know that number of free electrons in sodium,

$$n = \frac{\text{Avogadro's number} \times \text{Density}}{\text{Atomic weight}} = \frac{N \times d}{\text{Atomic weight}}$$

$$= \frac{(6.023 \times 10^{23}) \times 958.2}{(23 \times 10^{-3})} = 2.51 \times 10^{28}$$

$\therefore$ The conductivity,

$$\sigma = e.n.\mu = (1.602 \times 10^{-19}) \times (2.51 \times 10^{28}) \times (0.528 \times 10^8)$$
$$= 2.123 \times 10^{17} \text{ } (\Omega\text{-m})^{-1}$$

and resistivity,

$$\rho = \frac{1}{\sigma} = \frac{1}{2.123 \times 10^{17}} = 0.47 \times 10^{-17} \; \Omega\text{-m} \;\; \textbf{Ans.}$$

5.30 THERMOELECTRICITY

Fig. 5.18. shows two wires of different metals (say copper and iron) joined at their ends so as to form two junctions A and B. A sensitive galvanometer is included in the circuit as shown in the figure. Such an arrangement is known as thermocouple. When one of the junctions of the thermocouple is kept hot and the other as cold, the galvanometer gives deflection indicating the production of current in the arrangement. The current, so produced, is called the thermoelectric current. The flow of current indicates that an e.m.f. (electromotive force) is produced in the circuit. This e.m.f. is known as thermoelectric e.m.f.

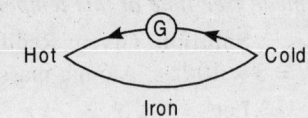

Fig. 5.18. Thermocouple.

This phenomenon of production of electricity, with the help of heat, is called thermo electricity. And this effect is called thermoelectric effect or seebeck effect. Thus Seebeck effect may be defined as the phenomenon of generation of an electric current in a thermocouple by keeping its two junctions at different temperatures.

5.31 ORIGIN OF THERMO E.M.F.

We know that the number of electrons per unit volume (called concentration) differs from one metal to another. When the wires of two different metal are placed in contact with each other, the electrons diffuse from one wire to another because of the concentration difference. As a result of this, one of the metallic wire becomes positively charged and the other negatively charged. This causes a potential difference to be set up across the contact. It is called contact potential. The contact potential is strongly affected by temperature. When one of the junctions of the thermocouple is heated (keeping the other junction cold), the potential difference set up at the hot junction is comparatively more than that of the cold junction. As a result of this, a net e.m.f. is produced in the thermocouple. It is called thermo e.m.f. and is the cause of thermoelectric current.

5.32 MAGNITUDE AND DIRECTION OF THERMO E.M.F.

The magnitude and direction of the thermo e.m.f. produced in a thermocouple depends upon the following two factors:

1. *Nature of the metals forming the thermocouple.* On the basis of experimental observations. Seebeck arranged a number of metals in the form of a series called thermoelectric series. Some of the important metals in this series are antimony, iron, zinc, silver, gold, molybdenum, chromium, tin, lead, mercury, magnesium, calcium, copper, cobalt, nickel and bismuth.

Suppose a thermocouple is formed with wires of any two metals from this series. Then direction of the current will be from a metal occurring earlier in this series to a metal occurring later in the series through the cold junction. For example, in an antimony-bismuth thermocouple, the current flows from antimony to bismuth through the cold junction and from bismuth to antimony through the hot junction. Similarly, in a copper-iron thermocouple, the current will flow from iron to copper through cold junction and from copper to iron through hot junction.

---

* Seeback was a scientist, who discovered thermoelectric effect in 1826. After his name, thermoelectric effect is called Seebeck effect.

Electron Theory of Metals

The magnitude of thermo *e.m.f.* is of the order of a few micro-volts per degree temperature difference between the two junctions. It depends upon how far the metals are separated in the series; As a general rule more the separation, greater will be the magnitude of thermo *e.m.f.* As a general rule, magnitude of thermo *e.m.f.* for a difference of 100°C temperature is about 1.3 mV for copper-iron thermocouple and about 8 mV for antimony bismuth thermocouple.

2. *Temperature difference between two junctions of the thermocouple.* To study the effect of differences of temperature between the two junctions, consider a copper-iron thermocouple as shown in Fig. 5.17 (*a*). One of its junctions is kept hot by immersing in oil bath and heated with a burner. The other junctions is kept cold by immersing it in powdered ice as shown in the figure. The temperature (*T*) of the hot junction can be recovered by a thermometer placed in the hot oil bath. A galvanometer is placed in the circuit for measuring the magnitude of the thermoelectric current.

We know that temperature of the cold junction (θ_c) is 0°C. The temperature of the hot junction (θ_h) can be varied by adjusting the amount of heat supplied by the burner. As the temperature of the hot junction is increased above 0°C, the deflection on the galvanometer is directly proportional to the thermoelectric current and hence the thermo *e.m.f.* As the temperature of the hot junction is further increased, a state reaches at which the thermo *e.m.f* becomes maximum. The temperature of the hot junction, at which thermo *e.m.f.* becomes maximum, is called a neutral temperature (θ_n). For a given thermocouple, the neutral temperature has a fixed value and does not depend upon the temperature of cold junction. For copper-iron thermocouple, the value of neutral temperature is 270°C.

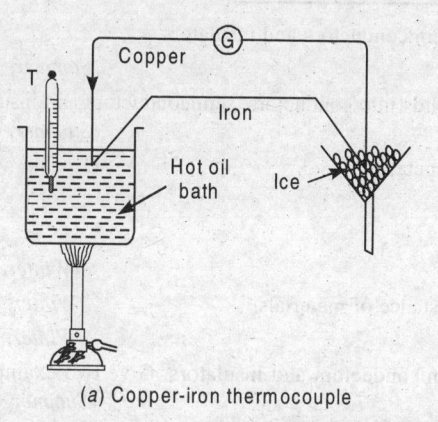

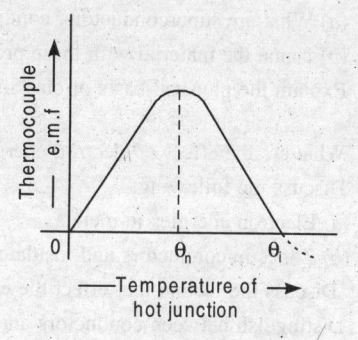

(a) Copper-iron thermocouple

(b) Variation of thermo e.m.f. with temperature difference between two junctions

Fig. 5.19.

As the temperature of the hot junction is further increased, the magnitude of thermo *e.m.f.* decreases, and ultimately, it becomes zero. If the temperature of the hot junction is still increased, the e.m.f. is produced in the reverse direction. The temperature of the hot junction, at which the direction of thermo e.m.f. reverses its direction is called the temperature of inversion (θ_i).

Now if we plot a graph with temperature of the hot junction (θ_h) along the horizontal axis and thermo e.m.f along the vertical axis, we shall obtain a curve as shown in Fig. 5.17 (*b*). It may be noted from this graph that the temperature of inversion is as much above the neutral temperature as the neutral temperature is above the temperature of the cold junction, i.e.,

$$\theta_n - \theta_c = \theta_i - \theta_n$$

$$\therefore \quad \theta_n = \frac{\theta_i + \theta_c}{2}$$

It is evident from the above relation that the neutral temperature is the mean of the temperature of inversion (θ_i) and temperature of cold junction (θ_c).

5.33 MEASUREMENT OF TEMPERATURE WITH THERMOCOUPLE

The principle of thermoelectricity may be used to measure the temperature of furnaces. In order to measure temperature, two suitable metals are taken in the form of thick uniform wires. They are joined together by electric or gas welding. So that they are in perfect contact with each other at two junctions and thus form a thermocouple. A very sensitive galvanometer is connected to read the value of thermoelectric current produced in the current. One of the junctions of a thermocouple is kept in melting ice (i.e., at 0°C) and the other is kept at known temperature. The scale of the galvanometer is calibrated to read directly on the scale to find the temperature of the furnace, the hot junction is placed in contact with the furnace and its temperature is read directly on the scale. It may be noted that we required different thermocouples for different values of temperature.

It is not essential for the thermocouple wires to be in contact with each other to form a cold junction. The wires can be separated by any conductor or an instrument, provided the temperature at all points of the conductor or an instrument is the same. In such a case, it is a common practice to call the junction as a reference junction rather than a cold junction. Thus, there is no need of keeping one of the function at 0°C as mentioned above.

A.M.I.E. (I) EXAMINATION QUESTIONS

1. (a) What are superconductors, conductors and semiconductors and insulators ?
   (b) Name the material with these properties.   *(Summer, 1993)*
2. Explain the physical basis of classification of solids into conductors, semiconductors and insulators.   *(Summer, 1994)*
3. What are the effect of electron energy levels in metals ?   *(Summer, 1994)*
4. Discuss the following:
   (a) Electron energies in metals.
   (b) Zones in conductors and insulators.   *(Winter, 1994)*
5. Discuss the factors that effect the electrical resistance of materials.   *(Winter, 1994)*
6. Distinguish between conductors and insulators.   *(Winter, 1993)*
7. Explain the difference between conductors, semiconductors and insulators. Give two examples of each category.   *(Summer, 1995)*
8. Discuss briefly the zone theory of solids.   *(Summer, 1995)*
9. Explain the formation of energy bonds in solids. Or on the energy bond diagram for metal, semiconductor and an insulator.   *(Summer, 1996)*
10. Draw the energy bond diagram of an insulator.   *(Summer, 1997)*
11. What do you understand by drifty velocity and mobility of a free electron ? *(Summer, 1998)*
12. Briefly discuss the factors affecting electrical resistance of materials.   *(Winter, 1998)*
13. Explain why metals will feel colder to the touch than ceramics and plastics although all are at room temperature.   *(Winter, 1999)*
14. Differentiate between electronic and ionic conduction. Illustrate with the help of suitable examples.   *(Summer, 2000)*
15. (a) A bimetallic strip is constructed from strips of two different metals that are bonded along their lengths. Explain how such a device may be used in a thermostat to regulate temperature ?
    (b) The thermal conductivity of a plain carbon steel is greater than that of stainless Steel. Explain why it is so ?   *(Summer, 2000)*

Electron Theory of Metals

16. Copper is a conductor but silicon is a semiconductor. Explain and illustrate with their electronic configurations. *(Winter, 2000)*
17. By applying the principle of Zone Theory, explain the difference between conductors, insulators and semiconductors. *(Summer, 2001)*
18. Write a note on Brillouin zones. *(Winter, 2002)*
19. In terms of electron energy bond structure, discuss reasons for the difference in electrical conductivity between metals, semiconductors and insulators. *(Winter, 2003)*

MULTIPLE CHOICE QUESTIONS

1. Which of the following metals has the lowest temperature coefficient of resistance ?
   (a) copper   (b) aluminium   (c) silver   (d) gold
2. The term 'Superconductivity may be defined as a state of a meterial in which it has
   (a) zero conductivity   (b) zero resistivity   (c) very high resistivity
   (d) resistivity depending upon the value of applied voltage.
3. Resistance of a metal wire is expressed in
   (a) ohms   (b) ohm-m   (c) (ohm-m)$^{-1}$ (d) ohm/m
4. The mobility of electrons (μ) is expressed in
   (a) m^2.volt-sec   (b) per m^2/volt-sec   (c) m^2/volt.sec   (d) volt-sec/m^2
5. "Thermoelectricity" is the phenomenon of producing electricity with the help of
   (a) light   (b) heat   (c) force   (d) pressure

ANSWERS

1. (c)   2. (b)   3. (a)   4. (c)   5. (b)

6
Mechanical Properties of Metals

1. Introduction. 2. Types of Mechanical Properties. 3. Elasticity. 4. Plasticity. 5. Ductility. 6. Brittleness. 7. Hardness. 8. Toughness. 9. Stiffness. 10. Resilience. 11. Creep, 12. Endurance. 13. Strength. 14. Types of Strengths. 15. Elastic Strength. 16. Plastic Strength. 17. Tensile Strength. 18. Compressive Strengths. 19. Shear Strength. 20. Bending Strength. 21. Torsional Strength. 22. Types of Technological Properties. 23. Malleability. 24. Machinability. 25. Factors Affecting Machinability. 26. Improving Machinability. 27. Weldability. 28. Formability or Workability. 29 Castability. 30. Factors Affecting Mechanical Properties of a Metal.

6.1 INTRODUCTION

The term 'property' in a broader sense, may be defined as the quality, which defines the specific characteristics of a metal. A detailed study of all the properties of a metal provides a sound basis for predicting its behaviour in manufacturing shop and also in actual use. As a matter of fact, the following properties of a metal are important for an engineer, to enable him in selecting suitable metal for his various jobs:

1. *Physical Properties.* These include shape, size, colour, lustre, specific gravity, porosity, structure, finish etc.
2. *Mechanical Properties.* These include elasticity, plasticity, ductility, brittleness, hardness, toughness, stiffness, resilience, creep, endurance, strength etc.
3. *Technological Properties.* Strictly speaking, all the technological properties of a metal are essentially its mechanical properties, which include properties like malleability, machinability, weldability, formability or workability, castability etc.
4. *Thermal Properties.* These include specific heat, thermal conductivity, thermal expansion, latent heat, thermal stresses, thermal shock etc.
5. *Electrical Properties.* These include conductivity, resistivity, relative capacity, dielectric strength etc.
6. *Chemical Properties.* These include atomic weight, equivalent weight, molecular weight, atomic number, acidity, alkalinity, chemical composition, corrosion, etc.

In this chapter, we shall discuss only the mechanical and technological properties of metals.

6.2 TYPES OF MECHANICAL PROPERTIES

The mechanical properties of a metal are those properties, which completely define its behaviour under the action of external loads or forces. Or in other words, mechanical properties are those properties which are associated with its ability to resist failure as well as behaviour under the action of the external forces. A sound knowledge of these properties is very essential for an engineer to enable him in selecting a suitable metal for his various structures or various

Mechanical Properties of Metals

components of a machine. Most of the mechanical properties of the metals are generally expressed in terms of *stress, strain or both.

Though there are many mechanical properties of a metal, an engineer should know, yet the following are important from the subject point of view.

1. Elasticity. 2. Plasticity. 3. Ductility. 4. Brittleness. 5. Malleability. 6. Weldability. 7. Castability. 8. Hardness. 9. Toughness. 10. Stiffness. 11. Resilience. 12. Creep. 13. Endurance. 14. Strength.

6.3 ELASTICITY

The term 'elasticity' may be defined as the property of a metal by virtue of which it is able to retain its original shape and size after the removal of the load. In nature, no material is perfectly elastic, over the entire range of stress, up to rupture. Steel and some other metals have a wide range over which they appear to be elastic.

The elasticity is always desirable in metals used in machine tools and other structural members.

6.4 PLASTICITY

The term 'plasticity' may be defined as the property of a metal by virtue of which a permanent deformation (without fracture) takes place, whenever it is subjected to the action of external forces. The plasticity of a metal depends upon its nature and the environmental conditions, *i.e.*, whether the metal is shaped red hot or in cold.

Most of the metals have been found to possess a good plasticity. Lead has good plasticity even at room temperature, which cast iron, does not possess any appreciable plasticity even when red hot. This property finds its use in forming, shaping and extruding operations of metals.

6.5 DUCTILITY

The term 'ductility' may be defined as the property of metal by virtue of which it can be drawn into wires or elongated before rupture takes place. It depends upon the grain size of the metal crystals. The measures of the ductility of a metal are its percentage elongation and percentage reduction in the cross-sectional area before rupture. The term percentage elongation is the maximum increase in the length expressed as percentage of original length. Mathematically, percentage of original length. Mathematically, percentage elongation

$$= \frac{\text{Increase in length}}{\text{Original length}} \times 100$$

Similarly, the term percentage reduction of cross-sectional area is the maximum decrease in cross-sectional area, expressed as the percentage of the original cross-sectional area. Mathematically, percentage reduction in cross-sectional area

---

* *Stress*. It is the internal resistance set up in a material under the action of the external forces. Mathematically, stress is expressed as forced divided by cross-sectional area.

Strain. It is the deformation per unit length under the action of the external forces. Mathematically, strain is expressed as change in length divided by original length. The strain may be classified into elastic strain and plastic strain. The elastic strain is a temporary strain, what appears so long as the external forces are applied. It disappears after the removal of the forces.

The plastic strain is a permanent strain caused by the external forces, when the stress exceeds the elastic limit. The plastic strain takes place as a result of permanent displacement of the atoms inside the material.

$$= \frac{Decrease\ in\ cross\text{-}sectional\ area}{Original\ cross\text{-}sectional\ area} \times 100$$

A little consideration will show, that a metal with a good percentage of elongation or reduction in cross-sectional area explains its high ductility. Metals with more than 15% elongation are considered as ductile. Metals with 5 to 15% elongation are considered of intermediate ductility. But the metal with less than 5% elongation are considered as brittle ones.

The following common metals have ductility in the decreasing order:

1. Gold. 2. Platinum. 3. Silver. 4. Iron. 5. Copper. 6. Aluminium. 7. Nickel. 8. Zinc. 9. Tin. 10. Lead.

6.6 BRITTLENESS

The term 'brittleness' may be defined as the property of a metal by virtue of which it will fracture without any appreciable deformation.

The property is opposite to the ductility of a metal. Cast iron, glass and concrete are the examples of brittle materials. This property finds its importance for the design of machine tools, which are subjected to sudden loads. We have already discussed in the last article that metals with less than 5% elongation are known to be brittle ones.

6.7 HARDNESS

The term 'hardness' may be defined as the property of a metal by virtue of which it is able to resist abrasion, indentation (or penetration) and scratching by harder bodies. It is measured by the resistance of the metal which it offers to scratching.

The hardness of a metal is determined by comparing its hardness with ten standard minerals. These minerals, in the increasing order of their hardness on the original Moh's Scale are:

1. Talc. 2. Gypsum. 3. Calcite. 4. Fluorite. 5. Apatite. 6. Orthoclase. 7. Quartz. 8. Topaz. 9. Corundum. 10. Diamond.

6.8 TOUGHNESS

The term 'toughness' may be defined as the property of a metal by virtue of which it can absorb maximum energy before fracture takes place. Tenacity and hardness of a metal are the measures of its toughness. It has been found that the value of toughness falls with the rise in temperature.

The important of toughness is in the selection of a material where load increases beyond the elastic limit or yield point e.g., power press punch and pneumatic hammer etc.

6.9 STIFFNESS

The term 'stiffness' may be defined as the property of a metal by virtue of which it resists deformation. Modulus of elasticity (*i.e.*, ratio of stress to the strain below elasticity limit) is a measure of stiffness of a metal. The stiffness of a metal is of great importance while selecting it for a member or a component of a machine or a structure. It is also used in graduating spring balances and spring controlled measuring instruments.

6.10 RESILIENCE

The term 'resilience' may be defined as the property of a metal by virtue of which it stores energy and resists shocks or impacts. The resilience of a metal is measured by the amount of energy that can be stored, per unit volume, after it is stressed up to the elastic limit.

The resilience of a metal is also of great importance in the selection of materials used for various types of springs.

Mechanical Properties of Metals

6.11 CREEP

The term 'creep' may be defined as the property of a metal by virtue of which it deforms continuously under a steady load. Generally, the creep occurs in steel at higher temperatures. The creep is always considered while designing I.C engines, boilers, turbines etc. The creep is of great importance in the following cases:

1. Soft meals used at about room temperature such as lead pipes and white metal bearings etc.
2. Steam and chemical plants operating at 450 to 550° C
3. Gas turbines working at higher temperatures.
4. Rockets, missiles, supersonic jets.
5. Nuclear reactor field.

6.12 ENDURANCE

The term 'endurance may be defined as the property of a metal by virtue of which it can withstand varying stresses (same or opposite nature). The maximum value of stress, that can be applied for an indefinite times without causing its failure, is known as endurance limit. It has been found that the endurance limit for ordinary steel is approximately half the tensile strength.

The endurance of a metal is of great importance in the design and production of parts in a reciprocating machines and components subjected to vibrations. It is always desirable to keep the working stress of material well within the endurance limit.

6.13 STRENGTH

The term 'strength' may be defined as the property of metal by virtue of which it can withstands or support an external force or load without rupture.

The strength of a metal is the most important property, which plays a decisive role in designing various structures and components.

6.14 TYPES OF STRENGTHS

A metal has an innumerable types of strengths. The strengths, which are important from the subject point of view, may be broadly grouped into the following two types:

1. Depending upon the value of stress, the strengths of a metal may be elastic or plastic.
2. Depending upon the nature of stress, the strengths of a metal may be tensile, compressive, shear, bending and torsional.

In the following pages, we shall discuss all these types of strengths one by one.

6.15 ELASTIC STRENGTH

The elastic strength of a metal is the value of load corresponding to transition from elastic to plastic range. The ideal stress values (i.e., proportional limit and elastic limit) are used to defined elastic strength of a material.

6.16 PLASTIC STRENGTH

The plastic strength of a material is the value of load corresponding to plastic range and rupture. It is also called ultimate strength. In actual practice, the specimen is subjected to a stress which is always less than the working stress. The ratio of ultimate stress, to the working stress of a metal, is called factor of safety. It is also known as the factor or ignorance, and greatly varies upon the nature of stresses or loads. Following values of factor of safety are generally kept for various loads:

1. Dead load   4 or 5
2. Live load   6
3. Alternating kind of load   8 to 12
4. Shock loading   12 to 15

The lower value of factor of safety are adopted by ensuring the metal to be without any defect, which is done through *non-destructive tests. The factor of safety is of great importance in determining the reliability of the design.

6.17 TENSILE STRENGTH

The tensile strength of a metal is the value of load applied to break it off by the pulling it outwards into two pieces. Mathematically, tensile stress

$$= \frac{Maximum\ tensile\ load}{Original\ cross\text{-}sectional\ area}$$

The tensile stress is always expressed in N/mm^2 or MN/m^2. In actual practice, the given specimen is subjected to a tensile stress less than the working tensile stress.

6.18 COMPRESSIVE STRENGTH

The comprehensive strength of a metal is the value of load applied to break it off by crushing. Mathematically, compressive stress

$$= \frac{Maximum\ compressive\ load}{Original\ cross\text{-}sectional\ area}$$

The compressive stress is also expressed in N/mm^2 or MN/m^2. In actual practice, the given specimen is also subjected to a compressive stress less than the working compressive stress.

6.19 SHEAR STRENGTH

The shear strength of a metal is the value of load applied tangentially to shear it off across the resisting section. Mathematically shear stress

$$= \frac{Maximum\ tangential\ load}{Original\ cross\text{-}sectional\ area}$$

The shear stress is also expressed in N/mm^2 or MN/m^2. In actual practice, the given specimen is also subjected to shear stress less than the working shear stress.

6.20 BENDING STRENGTH

The bending strength of a metal is the value of load to break it off by the bending it across the resisting section. Mathematically bending stress

$$= \frac{Maximum\ bending\ load}{Original\ cross\text{-}sectional\ area}$$

The bending stress is also expressed in N/mm^2 or MN/m^2. In actual practice, the given specimen is also subjected to bending stress less than the working bending stress.

6.21 TORSIONAL STRENGTH

The torsional strengths of a metal is the value of load applied to break it off by twisting across the resisting section. Mathematically, torsional stress

---

* These tests are discussed in Art. 9.13.

$$= \frac{Maximum\ twisting\ load}{Original\ cross\text{-}sectional\ area}$$

The torsional stress is also expressed in N/mm^2 or MN/m^2. In actual practice, the given specimen is also subjected to torsional stress less than the working torsional stress.

6.22 TYPES OF TECHNOLOGICAL PROPERTIES

The technological properties of a metal are those properties, which completely define its behaviour in shaping, forming and fabrication operation during the manufacturing processes. A sound knowledge of these properties is also essential for an engineer to enable him in selecting a suitable metal for his various structures or various components of a machine. Though there are many technological properties of a metal, an engineer should know, yet the following are important from the subject point of view:

1. Malleability.   2. Machineability.   3. Weldability.   4. Castability.

6.23 MALLEABILITY

The term 'malleability' may be defined as the property of a metal by virtue of which it can be deformed into thin sheets by rolling or hammering without rupture. It depends upon the crystal structure of the metal. All the metals, having small grain size, are used for very thin sheets, whereas the metal having large gain size are used for thick sheets.

The following common metals have malleability in the decreasing order:

1. Gold   2. Silver   3. Aluminium   4. Copper   5. Tin
6. Platinum 7. Lead   8. Zinc   9. Iron   10. Nickel.

8.24 MACHINABILITY

The term 'machinability' may be defined as the property of a metal, which indicates the ease with which it can be cut or removed by cutting tools in various machining operations such as turning, drilling, boring, milling etc. The machinability of a metal depends upon mechanical and physical properties of a metal, chemical composition of the metal, microstructure of the metal and the cutting condition.

The following common ferrous and non-ferrous alloys have machinability in the ascending order:

1. White cast iron	Not machinable
2. Sintered carbides	
3. Monel metal	Very poor
4. High speed steel	
5. Wrought iron	Poor
6. Low alloy steel	
7. Low carbon steel	Fair
8. Copper-aluminium alloy	
9. Grey cast iron	Good
10. Gun metal	
11. Zinc alloys	
12. Aluminium	Excellent
13. Magnesium alloys	

6.25 FACTORS AFFECTING MACHINABILITY

Though there are innumerable factors, which affect the machinability of a metal, yet the following are important from the subject point of view:

1. Composition of the metal.
2. Hardness of the metal.
3. Grain size and microstructure of the metal.
4. Work-hardening characteristics of the metal.
5. Type and quality of machine used.
6. Size, shape and velocity of cut.
7. Quality and lubricant used during operation.
8. Kind and type of cutting tool used.
9. Coefficient of friction between chip and tool.
10. Shearing strength of the work piece.

6.26 IMPROVING MACHINABILITY

The machinability may be improved by adding certain elements in the parent metal, *e.g.*, various steels, copper and its alloys. Lead is usually added to copper, brass, bronze, nickel etc. to improve their machinability. Sometimes selenium and tellurinium are also added for improving the machinability of the parent metal. Sulphur is the most suitable element for steels. Manganese (as manganese sulphide) or oxysulphide is effective for free machining steels. Sometimes, phosphorus is also added upto 0.1% to improve machinability of steel. It will be interesting to know that the addition of these elements helps in lowering the mechanical properties, e.g., low hardness, low ductility and low tensile strength. They also act as lubricants. It has been observed that the grey cast iron possesses very good machinability. But machinability of white cast iron is very poor. Mild steel has better machinability than medium carbon steel, high carbon steel, high speed and most of the alloy steels.

6.27 WELDABILITY

The term 'weldability' may be defined as the property of a metal, which indicates the ease with which two similar or dissimilar metals are joined by fusion (with or without the application of pressure) and with or without the use of filler metal.

Strictly speaking, a metal has good weldability, if it can be easily welded in a fabricated structure. The various factors affecting weldability or a metal are:

1. Composition of the metal.
2. Brittleness and strength of the metal at elevated temperatures.
3. Thermal properties of the metal.
4. Welding techniques, fluxing material and filler material.
5. Proper heat treatment before and after the deposition of the metal.

The following common metals have weldability in the descending order:

1. Iron.   2. Carbon.   3. Steel.   4. Cast iron.   5. Low alloy steels.
6. Stainless steel.

6.28 FORMABILITY OR WORKABILITY

The term 'formability or workability' may be defined as the property of a metal which indicates the ease with which it can be formed (*i.e.*, pressed by forging) into different shapes and sizes. The various factors affecting the formability of a metal are:

Mechanical Properties of Metals

1. Crystal structure of the metal.
2. Grain size of the metal.
3. Hot and cold working.
4. Alloying element present in the parent metal.

It will be interesting to know that the metals with small grain size are suitable for shallow forming, while metals with relatively large grain size are suitable for heavy forming. The distortion of grains is an important factor which affects the formability of a metal, as the distortion of grains takes place due to hot working to a metal. The cold working of a metal results in less ductility and more distortion than the hot working.

It has been observed that the addition of some alloying elements reduces the ductility and hence formability of the parent metal. Low carbon steels possess good formability, which does not interfere with the slip planes. Medium carbon steels, on the other hand, are not suitable for cold forming.

6.29 CASTABILITY

The term 'castability' may be defined as the property of a metal, which indicates the ease with which it is casted into different shapes and sizes from its liquid state. The various factors affecting the castability of a metal are:

1. Solidification rate (fludity of metal).
2. Gas porosity.
3. Segregation.
4. Shrinkage.

6.30 FACTORS AFFECTING MECHANICAL PROPERTIES OF A METAL

In the previous articles, we have discussed the important mechanical properties of metals. Following are the important factors, which affect the mechanical properties of a metal:

1. *Grain Size.* We know that the metals are composed of crystal or grains. If the grain size of a metal is small, it is called a fine-grained metal. And if the grain size of a metal is large, it is called a coarse-grained metal. It has been observed that a fine-grained metal has a greater tensile and fatigue strength, whereas a coarse-grained metal has greater hardenability and forgeability. Moreover, it has better creep resistance. It is less tough and has a greater tendency to cause distortions.

2. *Temperature.* The temperature of a metal greatly affects its mechanical properties. It has been observed that with the decrease in temperature, the tensile and yield strength of a metal increases. But the toughness and ductility decreases. In some of the metals such as copper, nickel, aluminium and austenitie steel, the toughness and ductility show a gradual decrease in their values with the decrease in temperature. In some other metals such as iron, there is a sudden decrease in these properties. The creep resistance of a metal also increases with the decrease in temperature.

On the other hand, increase of temperature decreases tensile and yield strengths. But the toughness and ductility increases.

3. *Heat treatment.* The heat treatment is an operation, or a combination or operations, involving heating and cooling of a metal or an alloy in its solid state. It is done in order to obtain certain desirable properties or conditions in a metal.

It has been observed that the heat treatment of a metal increases tensile, strengths, hardness, ductility and shock resistance etc. It improves the machinability of a metal.

4. *Atmospheric exposure.* Sometimes, a metal is exposed to a normal atmosphere containing wet air or to an industrial atmosphere for a considerable period. In such a case, a thin oxide film is formed on the surface of the metal. This oxide film breaks down due to crack or discontinuity in it. As a result of this, the condensed moisture on the surface of the metal absorbs carbon dioxide, sulphur dioxide, hydrogen sulphide etc. From the atmosphere and produceds dilute acids. These acids act as electrolyte. This results in the formation of electro-chemical cell in which electron current starts flowing from the exposed metal through the electrolyte to the other regions of the metal which are still protected. It has been observed that the atmospheric exposure of a metal decreases the tensile and fatigue strengths of a metal.

A.M.I.E. (I) EXAMINATION QUESTIONS

1. (a) Explain the term machinability.
   (b) Describe the method of improving machinability. *(Summer, 1993)*
2. Define ductility *(Winter, 1993)*
3. Which element is responsible for the high machinability of free cutting steels? *(Summer, 1994)*
4. Discuss the factors that affect the hardenability of steels. *(Winter, 1994)*
5. Differentiate between hardness and hardenability. *(Summer, 1995)*
6. What is meant by the term 'hardenability'? Describe how hardenability of steel can be estimated. *(Winter, 1996)*
7. What is compressive strength of a material? *(Summer, 1997)*
8. Differentiate between hardness and hardenability. How do you measure hardenability? *(Winter, 1999)*

MULTIPLE CHOICE QUESTIONS

1. Ductility is measured in terms of
   (a) Ultimate tensile strength
   (b) Percentage elongation
   (c) Modulus of toughness
   (d) Modulus of resilience *(Summer, 2000)*

ANSWERS

1. (b)

7

Mechanical Tests of Metals

1. Introduction. 2. Stress. 3. Strain. 4. Types of Stresses. 5. Hook's Law and Modulus of Material. 6. Poisson's Ratio. 7. Types of Mechanical Tests. 8. Tensile Test. 9. Stress-strain Curve for a low Carbon Steel (or Mild Steel). 10. Stress-strain Diagram for Different Metals. 11. Compressive Test. 12. Impact Test. 13. Charpy Test. 14. Izod Test. 15. Factors Affecting Impact Strength. 16. Hardness Test. 17. Brinell's Hardness Test. 18. Rockwell's Hardness Test. 19. Vicker's Hardness Test. 20. Knoop's Hardness Test. 21. Fatigue Test. 22 Creep Test. 23. Creep Curve. 24. Factors Affecting Creep Resistance.

7.1 INTRODUCTION

In the last chapter, we have discussed some of the important mechanical properties of metals. Almost all the mechanical properties of metal are established by conducting tests on various testing machines. In India, these tests are standardised by Bureau of Indian Standards [previously called Indian Standard Institute (ISI)]. Similarly in U.S.A., these tests are standardised by the American Society for Testing and Materials (ASTM).

7.2 STRESS

It is the resistance set up by the molecules of a material to the deformation due to the application of external force. Mathematically,

Stress, $$\sigma = \frac{p}{A_0}$$

where, p = Load or force acting on the body, and
  A_0 = Cross-sectional area of the body.

7.3 STRAIN

It is the deformation per unit length of the body. It is also called axial strain, linear strain or longitudinal strain. Mathematically

Strain, $$\varepsilon = \frac{l - l_0}{L_0} = \frac{\delta l}{l_0}$$

where, l_0 = Original length.
  l = Final length, and
  δl = Change in length of the body

7.4 TYPES OF STRESSES

Though there are many types of stresses, yet the following are important from the subject point of view:

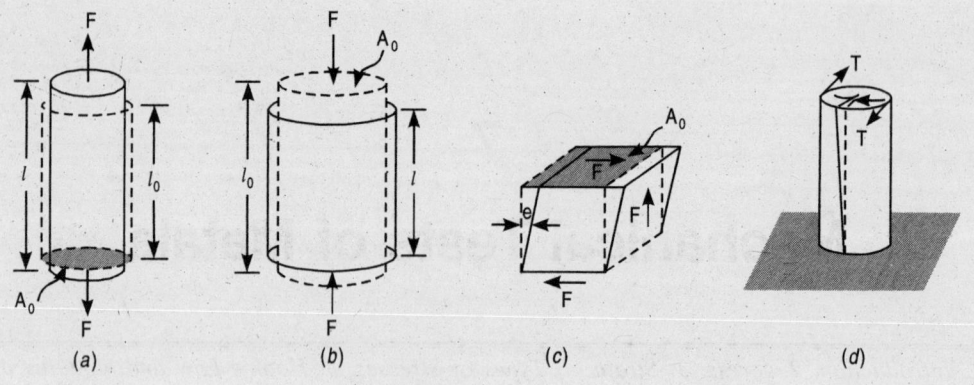

Fig. 7.1.

1. *Tensile stress.* When a section of any ductile material is subjected to two equal and opposite pulls, as a result of which the body tends increase its length as shown in Fig. 7.1 (*a*), the stress induced is called tensile stress. The corresponding strain is called tensile strain.
2. *Compressive stress.* When a section of any ductile material is subjected to two equal and opposite pushes, as a result of which the body tends to decrease its length, as shown in Fig. 7.1 (*b*), the stress induced is called compressive stress. The corresponding strain is called compressive strain.
3. *Shear stress.* When a section of any ductile material is subjected to two equal and opposite forces, acting tangentially across the resisting section, as a result of which the body tends to shear off across the section as shown in Fig. 7.1. (*c*), the stress induced is called shear stress. The corresponding strain is called shear strain its value is equal to tan θ. In engineering practice many loads are torsional rather than pure shear. This type of loading is illustrated in Fig. 7.1(*d*).

7.5 HOOK'S LAW AND MODULUS OF MATERIAL

The Hook's law states "when a material is loaded within its elastic limit, the stress is proportional to strain'. Mathematically,

$$\frac{Stress}{Strain} = E = a \text{ constant}$$

It may be noted that Hook's law holds good for tension as well as compression.

The proportionality constant (E) is called modulus of elasticity of material. It is different for different materials and different types of stresses, *e.g.*, for tensile and compressive stresses, it is called modulus of elasticity or Young's modulus (E). For shear stresses, the modulus of rigidity (G) and for volumetric distortion, bulk modulus (K). It is expressed in N/mm^2 MN/m^2 or GPa (*i.e.*, giga pascals)

Table 7.1 Shows the values of Young's modulus for some typical metals, ceramics and polymers.

Table 7.1

Material	Young's Modulus (Y)
1. Magnesium	45 GPa
2. Tungsten	407 GPa
3. Ceramics	70 to 500 GPa
4. Polymers	0.007 to 4 GPa

Mechanical Tests of Metals

As seen from the table, we find that the magnitude of Young's modulus for most of the metals lies in the range of 45 GPa to 407 GPa. On the other hand, the modulus of elasticity is slightly higher (70 to 500 GPa) than that for metals. Whereas for polymers, the modulus value is smaller (0.007 to 4 GPa) than that of metals and ceramics.

Deformation of a material in which stress and strain are **proportional** is called elastic deformation. For a elastic material, a plot between the stress versus strain results in a linear relationship. This situation is shown is Fig. 7.2.

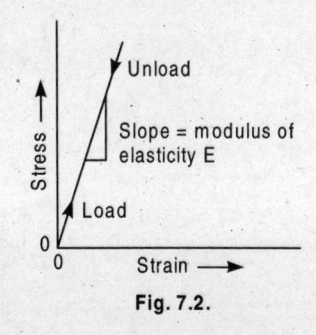

Fig. 7.2.

As seen from this diagram, the slope of the line segment corresponds to modulus of elasticity, E. This modulus may be thought of as stiffness, or a material's resistance to elastic deformation. As a matter of fact, the greater the modulus, the stiffer the material, or the smaller the elastic strain that results from the application of a given stress. The modulus of elasticity is an important design parameter for material engineers. It is used for computing elastic deflections.

It may be carefully noted that elastic deformation is *non-permanent*. This means that when the applied load is released, the piece returns to its original shape. This situation is indicated in Fig. 7.2 by an arrow marked upwards. Thus application of the load corresponds to moving from the origin up and along the straight line. Upon release of the load, the line is traversed in the direction back to the origin. This situation is shown by an arrow marked downwards in Fig. 7.2.

There are some materials (like gray cast iron, concrete and polymers) for which this elastic portion of the stress versus strain curve is not linear. Refer to Fig. 7.3 below. Hence, it is not possible to determine a modulus of elasticity for such materials with non-linear behaviour.

In order to determine, the modulus of elasticity for non-linear behaviour, there are two methods to do it:

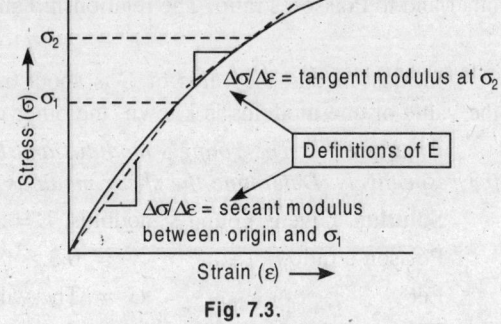

Fig. 7.3.

1. Find the slope of the stress versus strain curve at some specific level of stress (say σ_2 as indicated in Fig. 7.3.). The modulus of elasticity determined in this way is called tangent modulus.
2. Find the slope of a line drawn from the origin to some given point (say σ_1). The modulus of elasticity determined in this way is called secant modulus.

7.6 POISSON'S RATIO

When a tensile stress is imposed on virtually all materials, an elastic elongation and accompanying strain result in the direction of applied stress as shown in Fig. 7.4. Notice that the stress (σ_z) is applied along the Z-direction and hence the elongation is also produced in the same direction. The elongation is indicated by Δl_z.

It may be carefully noted that as a result of elongation in Z-direction, there will be constructions in the lateral (*i.e.* x and y directions perpendicular to the applied stress. From these contractions, we can determine the compressive strains along x-axis (ε_x) and along y-axis (ε_y). If the applied stress is uniaxial (*i.e.* in one direction only as in the present case shown in Fig. 7.4) and the material is isotropic, then $\varepsilon_x = \varepsilon_y$.

Fig. 7.4.

$$\nu = -\frac{\varepsilon_x}{\varepsilon_z} = -\frac{\varepsilon_y}{\varepsilon_z}$$

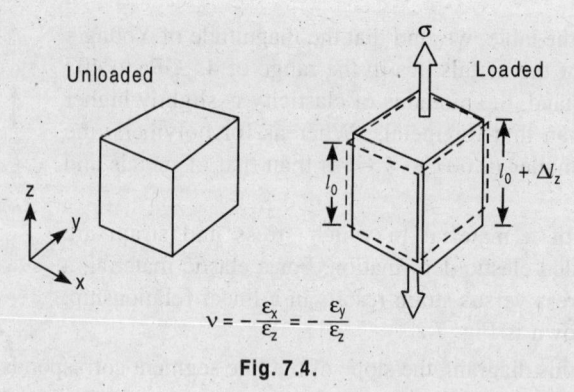

A parameter called Poisson's ratio is defined as the ratio of the lateral and axial strains. Mathematically, the Poisson's ratio.

$$\nu = -\frac{\varepsilon_x}{\varepsilon_z} \text{ or } -\frac{\varepsilon_y}{\varepsilon_z}$$

The negative sign is included in the expression so that Poisson's ratio (ν) is always be positive, since ε_x and ε_z will always be of opposite sign. For many metals and other alloys, values of Poisson's ratio range between 0.25 and 0.35.

For isotropic materials, shear modulus (G) and Young's modulus (E) are related to each other and to Poisson's ratio. The relationship among these three qualities is given by the equation,

$$E = 2G(1 + \nu)$$

In most metals, the value of G is about 0.4 E. It is evident from the above equation, that if the value of one modulus is known, the other may be calculated.

Example 7.1. *The Young's modulus and Poisson's ratio of a material are 210 GN/m^2 and 0.3 respectively. Determine the shear modulus of the material.* (A.M.I.E. Summer, 1995)

Solution. Given: Young's modulus, $E = 210$ GN/m^2;

Poisson's ratio, $\nu = 0.3$.

Let G = The value of shear modulus

We know that Young's modulus of elasticity (E)

$$210 = 2G(1 + \nu)$$
$$= 2G(1 + 0.3)$$

$$\therefore \quad G = \frac{210}{2(1+0.3)} = 80.77 \text{ GN/}m^2 \text{ **Ans.**}$$

7.7 TYPES OF MECHANICAL TESTS

Though there are a number of mechanical tests, which are carried out to determine the suitability of a metal, yet the following are important from the subject point of view:

1. Tensile test. 2. Compressive test. 3. Impact test. 4. Hardness test. 5. Fatigue test. 6. Creep test.

7.8 TENSILE TEST

The tensile test of a metal is generally performed to determine:
1. Proportional and elastic limit.
2. Yield point.

Mechanical Tests of Metals

3. Ultimate tensile strength.
4. Percentage elongation and reduction of area.

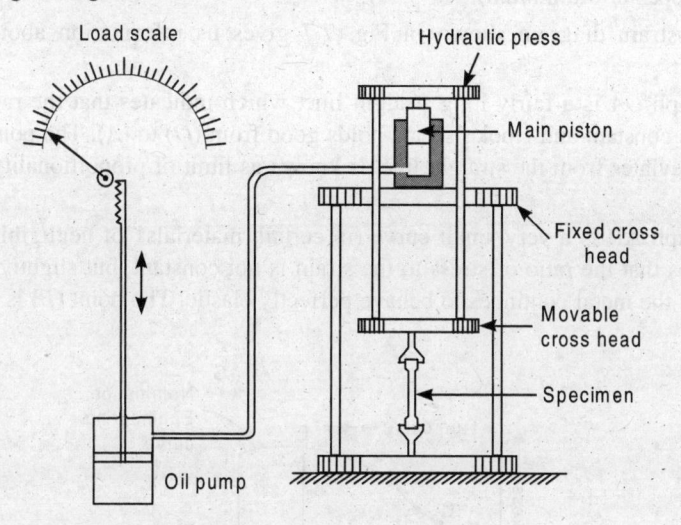

Fig. 7.5. Universal testing machine.

The results obtained by the tensile test are widely used in the design of materials for structures and other purposes. In this test, the test piece is pulled out at a constant rate by gradually increasing the axial pull, till the rupture takes place.

The tensile test for a ductile material is, generally, carried out with the help of a universal testing machine on the specimen made from the material to be tested.

The schematic working arrangement of a universal testing machine is shown in Fig. 7.5. The specimen is held in the jaws of the machine. And the load is applied gradually by a hydraulic press, which is measured from the pressure developed inside the cylinder. The function of the oil pump is to supply oil under pressure to the hydraulic cylinder. The load reading is noted directly from the load scale.

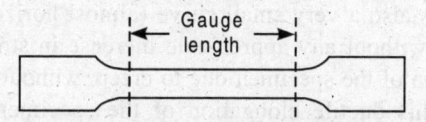

Fig. 7.6. Specimen for tensile test.

The dimension and form of the specimen varies according to the size and shape of the material to be tested and the main objective in view. The test is carried out on a specimen having uniform cross-section throughout the gauge length as shown in Fig. (7.6).

The Bureau of Indian Standards (previously Indian Standards Institute) the recommended several sizes of the specimen. But the standard practice is to use a specimen whose gauge length in mm,

$$l_0 = 5.65 \sqrt{A_0}$$
$$= 5.0 \, d_0$$

where, A_0 = Area of cross-section in mm^2, and

d_0 = Diameter of the specimen in mm

It means for a gauge length of 50 mm, the specimen diameter should be 10 mm.

The tensile load is gradually increased and the corresponding extensions are recorded. Now a graph is plotted with stresses along the vertical axis and the corresponding strains along the horizontal axis. If we draw a curve passing through the vicinity of all such points, we shall obtain a curve as shown in Fig. 7.7. Such a graph is known as stress-strain diagram for a given ductile material (say copper or aluminium).

The stress-strain diagram shown in Fig. 7.7 gives us information about the following important points:

1. The graph OA is a fairly long straight line, which indicates that the ratio of a stress to strain is constant and Hooke's Law holds good from (O) to (A). The point (A), where the curve deviates from the straight line, is known as limit of proportionality or proportional limit.

2. The graph AB is a very small curve (in certain materials* of negligible length), which indicates that the ratio of stress to the strain is not constant, but slightly changes. In this portion, the metal continues to behave perfectly elastic. The point (B) is known as elastic point.

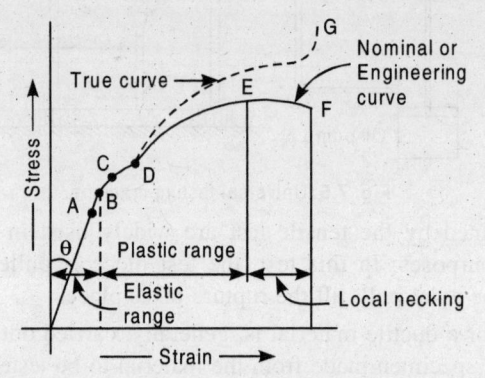

Fig. 7.7. Stress-strain curve for a ductile material.

3. The graph BC is another very small curve, which indicates that the strain increases more quickly than the stress (in comparison to OA or AB). The point (C) is called** yield point. It may be noted that if the load on the specimen is removed, then the elongation from (B) to (C) will not disappear. But it will remain as a permanent set.

4. The graph CD is also a very small curve (almost horizontal), which indicates that the strain increases without any appreciable increase in stress. This happens as there is a sudden elongation of the specimen due to creep, without any appreciable increase in the stress. During this ductile elongation of the specimen, its cross-sectional area gets reduced uniformly in proportion to its length.

5. The graph DE is an upward curve, which indicates that the specimen regains some strength, and higher values of stresses are required for higher strains. The graph rises up to the maximum limit indicated by the point (E). The stress, corresponding to the point (E), is known as ultimate tensile stress or tenacity, which is a measure of tensile srength of a material. The work done while stretching the specimen is transferred, largely, into heat and the specimen becomes hot.

6. The graph EF is a downward curve, which indicates that a neck is formed, which decreases the cross-sectional area of the specimen, Now it requires lesser load to continue extension till fracture takes place at F.

---

* At this point the elongation of a different materials is different. But in case of a mild steel specimen, the elongation is about 2%.
** In wrought Iron, the point A and B, are practically, the same point. But in rolled aluminium, these points exist separately.

Mechanical Tests of Metals

A little consideration will show, that less stress is necessary to break away the specimen. The stress corresponding to the point (F) is known as breaking stress. The breaking stress, *i.e.*, the stress at point F is less than that at point E. This statement appears to be somewhat misleading. But it is true because the formation of a neck takes place at point E, which reduces the cros-ssectional area of the specimen. It causes the specimen to fail suddenly at point F. If for each value of strain, between the points D and F, the tensile load is divided by the reduced cross-sectional area at the narrowest part of the neck, then the stress-strain curve will follow the line DG. The stress-strain curve shown by the dotted line DG is called true curve. Similarly the curve shown by the line DF is called the nominal curve or engineering curve. In actual practice, it is the engineering stress-strain curve, which is used for all applications.

As a matter of fact, the greater value of angle θ will represent the material to be more elastic. The higher position of yield point (C) will represent the material to be more hard. The higher position of the point (E) will represent the material to be more strong. Similarly, the position of the breaking load (point F) will represent the brittleness or toughness of the material.

It may be noted that the stress-strain curve gives a valuable information about mechanical properties of a metal. The following terms are important in the tensile test of a specimen:

1. *Elastic stress.* The elastic load (corresponding to point B) divided by the original cross-sectional area of the specimen is known as elastic stress. Mathematically the elastic stress,

$$\sigma_0 = \frac{\text{Elastic load}}{\text{Original cross-sectional area}} = \frac{p_e}{A_0}$$

The elastic stress is expressed in N/mm^2 or MN/m^2 etc.

2. *Working Stress.* It is the stress, which is used by the engineers for designing their structures or machine components. The working stress is kept much below the elastic stress, and is also known as safe stress.

3. *Yield stress.* The yield load (corresponding to point C) divided by the original cross-sectional area of the specimen is known as yield stress. Mathematically, the yield stress,

$$\sigma_s = \frac{\text{Yield load}}{\text{Original cross-sectional area}} = \frac{p_s}{A_0}$$

The yield stress is also expressed in N/mm^2 or MN/in^2

4. *Ultimate stress.* The ultimate load (corresponding to point E) divided by the original cross-sectional area of the specimen is known as ultimate stress. Mathematically the ultimate stress,

$$\sigma_u = \frac{\text{Ultimate load}}{\text{Original cross-sectional area}} = \frac{P_u}{A_o}$$

The ultimate stress is also called ultimate tensile stress, the maximum stress or tenacity. It is expressed in N/mm^2 or MN/m^2.

5. *Factor of safety.* The ratio of ultimate stress to the working stress is known as factor of safety. Mathematically the factory of safety,

$$= \frac{\text{Ultimate stress}}{\text{Working stress}}$$

These days, the value of factor of safety for structural steel is taken as 2 or 2.5. But in the case of cast iron, concrete, wood etc., the value of factor of safety is taken as 4 to 5.

6. *True Stress.* The load at any instant (called instantaneous load) divided by the cross-sectional area of the specimen at that instant (called instantaneous cross-sectional area of the specimen) is known as true stress. Mathematically the true stress,

$$\sigma_t = \frac{\text{Instantaneous load}}{\text{Instantaneous cross-sectional area}} = \frac{p_i}{A_i}$$

The true stress is expressed in N/mm² or MN/m²

7. *Nominal Stress.* The load at any instant (called instantaneous load) divided by the original cross-sectional area of the specimen is known as nominal stress. Mathematically the nominal stress,

$$\sigma_n = \frac{\text{Instantaneous load}}{\text{Original cross-sectional area}} = \frac{p_i}{A_i}$$

The nominal stress is expressed in N/mm² or MN/m². The value of nominal stress, in tension, is always smaller than the true stress. It is because of the fact that the instantaneous (or actual) cross-sectional area of the specimen becomes less than the original cross-sectional area on the application of load. It may be noted that it is the nominal stress, which is used in most of the engineering calculations for design purposes. Therefore nominal stress is also called engineering stress.

8. *True breaking Stress.* The breaking load (corresponding to point F) divided by the final cross-sectional area of the specimen is known as true breaking stress. Mathematically the true breaking stress

$$= \frac{\text{Breaking load}}{\text{Final cross-sectional area}} = \frac{P_b}{A_b}$$

The true breaking stress is expressed in N/mm² or MN/m².

9. *Nominal breaking stress.* The breaking load (corresponding to point f) divided by the original cross-sectional area of the specimen is known as true breaking stress. Mathematically the nominal breaking stress.

$$= \frac{\text{Breaking load}}{\text{Original cross-sectional area}} = \frac{P_f}{A_0}$$

10. *Percentage elongation.* The elongation or increase in length divided by the original length of the specimen, expressed as per cent, is known as percentage elongation. Mathematically the percentage elongation

$$= \frac{\text{Increase in length of the specimen}}{\text{Original in length of the specimen}} \times 100$$

Let, l = Length of the specimen at any load P, and
l_0 = Original length of the specimen.

∴ Increase in length

$$\Delta l = l - l_0$$

and percentage elongation

$$= \frac{\Delta l}{l_0} \times 100 = \frac{l - l_0}{l_0} \times 100$$

11. *Percentage reduction of area.* The reduction is cross-sectional area divided by the original cross-sectional area of the specimen, expressed as percent, is known as percentage reduction of area. Mathematically the percentage reduction of area

$$= \frac{\text{Reduction in cross-sectional area}}{\text{Original cross-sectional area}} \times 100$$

Mechanical Tests of Metals

Let A = Cross-sectional area of the specimen at any load P,
A_0 = Original cross-sectional area of the specimen,
d = Diameter of the specimen at the load P, and
d_0 = Original diameter of the specimen.

∴ Reduction in cross-sectional area,

$$\Delta A = A_0 - A = \frac{\pi d_0^2}{4} - \frac{\pi d^2}{4}$$

and percentage reduction of area

$$= \frac{A_0 - A}{A_0} \times 100 = \frac{\frac{\pi d_0^2}{4} - \frac{\pi d^2}{4}}{\frac{\pi d_0^2}{4}} \times 100$$

$$= \frac{d_0^2 - d^2}{d_0^2} \times 100 = \left[1 - \left(\frac{d}{d_0}\right)^2\right] \times 100$$

It is evident from the above relation that the percentage reduction of area depends upon the original diameter and diameter of the specimen at any instant. In other words, it does not depend upon the gauge length.

12. *Uniform Elongation.* The increase in length of the specimen, when it loaded up to the limit of proportionality (corresponding to point A) is uniform throughout its gauge length. However, as the applied load is increased, beyond the limit of proportionality, the area of cross-section starts reducing due to the permanent deformation of material. Therefore the elongation produced in the specimen is no longer uniform beyond the limit of proportionality.

13. *Total Elongation.* The total increase in length of the specimen measured immediately after its fracture, is known as total elongation. It may be noted that elongation of the specimen after fracture is not uniform throughout its gauge length. It is because of the fact that elongation in the region of neck is more than that in the remaining portion of the specimen.

14. *True Strain.* It is also called engineering strain. The change in length with reference to the instantaneous gauge length (rather than that of original length is called true strain.

True strain (or engineering strain),

$$\varepsilon_x = \frac{l - l_0}{L} = \int_{L_0}^{L} \frac{dl}{l} = \ln \frac{l}{l_0}$$

where $dl = l - l_0$

Since the value of the specimen remains constant,

∴ $$\frac{\pi}{4} d_0^2 l_0 = \frac{\pi}{4} d^2 l$$

where d_0 = original diameter and
d = diameter under the existing load,

From equation (*i*), we get,

$$\frac{l}{l_0} = \left(\frac{d_0}{d}\right)^2$$

∴ True strain (or engineering strain),

$$\varepsilon_x = ln\frac{l}{l_0} = ln\left(\frac{d_0}{d}\right)^2 = 2\,ln\left(\frac{d_0}{d}\right)$$

Notes: 1. For ductile materials, modulus of toughness is given by the equation:

$$= \left\{\frac{\text{yield strength + ultimate tensile strength}}{2}\right\} \times \text{strain at fracture}$$

$$= \left(\frac{\sigma_y + \sigma_u}{2}\right) \times \varepsilon_f$$

$$= \left(\frac{\sigma_y + \sigma_u}{2}\right) \times \left(\frac{l - l_0}{l_0}\right)$$

2. For brittle materials like cast iron, concrete etc, the modulus of toughness is determined by multiplying two-thirds of ultimate strength by the strain at fracture, *i.e.*,

$$\text{Modulus of toughness} = \frac{2}{3}\sigma_u \times \varepsilon_f$$

3. For a polycrystalline material, the variation of yield strength with grain size is given by the equation:

$$\sigma_y = \sigma_0 + \frac{k_y}{\sqrt{d}} \qquad \ldots (i)$$

where d = average grain diameter (in mm) and,
σ_0, k_y = constants for a given material.

Example 7.2. *The following observations were made during a tensile test on a mild steel specimen 40 mm in diameter and 200 mm long.*

Elongation with 40 kN load (within limit of proportionality),

$$\delta l = 0.0304 \text{ mm}.$$

Yield Point load = 161 kN ; Maximum load = 242 kN

Length of specimen at fracture = 249 mm.

Determine (a) Young's modulus of elasticity, (b) Yield point stress, (c) Ultimate stress and (d) Percentage elongation.

Solution. Given: Original length of the specimen, $L = 200$ mm $= 200 \times 10^{-3}$ m; Diameter, $d = 40$ mm $= 40 \times 10^{-3}$ m. Load W $= 40$ kN $= 40 \times 10^3\,N$; Elongation $\delta l = 0.0304$ mm $= 3.04 \times 10^{-5}$ m

Yield point load $= 161\,kN = 161 \times 10^3\,N$; Maximum load $= 242\,kN = 242 \times 10^3\,N$; Length of specimen at fracture (L) $= 249$ mm $= 249 \times 10^{-3}$ m.

(a) Young's modulus of elasticity

We know that area of cross-section of a specimen,

$$A = \frac{\pi d^2}{4} = \frac{\pi \times (40 \times 10^{-3})^2}{4} = 1.257 \times 10^{-3} \text{ m}^2$$

and the stress,

$$\sigma = \frac{P_e}{A} = \frac{40 \times 10^3}{1.257 \times 10^{-3}} = 3.18 \times 10^7 \text{ N/m}^2$$

We also know that strain,

$$\varepsilon = \frac{\Delta L}{L} = \frac{3.04 \times 10^{-5}}{200 \times 10^{-3}} = 1.52 \times 10^{-4}$$

Mechanical Tests of Metals

∴ Young's modulus,

$$E = \frac{\text{Stress}}{\text{Strain}} = \frac{\sigma}{\varepsilon} = \frac{3.18 \times 10^7}{1.52 \times 10^{-4}} = 2.09 \times 10^{11} \text{ N/m}^2 \text{ **Ans.**}$$

(b) *Yield Point Stress*

We know that yield point stress,

$$= \frac{\text{Yield point load}}{\text{Area}} = \frac{161 \times 10^3}{1.257 \times 10^{-3}}$$
$$= 12.8 \times 10^7 \text{ N/m}^2 \text{ **Ans.**}$$

(c) *Ultimate Stress*

We know that ultimate stress,

$$= \frac{\text{Maximum load}}{\text{Area}} = \frac{242 \times 10^3}{1.257 \times 10^{-3}}$$
$$= 19.26 \times 10^7 \text{ N/m}^2 \text{ **Ans.**}$$

(d) *Percentage elongation*

We know that percentage elongation,

$$= \frac{l - l_0}{l_0} \times 100 = \frac{(249 \times 10^{-3}) - (240 \times 10^{-3})}{(240 \times 10^{-3})} \times 100$$
$$= 0.245\% \text{ **Ans.**}$$

Example 7.3. *In order to evaluate various mechanical properties, a steel specimen of 12.5 mm diameter and 62.5 mm gauge was tested in a standard tension test. Following observations were made during the test:*

Yield load = 40 kN   Maximum load = 71.5 kN

Fracture load = 50.5 kN   Gauge length of fracture = 79.5 mm.

Strain at load of 20 kN = 7.75 × 10⁻⁴

Determine (a) Yield point stress (b) ultimate tensile, strength (c) percentage elongation (d) modulus of elasticity (v) modulus of resilience (vi) fracture stress (vii) percentage reduction in area (viii) modulus of toughness.

Solution. Given: Specimen diameter = 12.5 mm = 12.5 × 10⁻³ ;

Specimen length = 62.5 mm = 62.5 × 10⁻³ mm; Yield load = 40 kN = 40 × 10⁻³ N;

Maximum load = 71.5 × 10³ N; Fracture load = 50.5 kN = 50.5 × 10³ N;

Gauge length at fracture = 79.5 mm = 79.5 × 10⁻³ m;

Strain at load of 20 kN = 7.75 × 10⁻⁴

(a) *Yield point stress*

We know that area of cross-section,

$$A = \frac{\pi d^2}{4} = \frac{\pi \times (12.5 \times 10^{-3})^2}{4} = 1.227 \times 10^{-4} \text{ m}^2$$

and yield point stress,

$$\sigma_y = \frac{\text{Load at the lower yield point}}{\text{Original area}}$$

$$= \frac{40 \times 10^3}{1.227 \times 10^{-4}} = 325.95 \times 10^6 \text{ N/m}^2 \text{ **Ans.**}$$

(b) Ultimate tensile strength

We know that ultimate tensile strength,

$$\sigma_u = \frac{\text{Maximum load}}{\text{Original area}} = \frac{(71.5 \times 10^3)}{1.227 \times 10^{-4}}$$
$$= 582.6 \text{ N} \times 10^6 \text{ N/m}^2 \textbf{ Ans.}$$

(c) Percentage elongation

We know that percentage elongation,

$$= \frac{l - l_0}{l_0} \times 100 = \frac{(79.5 \times 10^{-3}) - (62.5 \times 10^{-3})}{(62.5 \times 10^{-3})}$$
$$= 27.2\% \textbf{ Ans.}$$

(d) Modulus of elasticity

We know that stress at 20 kN,

$$\sigma = \frac{20 \times 10^3}{1.227 \times 10^{-4}} = 162.97 \times 10^6 \text{ N/m}^2$$

and strain at 20 kN,

$$\varepsilon = 7.75 \times 10^{-4}$$

∴ Modulus of elasticity,

$$E = \frac{\sigma}{\varepsilon} = \frac{162.97 \times 10^6}{7.75 \times 10^{-4}} = 2.1 \times 10^{11} \text{ N/m}^2 \textbf{ Ans.}$$

(e) Modulus of resilience

We know that modulus of resilience,

$$= \frac{\sigma_y^2}{2E} = \frac{(325.95 \times 10^3)^2}{2 \times 2.1 \times 10^{11}} = 0.252 \textbf{ Ans.}$$

(f) Fracture stress

We know that fracture stress,

$$= \frac{\text{Fracture load}}{\text{Original area}} = \frac{50.5 \times 10^3}{1.227 \times 10^{-4}}$$
$$= 411.5 \times 10^6 \text{ N/m}^2 \textbf{ Ans.}$$

(g) Modulus of toughness

We know that modulus of toughness for steel,

$$= \frac{\sigma_y + \sigma_u}{2} \times \varepsilon_f = \frac{\sigma_y + \sigma_u}{2} \times \left(\frac{l - l_0}{l_0}\right)$$

$$= \left\{\frac{(325.95 \times 10^6) + (582.6 \times 10^6)}{2}\right\} \times \left\{\frac{(79.5 \times 10^{-3}) - (62.5 \times 10^{-3})}{62.5 \times 10^{-3}}\right\}$$

$$= 123.56 \times 10^6 \text{ N/m}^2 \textbf{ Ans.}$$

7.9 STRESS-STRAIN CURVE FOR A LOW CARBON STEEL (OR MILD STEEL)

Fig. 7.8 shows the stress-strain curve for a low carbon steel or mild steel. This curve is almost the same as that of a ductile material shown in Fig. 7.8, except one difference that it has

Mechanical Tests of Metals

two yield points *C* and *D* as indicated in the figure. The point *C* is called the upper yield-point and the point *D* as the lower yield point. The stresses, corresponding to these points, are the upper-yield point stress and lower yield-point stress respectively. The value of lower yield-point stress is about half of the tensile strength.

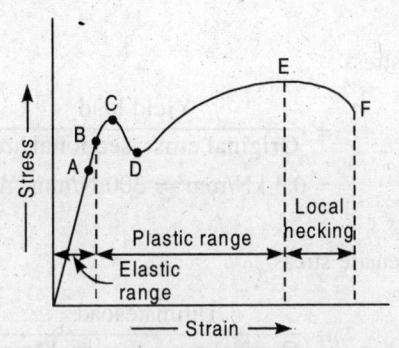

Fig. 7.8. Stress-strain curve for a low carbon steel or mild steel.

The behaviour of a low carbon steel (or mild steel) indicates that near the elastic limit, there is a sudden yield and fall-off of load. The material continues to deform at a lower load until, work hardening sets in.

The explanation for remaining part of the stress-strain curve for the low carbon steel (or mild steel) is the same as discussed in the last article.

Example 7.4. *A steel bar 12.7 mm diameter breaks with a load of 14 kN. Its final diameter is 7.87 mm. What is (i) true breaking stress, (ii) nominal breaking stress?*

Solution Given: Original dia. 12.7 mm or original cross-sectional area = $\frac{\pi}{4}(12.7)^2 = 126.7 \text{mm}^2$;

Breaking load = 14 kN; Final dia. = 7.87 mm or final cross-sectional area = $\frac{\pi}{4}(7.87)^2 = 48.65 \text{mm}^2$

(i) True breaking stress

We know that true breaking stress

$$= \frac{\text{Breaking load}}{\text{Final cross-sectional area}} = \frac{14}{48.65} \text{ kN/mm}^2$$

$= 28.78 \text{ kN/mm}^2 = 288 \text{ N/mm}^2$ **Ans.**

(ii) Nominal breaking stress

We also know that nominal breaking stress

$$= \frac{\text{Breaking load}}{\text{Original cross-sectional area}} = \frac{14}{126.7} \text{ kN/mm}^2$$

$= 0.11 \text{ kN/mm}^2 = 110 \text{ N/mm}^2$ **Ans.**

Example 7.5. *A mild steel rod of 12 mm diameter was tested for tensile strength, with a gauge length of 60 mm. Following were the observations:*

Final length = 78 mm

Final diameter = 7 mm

Yield load = 34 kN

Ultimate load = 61 kN

Calculate (a) yield stress, (b) ultimate tensile stress (c) percentage reduction, and (d) percentage elongation.

Solution. Given Original dia. = 12 mm or original cross-sectional area = $\frac{\pi}{4}(12)^2$ = 113.1 mm²; Original length = 60 mm; Final length = 78 mm; Final dia. = 7mm or final cross-sectional area = $\frac{\pi}{4}(7)^2$ = 38.5 mm²; Yield load = 34 kN; Ultimate load = 61 kN.

(a) Yield stress

We know that the yield stress

$$= \frac{\text{Yield load}}{\text{Original cross-sectional area}} = \frac{34}{113.1} \text{ kN/mm}^2$$

$$= 0.3 \text{ kN/mm}^2 = 300 \text{ N/mm}^2 \text{ \textbf{Ans.}}$$

(b) Ultimate tensile stress

We know that ultimate tensile stress

$$= \frac{\text{Ultimate load}}{\text{Original cross-sectional area}} = \frac{61}{113.1} \text{ kN/mm}^2$$

$$= 0.54 \text{ kN/mm}^2 = 540 \text{ N/mm}^2 \text{ \textbf{Ans.}}$$

(c) Percentage reduction

We know that percentage reduction in diameter.

$$= \frac{\text{Original area} - \text{Final area}}{\text{Original cross-sectional area}} \times 100$$

$$= \frac{113.1 - 38.5}{113.1} \times 100 = 66\% \text{ \textbf{Ans.}}$$

(d) Percentage elongation

We know that percentage elongation in length

$$= \frac{\text{Final length} - \text{Original length}}{\text{Original length}} \times 100$$

$$= \frac{78-60}{60} \times 100 = 30\% \text{ \textbf{Ans.}}$$

Example 7.6. *A specimen of copper having a rectangular cross-section of 15.2 mm × 19.1 mm is pulled in tension with 44.5 kN force, producing an elastic elongation. Calculate the resulting engineering strain. Given Young's modulus for copper is 110 GPa.*

(A.M.I.E., Summer, 1998)

Solution. Given: Area of cross-section of a specimen, A = 15.2 mm × 19.1 mm ;

Force ρ(or σ) = 44.5 × 10³ N ;

Young's modulus, E = 110 GP = 1.1 × 10⁵ N/mm².

We know that strain,

$$\varepsilon = \frac{\sigma}{AE} = \frac{44.5 \times 10^3}{(15.2 \times 19.1) \times (1.1 \times 10^5)} = 1.39 \times 10^{-3} \text{ mm \textbf{Ans.}}$$

Example 7.7. *The engineering stress and strain at fracture were found to be 450 MPa and 0.63 respectively. Determine true stress and true strain.* (A.M.I.E.; Winter, 2003)

Solution. Given: Engineering stress = 450 MPa and strain = 0.63

We know that true stress,
$$\sigma_t = \sigma(1+\varepsilon) = 450(1+0.63) = 733.5 \text{ MPa } \textbf{Ans.}$$
and the true strain,
$$\varepsilon_t = \ln(1+\varepsilon) = \ln(1+0.63) = 0.489 \textbf{ Ans.}$$

Example 7.8. *An aluminium bar of 24 mm × 30 mm cross-section is under a load of 7000 kg and a steel bar of diameter 10 mm is under a load of 5000 kg. Which part has a greater stress?*

(A.M.I.E.; Winter, 2001)

Solution. Given: Area of cross-section of aluminium bar = 24 mm × 30 mm = 720 mm² ; Load on aluminium bar = 7000 kg ; Diameter of steel bar = 10 mm ; Load on steel bar = 5000 kg.

We know that stress on aluminium bar,

$$\sigma_{Al} = \frac{P}{A_0} = \frac{7000}{720} = 9.72 \text{ kg/mm}^2 \qquad \ldots(i)$$

We also know that area of cross-section of steel bar,

$$A_{steel} = \frac{\pi d^2}{4} = \frac{\pi \times (10 \text{ mm})^2}{4} = 78.57 \text{ mm}^2$$

and the stress on steel bar,

$$\sigma_{steel} = \frac{P}{A_0} = \frac{5000}{78.57} = 63.64 \text{ kg/mm}^2 \qquad \ldots(ii)$$

On comparing the results from equations (*i*) and (*ii*), we find that stress is greater on steel bar. **Ans.**

7.10 STRESS-STRAIN DIAGRAM FOR DIFFERENT MATERIALS

The stress-strain diagrams for different materials may be grouped under the following two important heads: 1. For ferrous metals 2. For non-Ferrous metals.

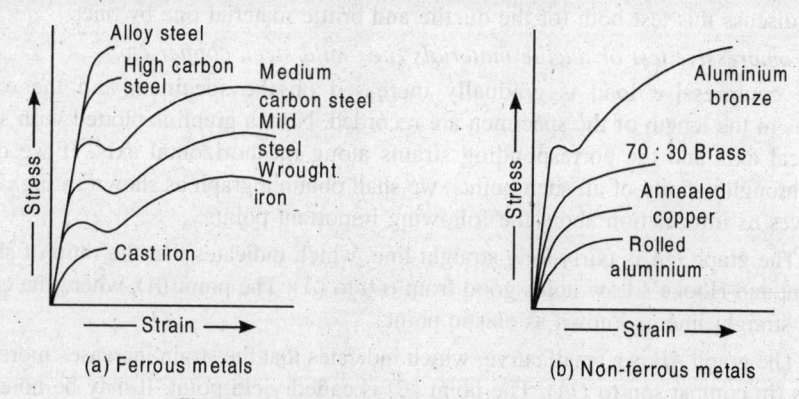

Fig. 7.9. Stress-strain curve for different metals.

(a) Ferrous metals

(b) Non-ferrous metals

1. Stress-strain diagram for ferrous metals.

Fig. 7.9. (*a*) show the stress-strain diagram for a few important types of steels, wrought iron and cast iron. It will be seen from the diagram, that the elastic portion for all the steel is nearly the same. Thus the Young's modulus for these metals are also approximately the same. As a matter of fact, the strength of a ferrous metal depends upon the percentage, of carbon content in it. This point is illustrated by the stress-strain curve of wrought iron, which indicates that the ductility decreases by increasing the percentage of carbon content.

It may be noted from the stress-strain curve of a cast iron (which is a brittle material), that there seems to be no perfectly defined location of the yield point. Therefore it becomes difficult to determine the yield strength of brittle materials. A common method to determine yield strength for such materials is to draw a straight line parallel to the elastic portion of the curve at a predetermined strain ordinate value (say 0.1%). The point, at which this line intersects the stress-strain curve, is called the yield strength or proof stress, which is at 0.1% of set strain.

2. *Stress-strain curve for non-ferrous metals.*

Fig. 7.9 (*b*) shows the stress-strain diagram for a few selected non-ferrous metal. It will be seen from the diagram, that there is no definite elastic portion of the curves. As a matter of fact, the elastic properties of non-ferrous metal vary considerably. In case of alloys, the stress-strain diagram depends upon their chemical composition.

7.11 COMPRESSIVE TEST

The compressive test is, merely the opposite of the tensile test. It is generally performed for testing brittle materials such as cast iron, concrete, stone etc. The specimens used in this test are, usually, made of cubical or cylindrical shape. It has been observed that some errors always creep in the compressive test due to the following practical difficulties:

1. Since the top and bottom faces of the given specimen are seldom absolutely parallel to each other, therefore it is very difficult to ensure axial loading on the specimen.

2. Since the length of the given specimen is kept short enough (not) more than twice off its diameter) to avoid its buckling, therefore, within the elastic limit, a small compression takes place which is difficult to measure accurately.

3. The friction between the ends of the given specimen and clutches of the machine prevents the dimensions of the specimen ends from increasing. This results in the lateral expansion to take place more in the centre instead of uniform increase in diameter throughout the whole length. Such an effect, which is called barrel effect, is not a case of an ideal compression.

Though the compressive test is, generally, performed for testing the brittle material only, yet we shall discuss this test both for the ductile and brittle material one by one.

1. *Compressive test or ductile materials (i.e., mild steel, copper etc.)*

The compressive load is gradually increased on the specimen and the corresponding reductions in the length of the specimen are recorded. Now a graph is plotted with stresses along the vertical axis and the corresponding strains along the horizontal axis. If we draw a curve passing through vicinity of all such points, we shall obtain a graph as shown in Fig. 7.10 (*a*). This graph gives us information about the following important points:

1. The graph *OA* is fairly long straight line, which indicates that the ratio of stress to strain is constant and Hooke's Law holds good from (*O*) to (*A*). The point (*A*), where the curve deviates from the straight line, is known as elastic point.

2. The graph *AB* is a small curve, which indicates that the strain increases more quickly than the stress (in comparison to *OA*). The point (*B*) is called yield point. It may be noted that if load from the specimen is removed, then the decrease in the length of the specimen from (*A*) to (*B*) will not disappear. But it will remain as a permanent set.

3. The graph *BC* is also a very small curve (almost horizontal), which indicates that the strain increases without any increase in the stress. This happens as there is a sudden decrease in the length of the specimen due to creep, without any appreciable increase in the stress. During this ductile reduction if the length of the specimen, its cross-sectional area gets increased in proportion to its length.

4. The graph *CD* is an upward curve, which indicates that the specimen regains some strength, and higher values of stresses are required for higher strains. The curve continues, almost

without any limit, as there is no failure of the material on account of its ductility. It may be noted that the cross-sectional area of the specimen goes on increasing, continuously with the increase in load. The specimen gets shortened and bulges out.

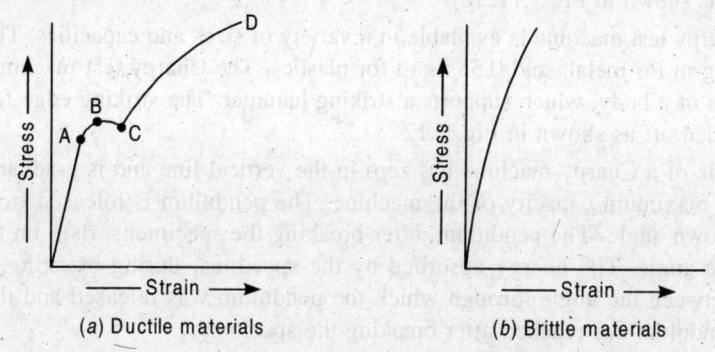

Fig. 7.10. Compressive test.

2. *Compression test for brittle materials (i.e. cast iron, concrete etc.)*

The compressive load is gradually increased on the specimen and the corresponding reduction in lengths of the specimen are recorded, Now a graph is plotted with stresses along the vertical axis and the corresponding strains along the horizontal axis. If we draw a curve passing through the vicinity of all such points, we shall obtain a graph as shown in Fig. 7.10 (*b*)

From the graph, we see that there is a little strain as compared to the stress. There is always a point, where the specimen will fail due to shear along a diagonal plane. The value of breaking stress for different materials is different. But the general pattern of the stress-strain diagram is approximately the same as shown in Fig. 7.10. (*b*).

7.12 IMPACT TEST

Many machine parts are subjected to suddenly applied load called impact blows. It has been observed that a metal may be hard, strong or of high tensile strength. But it may be unsuitable for uses where it is subjected to sharp blows. The capacity of a metal to withstand such blows without fracture is known as an impact resistance or impact strength. It is an indicative of the toughness of the metal *i.e.*, the amount of energy absorbed by the metal during plastic deformation. The S.I. Unit for expressing impact strength is meganewton per square metre, *i.e.*, MN/m^2 (known as MP_a).

There are many types of impact testing machines available in the market. But the basic principle, on which all of them are based, is the same. The following two types of impact testing machines are important from the subject point of view:

1. Charpy testing machine, and 2. Izod testing machine.

7.13 CHARPY TEST

The Charpy test is carries out on a specimen, which is 55 mm × 10 mm × 10 mm in size and has a 2 mm deep notch at its centre making an angle of 45° as shown in Fig. 7.11. (*a*).

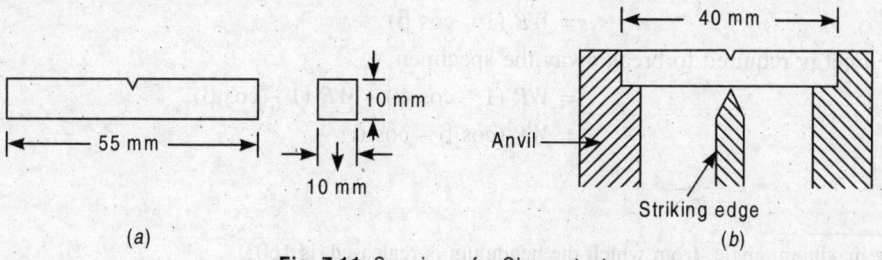

Fig. 7.11. Specimen for Charpy test.

The specimen is placed horizontally as a simply supported beam between two anvils 40 mm apart in such a way that the striking hammer strikes the specimen on the face which is opposite to the notch as shown in Fig 7.11. (b).

The Charpy test machine is available in a variety of sizes and capacities. The usual capacity is about 30 kg-m for metals and 0.55 kg.m for plastics. The Charpy test machine, in its simplest form, consists of a body, which supports a striking hammer. The striking edge for the hammer is slightly rounded off as shown in Fig. 7.12.

The scale of a Charpy machine has zero in the vertical line and is graduated on both sides from zero to maximum capacity of the machine. The pendulum is released from the right side from a * known angle. The pendulum, after breaking the specimens, rises on the opposite side through some angle. The energy absorbed by the specimen, during breaking, is given by the difference between the angle through which the pendulum was released and the angle through which the pendulum has reached after breaking the specimen.

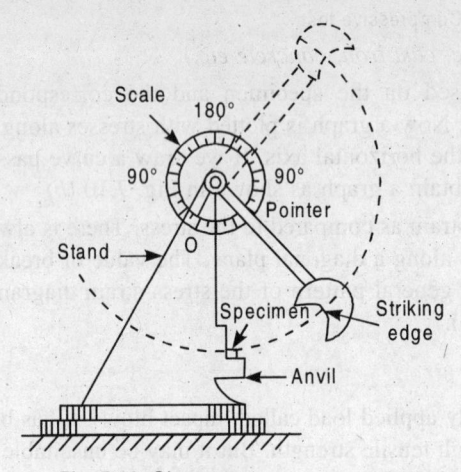

Fig. 7.12. Charpy test. 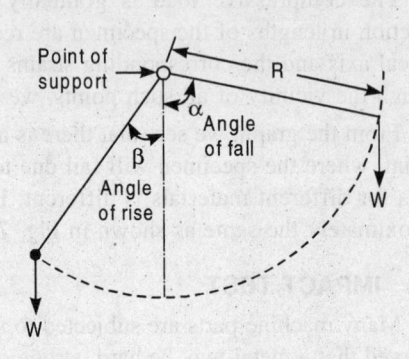   **Fig. 7.13.** Energy of pendulum.

Neglecting losses, the energy used in breaking away the specimen is determined is given below:

Let,   W = Weight of the pendulum,
        α = Angle through which the pendulum falls,
        β = Angle through which the pendulum rises,
        R = Distance between the centre of gravity of the pendulum and the axis of rotation.

From the geometry of the Fig. 7.13, we know that initial energy of the pendulum
$$= WR (1 - \cos \alpha)$$
and energy after breaking the specimen
$$= WR (1 - \cos \beta)$$
∴ Energy required to break away the specimen
$$= WR (1 - \cos \alpha) - WR (1 - \cos \beta)$$
$$= WR (\cos \beta - \cos \alpha)$$

---

*   The maximum angle, from which the pendulum is realeased, is 160°.

Mechanical Tests of Metals

7.14 IZOD TEST

The izod test is carried out on specimen, which is 75 mm × 10 mm × 10 mm in size and has a 2 mm deep notch making an angle of 45° as shown in Fig. 7.14. (*a*).

The specimen is held vertically as a cantilever between two jaws, in such a way that the striking hammer strikes the specimen on the same face as that of notch as shown in Fig. 7.14. (*b*).

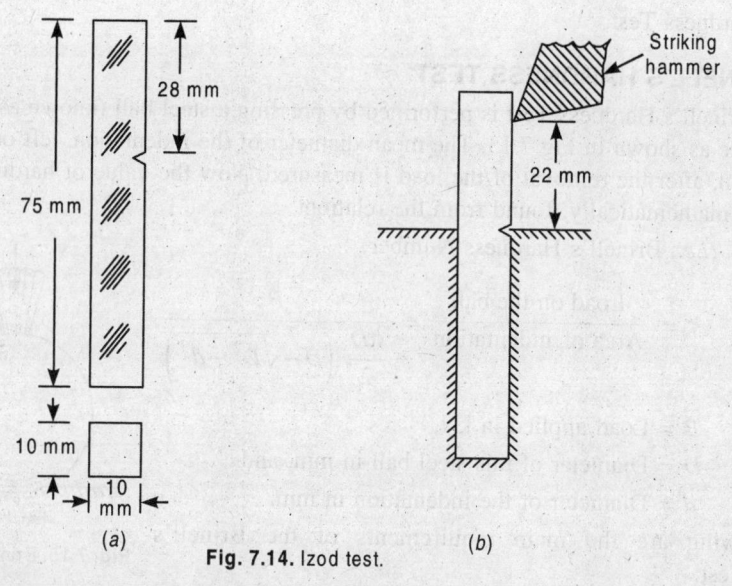

Fig. 7.14. Izod test.

The remaining procedure for obtaining the energy required to break away the specimen is the same as discussed in the Charpy test.

7.15 FACTORS AFFECTING IMPACT RESISTANCE

The resistance offered by specimen, to the impact load, is known as impact resistance. The following factors affecting the impact resistance are important from the subject point of view:

1. *Dimensions of the notch of the test specimen.* We know that the reduction in size of the specimen reduces the volume of the metal subjected to distortion. This, in turn, reduces the amount of energy required to rupture the specimen. However, by reducing the size and the tendency to cause brittle fracture reduces. And this, in turn, increases the energy absorbed the specimen.

2. *Impact velocity.* The impact resistance increases with the increase in impact velocity upto a limit, known as critical velocity, which is different for different metals. Beyond the critical velocity, there is a rapid decline in the value of impact resistance. The rate at which the impact resistance declines is also different for different metals.

3. *Temperature of specimen.* Most of the metal have brittle failure below a particular temperature, known as critical temperature, which is different for different metals. It has been observed that the energy required in a brittle fracture is also low. Beyond this critical temperature, the failure is ductile which requires very large energy.

4. *Angle and shape of notch.* It has been observed that if the angle of notch is below 60°, it does not affect the impact resistance. But if angle of notch is more than 60°, it greatly affects the impact resistance. For ductile material (like mild steel) the sharpness of the bottom a notch decreases the energy of **rupture**.

7.16 HARDNESS TEST

The hardness test of a metal is generally performed to know its resistance against indentation (*i.e.*, penetration) and abrasion. Though there are many tests to determine the hardness of different materials, yet the following are important from the subject point of view:

1. Brinell's Hardness Test. 2. Rockwell's Hardness Test. 3. Vicker's Hardness Test. 4. Knoop's Hardness Test.

7.17 BRINELL'S HARDNESS TEST

The Brinell's Hardness Test is performed by pressing a steel ball (known as indentator) into the test piece as shown in Fig 7.15. The mean diameter of the indentation, left on the surface of the specimen, after the removal of the load is measured. Now the value of hardness of the given material is, mathematically, found from the relation:

B.H.N. (*i.e.*, Brinell's Hardness Number)

$$= \frac{\text{Load on the ball}}{\text{Area of indentation}} = \frac{P}{\frac{\pi D}{2}\left(D - \sqrt{D^2 - d^2}\right)}$$

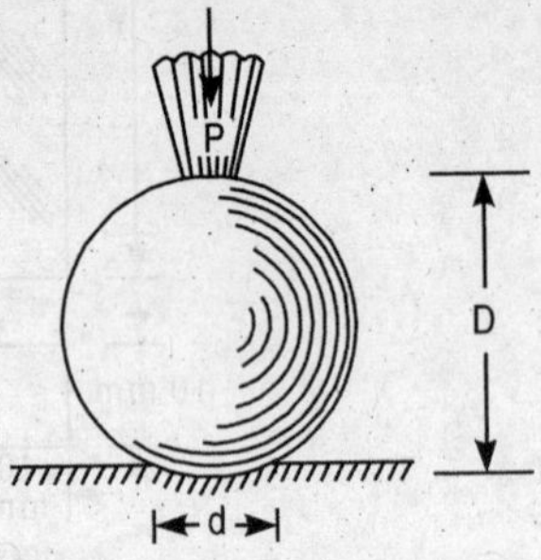

where, P = Load applied in kN,

D = Diameter of this steel ball in mm, and

d = Diameter of the indentation in mm.

Following are the main requirements of the Brinell's Hardness Test:

Fig. 7.15. Brinell's Hardness Test

1. *Steel ball.* The diameter of the steel ball is generally 10 mm ± 0.01 mm. For testing harder steels, tungsten carbide balls are widely used.
2. *Test specimen.* The thickness of the test specimen should always be more than 8 time the depth of indentation. The test surface should be free from any oxide film.
3. *Load.* For ferrous metals, the load should be applied for a period of 15 seconds, But for soft metals, the load should be applied for a minimum period of 33 seconds. Moreover, the load should be applied gradually and smoothly.

The Brinell's Hardness Test is mostly used for determining the hardness of metallic materials. This test is not recommended for materials having B.H.N. above 630 and also for very thin specimens. Moreover, it is a time consuming test and requires an expensive equipment. As the test leaves a large impression on the specimen, therefore it cannot be adopted in the industry.

Table 7.2. gives the value of B.H.N. of some important metals and alloys.

Table 7.2

S. No.	Name	B.H.N.
1.	Lead	4 to 8
2.	Zinc	25 to 40
3.	Copper	30 to 60
4.	Wrought iron	70 to 85
5.	Mild steel	80 to 105
6.	Cast iron, Sand cast	115 to 200
7.	Nickel steel (annealed)	130 to 160
8.	Nickel chrome steel	175 to 300

Mechanical Tests of Metals

7.18 ROCKWELL'S HARDNESS TEST

The Rockwell's Hardness Test is, generally, performed, when quick and direct reading is desirable. This test is also performed when the materials have hardness, beyond the range of Brinell's Hardness Test. It differs from the Brinell's test that in this test the loads for making indent are smaller, and thus make smaller and shallower indents. It is because of these reasons that the Rockwell' Hardness Test is widely used in the industry.

In this test a standard indentor either of 1.58 mm diameter loaded with 100 kN or a cone indentor with 120° cone and 150 kN is employed. The test has nine scales of hardness (A to H and K). But B and C scales are widely used.

The ball indentors are, generally, made of hardened tool steel or tungsten carbide. During the test, the specimen is placed on the anvil, and is raised till it comes in contact with the indentor. A minor load of 100 kN is applied on the specimen and the small pointer indicates 'set'. Now the main pointer is also brought to the 'set' position. The major load is then applied and is allowed to continue for one second. The depth of indentation in mm, is read the from small pointer. Now the Rockwell Number is obtained, mathematically, from the relation.

Rockwell B number

$$(R.H.B.) = 130 - \frac{\text{Depth of indentation in mm}}{0.002}$$

and Rockwell C number

$$(R.H.C) = 100 - \frac{\text{Depth of indentation in mm}}{0.002}$$

The following precautions should be taken in the Rockwell's Hardness Test:
1. The indentor and anvil should be clean and well placed.
2. The surface of the specimen should be flat, clean, dry, smooth and should be placed perpendicular to the indentors.
3. The thickness of the specimen should be more than 10 times the depth of indentation. If the specimen is curved in shape, then the indent should be made of the concave side.
4. The test should be carried out preferably at a temperature of 27°C ± 2°C.

7.19 VICKER'S HARDNESS TEST

The Vicker's hardness test is the most accurate test, which has a fairly continuous scale of hardness (Vicker's Hardness Number of 5 to 1500). The test makes the use of diamond square based pyramid indentor with 136° angle between the opposite faces. The load range is variable from 50 to 1200 in steps of 5 N.A. Piston and a dashpot of oil is used for controlling the rate and duration of the loading.

The test is performed by placing the specimen on an anvil and raised till it is close to the indentor point. The load is then gradually applied to the indentor and then removed. The diagonal of the square indentation is measured to 0.001 mm length. Now the Vicker's Hardness Number is obtained, mathematically, from the relation:

Vicker's Hardness Number

$$\text{V.H.N.} = \frac{P}{A} = \frac{2P \sin(\theta/2)}{L^2} = \frac{1.854 \, P}{L^2}$$

where P = Load applied in newton.
  A = Surface area of indentation in square mm.
  L = Length of the diagonal in mm, and
  θ = Angle between opposite faces of the diamond pyramid (equal to 136°).

This test is very suitable for testing polished and hardened material or nitrided surface due to small impression made on the test specimen. This test is accurate and is suitable for metal as thin as 0.15 mm. In spite of these advantages, the Vicker's Hardness Test is not widely adopted because of its being slow.

All the precautions necessary to perform the test are similar to those in a Rockwell's Hardness Test.

7.20 KNOOP'S HARDNESS TEST

The Knoop's Hardness Test is suitable for extremely thin metal plates exceptionally hard and brittle, very shallow carburised or nitrided surface. This test is also performed whenever the applied load is kept below 35N.

In this test, a diamond pyramidal indentor with short depth diagonals in the ratio of 7 : 1 is used to indent on the specimen. The length of the diagonal is read under microscope. Now the Knoop's Hardness Number is obtained, mathematically, from the relation:

$$\text{Knoop's Hardness Number} = \frac{\text{Load in newton}}{\text{Unrecovered projected area of the impression in square mm}}$$

7.21 FATIGUE TEST

The failure of a material, under repeatedly applied stress, is called fatigue. It has been observed that some of the machine parts such as axles, shafts, crankshafts, connecting rods springs, pinion teeth etc., are subjected to varying stresses. It includes the variation in the intensity of the same type of stress as well as different types of stresses (i.e., change of stress from tensile to compressive and vice versa). The varying stresses may be broadly classified into the following four types:

1. The stress varying between two limits of equal value, but of opposite sign.
2. The stress varying between two limits of unequal values, but of opposite sign.
3. The stress varying between zero and a definite value.
4. The stress varying between tow limits of unequal values, but of same sign.

Though there are a numerous ways of fatigue test is a laboratory, yet the basic principle is the same. The most common test is conducted on a rotating beam fatigue testing machine as shown in Fig. 7.16.

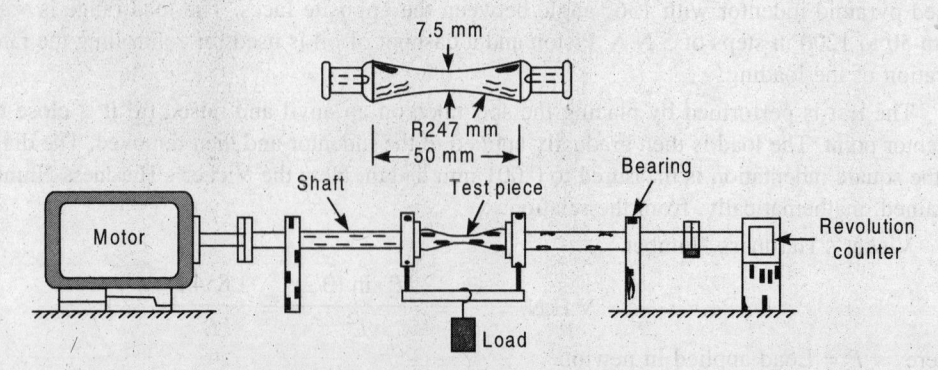

Fig. 7.16. Rotating beam fatigue testing machine.

Mechanical Tests of Metals

The test piece (or specimen) is loaded in pure bending. Now the test piece is rotated with the help of a motor. A little consideration will show, that with each rotation, all points on the circumference of the specimen will be alternately in tension and compression. Thus each revolution of the specimen will constitute a complete cycle of stress reversal. The speed of the motor will indicate the frequency of the stress reversal.

In this test, a number of identical test pieces or specimens (say 8 to 10) are made from the same material. The first test pieces is fixed in the machine, and is loaded as shown in the figure. Generally, this load is not less than that which can produce a stress equal to 3/4 of the tensile strength of the material upto its elastic limit. The speed of the motor shaft is kept constant. It has been found that after a sufficient number of stress reversals (or rotations) a crack is formed, on the outer surface, in the form of a ring. This crack goes on extending towards the centre of the test piece, till it breaks away. It has also been observed that the speed of the motor (or frequency of stress reversal) has no effect on the result. But it is the load or intensity of the stress, which controls the result.

After the first test piece breaks away, the second piece is tested with a decreased load. Similarly third, fourth ...etc., test pieces are then tested with still decreased loads. It well be interesting to know that the number of stress reversals will go on increasing each time with the decreasing in load. After some tests, a limit is reached, when the stress is not sufficient to break the test piece even after 10×10^6 stress reversals (in some countries this limit is 20×10^6). This safe stress, which after reversing for 10×10^6 times does not cause the test pieces to break, is called *endurance limit or fatigue limit*.

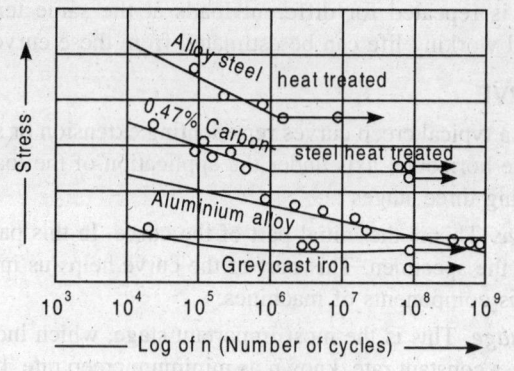

Fig. 7.17. Fatigue test curve.

Now the fatigue test results are plotted on diagrams known as *S-N* diagrams. In this diagram, the stress is plotted against vertical axis and log of N (*i.e.*, number of cycles) is plotted against horizontal axis as shown in Fig. 7.17.

The given figure shows the test results for steel, grey cast iron and aluminium alloy.

7.22 CREEP TEST

The continuous deformation of a metal, under a steady load, is known as creep. This test is very essential to predict the working life of some members or machine components which are subjected to creep. It is always exhibited in metals like iron, nickel, copper and their alloys at increased temperatures. But some metal like zinc, tin, lead and their alloys also creep at room temperatures. It has been observed that some organic metals such as plastics and rubber are very sensitive to creep.

Strictly speaking, the creep in metals is caused by slip occurring along crystallographic directions in the individual crystals together with some deformation of grain boundary. As a matter of fact, most of the deformation in case of metals, is non-recoverable. However, a small fraction of this plastic deformation is recovered with the time after the load is removed.

The creep test is generally performed by applying a static load to one end of the lever system. The other end is attached to the specimen, under test in the furnace, and held at constant temperature as shown in Fig. 7.18.

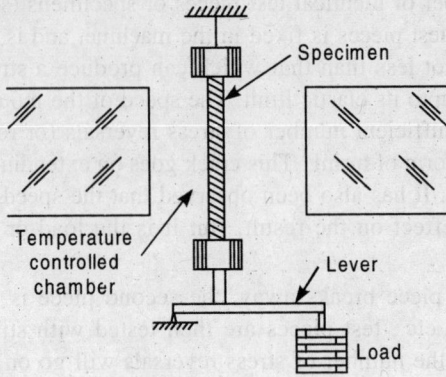

Fig. 7.18. Creep test.

The axial deformation is read, periodically, throughout the test. And a curve is plotted between extension (i.e., strain) along the vertical axis and time along the horizontal axis as shown in Fig. 7.19.

This procedure is repeated for different loads at the same temperatures. The maximum permissible strain and working life can be estimated from these curves.

7.23 CREEP CURVE

Fig. 7.19 shows a typical creep curves representing extension or strain along the vertical axis and the time along the horizontal axis under the application of the load. The creep curve may be studied in the following three stages :

1. *Primary stage.* This is the initial part of the curve. In this part, the curve indicates rapid extension of the specimen. This part of the curve helps us in providing proper clearance in the various components of machines.

2. *Secondary stage.* This is the most important stage, which indicates that the creep occurs at more or less constant rate, known as minimum creep rate. It will be interesting to know at it is the secondary stage of the creep, which has maximum practical utility.

3. *Tertiary stage.* This is the part of the curve, which indicates rapid extension of the specimen, which finally leads to rupture.

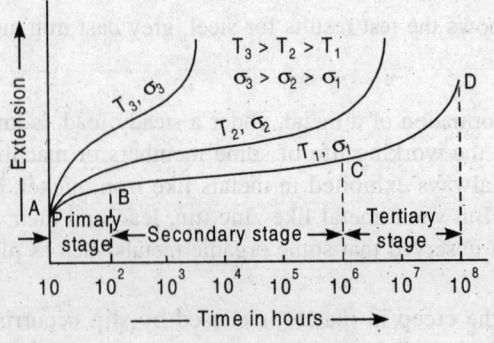

Fig. 7.19. Creep curves.

Mechanical Tests of Metals

It may be noted from the creep curve that the portions *AB* and *CD* (*i.e.*, primary stage and tertiary stage) are short time periods as compared to portion *BC* which represents the entire life period of the specimen. A little consideration will show, that this period is of great importance, because the life time of a machine component depends upon the rate of this extension. If any attempt is made to decrease the tensile strength, the slope *BC* will decrease accordingly. It is thus obvious, that in such cases, the design must be based on the assumption of a definite period of service and definite amount of permissible distortion. For example, in the design of moving parts in steam turbines and automobile industry, the creep should not exceed in 1 per cent in 10,000 hours.

It may also be noted that as the temperature (T) or the applied stress (σ) increases, the creep curve shifts upwards and the time duration of the various stages reduces significantly as shown in the figure. This indicates that the value of creep increases with the increase in temperature as well as stress. Therefore the material to be used at a high temperature must have high melting point.

7.24 FACTORS AFFECTING CREEP RESISTANCE

We have already discussed in the last article the creep test and creep curve. The resistance offered by a specimen, to the creep, is known as creep resistance. The following factors affecting the creep resistance are important from the subject of view:

1. *Effect of grain size.* The creep resistance of metal is greatly affected by its grain size. It has been observed that a fine-grained structure has better creep resistance at low temperature. But at high temperatures, coarse-grained structure has a better creep resistance. It is due to the fact, that in a fine-grained structure, its grain boundaries are stronger than the grains. As a result of this, the dislocations, piled up at grain boundaries, have less freedom to move. But at higher temperatures, they have a tendency to move easily. And in a coarse-grained structure, the grains are stronger than their grain boundaries. Therefore at higher temperatures, they do not allow the movement of dislocations. Thus at room temperature, a coarse-grained structure has a better creep resistance.

2. *Effect of strain hardening.* We know that strain hardening (or work hardening) is a process of deforming a metal at room temperature. Moreover, the strain hardening produces internal stresses in metals. It has been observed that at low temperatures, the failure of a metal takes place due to the fracture through its crystals. This type of fracture is known as intercrystalline. Whereas at higher temperatures, the failure takes place at crystal boundaries. This type of fracture is known as intercrystalline. The temperature, at which the fracture changes from intercrystalline to intercrystalline, is known as equicohesive temperature. The magnitude of creep, due to applied stress, depends upon the opposing effects of yielding of the materials and the strain hardening caused by such yielding. It has also been observed that at or below the equicohesive temperature, the creep takes place due to strain hardening. But the continuous creep does not take place unless the applied stress is large enough to overcome the resistance by strain hardening. It may be noted that the creep due to strain hardening takes place at a higher stress than that of yielding of a material. Therefore curve shifts upwards for higher temperatures.

3. *Effect of heat treatment.* The creep resistance of a metal is greatly affected after heat treatment. The maximum creep resistance is, usually, produced by normalising. In this process, the metal (say steel) is heated to a temperature of 550° or above. The metal is heated at this temperature for a period of about 15 minutes and then allowed to cool down in still air.

4. *Effect of alloying addition.* The creep resistance of a metal is also greatly affected by alloying the metal with regard to the temperature. It has been observed that at low

temperatures, the creep resistance is increased by the addition of cobalt, nickel, manganese, etc. But at high temperature, it is increased by the addition of chromium, tungsten, vanadium, molybdenum, etc.

5. *Effect of manufacturing process.* The creep resistance of a metal is also greatly affected by the kind of process adopted during the manufacturing process. The selection of the manufacturing process depends upon size and shape of the product, quantity of the production and type of heat treatment. It has been observed that the metal (say steel) manufactured by electric arc furnace, have better creep resistance than that of open hearth process. And the steel manufactured, by induction furnace has better creep resistance even than that of electric arc furnace.

A.M.I.E (I) EXAMINATION QUESTIONS

1. Draw stress-strain diagram of a ductile material and mark all the salient points. *(Summer, 1993)*
2. (*a*) Define creep. Explain its phases and mechanisms.
   (*b*) Describe the methods for minimizing creep of materials at low and high temperatures.
   (Summer, 1993)
3. What is the creep in metallic materials? Why is the knowledge of creep resistance important industrially? *(Winter, 1993)*
4. Describe the tensile testing of metals. What are the data one can get from such a test? *(Summer, 1994)*
5. Define yield stress and tensile strength. *(Summer, 1995)*
6. Draw the stress-strain diagram for a ductile material and show the salient points on it.
   (Summer, 1996)
7. (*a*) Differentiate between engineering strain and true strain.
   (*b*) Draw neat and labelled sketch of engineering strain and true strain.
   (*c*) Sketch a typical creep curve and label its important areas.
   (*d*) Explain the terms percentage elongation and proof stress. *(Summer, 1997)*
8. Sketch the stress-strain curve of mild steel and show the salient points. Define yield stress and proof stress. *(Winter, 1999)*
9. Define hardness. What is the principle of hardness-testing machines? Is hardness related to ultimate tensile strength and endurance limit? If so, what is the general relationship? *(Summer, 1998)*
10. Sketch the Stress-strain curve of mild steel and show the salient points. *(Summer, 1999)*
11. What is Hooke's Law? Draw and label the stress-strain diagram for mild steel and cast iron.
    (Winter, 1999)
12. Draw a nominal stress-strain diagram of a ductile material and indicate (*a*) proportional limit, (*b*) yield point (*c*) initiation of necking. Show how yield strength and the percentage of elongation can be measured from the diagram. *(Winter, 2000)*
13. Explain the phenomenon of yielding in mild steel. Why the yield point in copper is not distinct.
    (Summer, 2000)
14. How is the fatigue test performed in the laboratory? *(Summer 2001, Winter 1999)*
15. Explain the terms true stress, true strain, engineering stress and engineering strain.
    (Summer, 2001)
16. Draw a typical creep curve of strain versus time at constant stress and constant elevated temperature, clearly showing the instantaneous deformation, three stages of creep and rupture. At what temperature, does creep become significant? *(Summer 2002, Winter 2001, Summer 1998)*
17. Explain the effect of stress and temperature on the steady-state creep rate. *(Summer, 2003)*
18. How do you enhance creep resistance of steel? Which temperature is important for a creep resistance alloy. *(Summer, 2003)*

MULTIPLE CHOICE QUESTIONS

1. According to Hook's Law, "when a material is loaded within its elastic limit, the stress is ...."
   (a) directly proportional to the strain
   (b) inversely proportional to the strain
   (c) proportional to the square of the strain
   (d) proportional to the inverse of the square of the strain
2. Proof stress corresponds to
   (a) lower yield point (b) higher yield point
   (c) elastic limit (d) a specific strain
   (A.M.I.E., Summer, 2001)
3. Which hardness method can be used to measure hardness of this metal plates ?
   (a) Rockwell (b) Knoop (c) Vickers (d) Shore
4. For isotropic materials, the relationship between shear modulus (G), Young's modulus (E) and the Poisson's ratio (v) is,
   (a) $v = 2G(1 + E)$ (b) $E = 2G/(1 + v)$ (c) $G = 2E(1 + v)$ (d) $E = 2G(1 + v)$
5. The failure of a material under repeatedly applied stress is called:
   (a) creep (b) fatigue (c) corrosion (d) non of these.
6. The continuous deformation of a metal under a steady load is called:
   (a) creep (b) fatigue (c) corrosion (d) non of these.

ANSWERS

1. (a) 2. (b) 3. (b) 4. (d) 5. (b) 6. (a)

8
Deformation of Metals

1. Introduction. 2. Classification of Metal Deformations. 3. Elastic Deformation. 4. Plastic Deformation. 5. Comparison between Elastic and Plastic Deformations, 6. Modes of plastic Deformation. 7. Slip. 8. Mechanism of Slip. 9. Critical Resolved Shear Stress. 10. Twinning. 11. Mechanism of Twinning. 12. Comparison Between Slip and Twinning. 13. Dislocations. 14. Types of Dislocations. 15. Edge Dislocations. 16. Screw Dislocations, 17. Comparison between Edge and Screw Dislocations. 18. Dislocation Density. 19. Characteristics of Dislocations. 20. Burgers vector. 21. Dislocation Climb 22. Multiplication of Dislocations (Frank-Read Generator). 23. Elastic After Effect. 24. Bauschinger Effect. 25. Deformation of Polycrystalline Materials. 26. Mechanisms of strengthening in Metals. 27. Strengthening by Grain size Reduction 28. Solid Solution Hardening. 29. Work Hardening. 30. Preferred Orientation. 31. Cold Working. 32. Advantages and Disadvantages of Cold Working. 33. Annealing of a cold Worked Metal. 34. Recovery. 35. Recrystallisation. 36. Grain Growth 37. Variation of Mechanical properties in Annealing. 38. Comparison between Recovery and Recrystallisation. 39. Hot Working. 40. Advantages and Disadvantges of Hot Working. 41. Comparison between Cold Working and Hot Working.

8.1 INTRODUCTION

The changes, produced in the shape of a metal piece, under the action of a single force or a set of forces, is known as deformation or mechanical deformation. The mechanical deformation, of metals or their alloys, is essential to give a desired shape to the finished surfaces. It has been experimentally found that the deformed metals are more superior to the cast metals from which they are produced. The various operations used in the deformation of metals are rolling, forging, spinning, drawing etc.

8.2 CLASSIFICATION OF METAL DEFORMATION

The deformation of Metals may be classified into the following two types depending upon the nature of strain produced during deformation:

1. Elastic deformation. 2. Plastic deformation.

8.3 ELASTIC DEFORMATION

The term 'elastic' deformation' may be defined as the process of deformation, which appears and disappears simultaneously with the application and removal of stress.

It has been observed that whenever a stress of low magnitude is applied to a piece of metal, it causes displacement of atoms from their original positions. But on the removal of stress, the atoms spring back and occupy their original positions. In elastic deformation, the tensile strain is due to a slight elongation of the unit cell in the direction of the tensile load. Similarly, the

Deformation of Metals

compressive strain is due to a slight contraction of the unit cell in the direction of the compressive load.

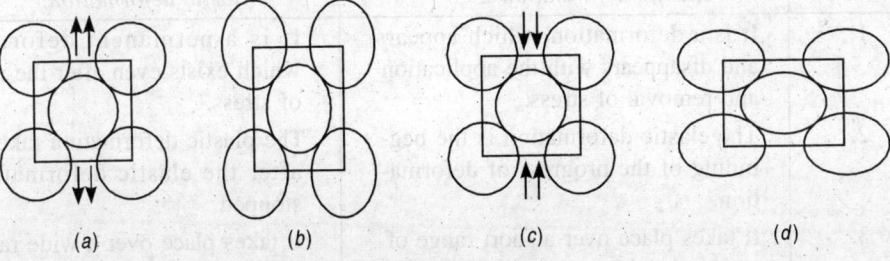

Fig. 8.1. Elastic deformation.

Fig. 8.1 shows the effect of tensile and compressive loads respectively on the atoms of the unit cell respectively. Fig. 8.1 (a) represents the front face of the F.C.C. unit cell before loading. In this figure, the circles represent the atoms of metal specimen. Fig. 8.1. (b) shows a slight elongation of the unit cell in the direction of tensile load. Similarly Fig. 8.1. (c) represents the front face of the F.C.C. unit cell before loading. And Fig. 8.1 (d) shows a slight contraction of the unit cell in the direction of compression load. The unit cell, after removal of the load, returns back to the normal positions as shown in Fig. 8.1. (a) and (c).

It may be noted, from the above discussion, that whenever there is an elongation or contraction, in the crystal structure of a metal, in one direction, due to uniaxial load, produces an adjustment in dimensions at right angles to the force. This type of change in dimension (i.e., at right angles to the applied load) is due to lateral strain. The ratio of lateral strain to the original strain is known as Poisson's Ratio.

It has been experimentally found that the strain is nearly proportional to the applied stress (or load). The ratio between stress and strain is known as Modulus of Elasticity or Young's Modulus. It is an important characteristic of a metal, which gives an idea about the amount of elasticity in it. It has been observed that the crystal structures of metals are also subjected to shearing loads. This produces a displacement of one plane of atoms, relative to the adjacent plane of atoms. This type of displacement in the crystal structure is known as shearing strain. The ratio between shearing stress and shearing strain is known as Shear Modulus or Modulus of Rigidity.

8.4 PLASTIC DEFORMATION

The term 'plastic deformation' may be defined as the process of permanent deformation, which exists in a metal, even after the removal of the stress. It is due to this property, that the metals may be subjected to various operations like rolling, forging, drawing, spinning etc.

The plastic deformation in crystalline materials occurs at temperatures lower than $0.4\ T_m$ (where T_m is the melting temperature in Kelvin). In this temperature range, the amount of deformation, which occurs after the application of stress is very small and generally ignored. However, the rate at which, the material is deformed plays some role in determining the deformation characteristics.

The plastic deformation may occur under the tensile, compressive or torsional stresses. It is a function of stress, temperature and rate of straining. There are two basic modes of plastic deformation, namely slip or gliding and twinning. The slip mode is common in many crystals at ambient and elevated temperatures. At low temperatures, the mode of deformation changes over to twinning in a number of cases.

8.5 COMPARISON BETWEEN ELASTIC AND PLASTIC DEFORMATIONS

The following table gives the important points of comparison between elastic and plastic deformations:

Table 8.1

S. No.	Elastic deformation	Plastic deformation.
1.	It is a deformation, which appears and disappears with the application and removal of stress.	It is a permanent deformation which exists even after the removal of stress.
2.	The elastic deformation is the beginning of the progress of deformation.	The plastic deformation takes place after the elastic deformation has stopped.
3.	It takes place over a short range of stress-strain curve.	It takes place over a wide range of stress-strain curve.
4.	In elastic deformation, the strain reaches its maximum value after the stress has reached its maximum value.	In plastic deformation, the strain occurs simultaneously with the application of stress.

8.6 MODES OF PLASTIC DEFORMATION

The plastic deformation may occur by any one (or both) of the following modes:
(i) Slip   (ii) Twinning

8.7 SLIP

The term 'slip' may be defined as a shear deformation, which moves the atoms through many inter eratomic distances, relative to their initial positions. The deformation by slip is as shown in Fig. 8.2.

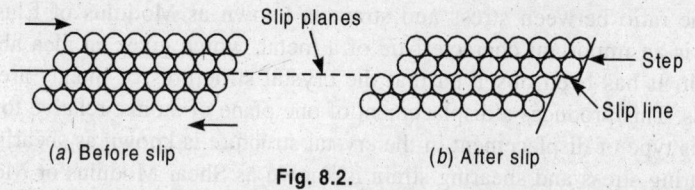

(a) Before slip   (b) After slip
Fig. 8.2.

Fig. 8.2 (a) shows the adjacent planes of a hypothetical crystal. A shearing stress, acting as indicated by the arrows, tends to move the atoms of the upper planes to the right. The movement of atoms or slip occurs only when the shear stress exceeds a critical values. The atoms move an integral number of atomic distances along the slip plane and a step is produced as shown in Fig. 8.2 (b).

A careful examination of the surface of a deformed crystal, under the microscope, shows a group of parallel lines, which correspond to a step on the surface. These are called slip lines. The slip occurs most readily in specific directions on certain crystallographic planes. Generally, the slip plane is the plane of greatest atomic density and the slip direction is the closest-packed direction within the slip plane. Since the planes, of greatest atomic density are also the most widely spaced planes in the crystal structure, therefore the resistance to slip is generally less for these planes than for any other set of planes. The slip plane together with the slip direction is called the slip system.

The common slip planes and slip directions for some simple crystals is as shown in Table 8.2. It is evident from this table that in a Face-Centred Cubic (F.C.C.) structure the {111} octahedral planes and the < 110> directions are the close-packed systems. Refer to Fig. 8.3 (a) and (b). In Fig. 8.3 (a) the arrows indicate that the slip occurs along < 110 > type directions within {111} planes. Hence, {111} < 100 > represents the slip plane and direction combination or the slip system for F.C.C. structure.

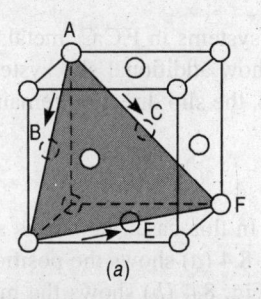

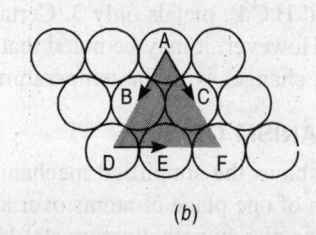

Fig. 8.3.

Fig. 8.3 (b) indicates that a given slip plane *i.e.* {111} may contain more than one slip direction. Thus several slip systems may exist for a particular crystal structure. In fact, the number of independent slip systems represent the different possible combinations of slip planes and directions.

The Base Centred Cubic (B.C.C.) structure is not a close-packed structure like F.C.C. structure. Accordingly, there is no one plane of predominant atomic density. The {110} planes have the highest atomic density in the B.C.C. structure. But they are no greatly superior in this respect to several other planes. However, in the B.C.C. structure, the <111> direction is just as close-packed as the <110> direction in the F.C.C. structure. Therefore the B.C.C. metal obeys the rule that slip direction is the close-packed direction, but not having a definite single slip planes. Thus slip in B.C.C. metals is found to occur on the {110}, {112} and {123} planes, while the slip direction is always the {111} direction.

Table 8.2

S. No.	Structure	Slip plane	Slip direction	Metals
1.	F.C.C.	{111}	<110>	Aluminium, Copper, Silver, Nickel etc.
2.	B.C.C.			
	More common:	{110}	<111>	Alpha iron
	Less common :	{110} {112} {123}		
3.	H.C.P.		<1120>	Cadmium, Magnesium, Zinc, Titanium etc.
	More common:	Basal plane (0001)		
	Less common:	Prismatic $(10\bar{1}0)$ and Pyramidal plane $(10\bar{1}1)$		

In the hexagonal close-packed (H.C.P). metals, the only plane with high atomic density is the basal plane (0001). The diagonal axes $<11\bar{2}0>$ are the close-packed directions. For Zinc. Cadmium, Magnesium and Cobalt, the slip occurs on the <0001> plane in the $<11\bar{2}0>$ directions, However, for Zirconium and Titanium, which have low c/a ratio, the slip occurs primarily on the prismatic $\{10\bar{1}0\}$ and pyramid $\{10\bar{1}1\}$ planes in the $<11\bar{2}0>$ direction.

It will be interesting to know that number of slip systems in F.C.C. metals are 12, in B.C.C. metals, 48 and H.C.P. metals only 3. Certain metals show additional slip systems with increased temperature. However, it may be noted that in all cases, the slip direction remains the same, while the slip plane changes with the temperature.

8.8 MECHANISM OF SLIP

Fig. 8.4 shows the simplified mechanism of slip. In this case, the slip is shown to occur by the translation of one plane of atoms over another. Fig. 8.4 (a) shows the position of atoms in two adjacent planes of a hypothetical crystal before slip. Fig. 8.4 (b) shows the movement of atoms of the upper plane towards right on the application of shear stress. Similarly, Fig. 8.3 (c) shows the new position of atoms after the slip has taken place.

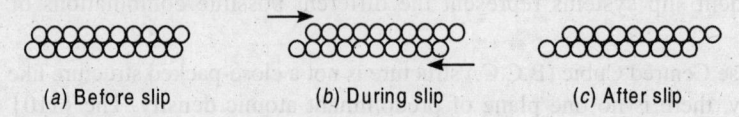

(a) Before slip (b) During slip (c) After slip

Fig. 8.4. Mechanism of slip.

The value of shear stress required for such a movement of atoms in a perfect lattice (called theoretical shear strength) is approximately equal to the value of shear modulus divided by 6 (i.e., G/6). However, in actual practice, the value of shear stress required to produce a slip is found to be at least 100 times smaller than the theoretical value. It means that a mechanism responsible for slip, is not due to the bodily shearing of planes of atoms but due to some other cause.

The experimental evidence shows that the mechanism of slip is actually due to the movement of dislocations in the crystal lattice. The shear stress required for producing a slip due to the movement of dislocations is a small fraction of theoretical value (i.e., G/6) and it matches the observed shear strengths of metals. Since the mechanism of slip requires the growth and movement of dislocation line, therefore energy is required for this purpose. The energy of a dislocation line is given by the relation,

$$E \propto l.\ G.\ b^2$$

Where l = Length of dislocation line,

G = Shear modulus, and

b = *Unit slip vector or Burgers vector.

The energy required will be minimum, when the unit slip vector (b) is the shortest one and the shear modulus (G) has the lowest value. It means that the dislocations having the shortest slip vector (b) are the easiest dislocations to generate and expand for plastic deformation. The directions in a metal, with the shortest slip vector, will be the directions with the greatest linear density of atoms. Similarly, the lowest value of shear modulus (G) corresponds to the planes, which are widely spaced and hence have the greatest planar density of atoms.

8.9 CRITICAL RESOLVED SHEAR STRESS

The slip in crystalline materials results from the action of a shear stress on the slip plane. Within the range of stresses in engineering application, the component of stress normal to the slip plane does not influence slip. Thus the slip process must be considered in terms of the shear stress resolved on the slip plane in the slip direction.

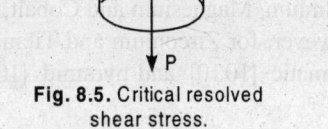

Fig. 8.5. Critical resolved shear stress.

---

* Please see Art 8.18.

Deformation of Metals

Consider a single crystal subjected to an axial load as shown in Fig. 8.5

Let
P = Load applied along the axis of the single crystal, and
A = Cross-sectional area of the crystal.

As a result of axial load let the slip takes place along the shaded plane as shown in the figure.

Now let
α = Angle, which the slip direction makes with the tensile axis (i.e., in the direction of the load), and
β = Angle which the slip plane makes with the normal to the tensile axis.

We know that component of the applied load acting in the slip direction.
$$= P \cos \alpha$$
and area of the slip plane
$$= A/\cos \beta$$

∴ Resolved shear stress,
$$\tau = \frac{\text{Load}}{\text{Area}} = \frac{P \cos \alpha}{A/\cos \beta} = \frac{P}{A} \cos \alpha \cos \beta$$
$$= \sigma \cos \alpha \cos \beta$$

Where σ is the applied tensile stress (equal to P/A).

The stress required to initiate slip in a pure and perfect single crystal is called critical resolved shear stress.

∴ Critical resolved shear stress,
$$\tau_{cr} = \sigma \cos \alpha \cos \beta$$

This equation is popularly known as Schmid's Law, and the term 'cos α cos β' as the Schmid's Factor. The following points are important for the values of critical resolved shear stress.

1. If the slip direction is at right angles to the tensile axis (i.e., $\alpha = 90°$), then $\cos \alpha = \cos 90° = 0$. Therefore $\tau_{cr} = 0$
2. If the slip plane is parallel to the tensile axis (i.e., $\beta = 90°$) then $\cos \beta = \cos 90° = 0$. Therefore $\tau_{cr} = 0$.
3. If both the slip plane and slip directions are inclined at an angle of 45° to the tensile axis, then

$$\tau_{cr} = \sigma \cos 45° \cos 45° = \sigma \times \frac{1}{\sqrt{2}} \times \frac{1}{\sqrt{2}} = \frac{\sigma}{2}$$

4. It will be seen that for all combinations of α and β, the critical resolved shear stress will always be less than $\sigma/2$ (i.e., half the tensile stress).

The value of critical resolved shear stress is a constant for a material at a given temperature. If many slip planes and slip directions of the same type are possible in a crystal, then the active plane is the one on which the critical resolved shear stress is reached first as the specimen is subjected to increasing applied stress.

Example 8.1. *A tensile stress of 15 MPa applied on $[1\bar{1}0]$ axis of a single crystal of silver is just sufficient to cause slip on the $[1\bar{1}1][0\bar{1}1]$ system. Calculate the critical resolved shear stress of silver.*
(A.M.I.E.; Summer, 2002)

Solution. Given: Tensile stress = 15 MPa; direction = $[1\bar{1}0]$.

For the given direction $[1\bar{1}0]$, we know that the angle which the slip direction makes with the tensile axis,

$$\cos \alpha = \frac{h_1 h_2 + k_1 k_2 + l_1 l_2}{\sqrt{h_1^2 + k_1^2 + l_1^2} \times \sqrt{h_2^2 + k_2^2 + l^2}}$$

$$= \frac{1.1 + \bar{1}.\bar{1} + 0.1}{\sqrt{1^2 + \bar{1}^2 + 0^2} \times \sqrt{1^2 + \bar{1}^2 + 1^2}} = \frac{2}{\sqrt{2} \times \sqrt{3}} = \frac{2}{\sqrt{6}}$$

and the angle which the slip plane makes with the normal to the tensile axis,

$$\cos \beta = \frac{1}{2}$$

∴ The critical resolved shear stress,

$$\tau_{cr} = \sigma \cdot \cos \alpha \cdot \cos \beta$$

$$= 15 \times \frac{2}{\sqrt{6}} \times \frac{1}{2}$$

$$= 6.12 \text{ MPa Ans.}$$

8.10 TWINNING

The term 'twinning' may be defined as the plastic deformation, which takes place along two planes due to a set of forces applied on a given metal piece. The process of twinning is shown in Fig. 8.5

Fig. 8.6 (a) shows the circles, which indicate the arrangement of atoms before twinning. Fig. 8.6. (b) shows the arrangement of atoms after twinning along the planes AB and CD. It may be noted that the process of deformation between the two planes AB and CD is similar to that of slip. Whereas the arrangement of atoms on either side of the twinning planes (i.e., towards left of AB and right of CD) remains unaffected.

(a) Before twinning   (b) After twinning

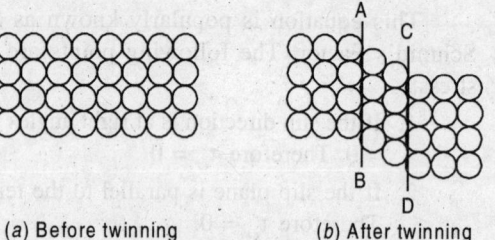

Fig. 8.6.

It has been observed that a metal, usually, deforms by twinning only if it is unable to slip Moreover, the deformation produced by twinning is small. But it places the slip planes in more favourable orientation causing the deformation to take place through slip.

It has been observed that the twinning may be produced either by mechanical deformation or annealing. The twins produced due to mechanical deformation, are known as mechanical twins. These are produced in B.C.C. (i.e., body centred cubic) and H.C.P. (i.e., hexagonal close packed) metals under the conditions of shock loading at room temperature. The twins produced due to annealing of cold work metal are known as annealing twins. These are produced in certain F.C.C. (i.e., face centred cubic) metals like copper, silver etc.

Like slipping, the twinning also occurs along certain crystallographic planes and directions. These are known as twin planes and twin directions respectively. Table 8.3 shows the twin planes and twin directions for common metals.

Deformation of Metals

Table 8.3

S. No.	Structure	Twin plane	Twin direction	Metals
1.	FCC	(111)	[112]	Copper, silver etc.
2.	BCC	(112)	[111]	Alpha iron
3.	HCP	(1012)	[1011]	Magnesium, zinc etc.

8.11 MECHANISM OF TWINNING

In twinning process, the movement of atoms is only a fraction of interatomic distance. Fig. 8.7 shows the dark circles, which indicate the arrangement of atoms. The lines. *AB* and *CD* represent the planes of symmetry, from where the twinning starts and ends respectively. These planes are known as twinning planes.

It has been observed that the crystals twin about the twinning planes. And the atoms in the regions to the left of the twinning plane *AB* and right of the twinning plane *CD* remain undisturbed whereas in the twinned region, each atom moves by a distance proportional to its distance from the twinning plane *AB*. The blank circles indicate the new position of the atoms.

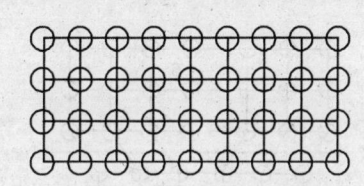

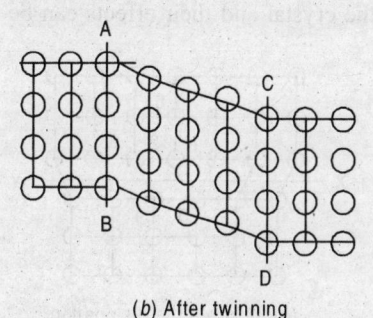

(a) Before twinning (b) After twinning
Fig. 8.7. Mechanism of twinning.

The twinning occurs due to the growth and movement of dislocations in the crystal lattice.

8.12 COMPARISON BETWEEN SLIP AND TWINNING

The following table gives the important points of comparison between slip and twinning:

Table 8.4

S. No.	Slip	Twinning
1.	In this process, the deformation takes place due to the sliding of atomic planes over the others. However, the orientation of the crystal above and below the slip plane is the same after deformation as before.	In this process, the deformation takes place due to orientation of one part of the crystal with respect to the other. The twinned portion is the mirror image of the original lattice.
2.	In this process, the atomic movements are over large distances.	In this process, the atomic movements are over a fraction of atomic spacing.
3.	It requires lower stress for atomic movements.	It requires higher stress for atomic movements.
4.	It occurs on widely spaced planes.	It occurs on every atomic plane involved in the deformation within the twinned region of the crystal.

8.13 DISLOCATIONS

The term 'dislocation' may be defined as a linear disturbance of the atomic arrangement in a crystal. It may occur in metallic crystals during their growth from its molten or vapour state. They may also be produced during slipping of the atomic planes over the others. The dislocation can move very easily on the slip planes through the crystals. This movement may be prevented by the presence of boundaries. It has been observed that the dislocations pile up at the grain boundaries, which act as barriers or obstacles for their further movement. The dislocations have a tendency to multiply during the process of deformation. It may be noted that the presence of dislocations, in a metal crystal, reduces its strength even lower than its theoretical value.

8.14 TYPES OF DISLOCATIONS

The following two types of dislocations are important from the subject point of view:

1. Edge dislocation. 2. Screw dislocation.

8.15 EDGE DISLOCATIONS

The term 'edge dislocation' may be defined as the plastic movement of atoms, which starts within the crystal and their effects can be seen on the edge of the crystal.

(a) Before edge dislocation

(b) Initial dislocation

(c) Before edge dislocation

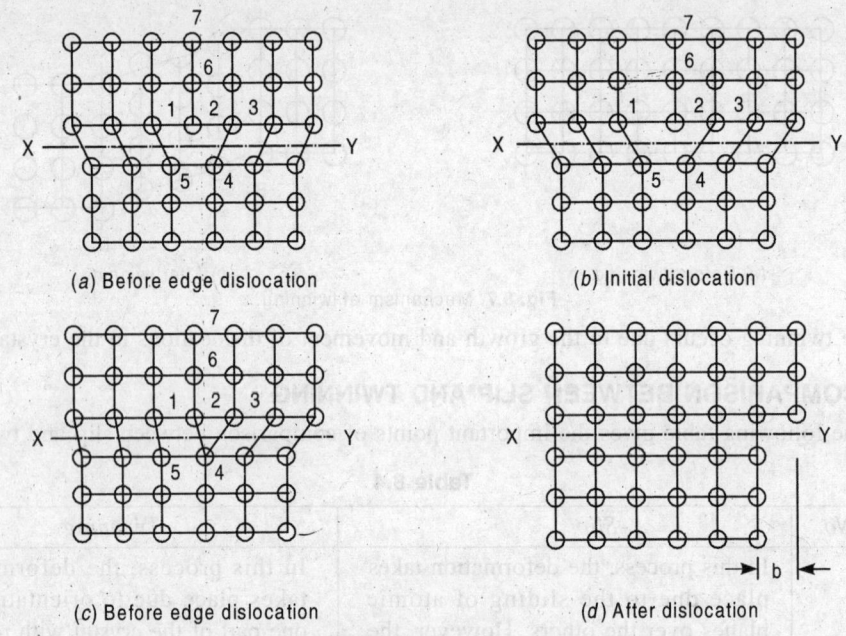

(d) After dislocation

Fig. 8.8. Edge dislocation.

Fig. 8.8 (a) shows the circles, which indicate the arrangement of atoms. In this figure, the dislocations line is shown by the atoms 2-6-7 and slip plane by X-Y Fig. 8.8 (b) and (c) show the process of dislocation and Fig. 8.8 (d) and arrangement after dislocation.

It may be noted that due to shear stress, the atom '2' moves towards the atom '3' and away from '1'. At the same time, the movement of atoms below the slip plane is such that the separation between atoms 1-5 decreases, whereas, the distance between atoms 3.4 increases as shown in Fig. 8.7 (b). On further increasing the shear stress, the atom '2' continues to move towards right, until it bonds with atom '4' as shown in Fig. 8.8 (c). It may be noted that at this stage, the dislocation has moved to the right side by one atomic distance. This procedure is repeated, until the dislocation has moved to extreme right corner giving a step of height 'b' as shown in Fig. 8.8 (d).

Deformation of Metals

From the above discussion, we see that in an edge dislocation, the movement of dislocation takes place in a direction perpendicular to the slip plane. It has been observed that an edge dislocation requires much lower yield stress to produce deformation than that required for slipping.

At the macroscopic level, plastic deformation simply corresponds to permanent deformation that result from the movement of dislocations or slip, in response to an applied shear stress as shown in Fig. 8.9.

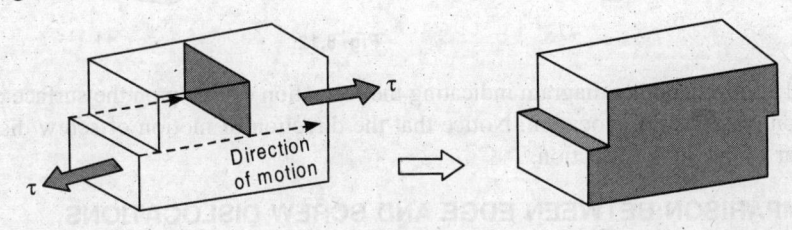

Fig. 8.9.

As seen from Fig. 8.9, there is a formation of step on the surface of a crystal by the motion of an edge dislocation. It may be noted carefully that for an edge dislocation, the dislocation line moves in the direction of applied shear stress.

Dislocation motion is analogos to the mode of locomotion observed in a caterpillar. Refer to Fig. 8.10. The caterpillar forms a hump near its posteror end by pulling in its last pair of legs. When the hump reaches the anterior end, the entire caterpillar has moved forward by the leg separation distance. The caterpillar hump and its motion correspond to the extra half-plane of atoms in the dislocation model of plastic deformation.

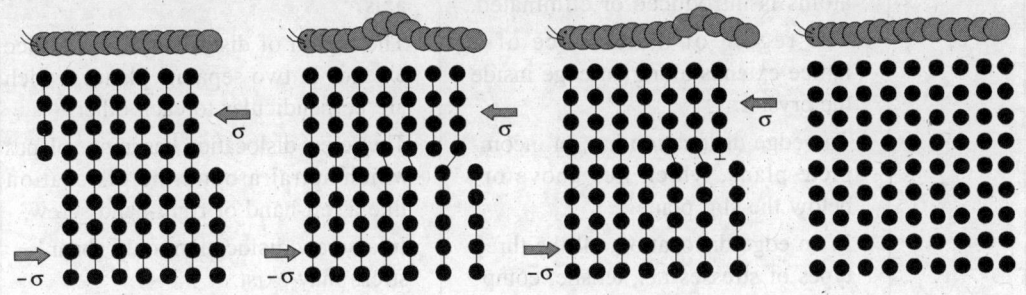

Fig. 8.10.

8.16 SCREW DISLOCATIONS

The term 'screw dislocation' may be defined as the plastic movement of atoms, within the crystal, along two separate planes perpendicular to each other. The effect of screw dislocation can be seen on the sides of the crystal. It has been observed that the effect on the crystal appears like that of a screw or a helical surface.

Fig. 8.11 shows the screw dislocation in a simple cubic crystal. In this figure, the dark circles indicate the arrangement of atoms in cubic crystal and X-Y represents slip plane. It will be interesting to know that in screw dislocation, the movement of dislocation takes place in a direction parallel to the slip plane.

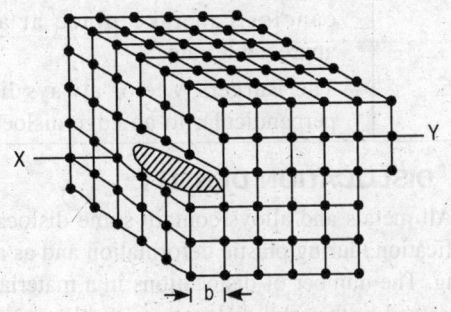

Fig. 8.11. Screw dislocation in a simple cubic structure.

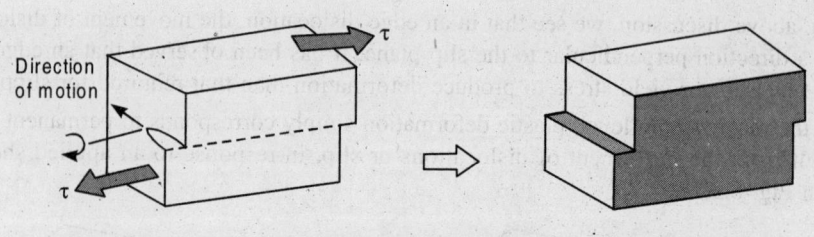

Fig. 8.12.

Fig. 8.12 shows another diagram indicating the formation of a step on the surface of a crystal by the motion of a screw dislocation. Notice that the direction of motion of screw dislocation is perpendicular to the stress direction.

8.17 COMPARISON BETWEEN EDGE AND SCREW DISLOCATIONS

The following tables gives the important points of comparison between the edge and screw dislocation.

Table 8.5

S. No	Edge dislocation	Screw dislocation
1.	It may originate when there is a slight mismatch in the orientation adjacent part of the growing crystal so that an extra row of atoms is introduced or eliminated.	It may originate during crystallization, when there is a twist in stacking sequences of atoms and unit cell formation of a step around a line known as screw axis.
2.	The region of disturbance of a lattice extends along an edge inside the crystal.	The region of disturbance of a lattice extends in two separate planes, which are perpendicular to each other.
3.	The edge dislocation has an incomplete plane, which lies above or below the slip plane.	The screw dislocation has lattice planes which spiral around the dislocation like a left-hand or right-hand screw.
4.	In an edge dislocation, all the three types of stresses i.e., tensile, compressive and shear may exist.	In a screw dislocation, only shear stress may exist.
5.	A pure edge dislocation can glide or slip in a direction perpendicular to its length. However, it may move vertically by a process known as climb, if diffusion of atoms or vacancies can take place at an appreciable rate,	The screw dislocation can move by slip. The movement by climb is not possible.
6.	The Burgers Vector always lies perpendicular to an edge dislocation.	The Burgers Vector always lies parallel to the screw dislocation.

8.18 DISLOCATION DENSITY

All metals and alloys contain some dislocation. These dislocations were introduced during solidification, during plastic deformation and as a consequence of thermal stresses that result rapid cooling. The number of dislocations in a material per unit area is known as dislocation density. It is expressed as the total dislocation length per unit volume. Alternatively dislocation density may also be expressed as the number of dislocations that intersect a unit area of a random section. The units of dislocation density are millimeters of dislocation per cm^3 or per mm^2.

Typically, dislocation densities may be *as low as 10^3 mm^{-2}*. However for heavily deformed metals, the dislocation density may run *as high as 10^9 to 10^{10} mm^{-2}*. On the other hand, *heat treating* a deformed specimen can *diminish* the dislocation density to an order of 10^5 to 10^6 mm^{-2}.

A typical dislocation density in ceramics is between 10^2 to 10^4 mm^{-2}. For silicon single crystals (used in manufacturing of integrated circuits), the value lies between 0.1 and 1 mm^{-2}.

8.19 CHARACTERISTICS OF DISLOCATIONS

Although there are several characteristics of dislocations that affect the mechanical properties of metals, yet the following two are important from the subject point of view.

1. strain fields that exist around dislocations
2. ability to multiply.

Both these characteristics are discussed in the following pages.

Strain Fields around Dislocations. When metals are plastically deformed, about 95% of the deformation energy is dissipated as heat while 5% is retained internally. The major portion of this retained energy is associated with dislocations.

In order to illustrate the above discussion further, let us consider the edge dislocation shown in Fig. 8.13. As seen from Fig. 8.13, some atomic lattice distortion exists along the dislocation line because of the presence of extra half-plane of atoms. As a consequence there are regions in which compressive, tensile, and shear lattice strains are imposed on the neighbouring atoms. The atoms immediately above and adjacent to the dislocation line are squeezed together. As a result, these atoms may be thought of as experiencing a compressive strain relative to atoms positioned in the perfect crystal and removed away from the dislocation line.

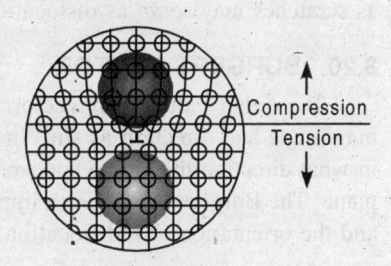

Fig. 8.13.

However, directly below the dislocation line, the situation is opposite. Here the lattice atoms experience tensile strain. Shear strains also exist in the vicinity of edge dislocation. For screw dislocation, lattice strains are pure shear only (*i.e.*, no tensile or compressive). These lattice distortions are considered to be strain fields that radiate from the dislocation line. The strains extend into the surrounding atoms. Their magnitude decrease with radial distance from the dislocation.

The strain fields surrounding dislocations in close proximity to one another may interact such that forces are imposed on each dislocation by the combined interactions of all its neighbouring dislocations. For example, let us consider too edge dislocations that have the same sign, and the identical slip plane as shown in Fig. 8.14.

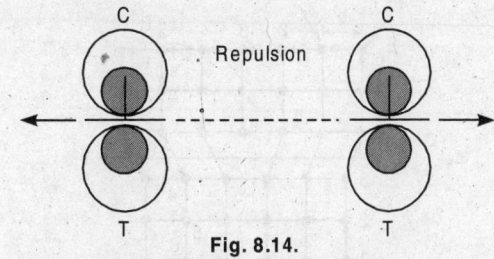

Fig. 8.14.

As seen from the diagram, the compressive and tensile strain fields for both lie on the same side of the slip plane. Because of this, the interaction between the strain fields is such that it produces a mutually repulsive force between the two isolated dislocations. This force of repulsion tends to move the two dislocations away from each other.

On the other hand, two dislocations of opposite sign and having the same slip plane will be attracted to one another as shown in Fig. 8.15. This situation leads to dislocation annihilation when they meet. That is two extra half-plane of atoms will allign and become a complete plane. As a matter of fact, dislocation interactions are possible between edge, screw and/or mixed dislocations and for a variety of orientations. These strain fields and associated forces are important in the strengthing mechanisms for metals.

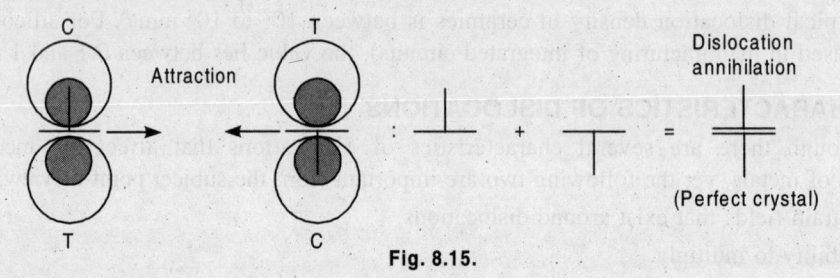

Fig. 8.15.

It will be 'interesting to know that during plastic deformation, the number of dislocations increases dramatically. The dislocation density in a metal that has been highly deformed may be as high as 10^{10} per mm^2. One major source of these new dislocations is existing dislocations which multiply. Besides this grain boundaries as well as internal defects and surface irregularities such as scratches may serve as dislocation formation sites during deformation.

8.20 BURGERS VECTOR

It is also called slip vector. The Burgers Vector 'b' is the vector, which defines the magnitude and direction as slip. In other words, it indicates how much above the slip plane and in what direction the lattice appears to have been shifted with respect to the lattice below the slip plane. The Burgers Vector is an important property of a dislocation, because if the Burgers Vector and the orientation of a dislocation line are known, the dislocation is completely described.

A convenient way of defining the Burgers Vector of a dislocation is by means of a Burgers Circuit. The Burgers Circuit may be obtained by considering atomic arrangement around an edge dislocation or screw dislocation. Starting from a lattice point, trace a path from atom to atom, an equal distance in each direction, always in the direction of one of the vectors of the unit cell. If the region enclosed by the path does not contain a dislocation, the Burgers Circuit will close. However, if the path encloses a dislocation the Burgers Circuit will not close. The closure failure of the Burgers Circuit is the Burgers's Vector 'b'.

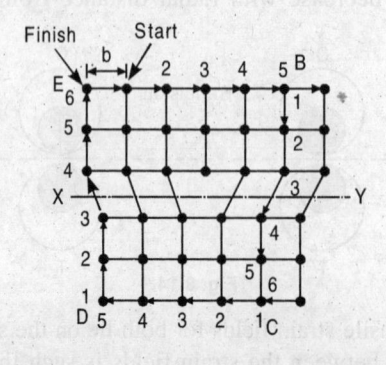

(a) Burgers circuit in edge dislocation   (b) Burgers circuit in screw dislocation

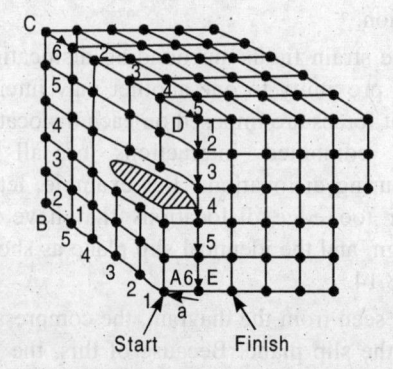

Fig. 8.16.

Deformation of Metals 183

Fig 8.16 (*a*) and (*b*) show the Burgers Circuit in an edge dislocation and screw dislocation respectively. The Burgers Circuit is the path traced by counting equal number of atoms in all the four directions. In Fig. 8.16 (*a*), let us start the Burger Circuit from one of the corner, say *A*. Let the circuit cover five unit cells and reach *B*. Similarly, from *B*, let the circuit cover another six units and reach *C*. Again from *C*, the circuit cover five units (*i.e.*, equal number of units as that of *AB*) and reach *D*. Finally, let the circuit covers six units (*i.e.*, equal number of units as that of *BC*) and reach *E*. The distance *EA* is known as Burgers Vector. It will be interesting to know that if the region enclosed by the circuit, does not contain any dislocation, then the Burgers Circuit must close.

It will be interesting to know that an edge dislocation lies perpendicular to its Burgers Vector. On the other hand, a screw dislocation lies parallel to its Burgers Vector. Similarly, an edge dislocation moves in the direction of Burgers Vector, while a screw dislocation moves in a direction perpendicular to the Burgers Vector.

8.21 DISLOCATION CLIMB

The term 'dislocation climb' may be defined as a process in which a dislocation moves in a direction perpendicular to the slip plane, Fig. 8.17 shows the circles, which indicate the arrangement of atoms in a crystal lattice,. In this figure, 1, 2 and 3 represent atoms in a dislocation plane and hatched region, a vacancy.

(*a*) Before dislocation   (*b*) After dislocation

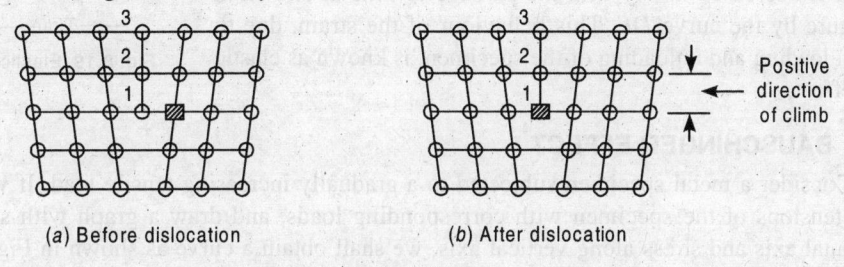

Fig. 8.17. Dislocation climb.

It has been observed that all the atoms surrounding vacancy have a tendency to replace it. But the atom '1' being in less equilibrium has a greater tendency to occupy the position of vacancy. The vacancy, in turn, occupies the position previously occupied by atom '1' as shown in Fig. 8.17 (*b*).

As a result of this, the length of dislocation plane is reduced from atoms 1-2-3 to 2-3 only. In other words, the length of dislocation plane has been pushed up one interatomic distance. This upward movement of dislocation is called positive direction of climb, Similarly, the downward movement of dislocation is called negative direction of climb.

8.22 MULTIPLICATION OF DISLOCATIONS (Frank-Read Generator)

The term 'multiplication of dislocation' may be defined as the process in which many dislocations are produced, automatically, one after the other. A number of mechanism have been proposed for multiplication of dislocations. But we shall discuss the most common mechanism, called Frank-Read Generator.

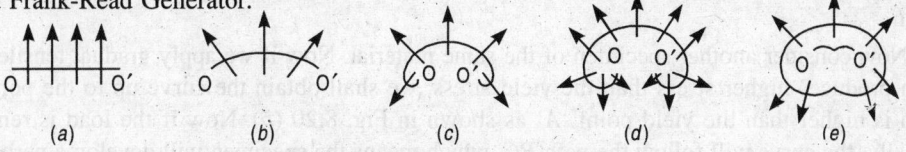

(*a*)   (*b*)   (*c*)   (*d*)   (*e*)

Fig. 8.18. Multiplication of dislocations.

Fig. 8.18 (*a*) indicates the presence of a dislocation line '0-0' in a slip plane at both the ends. Fig. 8.18 (*b*) shows that the dislocation line has bowed outwards due to the application of shear stress. As a result of this, the dislocation line produces a slip in the form of an arc. Fig. 8.18 (c) shows the continued movement of dislocation line first to form a semi-circular arc and then spiral about points '0-0'. Fig. 8.18 (d) shows further growth of the slipped region, which brings the two sections of the spiral together. Fig. 8.18 (e) shows further movement of the dislocation line, which produces an outerward loop around the entire slipped region and new dislocation line '0-0'. This process continues, and will begin another cycle.

8.23 ELASTIC AFTER EFFECT

The term 'elastic after effect' may be defined as the behaviour of the elastic strain due to sudden loading and unloading of a specimen. Consider a metal specimen, subjected to an impact load, well below its elastic limit. If we note the strains at corresponding times and plot a curve with time along the horizontal axis and strain along the vertical axis, we shall obtain a graph as shown in Fig. 8.19

The curve AB represents the sudden increase in strain due to impact load. If the load is maintained, the strain increases gradually as shown by the curve BC. Now if the load is removed, the strain drops immediately as shown by the curve CD. This drop in strain is equal to the initial increase in strain AB. The remaining strain DE decreases slowly with the passage of time as shown in the figure by the curve DF. This behaviour of the strain, due to sudden loading and unloading of the specimen, is known as elastic after effect.

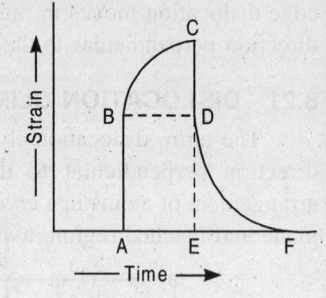

Fig. 8.19. Elastic after effect.

8.24 BAUSCHINGER EFFECT

Consider a metal specimen subjected to a gradually increasing tensile load. If we measure the extensions of the specimen with corresponding loads, and draw a graph with strain along horizontal axis and stress along vertical axis, we shall obtain a curve as shown in Fig. 8.20 (a).

(a) Before yield stress

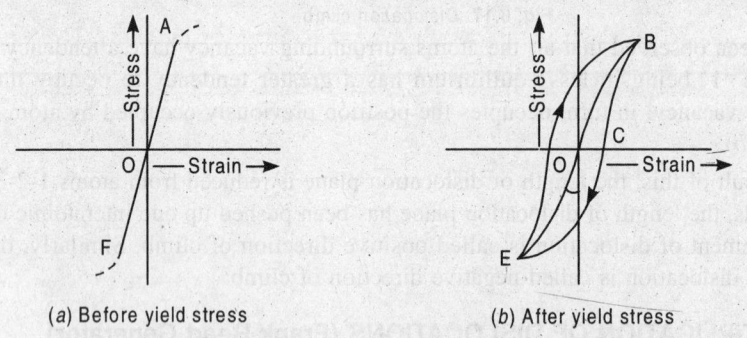

(b) After yield stress

Fig. 8.20. Stress-strain curve of a metal specimen.

The point 'A' on the curve represents the yield stress of the material loaded in tension. Similarly, the point 'F' represents the yield stress of the material when it is loaded in compression. It may be noted that the yield stresses at 'A' and 'F' will be equal in magnitude. But it is opposite in sign.

Now consider another specimen of the same material. Now if we apply gradual tensile load, which produces higher stress than the yield stress, we shall obtain the curve up to the point 'B' which is higher than the yield point 'A' as shown in Fig. 8.20 (b). Now if the load is removed gradually, the curve will follow the path BC, which means the specimen will develop a permanent strain OC. Now if the specimen is loaded with a gradually increasing compressive load, it will be seen that the plastic flow begins at 'D'. It will be seen that the compressive stress at D is lower

Deformation of Metals

than the original value of stress at 'F'. This reduction in-compressive yield stress, after the tensile loading of the specimen, is known as Bauschinger Effect. Similarly, if the specimen is loaded in compression initially, the Bauschinger Effect well be observed when the specimen is loaded in tension. The reduction in compressive (or tensile) stress is due to the presence of residual stress, even after the removal of the load. These stresses cause the dislocations to move more easily in a direction opposite to the original direction even at low stresses.

The degree of Bauschinger Effect is measured by the strain, which is known as Bauschinger Strain. This has been shown in the diagram as the difference in strains between the tension and compression curve.

8.25 DEFORMATION OF POLYCRYSTALLINE MATERIALS

As a matter of fact, the deformation of a polycrystalline material is different than that of single crystalline material. It is due to the reason that the deformation of a polycrystalline material takes place due to movement of dislocations along the slip system that has the most favourable orientation. This is illustrated in Fig. 8.21 by a photograph (at a micro level) of a polycrystalline copper specimen that has been plastically deformed.

Before the deformation, the copper surface was polished. As seen, slip lines are clearly visible. Notice that there are two sets of lines that are intersecting each other. This means that there were two slip systems that operated for most of the grains.

Fig. 8.21.

Moreover, the variation in grain orientation is indicated by the difference in alignment of the slip lines for the several grains.

It has been observed that the dislocations move under the action of applied load and caused slipping. If the movement of dislocations does not take place, the stress required to produce deformation will increase. This phenomenon may be better understood from the following example:

In a crystalline material, the grains have random orientations. On the application of a force, the grains which have their slip planes parallel to the direction of the force, deform more easily. But their deformation is restricted by the adjacent grains having different orientation. As a result of this, the grains develop a complex state of stresses.

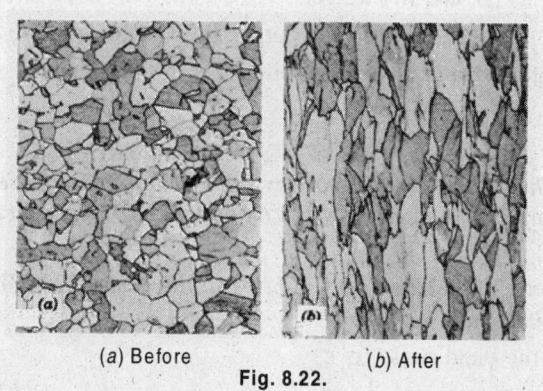

(a) Before (b) After
Fig. 8.22.

Fig. 8.22 shows the manner in which grains distort as a result of plastic deformation. Notice that the grain structure before deformation is equiaxed as shown in Fig. 8.22 (a). However after the deformation, the grains have elongated along the direction in which the specimen was extended. This is indicated in Fig. 8.22 (b).

The deformation of a polycrystalline material is also affected by the presence of grain boundaries. These areas act as physical barriers to the movement of dislocations. This results in piling up of dislocations at grain boundaries, which offer resistance to slipping. Thus a high stress is required to cause plastic deformation.

It will be interesting to know that greater the number of grain boundaries in a polycrystalline material, higher will be the resistance offered to the dislocation movement. Thus greater will be the stress required to produce plastic deformation. It is due to this fact, that a fine-grained material possess high tensile strength and better mechanical properties as compared to a coarse-grained material.

From the above discussion, we can conclude that the smaller the grains, closer are the grain boundaries and hence the strength is greater. The effect of grain size on the yield stress (σ_y) in steel is given by the relation,

$$\sigma_y = \sigma_i + \frac{k}{\sqrt{d}}$$

where d = Mean grain diameter in milimeters

k = A constant and

σ_i = A measure of the resistance of the material to dislocation motion due to effects other than grain boundaries in MP_a or MN/m^2.

The above relation is known as Patch equation.

Example 8.2. *The yield strength of a polycrystalline material increases from 115 MN m^{-2} to 215 MN m^{-2} on decreasing the grain diameter from 0.04 mm to 0.01 mm. Find the yield strength of the material when grain size is ASTM 9 for which the grain diameter is 0.016 mm.*

(A.M.I.E., Winter, 1999)

Solution. Given: Yield strength (σ_y) = 115 MN m^{-2} and 215 MN m^{-2} and Diameter of grains (d) = 0.04 mm and 0.01 mm.

We know that yield strength (σ_y),

$$115 = \sigma_i + \frac{k}{\sqrt{d}} = \sigma_i + \frac{k}{\sqrt{0.04}} = \sigma_i + 5k \qquad \text{... (i)}$$

and

$$215 = \sigma_i + \frac{k}{\sqrt{0.01}} = \sigma_i + 10k \qquad \text{... (ii)}$$

Solving equations (i), and (ii) we get,

$$100 = 5k \quad \text{or} \quad k = 20 \quad \text{and} \quad \sigma_i = 15 \text{ MN m}^{-2}$$

∴ Yield strength for a grain size (or diameter) of 0.016 mm,

$$\sigma_y = \sigma_i + \frac{k}{\sqrt{d}} = 15 + \frac{20}{\sqrt{0.016}} = 173.1 \text{ MN m}^{-2} \text{ **Ans.**}$$

Example 8.3. *The yield stress of a polycrystalline material increases from 120 MN m^{-2} to 220 MN m^{-2} on decreasing grain diameter from 0.04 mm to 0.01 mm. Find the yield stress for a grain size of 0.025 mm.*

(A.M.I.E, Summer, 1993)

Solution. Given: Yield stress (σ_y) = 120 MN m^{-2} and 220 MN m^{-2}. diameter of grains (d) = 0.04 mm and 0.01 mm.

We know that the yield stress (σ_y),

$$120 = \sigma_i + \frac{k}{\sqrt{d}} = \sigma_i + \frac{k}{\sqrt{0.04}} = \sigma_i + 5k \qquad \text{... (i)}$$

and

$$220 = \sigma_i + \frac{k}{\sqrt{0.01}} = \sigma_i + 10k \qquad \text{... (ii)}$$

Deformation of Metals

Solving equations (*i*) and (*ii*), we get
100 = 5k, or k = 20 and $\sigma_i = 20$
∴ Yield stress for a grain size of 0.025 mm.

$$\sigma_y = \sigma_i + \frac{k}{\sqrt{d}} = 20 + \frac{20}{\sqrt{0.025}} = 146.6 \text{ MN m}^{-2} \text{ Ans.}$$

8.26 Mechanisms of Strengthening in Metals

The engineers are often asked to design alloys having high strengths yet some ductility and toughness. Ordinarily, ductility is sacrificed when an alloy is strengthened. As a matter of fact, there are several hardening techniques at the disposal of an engineer. The selection of an alloy depends on the capacity of a material with the desired mechanical properties required for a particular application.

In order to understand the strengthening mechanisms, it is important to know the relation between dislocation motion and mechanical behaviour of metals. It is because of the fact that plastic deformation corresponds to the motion of large number of dislocations. The ability of a metal to plastically deform depends on the ability of dislocations to move.

Since hardness and strength are related to the case with which plastic deformation can be made to occur, therefore by reducing the mobility of dislocations, the mechanical strength can be enhanced. In other words, with the reduced mobility of dislocations, greater mechanical forces will be required to initiate plastic deformation.

On the other hand, the more unconstrained the dislocation motion is, the greater the ease with which a metal may deform, *i.e.* the metal will become softer and weaker. All the strengthening techniques are base on a simple principle: restricting or hindering dislocation motion renders a material harder and stronger.

We will keep our discussion for strengthening mechanisms limited to single-phase metals. Deformation and strengthening of multiphase alloys is more complicated and beyond the scope of this book.

In single-phase alloys, the commonly used mechanisms for strengthening are by :
1. Grain-size reduction
2. Solid-solution alloying and
3. Strain hardening or work hardening

Now we shall discuss all these methods one by one in the following pages.

8.27 Strengthening by Grain Size Reduction

It has been experimentally found that the size of the grains (or average grain diameter), in a polycrystalline metal affects the mechanical properties. Adjacent grains normally have different crystallographic orientations and of course, a common grain boundary as shown in Fig. 8.23.

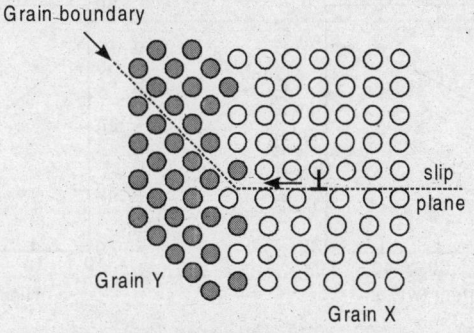

Fig. 8.23.

During plastic deformation, slip or dislocation motion must take place across the common boundary say from grain X to grain Y in Fig. 8.23. The grain boundary acts as a barrier to dislocation motion for the following two reasons:

1. Since the two grains are of different orientations, a dislocation moving into grain Y will have to change its direction of motion.
2. The atomic disorder within a grain boundary region will result in a discontinuity of slip planes from one grain into another.

It should be carefully noted that for high-angle grain boundaries, the dislocations may not traverse grain boundaries during deformation. Rather a stress concentration ahead of a slip plane in one grain may activate sources of new dislocations in an adjacent grain.

As mentioned earlier a fine-grained material is harder and stronger than the one that is coarse-grained. This is because of the fact that the fine-grained material has a greater total grain boundary area to resist dislocation motion.

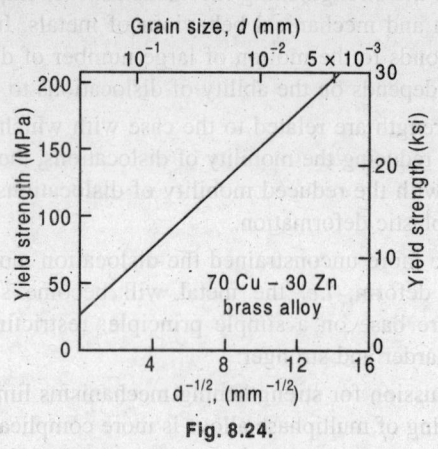

Fig. 8.24.

Fig. 8.24 shows the plot of yield strength versus grain size for a brass (70% Cu – 30% Zn) alloy. As seen from this figure, as the grain size is reduced from 0.1 mm to 0.005 mm, the yield strength of brass increases from 60 to 200 MPa.

8.28 Solid Solution Hardening

This is another technique to strengthen in and harden the metals. In this technique, the material is alloyed with impurity atoms that go into either substitutional or interstitial solid solution. Accordingly this method is called solid-solution strengthening. High-purity metals are always softer and weaker than alloys composed of the same base metal.

Increasing the concentration of the impurity atoms results in an increase in tensile and yield strengths. This fact is indicated in Fig. 8.25, for nickel in copper.

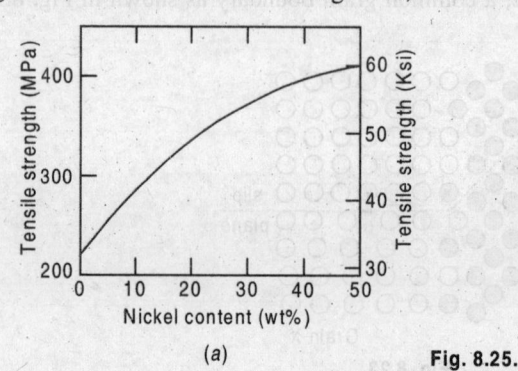

Fig. 8.25.
(a)   (b)

Deformation of Metals

Fig. 8.25 (a) shows the variation of tensile strength of the alloy versus the percentage of nickel content. Notice that as the nickel content is increased from 0 to 50%, tensile strength increases from 225 to 410 MPa. Fig. 8.25 (b) shows the variation of elongation or ductility with the increase in nickel content. As seen from this figure, ductility reduces from 55% to 30% with the increase in nickel content from 0 to 50%.

Alloys are stronger than pure metals because impurity atoms that go into solid solution impose lattice strains on the surrounding host atoms. These lattice strain field interactions between dislocations and these impurity atoms restrict the movement of dislocations.

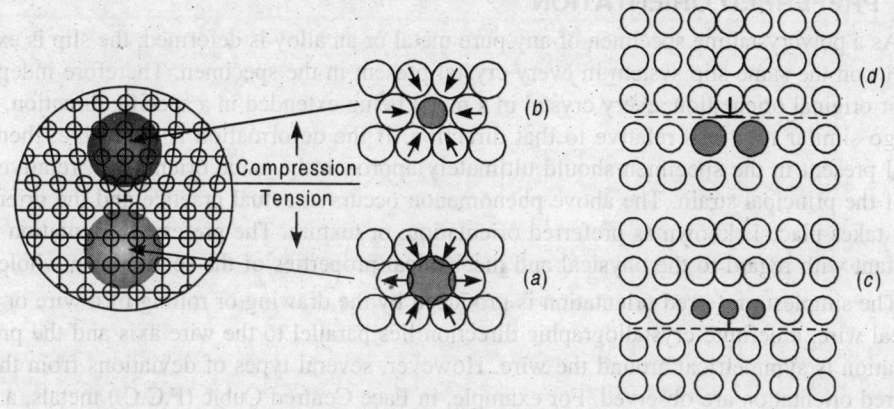

Fig. 8.26.

For example, an impurity atom that is smaller than the host atom for which it substitutes exerts tensile strains on the surrounding crystal lattice as shown in Fig. 8.26 (a). On the other hand, a large substitutional atom produces compressive strains in the vicinity as shown in Fig. 8.26 (b). These solute atoms tend to diffuse to and segregate around dislocations in a way so as to reduce the overall strain energy, *i.e.* to cancel some of the strain in the lattice surrounding a dislocation. To accomplish this, a smaller impurity atom is located where its tensile strain will partially nullify some of the dislocation's compressive strain. For the edge dislocation shown in Fig. 8.26 (c), this would be adjacent to the dislocation line and above the slip plane. A larger impurity atom would be situated as shown in Fig. 8.26 (d).

The resistance to slip is greater when the impurity atoms are present. This is because the overall strain must increase if a dislocation is torn away from them. Moreover, the same lattice strain interactions (indicated in Fig. 8.26 (c) and (d) will exist between impurity atoms and dislocations that are in motion during plastic deformation. As a result, a greater stress is necessary to first initiate and then continue plastic deformation for solid-solution alloys, as opposed to pure metals. This is evidenced by the enhancement of yield strength and hardness.

8.29 WORK HARDENING

It is also called strain hardening. The term 'work hardening' may be defined as a process of deforming a metal at room temperature to improve its hardness, tensile and fatigue strengths etc. We know that whenever a metallic piece is subjected to a load, beyond its elastic limit, some plastic deformation takes place. As a matter of fact, this deformation takes place due to slipping of the atomic planes. It has been observed that the slipping takes place more easily in those planes, which have favourable orientation, *i.e.*, the planes which are parallel to the direction of the load. As the deformation proceeds, the availability of slip planes decreases. This decreases the resistance of metal against plastic deformation. And the work piece is known to be work-hardened or strain-hardened.

It has been observed that the work hardening produces internal stresses in the metal. These stresses are produced due to:

1. Piling up of dislocations at grain boundaries.
2. Distortion of grains due to loading.

It will be interesting to know that the work hardening improves the hardness, yield point and strength of metals like iron, copper, aluminium nickel etc. But it decreases the ductility and electrical conductivity of the metals.

8.30 PREFERRED ORIENTATION

As a polycrystalline specimen of any pure metal or an alloy is deformed, the slip is expected to occur on the same slip system in every crystal present in the specimen. Therefore independent of their original orientation, every crystal in a metal being extended in a specific direction, should undergo similar rotations relative to that direction. If the deformation is extensive, then every crystal present in the specimen should ultimately approach the same orientation, relative to the axis of the principal strain. The above phenomenon occurs in actual practice and the orientation, which takes place is known as preferred orientation, or texture. The preferred orientation is very important with regard to the physical and mechanical properties of the section as a whole.

The simplest preferred orientation is produced by the drawing or rolling of a wire or rod. In an ideal wire, a definite crystallographic direction lies parallel to the wire axis and the preferred orientation is symmetrical around the wire. However, several types of deviations from the ideal preferred orientation are observed. For example, in Face Centred Cubic (F.C.C.) metals, a double preferred orientation is usually observed. The grains have either <111> or <100> parallel to the wire axis and have random orientations around the axis. The Body Centred Cubic (B.C.C.) metals have a simple <110> direction of preferred orientation. The preferred orientation in Hexagonal Close Placed (H.C.P) metals is not so simple.

The preferred orientation, resulting from deformation, is strongly dependent on the slip and twinning systems available for deformation. But it is not affected by processing variables such as die angle, roll diameter, roll speed etc. However, the direction of flow is the most important and process variable.

It has been observed that preferred orientation in metals or alloys results in an anisotropy in mechanical properties. The anisotropy means that the material properties vary in different directions. But when these crystals are combined in a random fashion, into a polycrystalline aggregate, the mechanical properties of the aggregate tend to be isotropic. However, the crystal alignment, which accounts for the preferred orientation, again introduces anisotropy in mechanical properties. As a result of this, different mechanical properties in different directions can result in an uneven response of the metal during forming and fabrication operations. The preferred orientation is of great advantage in magnetic applications, where it is used to produce grain-oriented or silicon-iron transformer sheets.

8.31 COLD WORKING

The term 'cold working' may be defined as the process of deforming plastically, a piece of metal below its recrystallisation temperature. The recrystallisation of a metal, usually, takes place within the temperature range of 0.3 to 0.5 times the melting temperature of a metal. Some of the important processes of cold working are drawing, squeezing, shearing, bending, extruding etc.

It has been observed that the process of cold working changes the various properties of metals such as hardness, ductility, tensile and fatigue strength etc. It increases the hardness, tensile and fatigue strength and the electrical resistance of a metal. But it decreases the ductility and creep resistance. The proportionality limit can also be increased with heavy cold working. The metals

Deformation of Metals

such as phosphor bronze and low carbon steel possess high proportionality limit, and are used as spring materials.

It has also been observed that during cold working, a certain amount of work done on the mental is stored internally in the form of strain-energy. This energy produces internal stresses in a cold worked metal. It is due to this fact, that these metals are liable to crack when subjected to heat treatment. The internal stresses can be removed by particular process of heat treatment called *annealing.

8.32 ADVANTAGES AND DISADVANTAGES OF COLD WORKING

Following are the advantages and disadvantages of cold working:

Advantages

1. It improves the mechanical properties like hardness, tensile and fatigue strength.
2. It prevents the loss of metal due to oxidation.
3. The cold worked metals have good surface finish.
4. It maintains a closer tolerance on dimension of a metal.

Disadvantages

1. It decreases ductility and creep resistance of a metal.
2. It requires greater energy to deform a metal plastically.
3. It produces internal stresses in a metal.
4. It produces distortion in the grain structure of a metal.

8.33 ANNEALING OF A COLD WORKED METAL

We have already discussed in Art. 8.26 that during cold working, a certain amount of work done on the metal is stored internally in the form of strain energy. This energy produces internal stresses in a cold worked metal, which leads to the cracking of metals. In order to relieve the metal from internal stresses, a particular process of heat treatment called *annealing is used. In this process, the metal is heated to a temperature below the melting point. In doing so the metal losses its stored energy and comes back to its strain-free condition. It has been observed that the metal losses its stored energy in the following three stages:

1. Recovery
2. Recrystallisation
3. Grain growth.

Now we shall discuss in detail the above mentioned three stages one by one.

8.34 RECOVERY

The term 'recovery' may be defined as the process of removing internal stresses, in a metal by heating it to a relatively low temperature, which is, usually, below the melting point.

It has been observed that the recovery process does not affect the grain structure. But it removes the internal stresses only. Moreover, the recovery process does not affect the hardness and strength. But it increases the ductility of the metal. As a result of cold working, the dislocations pile up at grain boundaries. During the process of recovery, these dislocations start reducing and rearrange themselves. They do so, through mechanism known as polygonization. In this mechanism, the dislocation climb out of their slip planes and rearrange themselves in a lower energy configuration.

---

* For details refer to Art. 11.6.

8.35 RECRYSTALLISATION

The term 'recrystallisation' may be defined as the process of forming strain-free new grains, in a metal, by heating it to a temperature known as recrystallisation temperature. It may be noted that the recrystallisation temperature is, usually, the temperature at which about 50% of the cold worked metal recrystallises in one hour. It has been observed that the formation of new grains in a recrystallisation takes place through the following three processes:

1. Nucleation. 2. Primary grain growth 3. Secondary grain growth.

It will be interesting to know, that during nucleation, small strain-free nuclei are formed at the grain boundaries where the deformation is very high. These nuclei grow into strain-free grains during primary grain growth. These grains meet each other and replace the old grains by the new ones. During the secondary grain-growth, these new grains grow at the expenses of others. At the end of this stage, the grains are, usually, of very small size but of the same shape. The factors, which affect recrystallisation process, are time of heating, temperature prior to deformation, impurity and alloy addition etc.

It has been observed that the increase in the time of heating or the annealing reduces the recrystallisation temperature. A certain amount of deformation is required before the crystallisation may take place. The addition if impurity or alloy increases the recrystallisation temperature. It may be noted that the recrystallisation process does not produce new structures. But it produces strain-free new grain and complete elimination of internal stresses. It results in a sharp decrease in the hardness and strength as well as increase in ductility.

8.36 GRAIN GROWTH

The term 'grain growth' may be defined as the process of forming strain-free grains larger in size in metal by heating it to a temperature above that of recrystallisation. It may be noted that the recrystallisation produces strain-free new grains. These grains are of smaller size, but of equal shape. When the temperature is increased above that of recrystallisation, these grains grow in size and the growth of grains takes place even during recrystallisation. But the growth rate is slow and becomes rapid with the increase of temperature. The grain growth takes place due to the combination of individual grains, thereby reducing their boundary area. As a result of this, the total energy decreases and the grains become stable. The factors, which affect the growth rate, are time of heating, temperature, degree of cold work and addition of impurities. It has been observed that the grain growth results in the decrease of hardness and strength as well as increase in ductility.

Note: The best way of indicating grain size is by specifying the number of grains per unit volume. But it is not a convenient way. As it is difficult to determine. A practical way of doing it is to count the number of grains that appear in unit length or unit area of a prepared section of a metal. In this case, the grain size is specified either by quoting a number of grains that appear per or square millimetre of cross-section or in a logarithmic manner as the A.S.T.M. (American Society for Testing Materials) grain size.

Let m = The number of grains per square millimetre of the actual section.

Now the grain size (as per A.S.T.M.) is given by the relation,

$$N = \frac{1}{0.693} \log_e \left(\frac{m}{8}\right) \quad \ldots (i)$$

Rearranging the above equation and simplifying, we get

$$m = 8 \times 2^N \quad \ldots (ii)$$

Deformation of Metals

For a given A.S.T.M. grain size, the above equation can be used to calculate the number of grains per square millimetre.

Example 8.4. *Determine the grain diameter of an A.S.T.M. number 9.*

(A.M.I.E. Summer, 1993)

Solution. Given: The grain diameter with A.S.T.M. number = 9

We know that the number of grains per square millimetre of the actual section,

$$m = 8 \times 2^N = 8 \times (2)^9 = 4096$$

∴ Average diameter of the grain

$$= \frac{1}{\sqrt{4096}} = 0.0156 \text{ mm Ans.}$$

8.37 VARIATION OF MECHANICAL PROPERTIES IN ANNEALING

We have already discussed in the previous articles the processes of recovery, recrystallisation and grain growth. It has been observed that the mechanical properties (such as strength, hardness, internal stresses, ductility and grain size) vary in all the three processes of annealing.

The variations in these mechanical properties as shown in Fig. 8.27 are described in detail as under:

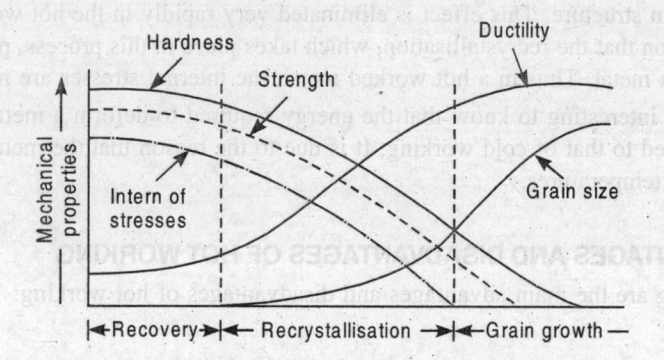

Fig. 8.27. Variation in mechanical properties.

1. *Hardness.* During recovery, it remains, somewhat, unaffected. But during recrystallisation, there is a sharp decrease. And during grain growth there is a slight decrease in hardness.
2. *Strength.* During recovery, it remains, somewhat, unaffected. But during recrystallisation, there is a sharp decrease. And during grain growth, there is also a slight decrease in strength.
3. *Internal stresses.* During recovery, they decrease. During recrystallisation there is further decrease in internal stresses and reach zero value.
4. *Ductility.* During recovery, it slightly increases. But during recrystallisation, there is a sharp increase in ductility. And during grain growth, there is slight increase in ductility. It may be noted that as the ductility of a metal increases, the internal stresses decrease.
5. *Grain size.* During recovery there is no change. But during recrystallisation there is a slight increase. And during grain growth there is a sharp increase in the grain size.

8.38 COMPARISON BETWEEN RECOVERY AND RECRYSTALLISATION

The following table gives the important points of comparison between recovery and recrystallisation:

Table 8.6

S. No.	Recovery	Recrystallisation
1.	It is a process of removing internal stresses in a metal by heating it to a relatively low temperature.	It is a process of forming strain-free new grains by heating it to a temperature known as recrystallisation temperature.
2.	It does not affect the grain size.	It affects the grain size.
3.	It reduces the internal stresses.	It removes the internal stresses completely.
4.	It does not affect the mechanical properties like hardness and strength.	It sharply decreases the mechanical properties like hardness and strength.

8.39 HOT WORKING

The term 'hot working' may be defined as the process of deforming plastically a metal piece at a temperature above the recrystallisation temperature. Some of the important processes of hot working are rolling, forging, hot spinning, hot extruding etc. It has been observed that the hot working improved the mechanical properties such as toughness, impact strength and ductility etc.

We have already discussed that the deformation produces strain hardening in a metal and distorts its grain structure. This effect is eliminated very rapidly in the hot working process. It is due to the reason that the recrystallisation, which takes place in this process, produces new strain-free grains in a metal. Thus in a hot worked metal, the internal stresses are not produced.

It will be interesting to know that the energy required to deform a metal in hot working is less as compared to that of cold working. It is due to the reason that the metal becomes soft and plastic at high temperatures.

8.40 ADVANTAGES AND DISADVANTAGES OF HOT WORKING

Following are the main advantages and disadvantages of hot working:

Advantages
1. It improves the mechanical properties like toughness, impact strength and ductility of metals.
2. It reduces the energy required to deform a metal.
3. It eliminates the blow holes and porosity.
4. It is rapid and economical for forming almost all commercial metals.

Disadvantages
1. A considerable amount of metal is lost due to oxidation at high temperatures.
2. The hot worked metals have poor surface finish due to the rapid oxidation.
3. The dimensional tolerances are greater than that in cold worked metals.
4. The structure and properties of metals are not uniform over the whole cross-section.

8.41 COMPARISON BETWEEN COLD WORKING AND HOT WORKING

The following table gives the important points of comparison between cold working and hot working of metals:

Deformation of Metals

Table 8.7

S. No.	Cold Working	Hot Working
1.	It takes place at a temperature below the recrystallisation temperature.	It takes place at a temperature above the recrystallisation temperature.
2.	In this process, no appreciable recovery takes place during deformation	In this process, the deformation and recovery takes place simultaneously.
3.	It produces internal stresses in a metal.	It does not produce internal stresses in a metal.
4.	It improves hardness, tensile and fatigue strengths.	It does not affect hardness, tensile and fatigue strengths.
5.	It reduces toughness, impact strength and ductility.	It improves toughness, impact strength and ductility.
6.	It produces a good surface finish on the metals.	It does not produce a good surface finish on the metals.

A.M.I.E. (I) EXAMINATION QUESTIONS

1. Explain the advantages and disadvantages of hot working of materials. *(Summer, 1993)*
2. What do you mean by 'preferred orientation? Discuss the effect of preferred orientation upon elastic properties. *(Winter, 1993)*
3. Distinguish between elastic deformation and elastic after effect. *(Winter, 1993)*
4. (a) Distinguish between slip and twin mechanisms of plastic deformation in metals.
   (b) Explain the phenomenon of recrystallisation. *(Summer, 1994)*
5. Distinguish between hot and cold working. *(Winter, 1994; Winter, 1993)*
6. Distinguish between elastic and plastic deformation of a solid. *(Summer, 1995)*
7. (a) Discuss the difference between slip and twinning.
   (b) Define critical resolved shear stress and discuss its dependence on temperature.
   (c) What is strain hardening? Draw shear stress versus shear strain curves for single crystals of iron, copper and magnesium. *(Summer, 1995)*
8. What is Bauschinger's effect? *(Summer, 1996)*
9. (a) Explain two modes of plastic deformation using sketches.
   (b) Draw neat sketches of (i) an edge dislocation and (ii) screw dislocation Derive the relationship between the Burger's vector and dislocation line in each case. *(Winter, 1996)*
10. (a) Distinguish between cold working and hot working.
    (b) Explain Bauschinger's effect with a sketch. What is its significance?
    (c) Derive an expression for critical resolved shear stress in a material subjected to uniaxial tensile loading. *(Winter, 1996)*
11. (a) What is dislocation and how its if formed.
    (b) What is Bauschinger's effect ? Explain to effect in deformation of metals.
    (c) Differentiate between hot working and cold working. Explain the effect of each of them on material properties. *(Summer, 1997)*
12. Distinguish between recovery and recrystallization. *(Summer, 1997)*
13. Distinguish between shear stress and critical resolved shear stress. *(Summer, 1997)*
14. What is critical resolved shear stress? Derive an expression for it, *(Winter, 1995)*
15. Distinguish between elastic and plastic deformation of metals. Define yield stress and **uniform elongation**. *(Winter, 1998)*

16. What is Bauschinger effect? What do you mean by critical resolved shear stress ? *(Summer, 1999)*
17. Why metals are polycrystalline ? What is whisker ? Explain briefly. *(Winter, 1999)*
18. Show how two edge dislocations of opposite sign on the same slip plane can innihilate each other. Can two screw dislocations of opposite sign also innihilate each other ? *(Winter, 1999)*
19. What is the difference between elastic and plastic deformation. *(Winter, 1999)*
20. Explain the vector quantity known as "Burger's vector". Show with sketch two vector quantities there on. *(Winter, 1999)*
21. (a) What is recovery, recrystallization and grain growth ?
    (b) Explain the term work hardening or strain hardening with its reasons. How does it differ from strain aging ?
    (c) What is Bauschinger effect ? What do you mean by critical resolved shear stress ?
    (Winter, 1999)
22. How does dislocation density influence mechanical properties? Is dislocation density in materials influenced by annealing ? *(Summer, 2000)*
23. What is slip? How it is measured? Distinguish slip and turning. Why stress required for slop in a actual metals is considerable less than the theoretically calculate stress. *(Summer, 2000)*
24. Why are most metals and alloys used in common applications polycrystalline in nature ? Is it possible to form single crystals of metals and alloys ? Describe a common method for measuring grain size in metals and alloys. *(Summer, 2000)*
25. (a) What is Burger's Vector ? How does dislocation density influence mechanical properties ? Is dislocation density in materials influenced by annealing ?
    (b) Explain phenomenon of yielding in mild steel. Why is the yield point in copper is not distinct ?
    (Summer, 2000)
26. Explain the phenomenon of :
    (i) Recovery   (ii) Recrystallization and   (iii) Grain growth
    What two parameters constitute a slip system. *(Summer, 2000)*
27. What is meant by critical resolved shear stress ? Derive its expression. *(Summer, 2000)*
28. (a) Define (a) screw dislocation (b) jog (c) stacking fault energy (d) shockley partial, (e) low-angle grain boundary, and (f) critical resolved shear stress. *(Winter, 2000)*
    (b) Discuss the mechanism of dislocation multiplication by the Frank-Read Source. *(Winter, 2000)*
29. Discuss the mechanism of dislocation multiplication by Frank-Read Source. *(Winter, 2000)*
30. Why does continuous cold working make a material harder ? How can its softness be recovered ? *(Winter, 2000)*
31. What is a slip ? How it is measured ? Distinguish slip with twinning ? Why stress required for slip in actual metals is considerably less than the theoretical calculated stress. *(Winter, 2000)*
32. (a) Why does Continuous cold working make a material harder ? How can its softness be recovered ?
    (b) Rolling of pure lead at room temperature can be called hot rolling – explain. *(Winter, 2000)*
33. Explain the phenomenon of work hardening in metals. *(Winter, 2001)*
34. What are mechanical twins ? Why HCP metals deform by twinning ? *(Winter, 2002)*
35. What do you mean by dislocation density ? How does it affect various mechanical properties ?
    (Summer, 2003)
36. Briefly explain the following : (i) season cracking and (ii) polarization. *(Winter, 2003)*

9
Fracture of Metals

1. Introduction. 2. Causes of Fracture. 3. Classification of Fractures. 4. Brittle Fracture. 5. Mechanism of Brittle Fracture: Griffith's Theory. 6. Ductile Fracture. 7. Mechanism of Ductile Fracture. 8. Ductile to Brittle Transition. 9. Creep Fracture. 10. Mechanism of Creep Fracture. 11. Fatigue Fracture. 12. Mechanism of Fatigue Fracture. 13. Non-destructive Tests. 14. Types of Non-destructive Tests. 15. Ultrasonic Test. 16. Radiographic Test. 17. Magnetic Particle Test.

9.1 INTRODUCTION

The term 'fracture' of a material may be defined as its fragmentation or separation, under the action of an external force, into two or more parts. It may occur as a sudden breaking up of a material either as the end result of extensive plastic deformation or as a result of fatigue in a part of the material. In actual practice, the fracture of materials is a serious problem, and should always be avoided.

9.2 CAUSES OF FRACTURE

It has been observed that most of the materials possess some or the other weakness due to the presence of submicroscopic defects known as cracks. These cracks act as the points of stress concentration. The actual fracture starts at these points, and propagates to cause complete failure. There are a number of potential sources of such cracks, Moreover, surface roughness and surface scratches also serve as the notches for stress concentration. It has been found that the maximum stress concentration occurs at the corners of the notch. When the stress exceeds the cohesive strength in the region, the crack itself propagates. It has been observed that, usually, the magnitude of stress concentration at these cracks is much higher than that at normal cross-section. Internal cracks or voids may also develop during the formation of the solids. In amorphous solids, the particles of dirt, dissolved gases etc. also serve as potential sources of submicroscopic defects. In crystalline materials, the impurities, blow holes, clusters of vacancies, dislocation clusters etc. also serve as the potential source of defects, which prove to be responsible for fracture.

9.3 CLASSIFICATION FRACTURES

The fractures may be broadly classified into the following four types:

1. Brittle fracture. 2. Ductile fracture. 3. Creep fracture. 4. Fatigue fracture.

It will be interesting to know that a brittle fracture takes place by a rapid propagation of crack with negligible plastic deformation. The ductile fracture takes place by a slow propagation of the crack after extensive plastic deformation. The creep fracture takes place due to continuous deformation of materials, under constant stress at elevated temperatures. And the fatigue fracture takes place due to repeatedly applied stresses.

In the following pages, we shall discuss all the above mentioned types of fractures one by one.

9.4 BRITTLE FRACTURE

The term 'brittle' fracture' may be defined as a fracture which takes place by the rapid propagation of a crack with negligible deformation. It has been observed that in amorphous materials, the fracture is completely brittle. But in crystalline materials, it occurs after a small deformation.

It may be noted that in crystalline materials, the fracture takes place normal to the specific crystallographic planes, called cleavage planes. The brittle fracture, generally, occurs in body centred cubic and hexagonal close packed single crystals at very low temperatures. It may also occur in crystals of graphite, talc and mica. In Polycrystalline materials, the fracture takes place along the grain boundaries, because of the tendency of brittle fracture increases with the decrease in temperature and higher rate of straining.

Fig. 9.1. Brittle fracture in mild steel.

Fig. 9.1 shows the photograph taken using scanning electron microscope (SEM) of a brittle fracture in mild steel. Notice that the fracture surface has a grainy texture. The grainy texture is due to the result of changes in orientation of the cleavage planes from grain to grain. This feature is more evident in the scanning electron micrograph shown in Fig. 9.2 (a). This type of fracture is called transgranular (or transcrystalline) because the fracture cracks pass through the grains.

However in some alloys, crack propagation is along grain boundaries. This type of brittle fracture is called intergranular. Fig. 9.2 (b) shows a scanning electron microphotograph showing a typical intergranular fracture. Notice that the shape of the grains is 3-dimensional. This type of fracture generally results subsequent to the occurrence of processes that weaken or embrittle grain boundary regions.

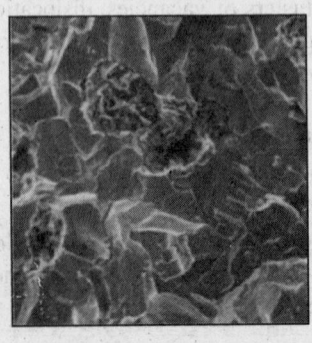

(a) Transgranular   (b) Intergranular
Fig. 9.2. Brittle fracture.

9.5 MECHANISM OF BRITTLE FRACTURE: GRIFFITH'S THEORY

It has been observed that the stress required for a material, at which it fractures, is only a small fraction of the cohesive strength. The discrepancy led Griffith to suggest that the low observed strengths were due to the presence of microcracks, which act as the points of stress concentration.

In order to explain the mechanism of brittle fracture, let us consider the stress distribution in the vicinity of crack and the conditions, under which it propagates with constant velocity.

Now consider a crack of an eliptical cross-section in a rectangular specimen as shown in Fig. 9.3

Let
- σ = Tensile stress applied to the specimen,
- C = Half length of the crack, and
- r = Radius of the curvature at its tips

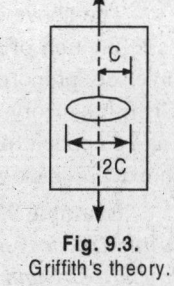

Fig. 9.3. Griffith's theory.

It has been observed that when a tensile stress is applied to the specimen, the stress is distributed about the crack in such a way that the maximum stress occurs at its tips. The expression for maximum stress at the tip of the crack is given by the relation.

$$\sigma_{max} = 2\sigma\sqrt{\frac{C}{r}}$$

We know that a certain amount of energy is always stored in a material before the propagation of crack, is known as elastic strain energy. The energy is released when the crack begins to propagate. We also know that as the crack propagates, new surfaces are created and a certain amount of energy must be provided to create them. According to Griffith's theory such a crack will propagate, when the released strain energy is just sufficient to provide the surface energy for the creation of new surfaces. The expression for total elastic strain energy is given by the relation:

$$U_E = -\frac{\pi \cdot C^2 \cdot \sigma^2}{E}$$

where E is the Young's modulus of elasticity. The negative sign indicates that the elastic strain energy stored in the material is released as the crack formation takes place.

Now if γ is the surface energy per unit area in joules/m^2. then the surface energy due to the presence of a crack of length $2C$ is given by the relation.

$$U_S = 4C \cdot \gamma$$

The above expression has been multiplied by 2, because there are two surfaces. The total change in potential energy, resulting from the creation of the crack,

$$U = U_E + U_S = -\frac{\pi \cdot C^2 \cdot \sigma^2}{E} + 4C \cdot \gamma$$

According to the Griffith's criterion, the crack will propagate under the effect of a constant applied stress (σ) if an incremental increase in length produces no change in the total energy of the system. In other words, the increased surface energy is compensated by a decrease in elastic strain energy. Mathematically, the above criterion may be expressed as follows:

$$\frac{dU}{dC} = \frac{d}{dC}\left(-\frac{2\pi \cdot C^2 \cdot \sigma^2}{E} + 4C \cdot \gamma\right) = 0$$

or
$$-\frac{2\pi \cdot C^2 \cdot \sigma^2}{E} + 4C \cdot \gamma = 0$$

$$\therefore \quad \sigma^2 = \frac{2 \cdot E \cdot \gamma}{\pi C} \quad \text{or} \quad \sigma = \sqrt{\frac{2E \cdot \gamma}{\pi C}}$$

The above expression gives us the stress required to propagate a crack in a brittle material as a function of the size of the microcrack. The expression indicates that the fracture stress is inversely proportional to the square root of the crack length. It will be interesting to know that the Griffith's theory, discussed above, is valid only for a perfect brittle material like glass. The theory can be applied for metals with certain modifications such as the inclusion of plastic work required to extend crack wall (p) and crack extension force (G).

Example 9.1. *The length of a crack in a steel is 4 μm. Taking E = 200 GN/m², estimate the brittle fracture strength at low temperatures, if the true surface energy is 1.48 J/m². The actual fracture strength is found to be 1230 MN/m². Explain the difference if any between this and your result.*

Solution. Given: Full-length of the crack ($2C$) = 4 μm = 4 × 10⁻⁶ m ; E = 200 GN/m² = 200 × 10⁹ N/m² ; γ = 1.48 J/m², Actual fracture strength = 1230 MN/m² = 1230 × 10⁶ N/m²

We know that fracture strength,

$$\sigma = \sqrt{\frac{2E \cdot \gamma}{\pi C}} = \sqrt{\frac{2 \times (200 \times 10^9) \times 1.48}{\pi \times \{(4 \times 10^{-6})/2\}}}$$

$$= 3.07 \times 10^8 \text{ N/m}^2 = 307 \text{ MN/m}^2 \text{ Ans.}$$

It is evident from the above result that the actual fracture strength (1230 MN/m²) is 4 times larger than the calculated value. As the iron plastically deforms, the higher observed strength can be attributed to the plastic work done by the crack as it propagates.

Example 9.2. *A glass sample has a crack length of 4.2 μ m. If the Young's modulus of the glass is 70 GN/m² and the specific energy is 1.1 J/m². Estimate the fracture strength and compare it with Young's modulus.*

Solution. Given: full-length of the crack ($2C$) = 4.2 μ m = 4.2 × 10⁻⁶ m; E = 70 GN/m² = 70 × 10⁹ N/m² and γ = 1.1 J/m².

We know that fracture strength,

$$\sigma = \sqrt{\frac{2E \cdot \gamma}{\pi C}} = \sqrt{\frac{2 \times (70 \times 10^9) \times 1.1}{\pi \times [(4.2 \times 10^{-6})/2]}}$$

$$= 1.528 \times 10^8 \text{ N/m}^2 \text{ Ans.}$$

Note. The fracture strength of 1.528 × 10⁸ N/m² or 152.8 MN/m² is 1/450ᵗʰ of the Young's modulus which is 70 GN/m². Thus the Griffith's criteria bridges the gap between the observed and ideal strengths of brittle materials.

Example 9.3. *A relatively large plate of glass is subjected to a tensile stress of 36 MN/m². If the specific surface energy and modulus of elasticity for this glass are 0.27 J/m² and 70 GN/m² respectively, determine the maximum length of a surface flow (or crack) that is possible without fracture.*

Solution. Given: σ = 36 MN/m² = 36 × 10⁶ N/m² ; γ = 0.27 J/m² and E = 70 GN/m² = 70 × 10⁹ N/m²

We know that fracture stress, (σ)

Fracture of Metals

$$36 \times 10^6 = \sqrt{\frac{2E \cdot \gamma}{\pi C}} = \sqrt{\frac{2 \times (70 \times 10^9) \times 0.27}{\pi C}}$$

or

$$(36 \times 10^6)^2 = \frac{2 \times (70 \times 10^9) \times 0.27}{\pi C}$$

$$C = 9.284 \times 10^{-6} \text{ m} \quad \text{or} \quad 9.284 \text{ } \mu\text{m}$$

and maximum length of surface flow, (or crack),

$$2C = 2 \times 9.284 = 18.568 \text{ } \mu\text{m} \textbf{ Ans.}$$

9.6 DUCTILE FRACTURES

The term 'ductile' fracture' may be defined as the fracture, which takes place by a slow propagation of crack with appreciable plastic deformation. It has been observed that the ductile fracture, generally, takes place in metals which do not work-harden much.

It has been seen that the ductile fracture is the end result of the extensive plastic deformation of a specimen in a tensile test. If we note the deformation of the specimen and the corresponding force (*i.e.,* load) and plot a graph with strain along the horizontal axis and stress along the vertical axis, we shall obtain a * curve as shown in Fig. 9.4.

The curve *OA* indicates that the ratio of stress to the strain is constant. The curve *AB* indicates that the ratio of stress to the strain is not constant. The point '*B*' is known as elastic point. The curve *BC* indicates that the strain increases more quickly than the stress (in comparison to *OA* or *OB*). The point '*C*' is known as yield point. The curve *CD* indicates that strain increases without any increase in stress. The curve *DE* indicates that higher stresses are required for higher values of strain. The stress, corresponding to the point '*E*', is known as ultimate tensile stress. As the stress is increased beyond this point, the fracture takes place.

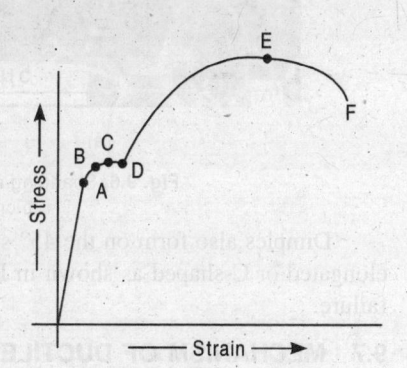

Fig. 9.4. Ductile Fracture.

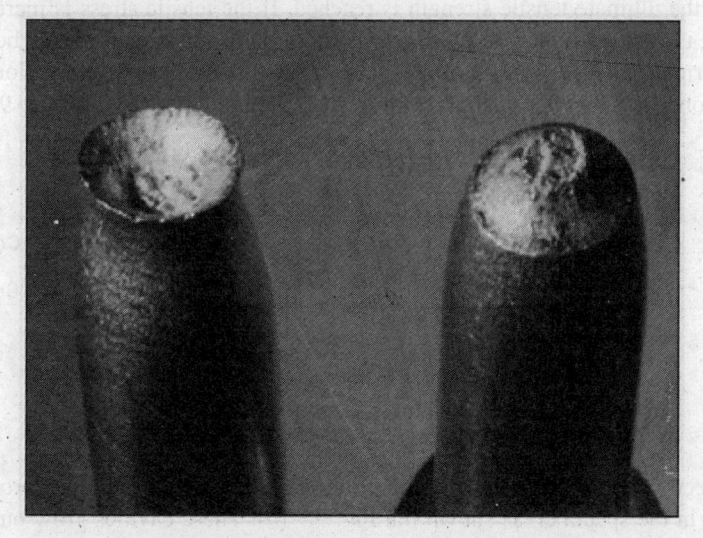

Fig. 9.5. Ductile fracture-cup and cone fracture in Al.

---

* For details, please refer to Art. 7.3.

Fig. 9.5 shows the ductile fracture in aluminium. In this type of fractured specimen, the central interior region of the surface has an irregular and fibrous appearance. This appearance is indicative of plastic deformation.

Fig. 9.6 (*a*) shows the scanning tunnel micrograph of a ductile fracture. Notice that the fracture consists of numerous spherical "dimples". This structure is characteristic of fracture resulting from uniaxial tensile failure. Each dimple is one-half of the microvoid that formed and then separated during the fracture process.

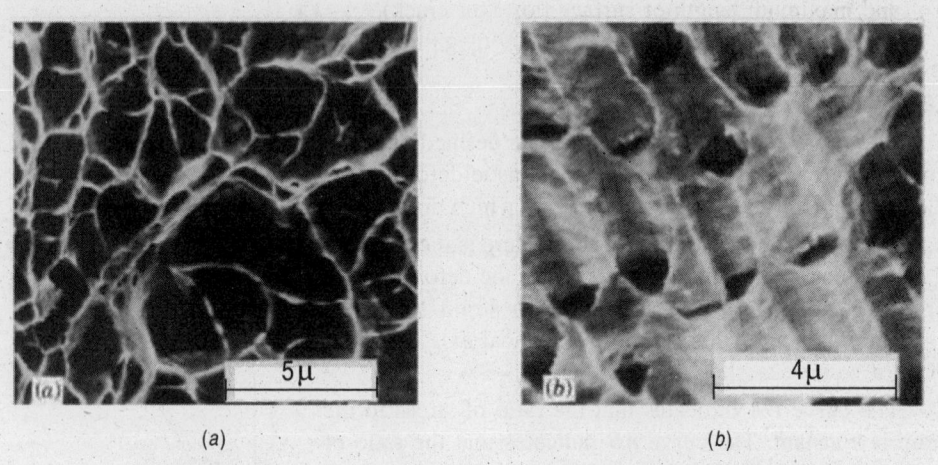

(a) (b)

Fig. 9.6. Scanning electron microscopy-spherical dimples corresponds to micro-cavities that initiate crack formation.

Dimples also form on the 45° shear lip of the cup-and-cone fracture. However these will be elongated or C-shaped as shown in Fig. 9.6 (*b*). This parabolic shape may be indicative of shear failure.

9.7 MECHANISM OF DUCTILE FRACTURE

We have already discussed that when the tensile stress, across the specimen, is increased, beyond the elastic limit, there is a uniform reduction in its cross-sectional area. The precess continues till the ultimate tensile strength is reached. If the tensile stress is increased beyond this value, a neck is formed somewhere near the middle of the specimen. It has been observed that during the formation of neck, small cracks are formed due to the combination of dislocations, which were formed during the manufacture of the specimen as shown in Fig. 9.7 (*a*).

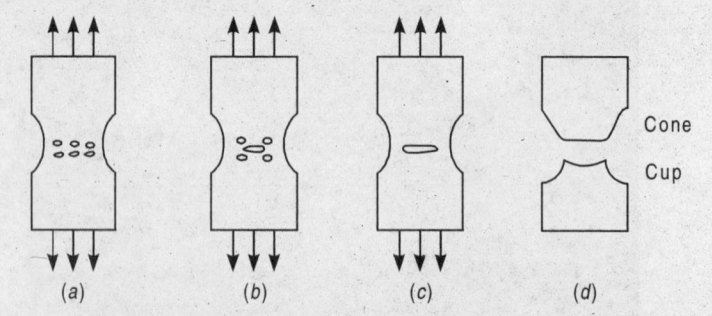

(a)   (b)   (c)   (d)

Fig. 9.7. Mechanism of ductile fracture.

It has been observed that the continuation of the plastic deformation produces cavities or bigger cracks in the specimen as shown inn Fig. 9.7 (*b*). These cavities grow outward and form a central crack as shown in Fig. 9.7 (*c*). On further increase of stress, the crack propagates on the surface of the specimen, which results in a cup and cone type of fracture as shown in Fig 9.7. (*d*).

Fracture of Metals

It may be noted that a ductile fracture occurs by a slow propagation of crack (*i.e.*, slow tearing of the metal) with a considerable amount of energy expenditure. The appearance of the fractured surface is rough and dull. Moreover, a considerable amount of slip is apparent. It will be interesting to know that in a ductile fracture, the crack continues to propagate, so long as the material is strained. If we stop the straining of material, the development of crack also stops. In actual practice, the ductile fracture is not of much significance.

9.8 DUCTILE TO BRITTLE TRANSITION

The ductile to brittle transition is commonly observed in B.C.C. metals. It is not observed in most of the F.C.C. metals. The ductile to brittle transition is observed at low temperatures, extremely high rates of strain or notching the material. The notched bar impact test described in the chapter on Mechanical Tests of Metals can be used to determine the temperature range over which the transition from ductile to brittle takes place. Such a temperature is called transition temperature.

The B.C.C. metal require a high stress to move dislocations and this stress increases rapidly as the temperature is lowered. On the other hand, the stress required to propagate a crack is least affected by temperature. At the transition temperature, the stress to propagate a crack (called fracture stress) is equal to the stress to move dislocations (called yield stress). At temperatures higher than the transition temperature, the yield stress reaches earlier than the fracture stress and the material first yields plastically. However, at temperatures lower than the transition temperature, the fracture stress reaches earlier and the material becomes brittle. Thus at all temperatures, below the transition temperature, the fracture stress is smaller than that of yield stress. It means that fracture stress may be controlled by the yield stress. As soon as the applied stress reaches a value equal to the yield stress, the crack is nucleated at the intersection of the slip planes and propagates rapidly.

It will be interesting to know that the yield stress as well as the fracture stress is a function of grain size. These stresses increase with the decreasing grain size. Therefore the fine-grained metals have a lower transition temperature as compared to the course grained metals.

Example 9.4. *For molybdenum, the temperature and strain rate dependence of yield stress in MN/m² is given by:*

$$\sigma_y = 20.6 + \frac{173600}{T} + 61.3 \log_{10} \varepsilon$$

where 'T' is the temperature in K and ε is the strain rate in sec⁻¹. Sharp cracks of half-length 2 μm are present in the metal. If E = 350 GN/m² and specific energy is 2 J/m², estimate the temperature at which the ductile-to-brittle transition occurs at a strain rate of (a) 10^{-2} per second and (b) 10^{-5} per sec.

Solution. Given: $C = 2 \, \mu\text{m} = 2 \times 10^{-6}$ m; $E = 350$ GN/m² $= 350 \times 10^9$ N/m²; $\gamma = 2$ J/m² ;

We know that fracture stress,

$$\sigma = \sqrt{\frac{2E \cdot \gamma}{\pi C}} = \sqrt{\frac{2 \times (350 \times 10^9) \times 2}{\pi \times 2 \times 10^{-6}}} = 472 \text{ MN/m}^2$$

(a) Temperature for Ductile-to-brittle transition at a strain rate of 10^{-2} per second

The yield stress

$$\sigma_y = 20.6 + \frac{173600}{T} + 61.3 \log_{10} \varepsilon$$

and the yield stress at *a* strain rate of 10^{-2} per second,

$$\sigma_y = 20.6 + \frac{173600}{T} + 61.3 \log_{10}(10^{-2})$$

Substituting for σ_y ($= \sigma$) in the above equation, we get,

$$472 = 20.6 + \frac{173600}{T} + 61.3 \times (-2)$$

$$\therefore \quad T = \frac{173600}{472 - 20.6 + (61.3 \times 2)} = 302.4 \text{ K Ans.}$$

(b) Temperature for Ductile-to-brittle transition at a strain rate of 10^{-5} per second

We know that the yield stress,

$$472 = 20.6 + \frac{173600}{T} + 61.3 \log_{10}(10^{-5})$$

or $$472 = 20.6 + \frac{173600}{T} + 61.3 \times (-5)$$

$$\therefore \quad T = \frac{173600}{472 - 20.6 + (61.3 \times 5)} = 229 \text{ K Ans.}$$

9.9 CREEP FRACTURE

The term 'creep fracture' may be defined as the fracture, which takes place due to excessive creeping of materials, under steady loading. It is always exhibited in metals like iron, nickel, copper and their alloys at higher temperatures. But some metals like zinc, tin, lead and their alloys also creep at room temperature. It has been observed that the tendency of creep fracture increases with the increase in temperature, higher rate of the straining etc. It has been observed that the creep resistance may be increased by the addition of certain elements such as cobalt, nickel manganese, tungsten etc.

9.10 MECHANISM OF CREEP FRACTURE

The creep fracture may take place due to the following two processes:
1. Sliding of grain boundaries.
2. Movement of dislocations, from one slip to another, by climbing.

It has been observed that at moderate stresses and temperatures, the creep occurs due to the sliding of grain boundaries only. But at higher stresses and temperatures, it occurs due to the movement of dislocations by climbing. It has also been observed that a creep fracture, which takes place due to continuation creep, is known as intercrystalline fracture, as it takes place along the grain boundaries.

It will be interesting to know that some fractures initiate along the grain boundaries, which act as points of high stress concentration. In such cases, the crack generally initiates along the lines where three or more boundaries meet or intersect. But sometimes, the fractures initiate due to the growth of voids along grain boundaries. With the passage of time, these voids grow and combine with each other to form cracks. This type of fracture, usually, occurs when small stresses are applied for a longer period.

The factors, which affect creep fracture are grain size, strain hardening, heat treatment alloying etc. But by controlling all these factors, the tendency to creep fracture may be greatly avoided.

9.11 FATIGUE FRACTURE

The term 'fatigue fracture, may be defined as the fracture which takes place under repeatedly applied fatigue stresses. It occurs at stresses well below the tensile strength of the materials. The fatigue stress is always exhibited in some parts of the machine parts such as axles, shafts, crank shafts etc.

9.12 MECHANISM OF FATIGUE FRACTURE

The fatigue fracture takes place due to the initiation of microcracks at the surface of the materials. It results in to and fro motion of dislocations near the surface. The microcracks act as the points of stress concentration. The excessive stress helps the propagation of crack with every cycle of stress application. The extent, to which a crack may propagate, depends upon the brittleness of the material. In brittle materials, the crack first grows to a critical size and then propagates rapidly through the material. But in ductile materials, the crack grows slowly, until the remaining cross-sectional area cannot support the applied stress. The fracture then takes place, rapidly, in a ductile manner.

It has been observed that the tendency of fatigue fracture increases with the increases in temperature, higher rate of straining etc. The fatigue resistance of a material may be increased by a careful design of the machine parts, so that the points of stress concentration are avoided.

9.13 NON-DESTRUCTIVE TESTS (NDT)

We have already discussed in chapter 8 the various types of mechanical tests. These tests are performed on the specimens made from the materials to determine the values of their different mechanical properties such as tensile strength, compressive strength, hardness, toughness etc.

The metal specimens, after tests, in all the above cases are destroyed. Or in other words, the specimens, after tests cannot be reused. That is why, these tests are known as destructive tests. There is another important class of tests, in which the specimen is not destroyed and can be reused after test. These tests are known as non-destructive tests. As a matter of fact, the non-destructive tests are performed to inspect the materials against any defects such as gas porosity, blow holes etc.

9.14 TYPES OF NON-DESTRUCTIVE TESTS

Though there are many non-destructive tests, which are carried out these days, yet the following are important from the subject point of view:

1. Ultrasonic test. 2. Radiographic test. 3. Magnetic particle test.

Now we shall discuss all the above mentioned non-destructive tests one by one.

9.15 ULTRASONIC TEST

In this test, the *ultrasonic radiations are made to fall on the material to be tested. While passing through the material, the ultrasonic radiations are absorbed and scattered along different directions. It will be interesting to know that ultrasonic radiations are attenuated (*i.e.*, weakened) while passing through the defective area. And they remain unattenuated, while passing through the perfect area. This leads to the variation in the intensity of transmitted radiations. Now the received radiations are converted to an oscilloscope beam for interpretations.

There are two methods for transmitting and receiving the ultrasonic radiations. In the first method, separate transducers are used for transmitting and receiving radiations. And in the second method the same transducer is used both for transmitting and receiving radiations. The ultrasonic

---

* These are the radiations having frequency more than 200000 cycles/sec.

testing is very useful for detecting voids, cracks and other defects near or far below the surface of any material. This method has the advantage of detecting air gaps even of the order of 0.003 mm.

9.16 RADIOGRAPHIC TEST

In this test, high frequency (or short wavelength) radiations of constant intensity are made to fall on the material to be tested. While passing through the material, the radiograph radiations are also absorbed and scattered along different directions. This also leads to the variation in the intensity of the transmitted radiations, which are usually recorded on a sensitive photographic film. It will be interesting to know that the darker region, on the photographic film, corresponds to a defective region. And less dark region corresponds to the perfect regions. It is due to the fact that more radiations pass through the defective area than the perfect one. The developed image of the photographic film is called radiography. And the various techniques to determine the flaws in the material are known as radiographic techniques. The radiographic techniques may be broadly classified into the following three categories, depending upon the nature of high frequency radiation sources:

1. X-radiography. 2. Gamma-radiography. 3. Neutron-radiography.

It will be interesting to know that in X-radiography, the X-rays are used as a high frequency radiations. This technique is useful to detect defects such as gas porosity in castings. But this technique has a drawback that it cannot be used to detect smaller defects. This is due to the fact that due to scattering of X-ray, the small defects are not detected.

In Gamma-radiography, the radioactive elements like radium, cobalt etc., are used as a source of producing radiations of very high frequency (even higher than X-rays). This technique is very useful for inspection of medium sized-parts and also for the jobs which require stronger radiations and higher penetrations. The Gamma radiography is less expensive as compared to X-radiography. But it requires a longer period to obtain radiography as compared to that of X-radiography.

In Neutron-radiography, a neutron beam is used as a source of high frequency radiations. This technique is very useful for inspection of very light material such as plastics, explosive, rubber components etc., where X-radiography and gamma-radiography cannot be used.

The Neutron-radiography may also be used for the detection of voids and detonators as well as inspection of rubber and paper products etc.

9.17 MAGNETIC PARTICLE TEST

In this test, the material to be tested, is magnetised and then fine magnetic particles are spread over the whole surface. We know that as the presence of a crack causes some leakage in the magnetic field, the particles will not be able to spread in the crack area. But they adhere to the surrounding surface, thus outlining the presence of a crack. In order to increase the visibility of a crack, flourescent magnetic particles or ultraviolet light may be used.

The magnetic particle testing is useful for ferrous alloys or magnetic materials such as iron, steel, nickel alloys etc. This technique is also used for detecting blow holes, grinding and fatigue cracks etc. This technique can only be used to detect cracks on the surface or just below it.

A.M.I.E. (I) EXAMINATION QUESTIONS

1. Explain the role of fatigue fracture behaviour of materials. *(Summer, 1993)*
2. Discuss Orowan theory for failure of material due to fatigue. *(Winter, 1993)*
3. Discuss with examples the ductile and brittle fracture. *(Summer, 1995)*

Fracture of Metals

4. State and explain Griffith's theory of fracture. *(Summer, 1996)*
5. What do you mean by metal fatigue? How does it differ from creep? *(Winter, 1999)*
6. (a) Briefly explain the mechanism of fatigue crack initiation in metals.
   (b) How does creep differ from high temperature fatigue? Explain different stages of creep.
   (c) What is thermal fatigue? Explain with the help of an example. *(Summer, 2000)*
7. What is the essential difference between brittle fracture and ductile fracture? *(Summer, 2000)*
8. Distinguish between ductile and brittle fracture. *(Summer 2001, Winter 1999, 1998, 1994, 1993,)*
9. Describe the mechanism of brittle fracture in materials. *(Winter, 2002)*
10. What is meant by brittle transition temperature of the material? How it can be estimated?
    (Winter, 2004)
11. Explain the mechanism of crack initiation and growth when a metal is subjected to cyclic stress.
    (Summer, 2002)

MULTIPLE CHOICE QUESTIONS

1. Creep of solid occurs at
   (a) half the melting point or above an absolute scale
   (b) temperature above 400 °C
   (c) any temperature
   (d) any temperature but not below °C

2. The maximum stress at the tip of the crack is given by:
   (a) $\sigma_{max} = 2\sigma\sqrt{\dfrac{C}{r}}$
   (b) $\sigma_{max} = \dfrac{\sigma}{2}\sqrt{\dfrac{C}{r}}$
   (c) $\sigma_{max} = 2\sigma\sqrt{\dfrac{r}{C}}$
   (d) $\sigma_{max} = \dfrac{1}{2\sigma}\cdot\sqrt{\dfrac{C}{r}}$

3. The stress required to propagate a crack is a brittle material as a function of size of the crack, is
   (a) $\sigma = \sqrt{\dfrac{2\pi C}{E.\gamma}}$
   (b) $\sigma = \sqrt{\dfrac{2E.C}{\pi\gamma}}$
   (c) $\sigma = \sqrt{\dfrac{\pi C}{2E.\gamma}}$
   (d) $\sigma = \sqrt{\dfrac{2E.\gamma}{\pi C}}$

4. The creep fracture is a metal may take place due to
   (a) sliding of grain boundaries
   (b) the initiation of microcracks
   (c) combination of dislocations
   (d) none of these

5. The fatigue fracture in a metal may take place due to
   (a) combination of dislocations
   (b) movement of dislocations, from one slip to another, by climbing
   (c) initiation of microcracks
   (d) none of these

ANSWERS

1. (c)   2. (d)   3. (a)   4. (a)   5. (c)

10

Iron-Carbon Alloy System

1. Introduction. 2. Allotropic Forms of Pure Iron. 3. Critical Points. 4. Iron-Carbon System. 5. Phase diagram of Iron-Iron Carbon System. 6. Solid Phases in Iron-Iron carbide Phase Diagram. 7. Critical Temperatures 8. Eutectoid, Hypoeutectoid and Hypereutectoid Steels. 9. Modified Iron-Iron Carbide Phase diagram. 10. Primary Solidification. 11. Secondary Solidification. 12. Decomposition of Austenite. 13. Transformation of Steel on Cooling. 14. Transformation of Austenite at Constant Temperature 15. Isothermal Transformation Diagram. 16. Changes in Steel Structure below Lower Critical Temperature. 17. Times.-Temperature Transformation Diagram. 18. Salient Points of a Time-Temperature Transformation Diagram. 19. Transformation of Austenite upon Continuous Cooling. 20. Transformation of Austenite to Martensite. 21. Microstructure of Martensite. 22. Retained Austenite.

10.1 INTRODUCTION

It is a well known fact that iron is an element of great importance these days. The carbon is added into iron in varying amounts to produce a number of useful alloys such as mild steel, stainless steel, white cast iron, grey cast iron etc. In order to understand the microstructure of these alloys, we shall first discuss the phase transformation occurring at different temperatures in the iron carbon system.

10.2 ALLOTROPIC FORMS OF PURE IRON

We know that pure substances may exist in more than one crystalline form. Each such crystalline forms is stable over more or less well defined limits of temperature and pressure. This is known as allotropy or polymorphism. The pure iron exists in three allotropic forms *i.e.*, alpha (α), gamma (γ) and delta (δ). Fig. 10.1 shows in ideal curve for pure iron indicating the temperature ranges over which each of these crystallographic forms are stable at atmospheric pressure.

It is evident from this diagram that from room temperature to 910°C, pure iron has a body-centred cubic (BCC) structure and is called α-iron. The α-iron is ferromagnetic at room temperature. But on heating to 786°C, (called curie point), the ferromagnetism disappears, However, the crystal structure remains BCC. Non-magnetic α-iron is stable up to 910°C. Above 910°C, it is transformed into face centred cubic (FCC) structure called γ-iron. Upon heating to 1404°C, the γ-iron is transformed back into the BCC structure called γ-iron. It is stable upto the melting point (1539°C) of pure iron. The BCC structure of δ-iron has a longer cube edge (*i.e.*, lattice parameter) than BCC structure of α-iron.

Note: The non magnetic α-iron was formerly known as β-iron. But later, β-iron was disapproved, because X-ray crystallographic methods showed that there was no change crystal structure at 768°C. To avoid confusion, the original naming of the sequence has been retained with the β phase deleted.

Iron-Carbon Alloy System

10.3 CRITICAL POINTS

We have already discussed in the last article the allotropic forms of pure iron. In this article, we discussed that the crystal structure of pure iron changes when it is heated from room temperature to its molten state or it is cooled from its molten state to room temperature. The temperatures at which the structural changes take place, are called critical points or arrest points. These points are designated by the symbol A_r for cooling and A_c for heating. The symbol A stands for arrest 'r' for *refroidissment* (a French word used for cooling) and 'c' *for chauflage* (a french word used for heating). The critical points correspond to temperatures, 910°C and 1404°C and are referred to as A_{r_3} and A_{r_4} respectively.

Note: As a matter of fact, it is possible to distinguish between the points A_{r_4} and A_{C_4} as well as A_{r_3} and A_{C_3} respectively. It has been observed that there exists approximately 30°C difference between a pair of similar points.

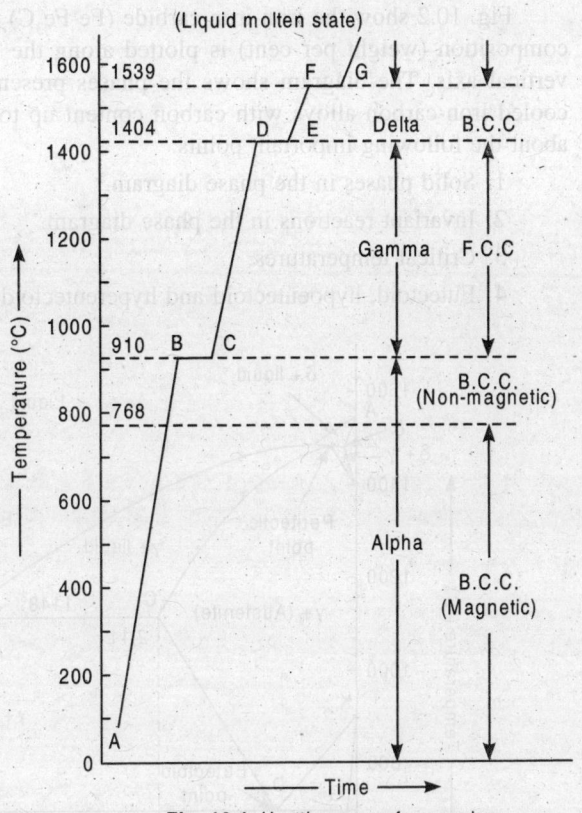

Fig. 10.1. Heating curve for pure iron.

10.4 IRON-CARBON SYSTEM

It is the most important binary system in engineering alloys. The alloys of iron-carbon system containing from 0 to 2.0% carbon are called steels and those containing from 2.0% to 6.7% are called cast irons. However in practice, the steels are manufactured with carbon content upto 1.4%. It is due to the fact that steels with carbon content more than 1.4% are brittle and hence are not useful. Similarly, the cast irons that are manufactured in practice contain carbon from 2.0% to 4.5% only.

It will be interesting to know that iron-carbon alloys exist in different phases in steels and cast irons. In steels, the iron and carbon exists as two separate phases, ferrite and cementite. The ferrite is a solid solution of carbon in α-iron, with negligible amount of carbon and cementite is an intermetallic compound called iron carbide (Fe_3C). Cementite is a stable phase in steels only. But it is not stable in cast irons under all conditions and hence cementite is called a metastable phase. Under certain conditions, cementite decomposes into the more stable phases of graphite and iron. However, once cementite is formed, it is very stable for all practical purposes and therefore can be treated as an 'equilibrium phase'. For this reason, we have the following two phase diagrams of iron-carbon system:

1. Iron-Iron carbide (Fe-Fe_3C) phase diagram.
2. Iron-Carbon (Fe-C) phase diagram.

We shall now discuss the iron-iron carbide phase diagram. The study or iron-carbon phase diagram is beyond the scope of this book.

10.5 PHASE DIAGRAM OF IRON-CARBON SYSTEM

Fig. 10.2 show the iron-iron carbide (Fe-Fe$_3$C) phase diagram. In this diagram, the carbon composition (weight per cent) is plotted along the horizontal axis and temperature along the vertical axis. The diagram shows the phases present at various temperatures for very slowly cooled iron-carbon alloys with carbon content up to 6.7%. This diagram gives us information about the following important points:

1. Solid phases in the phase diagram.
2. Invariant reactions in the phase diagram.
3. Critical temperatures.
4. Eutectoid, hypoeutectoid and hypereutectoid steels.

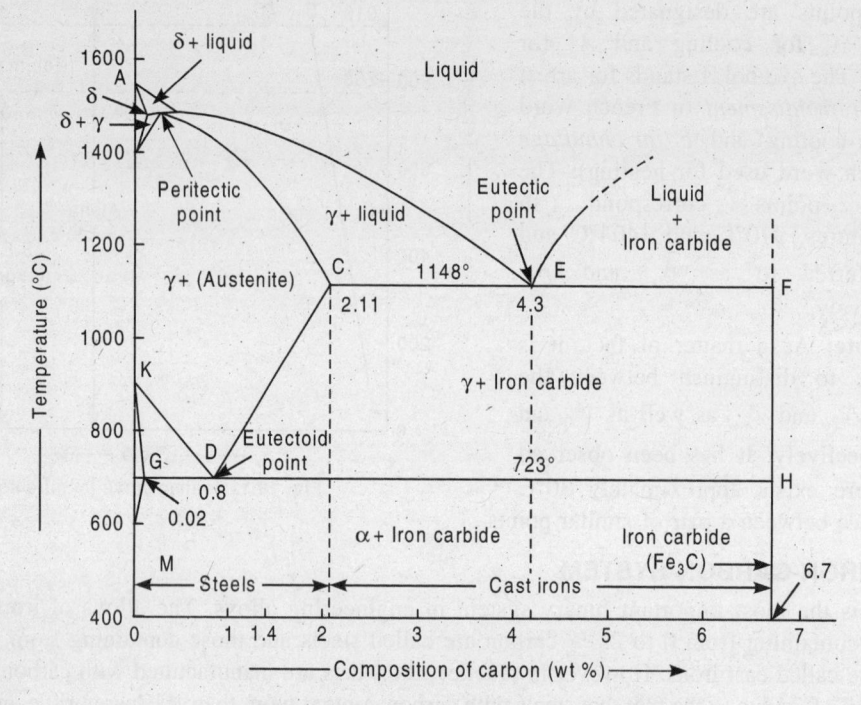

Fig. 10.2. Iron-Iron carbide phase diagram.

The low-carbon-region found 1400° in the phase-diagram is not of any practical importance. However, the region lying in the 700-900°C temperature range and 0-1% carbon range is the most important region in the phase diagram. In this region, an engineer can develop within steel, those microstructures which are required for desired properties.

10.6 SOLID PHASES IN IRON-IRON CARBIDE PHASE DIAGRAM

The iron-iron carbide phase diagram shown in Fig. 10.2 contains four-solid phases, *ie.*, α-ferrite, austenite (γ), cementite (Fe$_3$C) and δ-ferrite. A brief description of each of these phases is given as below:

1. **α-*ferrite*.** The solid solution of carbon in α-iron is called α-ferrite or simple ferrite. This phase has a body-centred cubic (BCC) structure. And at 0% carbon, in corresponds to α-iron. The phase diagram indicates that the carbon is slightly soluble in ferrite. It is due to the fact that maximum solid solubility of the carbon in α-ferrite to 0.02% per cent at 723°C. The solubility of carbon in α-ferrite decreases with the decrease in temperature,

until it is about 0.008% at 0°C as shown by the line GM in the phase diagram. The carbon atoms, because of their small size, are located in interstitial spaces or voids in the crystal lattice. The α-ferrite is soft ductile and highly magnetic. Its density is 7.88 gm/cm^3 and tensile strength is about, 310 MPa.

2. *Austenite.* The solid solution of carbon in γ-iron is called austenite. It has a face-centred cubic (FCC) structure and has a much greater solid solubility for carbon than α-ferrite. The solubility of carbon in austenite reaches a maximum of 2.11% at 1148°C and then decreases to 0.8% at 723°C as shown by the line CD in the phase diagram. The carbon atoms are dissolved interstitially (in the same way as in ferrite) but to a much greater extent in the FCC lattice. The difference in the solid solubility in austenite and α-ferrite is the basis for the hardening of most steels. The austenite is soft and ductile. It is not ferromagnetic at any temperature.

3. *Cementite.* The intermetallic iron-carbon compound is called iron carbide or cementite. Its chemical formula is Fe$_3$C. This means that in a cementite crystal lattice, the number of iron atoms are 3 time more that those of carbon atoms. Cementite has negligible solubility limits and contains 6.7% carbon and 93.3% iron. Cementite has an orthorhombic crystal structure with 12 iron atoms and 4 carbon atoms per unit cell. Its density is 7.6 gm/cm^3. As compared to austenite and ferrite, cementite is extremely hard and brittle. It is magnetic below 210°C.

4. *δ-ferrite.* The solid solution of carbon in δ-iron is called δ-ferrite. It has a BCC crystal structure but with a different lattice parameter than α-ferrite. The maximum solid solubility of carbon in δ-ferrite is 0.09% at 1495°C.

10.7 CRITICAL TEMPERATURES

We have already discussed in Art. 10.3 the critical points or arrest points of pure iron. With the addition of carbon in iron, these critical points do not occur at those temperatures as mentioned in Art. 10.3. As a matter of fact, these critical points vary with the composition of carbon in iron. The A_{r_3} (or A_{C_3}) critical points for steels with carbon content up to 0.8% are given by the line BD of Fig. 10.2. However, for steels with carbon content from 0.8% to 2.8%, the A_{r_2} (or A_{C_2}) critical points are given the line CD (also called A_{cm} line) of Fig. 10.2. In addition to this there is another temperature line which represents a critical temperature designated as A_{r_1} or A_{C_1}. This critical temperature is called lower critical temperature and corresponds to a line GDH of Fig. 10.2 which is at a temperature of 723°C.

10.8 EUTECTOID, HYPOEUTECTOID AND HYPEREUTECTOID STEELS

We have already discussed that iron-carbon alloys containing carbon from 0 to 1.4% are called steels. These steels are quite often referred as plain carbon steels, when they do not contain any alloying element. A plain carbon steel containing 0.8% carbon is known as eutectoid steel. If the carbon content of the steel is less than 0.8%, it is called hypoeutectoid steel. Most of the steels produced, commercially, are hypoeutectoid steels.

The steels, which contain more than 0.8% of carbon are called hypereutectoid steels. Hypereutectoid steels with carbon content upto 1.4% are produced commercially. When the carbon content to steel is more than 1.4%, it becomes very brittle. Thus very few steels are made with carbon content more than 1.4%. In order to increase the strength of steels, other alloying elements are added. These elements increase the strength as well as maintain ductility and toughness.

The iron-carbon alloys containing carbon above 2% are called cast irons. The cast iron containing 4.3% carbon is called eutectic steel. If the carbon content of cast iron is less than 4.3% it is called hypoeutectic cast iron. Most of the cast irons produced commercially are hypoeutectic cast irons. The cast irons containing more than 4.3% carbon are called Hypereutectic cast irons. Hypereutectic cast irons with carbon content up to 4.5% are produced commercially.

10.9 MODIFIED IRON-IRON CARBIDE PHASE DIAGRAM

We have already discussed in the Art. 10.5 that the low carbon region found above 1400°C in the iron-iron carbide diagram of Fig. 10.2 is not of any practical importance. Therefore, it is possible to modify the iron-iron carbide phase diagram as shown in Fig. 10.3 by omitting the low carbon region above 1400°C. Moreover, the various solubility curves, in the actual phase diagram, are taken as straight lines. We shall use this diagram to study transformation of various alloys from the liquid to the solid state. The transformation of various alloys may be discussed under the following two heads:

1. *Primary solidification.* This corresponds to the transformation of alloys from its liquid state to a temperature just below the eutectic temperature. It is also called transformation from liquid to solid state.

2. *Secondary solidification.* This corresponds to the transformation of alloys from the solid state below the eutectic temperature to the solid state below the eutectoid temperature. It is also called transformation of alloy from solid-to-solid state or secondary crystallisation.

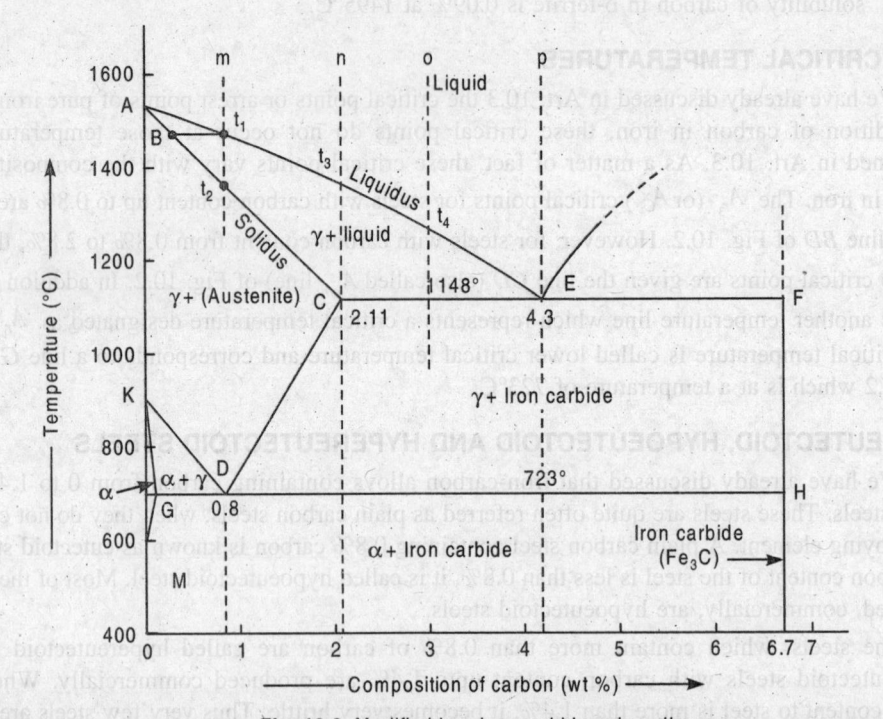

Fig. 10.3. Modified iron-iron carbide pahse diagram.

We shall now discuss the primary and secondary solidifications in detail in the following pages.

Iron-Carbon Alloy System

10.10 PRIMARY SOLIDIFICATION

We have already discussed in the last article that the transformations which occurs on cooling iron-carbon alloys from the liquid state below the eutectic temperature, i.e., 1148°C are called primary solidification. To study these transformations, consider the sequence of events when liquid alloys of various carbon contents are cooled to a temperature just below the eutectic temperature 1148°C, as shown by a line *CEF* in Fig. 10.3. If an alloy containing 0.8% carbon is cooled from a point '*m*' (in the liquid state) lying above the liquids line *AE*, it will remain in its liquid state upto a temperature t_1 (about 1463°C). At this temperature, the crystals of austenite (or γ-phase) begin to precipitate from the liquid alloy. The composition of austenite crystals at t_1 may be determined by drawing a horizontal line (called tie-line) at t_1. The intersection of the tie-line with the solids line *AC* (point *B*) gives the composition of austenite crystal at t_1. As the alloy is further cooled below the liquids line *AE*, the amount of austenite increases continuously. The composition of austenite, formed during cooling, varies along the solidus line *AB* and that of the liquid phase along the liquidus line *AE*.

When the temperature decreases to a temperature t_2, the alloy will become completely solid and will consist entirely of crystal of austenite. The alloys with carbon content varying from 0.2% will solidify exactly in the same manner as discussed above. All these alloys, at the end of solidification, will consist of only austenite.

Now consider an alloy containing 2.11% carbon. When this alloy is cooled from point '*n*' a process similar to the one discussed above will take place. The austenite crystals will begin to form a temperature t_3 lying on the liquidus line *AE*. As the crystals separate, the liquid gets richer in carbon and the last drop of the liquid containing 4.3% carbon will solidify when the eutectic temperature of 1148°C is reached. At this temperature, the alloy will consist entirely of austenite crystals containing 2.11% carbon dissolved in γ-iron. Thus austenite of 2.11% carbon composition is a solid solution of carbon in γ-iron.

When a liquid alloy containing 3.0% carbon is cooled from point '*O*', it begins to precipitate austenite crystals from the liquid alloy at the temperature t_4 lying on the liquidus line *AE*. In this case, also the austenite crystals also increase continuously as the temperature falls. When the temperature falls to 1148°C (*i.e.*, eutectic temperature), the remaining liquid of eutectic composition decomposes into a mixture of saturated austenite (or composition 2.11% carbon) and cementite (or composition 6.7% carbon). This eutectic mixture is called ledeburite. The cast iron alloys of any composition between 2.11% and 4.3% carbon will solidify in this manner only. It may be noted that the microstructure of all these alloys is composed of a proeutectic austenite (*i.e.*, austenite formed above the eutectic temperature) in a matrix of the eutectic mixture (*i.e.*, ledeburite).

Now, suppose an alloy of exactly the eutectic composition *i.e.*, 4.3% carbon is cooled from a temperature corresponding to point '*p*' in Fig. 10.3. This alloy will remain in the liquid phase, until the eutectic temperature is reached. At this temperature, the alloy will solidify completely to a eutectic mixture called ledeburite.

10.11 SECONDARY SOLIDIFICATION

We have already discussed in Art. 10.9 that the transformations, which occur on cooling iron-carbon alloys below the eutectoid temperature *i.e.*, 723°C, are called secondary solidification or secondary crystallisation.

To study these transformations consider the sequence of events when the solid alloys of various carbon contents are cooled below the eutectoid temperature (723°C). Now consider an alloy with 0.3% carbon cooled from a temperature above the line *KD*, where the steel is entirely austenite as shown in Fig. 10.3. A little consideration will show, that nothing will happen to this until, a temperature of about 800°C, on the line KD, is reached. At this temperature γ-iron is austenite. It will begin to transform into α-iron. As the alloy is further cooled, the carbon content in the austenite increases along the line KD. When the alloy is cooled to 723°C, the remaining austenite containing 0.8% carbon decomposes into eutectoid, a mixture of ferrite and cementite. This eutectoid mixture is called pearlite. The point D is called the eutectoid point and the line GDH as eutectoid line.

Now consider and alloy with 0.8% carbon being cooled from austenite state. When this alloy is cooled, no change occurs until the eutectoid point D is reached. At this point, the entire austenite will decompose into pearlite.

When an austenite containing 1.3%, carbon is cooled from the austenite state, no change occurs until it is cooled to 960°C (a temperature on the line CD). At this temperature, austenite begins to decompose with the precipitation of excess carbon as cementite. Since cementite contains 6.7% carbon, its separation will cause a progressive decrease in carbon content of the austenite with the fall in temperature along the line CD. When the eutectoid temperature, (*i.e.*, 723°C) is reached, the remaining austenite will have eutectoid composition (*i.e.*, 0.8% carbon). This will transform completely into pearlite at this constant temperature.

10.12 DECOMPOSITION OF AUSTENITE

We have already discussed in the last article that if an austenite with 0.8% carbon (called eutectoid composition) is cooled below the eutectoid temperature *i.e.*, 723°C, it decomposes into a mixture of ferrite and cementite (or iron carbide). This eutectoid mixture is called *pearlite*. Its microstructure is composed of alternate layers of ferrite and cementite as shown in Fig. 10.4. In this figure, the light region represents a ferrite and a dark region represents cementite.

Fig. 10.4. Microstructure of pearlite (eutectoid steel).

The fraction of ferrite and cementite present in the pearlite may be obtained as discussed below:

Fraction of ferrite in pearlite,

$$f_\alpha = \frac{6.7 - 0.8}{6.7 - 0.0} = 0.88$$

and the fraction of cementite,

$$f_{Fe_3C} = \frac{0.8 - 0.02}{6.7 - 0.02} = 0.12$$

This means that a microstructure of pearlite, is composed of 88% ferrite and 12% cementite. The structure of any alloy with carbon content below 0.8%, at room temperature will consist of partly proeutectoid ferrite and partly pearlite. Similarly, the structure of any alloy with carbon content above 0.8% at room temperature will consist of partly pearlite and partly proeutectoid cementite. For example, steel with 0.2% carbon consists of about 75% of proeutectoid, ferrite and about 25% of pearlite. On the other hand, a steel with 0.6% carbon consists of about 38% proeutectoid ferrite and 62% pearlite.

It is evident from the above discussion, that as the carbon content in the steel is increased, the amount of pearlite increases, until we get a fully pearlite structure at 0.8% carbon. However, if the carbon content in the steel is increased beyond 0.8%, the amount of pearlite decreases, because of the presence of proeutectoid cementite.

Notes: 1. The ferrite that forms above 723°C (*i.e.*, before the eutectoid reaction) is called proeutectoid ferrite. On the other hand, the ferrite that is part of a pearlite (which is formed from austenite with eutectoid composition is called eutectoid ferrite.

2. The cementite that forms above 723°C *i.e.*, before the eutectoid reaction) is called proeutectoid cementite. On the other hand, the cementite that is a part of pearlite, is called eutectoid cementite.

10.13 TRANSFORMATION OF STEEL ON COOLING

We have already discussed in the article 10.9 the modified iron-carbon equilibrium diagram. It gives us an important information about the phase relations and the resulting microstructure of various iron-carbon alloys at equilibrium conditions. The equilibrium conditions are obtained by assuming that the transformation (or solidification) of iron-carbon steel alloys takes place at very slow cooling rates. But in actual practice, the transformation takes place at rapid cooling rates. It has been observed that when a specimen of steel, in the austenite phase, is cooled down to room temperature, at different cooling rates, the structure and properties of the specimen are also different. We shall discuss the effect of different cooling rates, on the transformation of austenite, under the following two heads:

1. Transformation of austenite at constant temperature.
2. Transformation of austenite upon continuous cooling.

10.14 TRANSFORMATION OF AUSTENITE AT CONSTANT TEMPERATURE

Consider a small specimen of steel heated continuously form room temperature to a temperature above the critical point A_3. We know that at this temperature, the steel is present in the form of stable austenite. Let the specimen be held at this temperature for some time, in order to obtain a constant temperature throughout. Now let this specimen be cooled isothermally (*i.e.*, at constant temperature) in a bath, which is maintained to a temperature below the lower critical point A_1 (723°C). At this temperature, the austenite is unstable and therefore it transforms into a mixture of ferrite and cementite. The degree of transformation is determined by methods such as microscopic, magnetometric, dilatometric etc. Now, if we plot the results of this experiment on a graph with amount of austenite transformed along vertical axis and time elapsed along horizontal axis, we shall obtain a curve as shown in Fig. 10.5. This curve is known as isothermal curve.

It may be noted from this curve that the rate of transformation increases rapidly at the beginning, and gradually slows down at the end of the transformation. If may also be noted from the figure that the transformation process does not begin as soon as the specimen has attained the temperature of the bath. But it takes a certain time (indicated by the length (0*a*) before the transformation actually begins. This time is known as incubation period. The transformation, which begins at point '*a*' and ends at point '*b*' after a lapse of time is indicated by the length '*ab*'. The steel at point '*a*' has austenic structure and at '*b*' its structure is a mixture of ferrite and cementite.

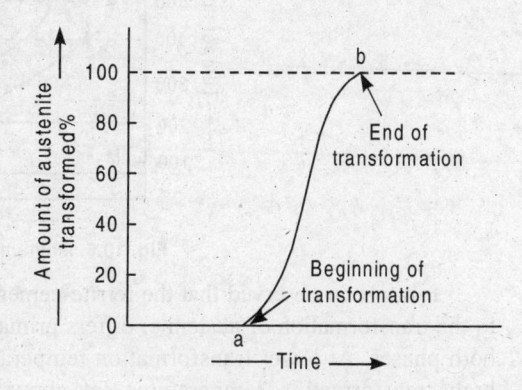

Fig. 10.5. Isothermal curve.

10.15 ISOTHERMAL TRANSFORMATION DIAGRAM

We have already discussed in the last article the isothermal curve for a steel specimen, when it is cooled from the upper critical temperature to a temperature below the lower critical point (723°C). Now, if we cool the steel specimen from the upper critical temperature to different temperatures (say 650°C, 500°C, 400°C and 250°C) and plot the graphs is the same way as discussed above. We shall obtain different curves as shown in Fig. 10.6 (*a*).

Now let us draw another graph with temperature along the vertical axis and time along the horizontal axis. Now let the time elapse before the beginning to the end of transformation be marked on the isotherms (*i.e.*, constant temperature lines) by projecting them from the isothermal curve. Now, if we draw a curve connecting all these points of the beginning and end of transformation, we shall obtain a curve as shown in Fig. 10.6 (*b*). The diagram, so obtained, is known as isothermal transformation diagram..

10.16 CHANGES IN STEEL STRUCTURE BELOW LOWER CRITICAL TEMPERATURE

We have already discussed that the isothermal transformation of austenite at a temperature below the lower critical point (723° C) gives rise to the formation of a mixture composed of the ferrite and cementite. But depending upon the transformation temperature, the structures and properties of this ferrite-cementite mixture are substantially different. The various changes in the structures due to the transformation of austenite at different temperatures, are shown in Fig. 10.6 (*b*).

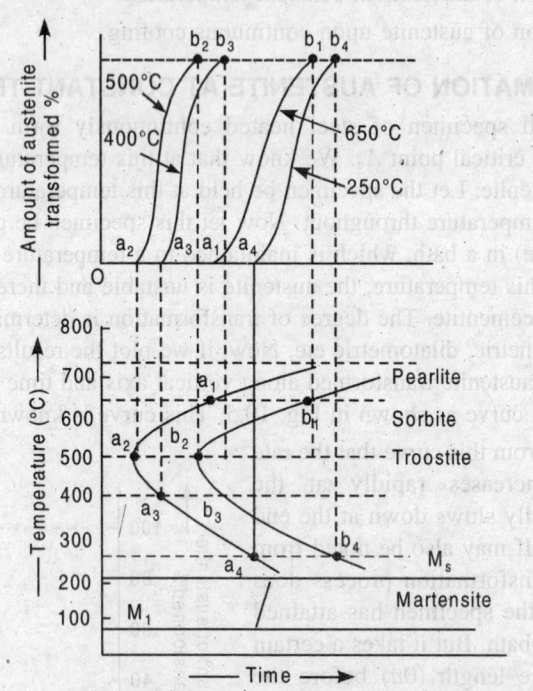

Fig. 10.6. Isothermal transformation diagram.

It has been observed that the ferrite-cementite mixture, formed at different temperatures, due to the transformation of austenite, differs primarily in the degree of dispersion (or refinement) of both phases. At lower transformation temperatures, both the phases are more dispersed. But at high transformation temperatures (at about 700°C) the ferrite-cementite mixture becomes sufficiently distinct and is known as pearlite.

Iron-Carbon Alloy System

It has also been observed that at somewhat lower transformation temperatures (at about 650°C) the ferrite-cementite mixture is more dispersed and is known as sorbite. At still lower transformation temperatures (at about 600°C) the ferrite cementite mixture becomes so dispersed that its structure can only be determined by an electronic microscope. Its structure is similar to pearlite and sorbite and is known as * troosite. At still more lower transformation temperature (from about 550°C to 200°C) the ferrite-cementite mixture is more dispersed and is known as accicular troostite. It has a needle like structure. At still more lower transformation temperatures (below 200°C) the austenite does not change into ferrite-cementite mixture. But it changes into martensite. The formation of martensite does not take place instantaneously. It depends upon the temperature and occurs over a wide range. The upper and lower limits of the transformation range are known as M_s and M_f respectively.

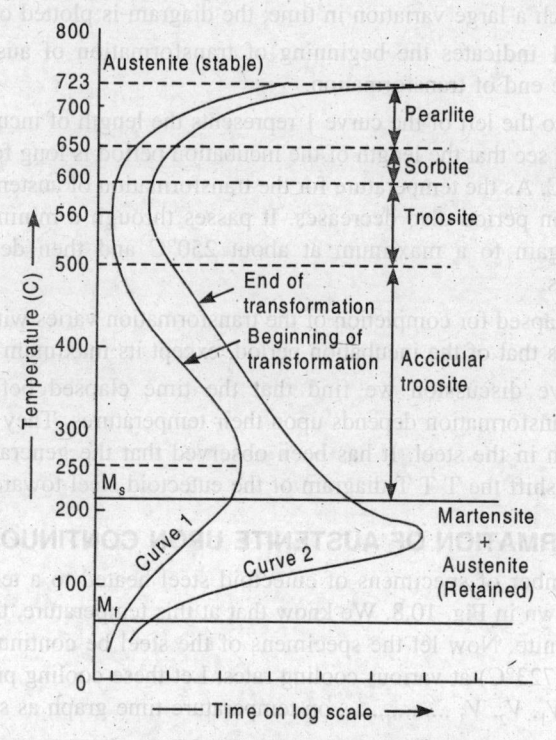

Fig. 10.7. T T T-diagram for eutectoid steel.

Note. The symbol M_s stands for the temperature at which the formation of martensite begins and the symbol M_f stands for the temperature at which the formation of martensite ends.

10.17 TIME-TEMPERATURE TRANSFORMATION DIAGRAM

We have already discussed in the Art. 10.16 about the cooling of, small specimens of steel, from the critical temperature (723°C) to 250°C. In this article, we have also discussed the construction isothermal transformation diagram. Now if we further cool the specimen upto the room temperature, the extend the curve (as discussed above), we shall obtain the curves as shown in Fig. 10.7. This complete isothermal transformation diagram is popularly known as Time-

---

* It is also known as bainite in the literature. Then sorbite is called upper bainite and accicular troostite as lower bainite.

Temperature Transformation Diagram, T T T diagram, Triple T-diagram or S-curves because of their shape.

The complete isothermal or T T T diagram for an eutectoid steel is shown in the figure.

10.18 SALIENT POINTS OF A TIME-TEMPERATURE TRANSFORMATION DIAGRAM

We have already drawn a complete time-temperature transformation diagram in the last article. Though this diagram has so many salient points, yet the following are important from the subject point of view:

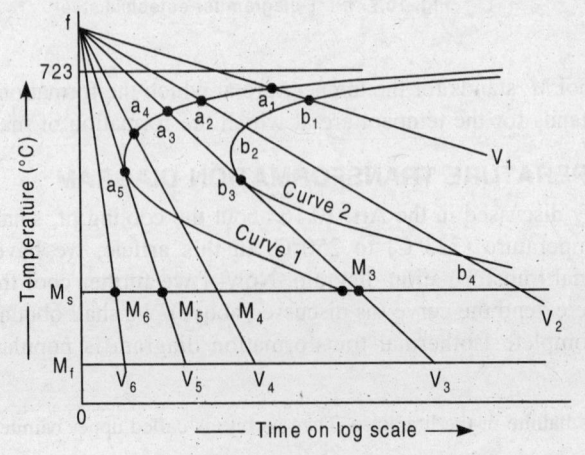

1. The complete transformation of austenite on cooling takes a few seconds only. But in case of some other metals, the complete transformation may take several hours, In order to avoid, such a large variation in time, the diagram is plotted on a logarithmic scale.
2. The curve 1 indicates the beginning of transformation of austenite and the curve 2 indicates the end of transformation.
3. The region to the left of the curve 1 represents the length of incubation period. From the diagram, we see that the length of the incubation period is long for a temperature slightly below 723°C. As the temperature for the transformation of austenite decreases, the length of incubation period also decreases. It passes through a minimum at about 560°C. It increases again to a maximum at about 250°C and then decreases again at lower temperatures.
4. The time elapsed for completion of the transformation varies with the temperature in the same way as that of the incubation period, except its maximum which occurs at 150°C.

From the above discussion we find that the time elapsed before the beginning and completion of the transformation depends upon their temperatures. They are also affected by the percentage of carbon in the steel. It has been observed that the general effect of reducing the carbon content is to shift the T T T diagram of the eutectoid steel towards left.

10.19 TRANSFORMATION OF AUSTENITE UPON CONTINUOUS COOLING

Consider a number of specimens of eutectoid steel heated to a temperature (t) above the critical points as shown in Fig. 10.8. We know that at this temperature, the steel is present in the form of stable austenite. Now let the specimens of the steel be continuously cooled below the lower critical point (723°C) at various cooling rates. Let these cooling process be represented by the inclined curves V_1, V_2, V_3 .... on temperature-time graph as shown in Fig. 10.8.

Fig. 10.8. Transformation of austenite.

It may be noted that the slowest cooling rate is represented by the curve V_1. The slightly more rapid cooling is represented by the curve V_2. The still more rapid cooling rates are represent by the curves V_3, V_4 and V_5. In fact, these curves are straight lines.

Now let us superimpose these cooling curves (V_1, V_2, V_3 ....) on the Time-Temperature transformation diagrams as shown in the Fig. 10.8. From this figure we find that the curve V_1 crosses transformation curves 1 and 2 at point a_1 and b_1 respectively. This means that on slow cooling, the austenite completely transforms into a ferrite-cementite mixture. Since the transformation occurs at the highest temperature, therefore the ferrite cementite mixture is pearlite. The curve V_2 also intersects both the transformation curves at points a_2 and b_2 respectively. Thus at this cooling rate also the austenite completely transforms into ferrite-cementite mixture. But as the transformation occurs at lower temperature (as compared to the V_1), the resulting ferrite-cementite is sorbite. Similarly, the curve V_3 also intersects both the curves at points a_3 and b_3 respectively. The resulting ferrite-cementite mixture is troostite.

It will be interesting to know that the curve V_4 does not cross both the transformation curves. It intersects only the curve 1 at point a_4. And it does not reach the stage of completion. It means that a part of austentic grains transform into ferrite-cementite mixture, while the other does not transform due to the insufficient time. It has been observed that the remaining part of austenite, which has not been transformed, undergoes transformation into martensite on reaching the temperature M_s. It is shown by the intersection of the curve V_4 and M_s temperature at point m_4. Thus the structure of steel, cooled at the rate of V_4, consists partly of troostite and partly of martensite. This type of structure is common to all steels, which are cooled at a rate faster than those represented by V_3, but slower than V_5. It has been observed that for carbon steels, this cooling rate is achieved by quenching in oil.

At any cooling rate, higher than V_5, *e.g.*, curve V_6, of austenite does not transform into ferrite-cementite mixture. But the austenite is transformed into martensite. It is shown by the points m_5 and m_6 in Fig. 10.8. For carbon steel, this cooling rate corresponds to the quenching in water. It will be interesting to know that the austenite is never completely transformed into martensite. This untransformed austenite is known as retained austenite. The minimum cooling rate, at which all the austenite is rapidly cooled to temperature M_s and is transformed into martensite, is known as critical cooling rate. It is represent by straight line V_5 *i.e.*, the tangent drawn to the curve 1.

It may be noted that the curves V_2 and V_3 and others between them, which have more slope also intersect the line M_s. This shows that the martensite is formed at the end of transformation. But it has been obsorbed that martensite is never formed at such cooling rates. It is due to the fact that the curves V_2 and V_3 and others intersect both the transformation curves. Thus the complete transformation of the austenite takes place at points b_2 and b_3 respectively. Beyond these points, there is no austenite left in the steel. Therefore nothing is to be transformed into martensite. As a result of this, the point m_3 in the figure has no physical sense.

10.20 TRANSFORMATION OF AUSTENITE TO MARTENSITE

We have already discussed in the last article that when the specimen of eutectoid steel is rapidly cooled (*i.e.*, cooled at a rate higher than critical cooling) from the austentic region, a new phase called martensite is formed at temperatures below about 220°C (but still above room temperature). The martensite phase is a metastable structure, because if an opportunity is given, It will proceed to form ferrite an cementite. It is because of this fact that the phase diagram of Fig. 10.2 does not show the martensite anywhere. We know that transformation of austenite into a ferrite and cementite consists of the following two processes:

1. Change of space lattice of austenite to the space lattice of ferrite *i.e.*, it changes from FCC to BCC structure.

2. Separation of the majority of the carbon atoms from the space lattice of austenite and their precipitation into an isolated cementite phase.

In the formation of martensite, the FCC structure of austenite changes to body centred in a special way which does not involve diffusion, but results from a shearing action. All the atoms shift together and no individual atom moves more than a fraction of a nanometer from its previous neighbours. As a result of this, the diffusion does not take place and carbon remains in the solid solution. With the carbon retained in the solid solution, the resulting structure is body centred tetragonal (BCT) rather than body centred cubic (BCC), as shown in Fig. 10.9. Since the carbon is present in the martensite, therefore it is hard, strong and brittle.

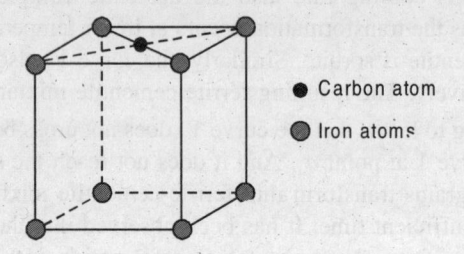

Fig. 10.9. Space lattice of martensite.

The hardness of martensite is much higher than that of pearlite. This enhanced hardness is of great importance, because it provides an extremely high resistance to abrasion and deformation. However, martensite has a drawback that it is brittle. Martensite is reheated to reduce its brittleness without loss of its hardness. The process is called tempering. During tempering, metastable martensite decomposes into more stable phases of ferrite and cementite.

10.21 MICROSTRUCTURE OF MARTENSITE

It has been observed that the martensite has characteristic needle like structure as shown in Fig. 10.10. The needles, in fact, are the plates of martensite, which appear in the microsection as long and thin needle. The rate of formation of martensite needles is very high (of the order of 50 milli-seconds). The formed martensite needles do not grow in length or breadth. But as the transformation proceeds further the subsequent needles become shorter and shorter. It will be interesting to know that the martensite transformation cannot proceed isothermally like the pearlite, sorbit transformation. Thus when the steel is held a constant temperature (below the M_s point) the martensite transformation ceases very rapidly. As the temperature is further lowered, the martensite transformation begins again. The temperature, at which the transformation is finally ceased, is denoted by M_f. This temperature is different for each type of steel. It is thus obvious, that the martensite transformation occurs over a wide range of temperatures from M_s to M_f, where M_s corresponds to the beginning and M_f to the end of martensite transformation.

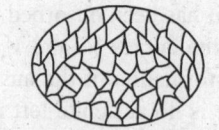

Fig. 10.10. Martensite structure.

10.22 RETAINED AUSTENITE

We have already discussed in the Art. 10.21 the transformation of austenite into martensite. In fact, the martensite transformation range is determined by the percentage of carbon in the steel. It has been observed that higher the percentage of carbon in steel, lower is the temperature of beginning and end of the martensite transformation. The variation of the martensite temperature range with the percentage of carbon in steel is shown in Fig. 10.11

From the figure, we find that above 0.7 per cent carbon, the M_f temperature is below 0°C. This means 100 per cent martensite cannot be produced even if the high carbon steels are

Iron-Carbon Alloy System

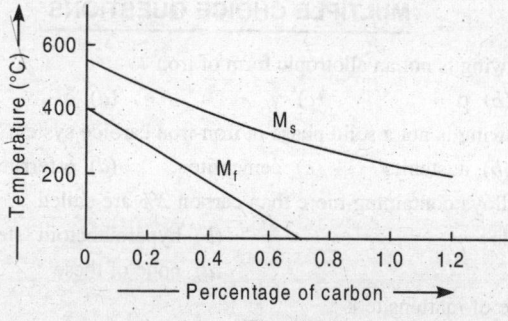

Fig. 10.11.

quenched in ice cold water. Thus some austenite is always left untransformed. This austenite is known as retained austenite or residual austenite. It has been observed that higher the percentage of carbon in steel, great will be the amount of retained austenite. It will be interesting to know that some of the retained austenite still remains in the structure of steel, even if this is cooled below the M_f temperature. The characteristic properties of martensite are its high hardness and extremely low impact strength.

A.M.I.E. (I) EXAMINATION QUESTIONS

1. Explain the term 'martensitic transformation.' *(Winter, 1994)*
2. Write a short note on critical cooling rate. *(Winter, 1994)*
3. Describe the main features of martensitic transformation. *(Summer, 1995)*
4. Write short note on critical cooling rate and heat resisting alloys. *(Summer, 1996)*
5. Define 'recrystallization temperature'. *(Summer, 1996)*
6. What is δ-iron ? *(Summer, 1996)*
7. (a) Draw a neat sketch of the T T T diagram for a eutectoid steel and label the regions. Mark the costing rate corresponding to (i) critical costing rate, (ii) normalising.
   (b) Discuss the important characteristics of martensite transformation. *(Winter, 1996)*
8. (a) What is meant by the term 'high carbon steel' ?
   (b) Define the term 'recrystallization temperature' ? *(Winter, 1996)*
9. Explain TTT curves. *(Summer, 1997)*
10. Distinguish between eutechtic and eutectoid steels. *Summer, 1997)*
11. (a) Explain TTT curves for eutectoid steel.
    (b) How do the alloying elements influence TTT curves of plain carbon steel ?
    (c) Give micro structure of mild steel and grey cast iron and explain their applications with justification. Do these alloys are heat treated? Justify your answers. *(Winter, 1999)*
12. Sketch the microstructure and label the phases for annealed condition of the following:
    (a) Mild steel (b) Grey cast iron
    (c) 0.8% carbon steel (d) White cast iron *(Winter, 2000)*
13. (a) What is martensite ? What are M_s and M_{90} in T-T-T diagram ? In dual phase steels, why the martensite is not hard ?
    (b) Differentiate between pearlitic reaction and Bainitic reaction. *(Winter, 2000)*

MULTIPLE CHOICE QUESTIONS

1. Which of the following is not an allotropic form of iron ?
   (a) α (b) ρ (c) γ (d) δ (e) θ
2. Which of the following is not a solid phase in iron-iron carbide system ?
   (a) α-ferrite (b) austenite (c) cementite (d) δ-ferrite (e) martensite
3. The iron-carbon alloys containing more than carbon 2% are called
   (a) eutectoid steels
   (b) hypereutectoid steels
   (c) cast irons
   (d) none of these
4. The microstructure of martensite is
   (a) needle like structure
   (b) BCC structure
   (c) FCC structure
   (d) HCP structure
5. The solid solution of carbon in γ-iron (called austenite) has
   (a) BCC structure
   (b) FCC structure
   (c) HCP structure
   (d) needle like structure

ANSWERS

1. (θ) 2. (e) 3. (c) 4. (a) 5. (b)

11

Heat Treatment

1. Introduction. 2. Objectives of Heat Treatment. 3. Process of Heat Treatment. 4. Types of Heat Treatment Processes. 5. Normalising 6. Annealing 7. Spheroidising. 8. Hardening. 9. Hardenability. 10. Measurement of hardenability 11. Comparison between Hardness and Hardenability. 12. Tempering. 13. Surface Hardening or Case Hardening. 14. Carburising 15. Nitriding. 16. Cyaniding. 17. Induction Hardening. 18. Flame Hardening. 19. Major Defects in a Metal or Alloy due to Faulty Heat Treatment.

11.1 INTRODUCTION

The term 'heat treatment' may be defined as an operation or a combination of operations, in the manufacturing process of machine parts and tools. As a matter of fact, the heat treatment of a metal or an alloy is carried out first by heating it in solid state and then cooling it. It is possible to impart the required or desirable mechanical properties to steel or alloys for normal operations by heat treatment.

11.2 OBJECTIVES OF HEAT TREATMENT

Though there are innumerable objectives, which are achieved by heat treatment, yet the following are important from the subject point of view:

1. To relieve internal stresses, which are set up in the metal due to cold or hot working.
2. To soften the metal.
3. To improve hardness of the metal surface.
4. To improve machinability.
5. To refine grain structure.
6. To improve mechanical properties like tensile strength, ductility and shock resistance etc.
7. To improve electrical and magnetic properties.
8. To increase the resistance to wear, tear, heat and corrosion etc.

11.3 PROCESS OF HEAT TREATMENT

We have already discussed in Art. 11.1 that the process of heat treatment is carried out first by heating the metal and then cooling it. In fact, this process consists of:

1. Heating the metal to a specified temperature.
2. Holding the metal at increased temperature for a specified period.
3. Cooling the metal (*i.e.*, quenching) according to specified process.

It will be interesting to know that the true heat treatment does not involve any chemical change. But it essentially involves time-temperature cycle, which means the range and rate of heating and cooling.

The temperature to which a metal or an alloy is heated depends upon its grade, grain size as well as type and shape. In general, the metal is never heated much beyond its * upper critical temperature. As a result of this, some plastic deformation of the grains of the metal takes place. This deformation depends upon the chemical composition of the metal or alloy and the temperature to which it is heated.

The metal or alloy is now held at the increased temperature to ensure uniformity of temperature throughout the mass. The period of heating depends upon the size and shape of the component. In general, time required for heating a component is approximately 1 minute to 3 minutes for 1 mm thickness of the largest section.

The main transformation in the properties of a component takes place in the cooling process of the metal. It depends mostly upon the rate at which the cooling takes place. It also depends upon the medium in which the cooling takes place. It may be noted that these days only five quenching medias adopted are caustic soda solution, brine, water, oil and air.

11.4 TYPES OF HEAT TREATMENT PROCESSES

Though there are many types of heat treatment processes, yet the following are important from the subject point of view:

1. Normalising
2. Annealing
3. Spheroidising
4. Hardening
5. Tempering
6. Carburising
7. Nitriding
8. Cyaniding
9. Induction hardening
10. Flame hardening

Now we shall discuss, in detail, all the above mentioned processes one by one.

11.5 NORMALISING

To main objects of normalising in heat treatment are:

1. To refine the grain structure of the steel and to improve machinability, tensile strength and structure of the weld.
2. To remove strains caused by cold working processes like hammering, rolling, bending etc., which makes the metal brittle and unreliable.
3. To remove dislocations caused in the internal structure of the steel due to hot working.
4. To improve certain mechanical and electrical properties.

The process of normalising consists of heating the steel 30°C to 50°C above its upper critical temperature for hypoeutectoid steel or (Acm) line for hypereutectoid steels. It is held at this temperature for about fifteen minutes and then allowed to cool down in still air.

The process provides a homogeneous structure consisting of ferrite and pearlite for hypoeutectoid steels, and pearlite and cementite for hypereutectoid steels. The homogeneous structure provides a higher yield point, ultimate tensile strength and impact strength with lower

---

* It is the temperature at which the carbon is completely dissolved into the iron. It is generally denoted as A_2 in the iron carbon equilibrium diagram.

ductility to steels. The process of normalising is frequently applied to castings and forgings etc. The alloy steels may also be normalised. But they should be held for two hours, at a specified elevated temperature and then cooled in the furnace.

11.6 ANNEALING

The main objects of annealing in heat treatment are:
1. To soften the steel, so that it may be more easily machined or cold worked
2. To refine the grain-size and structure to improve mechanical properties like strength and ductility.
3. To relieve internal stresses which may have been caused by hot or cold working or by unequal contraction in casting.
4. To alter electrical, magnetic or other physical properties.
5. To remove the gases, trapped in the metal, during initial casting.

The process of annealing is of the following two types:

1. *Full annealing*. The main object of full annealing is to soften the metal, to refine its gain structure, to relieve the stresses and to remove gases trapped in the metal. This process consists of heating the steel 30°C to 50°C above the upper critical temperature for hypoeutectoid steel and by the same temperature above the lower critical temperature for hypereutectoid steels. The steel is then held at this temperature for sometime to enable the internal changes to take place. The time allowed is approximately 3 to 4 minutes for each millimeter of thickness of the largest section, and then slowly cooled in the furnace.

The rate of cooling varies from 30°C to 200°C per hour, depending upon the composition of steel. The cooling, usually, carried out in the furnace. The objects may also be taken out of the furnace and cooled in ashes so as to prolong the cooling time.

In order the avoid decarburisation of the steel, during full annealing, the steel is packed in a cast iron box containing a mixture of cast iron borings, charcoal, lime, sand or ground mica. The box, along with its contents, is generally allowed to cool slowly in the furnace after the proper heating has been completed.

2. *Process annealing*. The main object of process annealing is to relieve the internal stresses set up in the metal and for increasing the machinability of the steel. In this process, the steel is heated to a temperature below or close to the lower critical temperature, held at this temperature for sometime and then cooled slowly. This causes complete recrystallisation in steels, which have been severely cold worked and a new grain structure is formed. The process of annealing is commonly used in the sheet and wire industries.

The approximate temperatures for annealing depending upon the carbon content in steel, are given in Table 11.1

Table 11.1

S. No.	Carbon content (in per cent)	Annealing temperature in °C
1.	Less than 0.12 (Dead mild steel)	875-925
2.	0.12 to 0.25 (Mild steel)	840-970
3.	0.25 to 0.55 (Medium carbon steel)	815-840
4.	0.55 to 0.80 (High carbon steel)	780-810
5.	0.80 to 1.40 (High carbon or tool steel)	760-780

11.7 SPHEROIDISING

It is a particular type of annealing in which cementite in the granular form is produced in the structure of steel. This is, usually, applied to high carbon tool steels, which are difficult to machine. The operation consists of heating steel upto a temperature slightly above the lower critical temperature (730°C to 770°C). It is held at this temperature for some time and then cooled slowly to temperature of 600°C. The rate of cooling is from 25°C to 30°C per hour.

The spheroidising improves the machinability of steels, but lowers the hardness and tensile strength. These steels have better elongation properties than normally annealed steel.

11.8 HARDENING

The main object of hardening are:
1. To increase the hardness of the metal, so that it can resist wear.
2. To enable it to cut other metals, *i.e.*, to make it suitable for cutting tools.

The process of hardening consists of heating the metal upto a temperature of 30°C to 50°C above the upper critical temperature for the hypoeutectoid steels and by the same temperature above the lower critical point for hypereutectoid steels.

The metal is held at this temperature for a considerable time, depending upon its thickness and then quenched (cooled suddenly) in a suitable cooling medium.

The hardness obtained from a given treatment depends upon the rate of cooling, the carbon content and the work size. A very rapid cooling is necessary to harden low and medium plain carbon steels, The quenching in a water or brine solution in a method of rapid cooling, which is commonly used. For high carbon and alloy steels, mineral oil is generally used as the quenching medium, because its action is not so severe as that of water. Certain alloy steels can be hardened by air cooling. But for ordinary steel, such a cooling rate is too slow to give an appreciable hardening effect. Large parts are, usually, quenched in an oil bath. The temperature of the quenching medium must be kept uniform to achieve uniform results. Any quenching bath, used in production work, should be provided with some means for cooling.

A rapid cooling from the hardening temperature cases the austenite to be transformed into another constituent called martensite, which is very hard and brittle. The hardening of steel depends entirely upon the formation of martensite, because austenite is comparatively soft and ductile.

It may be noted that the low carbon steels cannot be hardened appreciably, because of the presence of ferrite which is soft and is not changed by heat treatment. As the carbon content goes on increasing, the possible obtainable hardness also increases. The process of hardening is of the following four types:

1. *Work hardening*. It is the process of hardening a metal, while working on it.
2. *Age hardening*. It is also called precipitation hardening. The age hardening is a process of hardening a metal when allowed to remain or age after heat treatment. It is mostly applicable to non-ferrous metals such as alloys of aluminium, magnesium and nickel However, the effect of age hardening shows a marked increase in strength and hardness for duralumin. The duralumin is an alloy of aluminium with 4% copper and smaller quantities, of other alloying elements. The process of age-hardening consists of two steps namely solution treatment and precipitation treatment. In the solution treatment, the alloy is heated into the single phase region, held there long enough to dissolve all existing soluble particles and is then rapidly quenched into the two phase region. This produces a supersaturated solid solution. In the precipitation treatment, the alloy is allowed to age at or above the room temperature for a specified time. This produces a very fine precipitate particles, which increase the strength and hardness of the alloy.

Heat Treatment

3. *Air hardening*. It is the process of hardening a metal, when it is cooled slowly in air blast. The effect of air hardening is, usually, seen in high speed steels and some of the tungsten alloys.
4. *Hardening by heating and quenching*. It is the most common process of hardening (as the name indicates), which is generally employed for iron base alloys having a low carbon content.

11.9 HARDENABILITY

The ability of an iron metal or alloy to form martensite when cooled at various rates is measured by its hardenability. A material with higher hardenability will form martensite in large sections on quenching. It is not related to actual hardness of the martensite formed, which is almost solely dependent upon the carbon content of the iron metal or alloy. As a matter of fact hardenability is in effect reciprocal of the critical cooling rate.

11.10 MEASUREMENT OF HARDENABILITY (JOMINY END-QUENCH TEST)

The hardenability of an iron metal or alloy is measured by the Jominy end-quench test. In this test, a specimen of standard dimensions as shown in Fig 11.1(a) is first austenized. Then it is transferred very quickly from the furnace to an apparatus in which it is supported by one end and the other end is subjected to a jet of water as shown in Fig. 11.1 (b). The water can be supplied via a quick-acting valve immediately the specimen as shown in the figure.

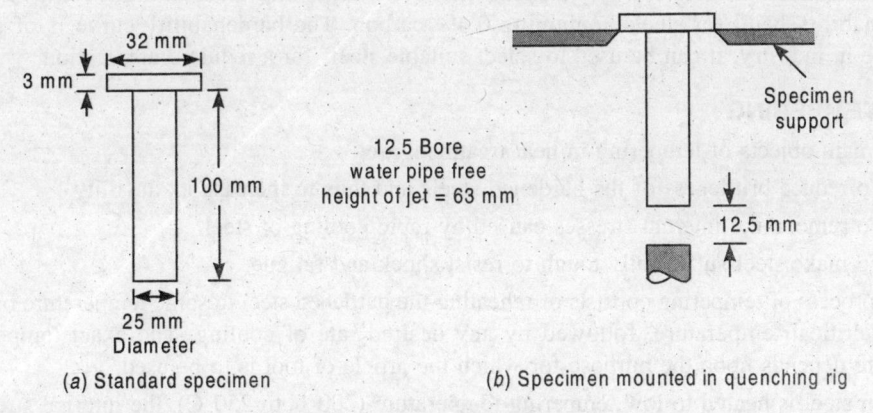

Fig. 11.1. Jominy end-quench test.

When the specimen is cold, a 0.38 mm deep flat region is made along the length of the specimen. Hardness measurements are made along this flat region. These are present in a graph plotted against distance from the quenched end.

11.11 COMPARISON BETWEEN HARDNESS AND HARDENABILITY

The term hardness is the property of a material by virtue of which it is able to resist abrasion, indentation and scratching by hard bodies. It is a surface property and can be measured by the methods like Brinelle, Rockwell, Vicker, Knoop's etc. The hardness of a material can be increased by cold working, surface diffusion (such as carburizing, nitriding etc.) or by age hardening.

On the other hand, hardenability is the susceptibility of a material to get hardened. It is tested by Jominy-end-quench test method. Hardenability is affected by he composition of an alloy, its percentage contents and the grain size.

Fig. 11.1 (*a*) shows the variation of hardness versus the percentage of carbon content for quenched steel. As seen from this figure, the hardness of a steel increases with an increase in the carbon content.

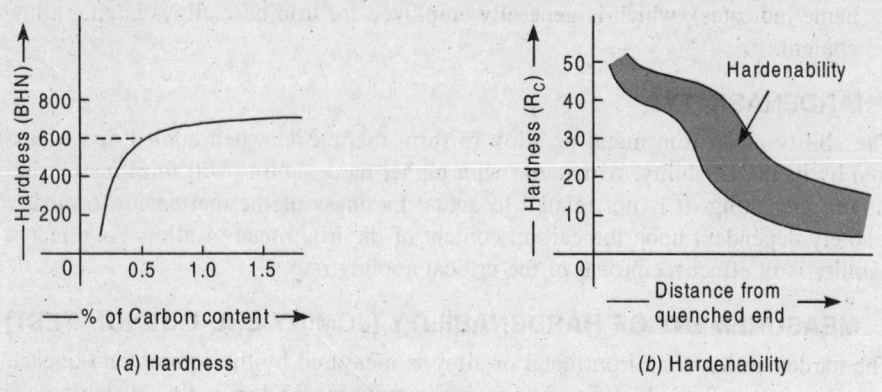

(*a*) Hardness

(*b*) Hardenability

Fig. 11.2. Hardness and hardenability curves.

Fig. 11.2 (*b*) shows the variation of hardness versus the distance from the quenched end. Since the hardenability varies with the alloying constituents and the grain size, consequently the hardenability of steel is indicated by a hardenability band rather than a curve. Fig. 11.2 (*b*) shows the hardenability band for steels containing 0.4% carbon. The hardenability curve is of great importance in industry. It can be used to select suitable steels for a required application.

11.12 TEMPERING

The main objects of tempering in heat treatment are:
1. To reduce brittleness of the hardened steels and thus to increase its ductility.
2. To remove the internal stresses caused by rapid cooling of steel.
3. To make steel sufficiently tough to resist shock and fatigue.

The process of tempering consists of reheating the hardened steel to some temperature below the lower critical temperature, followed by any desired rate of cooling. The exact tempering temperature depends upon the purpose for which the article or tool is to be used.

When steel is heated to low tempering temperature (200°C to 250°C), the internal stresses are removed and ductility increases without changing the structure of steel from martensite or reducing its hardness. On heating it to about 300°C, troostite-martensite mixture is obtained which imparts some ductility to the metal. On heating it to 400°C, martensite begins to change into fine pearlite or sorbite and the transformation of sorbite is completed on reaching a temperature of 600°C. The sorbitic steel is employed for making highly stressed parts, because it has a remarkable mechanical properties like ductility, strength and shock, resistance.

The tempering temperatures may be judged by the colour formed on the surface of the steel being tempered. The colours are caused by surface oxidation of the steel with the formation of thin films of iron oxide. The various tempering temperatures and colours, for different types of tools made from carbon steels, are given in Table 11.2.

Table 11.2

S. No.	Type of work	Tempering Temperature, °C	Tempering colour
1.	Scrapers and lathe tools for brass.	220	Light straws or pale yellow
2.	Hacksaws, steel-engraving tools circular saws for steel, light turning and parting tools.	225	—
3.	Hammer faces, planers for steel, screwing dies for brass.	230	Straw or dark straw
4.	Wood-engraving tools, paper cutters, planers for iron.	235	—
5.	Shear blades, milling cutters and hollow mills, bone cutting tools.	240	Golden yellow
6.	Rock drills, screw cutting dies, boring cutters and reamers.	245	—
7.	Taps and chasers, knurls, pen-knives, mill chisels and picks.	250	—
8.	Moulding and planing cutters for hard wood, punches and dies.	255	Yellowish brown or brown
9.	Plane irons, gauges and brace bits, stone-cutting tools, planing and moulding cutters.	260	—
10.	Twist drills for wood.	265	Reddish brown dappled with purple
11.	Dental and surgical instruments, augers and pressing cutters.	270	—
12.	Gimlets, axes and hot sets.	275	—
13.	Gold chisels and sets for steel, cold chisels for cast iron, chisels for wood.	280	Purple
14.	Planing cutters for soft wood.	285	—
15.	Cold chisels for wrought iron.	290	Violet
16.	Screw drivers.	295	Blue
17.	Springs.	300	Dark blue

The baths using tempering oils may be employed for temperatures upto approximately 230°C. The tempering oils are, usually, mineral oils having flash points of the order of 300°C. An adequate quantity of oil should be employed and the baths should be provided with wire baskets which when loaded with work may be lowered into the tempering bath. For temperatures above about 230°C, liquid salt baths are preferred. These salt baths, usually, consist of mixtures of nitrates and nitrites. The chlorides and fluorides are, usually, employed for higher temperatures. The process of tempering is of the following two types;

1. *Austempering.* It is a process of tempering in which steel is heated, above the upper critical temperature, at about 875°C, where the structure consists entirely of austenite. It is then suddenly cooled by quenching it in a salt bath or lead bath maintained at a temperature of about

250°C to 525°C, so as to facilitate the transformation of austenite into bainite. After complete transformation, the steel is cooled air. In this process, a good impact strength is obtained and the degree of cracking is also reduced.

2. *Martempering.* It is a process of tempering in which the steel is heated above the upper critical point and then quenched in a bath kept at a suitable temperature, so that it is in the upper martensite range. After the temperature becomes uniform throughout the steel structure, without the formation of bainite, it is further cooled in air. The steel is then tempered. The martempered steel is free from internal stresses. It avoids cracks and warping etc., which are usually caused by ordinary hardening. The martempering is mostly used in case of alloy steels.

11.13 SURFACE HARDENING OR CASE HARDENING

In many engineering applications, it is desirable that a steel to be used should have a hardened surface to resist wear and tear. At the same time, it should have soft and tough interior or core so that it is able to absorb any shocks etc. This is achieved by hardening the surface layers of the article, while the rest of it is left as such. This type of treatment is applied to gears, ball bearing, railway wheels, etc.

Following are the various surface or case hardening processes by means of which the surface layer is hardened:

1. Carburising. 2. Cyaniding. 3. Nitriding. 4. Induction hardening. 5. Flame hardening.

Now we shall discuss, in detail, these processes in the following pages one by one.

11.14 CARBURISING

We have discussed in Art. 11.8, that a low carbon steel (containing carbon content upto 0.25 per cent) cannot be hardened appreciably by any one of the hardening processes. Such steels are enriched in carbon on their surface before hardening by quenching. The process of introducing carbon, to low carbon steels, in order to give it a hard surface, is called carburising. The surface is made hard only upto a certain depth. Following two methods are commonly used for carburising;

1. Pack or solid carburising. 2. Gas carburising.

1. *Pack or solid carburising.* In this process, the article to be carburised is placed in a carburising box of proper design made of special heat resisting alloy, cast steel or steel sheet. The space between the article and the box is filled with a solid * carburising compound.

The layer of compound between the article and the box should be as uniform as possible to give uniform transfer of heat from the under surface of the box to the article. The box is covered with a lid and sealed with clay to eliminate the entry of air. The box is now gradually heated in a furnace to the selected carburising temperature and maintained there for a specified period. During heating, carbon monoxide gas is formed which reacts with the article to form carbon and carbon dioxide gas. The surface of the article absorbs the carbon and gets rich in its carbon content. The carbon thus absorbed impregnates into the body of the article by the diffusion process. The temperature, time and carburising compound composition used depends upon the depth of case (carbon rich surface) desired. For example, with a compound composition (as given in footnote), a case depth of 1 to 1.25 mm can be obtained at 925°C in an overall carburising time of nine hours. Out of these nine hours, five hours are required for heating up the work and four

---

* There are numerous varieties of carburising compounds. But the one with the following composition is mostly used:

Hardwood charcoal – 53 to 55 per cent. coke–30 to 32 per cent, sodium carbonate – 2 to 3 per cent, barium carbonate – 10 to 12 per cent and calcium carbonate –3 to 4 per cent.

hours represent the time at 'temperature'. After heating, the article may be quenched in oil directly or it may be slowly cooled and then subjected to proper heat treatment.

Note: If certain parts of the article are not to be carburised, then those parts may be protected by copper plating or by covering with a non-carburising material such as sand and fire clay.

2. *Gas carburising.* This process has gained popularity now-a-days, especially for the production of hard surfaces on light and small articles of low carbon steel. In this process, the articles to be case hardened are heated to the proper temperature in an oven having an atmosphere of gases rich in hydrocarbons. The gases used for this purpose may be methane, ethane, coal gas etc. The operation is performed in the oven at 900° to 950°C. The carbon in the gas combines with the surface of the article and makes it hard. This is a quicker process as compared to the pack carburising. It is useful only when mass production of the article is desired.

11.15 NITRIDING

It is a process of case or surface hardening in which nitrogen gas is employed in order to obtain hard surface of the steel. This process is commonly used for those steels, which are alloyed with chromium, molybdenum, aluminium, manganese etc. The steel article, usually, well machined and finished are placed in an air tight container made of high nickel chromium, steel and provided with inlet and outlet tubes through which the ammonia gas is circulated. The nitriding process is generally carried out in the electric furnace, where the temperature in the range of 450°C to 550°C is maintained. The container with the articles is placed in the furnace and ammonia gas is passed through it. The ammonia gas, when comes in contact with steel articles, gets dissociated in the form of nascent nitrogen, which reacts with the surface of the articles and form nitrides which is very hard. This process can give surface hardness upto a depth of 0.8 mm.

The nitriding process is used in the production of machine parts, which require high wear resistance at elevated temperatures such as automobile and air valves and valve parts, piston pins, crank shafts and cylinder liners. It also finds some application in the production of ball and roller bearing parts, or other parts to withstand high pressure steam services die-casting dies, wire-drawing dies etc.

Note: The core should be brought to its original toughness before nitriding by quenching in oil from about 900°C and tempering from about 600°C to 650 °C.

11.16 CYANIDING

The cyaniding (also called liquid carburising) is a case of surface hardening process in which both carbon and nitrogen are absorbed by the metal surface to get it hardened. In this process, the piece of low carbon steel is immersed in a bath of cyanide salt, such as sodium cyanide or potassium cyanide maintained at 850°C to 950°C. The immersed steel piece is left in the molten cyanide salt bath, at the above temperature, for about 15 to 20 minutes. It is then taken out of the bath and quenched in water or oil. The cyanide yields carbon monoxide and nitrogen, which behaves as active carburising agents in hardening the surface of steel. The process can give surface hardness upto a depth of 0.8 mm.

This process is mainly applied to the low carbon steel parts of automobiles (oil pump gears, drive worm screws, change over switch shafts, steps, sleeves, brake cam, speed box gears) some parts of motor cycles (gears, shafts, pins) and agricultural machinery (self propelled gears) etc.

11.17 INDUCTION HARDENING

It is a process of surface hardening in which the surface, to be hardened, is surrounded by an inductor block which acts as a primary coil of a transformer. The inductor block, should not touch the surface to be hardened. A high frequency current is passed through this block. The heating effect is due to induced eddy currents and hysteresis losses in the surface material. The inductor block, surrounding the heated surface, has water connections and numerous small holes on its inside surface. As soon as the surface reaches to the proper temperature (750°C to 760°C for 0.5 per cent carbon steel and 790°C to 800°C for alloy steel), it is automatically spray-quenched under pressure.

An important feature of this method of hardening is its rapidity of action, because it requires only a few seconds to heat the steel to a depth of 3 mm. The actual time depends upon the frequency used, power input and depth of hardening required.

The induction hardening is widely used for wearing surfaces of crankshafts, camshafts, gear teeth etc.

11.18 FLAME HARDENING

Sometimes, a particular portion (or portions) of an article is required to be hardened. This is generally done in case of a portion subject to wear, abrasion or shocks. This type of local hardening is done by a process, known as flame hardening. In this process, the portion, to be hardened, is heated with the help of a flame of oxyacetylene torch above its critical temperature. The heated portion is then immediately quenched by means of spray of water, which is directed towards heated portion. Since the heating is localised, therefore no stresses are developed. As a result of this, the chances of distortion or cracking are reduced.

The flame hardening process is, mostly, used for articles like gears, or wheels, which cannot be quenched as a whole. The main advantage of this method is that the time taken for heating is much less as compared to the methods when the metal as a whole is to be heated in a furnace.

11.19 MAJOR DEFECTS IN A METAL OR ALLOY DUE TO FAULTY HEAT TREATMENT

Following are some of the major defects found in metal or alloy due to faulty heat treatment.

1. *Overheating*. Prolonged heating of a metal or alloy above a temperature marked by the line *DK* in the iron-iron carbide phase diagram (called A_3 line), leads to the formation of very large actual grains. It is called overheating. On cooling a metal, this yields a structure containing coarse crystalline martensite (known as Widman tatten structure). Such a structure has a reduced ductility and toughness. It is possible to retrieve an overheated metal or alloy by usual annealing. For considerable overheating, double annealing may be used.

2. *Burning*. Heating a metal or an alloy to still higher temperatures near melting point for a longer time leads to burning. This leads to the formation of iron oxide inclusions along the grain boundaries. Burnt metal or alloy has a stoney fracture and such a metal or alloy is irremediable and is rejected.

3. *Oxidation*. Sometimes a metal or alloy is oxidized due to oxidizing atmosphere in the furnace. It is characterized by a thick layer of scale on the surface of a metal or alloy. It can be prevented by using controlled atmosphere in the furnace or using molten salt baths.

4. *Decarburization*. It is the loss of carbon in the surface layers of the metal or alloy. Decarburization results in lower hardness and lower fatigue limit. It is caused by the oxidizing furnace atmosphere. In order to prevent decarburization, the metal or alloy should be heated in a neutral or reducing atmosphere or in boxes with cast iron chips or in molten salts baths.

5. *Cracks*. The cracks occur in quenching when the internal tensile stresses exceed the resistance of metal or alloy to separation. The tendency of a metal or alloy to crack formation increases with carbon content, hardening temperature and cooling rate in the temperature interval

of martensite transformation. It also increases with the hardenability of metal or alloy. Another reason of crack formation is the concentration of local stresses.

It will be interesting to know that the cracks are irremediable defects. To prevent their formation, it is advisable to (*i*) avoid stress concentrations such as sharp corners on projections acute angles, abrupt changes from thicker to thinner cross-sections etc. in the component design, (*ii*) conduct quenching from the lowest possible temperature, (*iii*) cool the metal or alloy slowly in the martensite interval of temperatures by quenching in two media and stepped quenching (*i.e.*, martempering), (*iv*) apply isothermal quenching.

6. *Distortion and warping.* Distortion or deformation consisting in changes in the size and shape of heat-treated work, is due to thermal and structural stresses. Asymmetrical distortion of work is often called warping in heat-treating practice. It is usually observed in case of non-uniform heating or overheating for hardening, when the work is quenched in the wrong position and when the cooling rate is too high in the temperature interval of martensite transformation. An elimination of these causes should substantially reduce warping.

A.M.I.E. (I) EXAMINATION QUESTIONS

1. (*a*) Explain the various purposes of heat treatment.
   (*b*) Describe the flame hardening process and its application.
   (*c*) Explain the process of annealing. *(Summer, 1993)*
2. Distinguish between the hardness and hardenability of steel. State the factors which effect hardenability. *(Winter, 1993)*
3. (*a*) Define the term "heat treatment". Why are the steels heat treated ?
   (*b*) Discuss the major defects in the steel due to faulty heat treatment. *(Winter, 1993)*
4. Distinguish between the full annealing and process annealing. *(Winter, 1993)*
5. Discuss the factors that affect the hardenability of steels. *(Winter, 1994)*
6. Write a short note on
   (*a*) Cyaniding (*b*) Case hardening
   (*c*) Normalizing (*d*) Annealing. *(Winter, 1994)*
7. (*a*) Define the terms (*i*) annealing and (*ii*) tempering.
   (*b*) Differentiate between hardness and hardenability. Draw hardness and hardenability curves for steel. *(Summer, 1995)*
8. Discus the use of Hardenability curves. *(Summer, 1995)*
9. (*a*) Describe the flame hardening process with the aid of next sketch.
   (*b*) Describe the main features of martensite transformation.
   (*c*) State the objectives of heat treatment of metals. State the process of tempering. *(Summer, 1996)*
10. What is meant by the term 'hardenability' ? Describe how hardenability of a steel can be estimated. *(Winter, 1996)*
11. What type of heat treated is given for die steels ?
12. Explain the different types of annealing treatments and their objects ? *(Summer, 1997)*
13. Explain type of heat treatment required with justification for welded LPG cylinders. *(AMIE; Winter, 1999)*
14. How do the mechanical properties are controlled by hardening followed by suitable tempering ? Differentiate between martempering and austempering. *(AMIE., Winter, 1999)*
15. (*a*) Describe following heat treatment procedures :
    (*i*) full annealing (*ii*) normalizing (*iii*) quenching and tempering
    Assuming a medium carbon steel, explain for each heat treatment the final microstructure.

(b) Cite three sources of internal residual stress in metal components. What are the possible adverse consequences of these stresses ?

(c) Briefly explain the difference between hardness and hardenability. (*AMIE., Summer, 2000*)

16. What is precipitation hardening ? How does it differ with dispersion hardening ?

(*AMIE., Winter, 2000*)

17. Explain temper brittleness and its problem. (*AMIE., Winter, 2000*)

MULTIPLE CHOICE QUESTIONS

1. Which of the following is not the objective of heat treatment
   (a) to relieve internal stresses which are set up into metal due to hot or cold working
   (b) to improve tensile strength, ductility, shock resistance, hardness, resistance to wear, tear and heat and corrosion resistance
   (c) to improve electrical and magnetic properties of metal
   (d) to change the atomic structure

2. Hardness is
   (a) the property of a material due to which it is able to resist abrasion indentation and scratching
   (b) the susceptability of a material to get hardened.
   (c) the same as hardenability.
   (d) none of the above.

3. Nitriding is a process used to
   (a) reduce the wear resistance
   (b) increase the wear resistance
   (c) increase the surface hardness
   (d) none of the above.

ANSWERS

1. (d)   2. (a)   3. (b)

12
Corrosion of Metals

1. Introduction. 2. Electrode Potential. 3. Galvanic Series. 4. Classification of Corrosion. 5. Direct Chemical Corrosion (Dry Corrosion). 6. Electrochemical Corrosion (Wet Corrosion). 7. Mechanism of Electrochemical Corrosion. 8. Hydrogen Evolution Corrosion Reaction. 9. Oxygen Absorption Corrosion Reaction. 10. Passivity. 11. Galvanic Cell. 12. Types of Galvanic Cells. 13. Composition Cell. 14. Stress Cells. 15. Concentration Cells. 16. Types of Corrosions. 17. Uniform Corrosion. 18. Pitting Corrosion. 19. Intergranular Corrosion. 20. Stress Corrosion 21. Season Corrosion. 22. Crevice Corrosion. 23. Corrosion Fatigue. 24. Errosion Corrosion. 25. Atmospheric Corrosion. 26. Underground Corrosion. 27. Fretting Corrosion. 28. Selective Corrosion. 29. Prevention and Control of Corrosion. 30. Suitable Design and Fabrication Procedure. 31. Use of Inhibitors. 32. Modification of Corrosive Environment. 33. Use of Protective Coating. 34. Metallic Coatings. 35. Electroplating. 36. Dipping. 37. Spraying. 38. Cladding. 39. Cementation. 40. Non-metallic Coatings. 41. Paints and Lacquering. 42. Plastic Coatings. 43. Vitreous Coatings. 44. Oxide Coatings. 45. Chemical-dip Coatings. 46. Alloying of metals. 47. Cathodic Protection. 48. Anodic or Sacrificial Anode Method. 49. External Voltage Method.

12.1 INTRODUCTION

As a matter of fact, metals (except gold and platinum) exist in nature in the form of their oxides, carbonates, sulphides and silicates. These are reduced to their metallic states from ores during their extraction process. A considerable amount of energy is spent during their extraction process. As a result of this, the pure metals can be regarded in a high energy state as compared to their ores. Since the high energy state is not a stable state, therefore the pure metals have a tendency to revert back to their natural state. Thus when metals are put into use in various forms, these are exposed to the environment containing liquids and gases etc. As a result of this, the surface of the metal starts deteriorating. This type of deterioration or destruction may be due to direct chemical attack or electrochemical attack.

This destruction (or deterioration) of a metal and unwanted chemical or electrochemical attack, by its environment starting at the surface, is called corrosion. In the word of science, the corrosion of a metal is sometimes regarded as a process reverse to that of producing metal from its ore. Two most familiar examples of corrosion are the rusting of iron and the formation of green film on the surface of copper. The rusting of iron takes place, when the iron is exposed to atmospheric conditions. During this exposure, a layer of reddish scale and powder of oxide is formed and the iron becomes weak. The formation of green film on the surface of copper takes place, when it is exposed to moist-air containing carbon dioxide.

It will be interesting to know that the annual loss to the whole world, through the corrosion of different materials, is about 2 billion dollars. As a result of this colossal loss, research and

12.2 ELECTRODE POTENTIAL

It is an important term in the field of corrosion, which signifies the rate at which the corrosion is taking place. The electrode potential may be defined as the voltage developed at an electrode, with reference to a standard electrode the electrode potential is built up due to the dissociation of metals to irons and electrons. It depends upon the following two factors:

1. Nature of the metal, and
2. Nature and concentration of solution.

There is no method of measuring the absolute value of an electrode potential. Hence all electrode potentials are determined under a standard condition with reference to a standard hydrogen electrode, whose electrode potential is assumed to be zero.

Table 12.1 lists electrode potential at 25°C of some important metals with reference to the later part of this chapter.

The elements arranged in the order to their increasing electrode potential constitute a series called electrochemical series. Table 12.2 gives us the information about the following points:

1. The metals having electrode potential lower than that of the hydrogen (taken as reference electrode *i.e.*, having negative values are known as anodic metals, whereas the metals having electrode potential higher than that of the hydrogen *i.e.*, having positive values are called cathodic metals.

Table 12.1

S. No.	Metal iron	Symbol	Electrode Potential in volts.
1.	Sodium	Na	– 2.71 (Anodic)
2.	Magnesium	Mg	– 2.40
3.	Aluminium	Al	– 1.66
4.	Zinc	Zn	– 0.76
5.	Chromium	Cr	– 0.74
6.	Iron (Ferrous)	Fe	– 0.44
7.	Nickel	Ni	– 0.25
8.	Tin	Sn	– 0.14
9.	Lead	Pb	– 0.13
10.	Iron (Ferric)	Fe	– 0.045
11.	Hydrogen	H	+ 0.00 (Reference)
12.	Copper (cuprous)	Cu	0.34
13.	Copper (cupric)	Cu	0.47
14.	Silver	Ag	0.80
15.	Platinum	Pt	1.20
16.	Gold	Au	+ 1.50 Cathodic)

2. The metals with higher negative electrode potential are more active to corrosion. Thus zinc is more corrosive than copper.

12.3 GALVANIC SERIES

The electrochemical series (Table 12.1) provides valuable information regarding the chemical activity of the metal. But it does not provide sufficient information to predict the behaviour of corrosion in actual practice. Therefore electrode potential of various metals and alloys in common use have been measured by immersing them in sea water.

The values so obtained are arranged in the decreasing order of their activity as shown in Table 12.2. The metals or alloys arranged in this order constitute a galvanic series.

Table 12.2

S. No.	Corroded or Anodic end
1.	Magnesium
2.	Magnesium alloys
3.	Zinc
4.	Aluminium
5.	Aluminium alloys
6.	Low-carbon steel
7.	Cast iron
8.	Stainless steel (active)
9.	Lead-tin alloys
10.	Lead
11.	Tin
12.	Brass
13.	Copper
14.	Bronze
15.	Copper-nickel alloys
16.	Silver
17.	Stainless steel (passive)
18.	Monel
19.	Graphite
20.	Titanium
21.	Gold
22.	Platinum
	Protected or Cathodic end

The galvanic series provide more accurate information regarding the relative tendency of common metals and alloys to undergo corrosion.

12.4 CLASSIFICATION OF CORROSION

The corrosion may be broadly classified into the following two categories:
1. Direct chemical corrosion (Dry corrosion), and
2. Electrochemical corrosion (Wet corrosion).

12.5 DIRECT CHEMICAL CORROSION (DRY CORROSION)

The corrosion, which involves direct combination between metals and dry gases is known as direct chemical corrosion or dry corrosion. Chemical reactions of dry chlorine, hydrogen sulphide, oxygen etc. with dry metal are few examples of direct chemical corrosion. It has been experienced that under ordinary conditions, the oxygen is the most commonly encountered reacting gas, due to which direct oxidation takes place. The oxidation of all metals, except gold, silver and platinum, takes place at high temperature. However, in case of alkaline earth metals, the oxidation takes place at low temperature.

It has been observed that whenever a clean oxide free surface of a metal is exposed to the air, oxygen gets absorbed instantaneously on the metal surface. This results in the formation of scale (oxide film), which results further oxidation. Now for the oxidation of continue, either the metal must diffuse the oxide film to the surface or the oxygen must diffuse through the oxide film to the underlying metal as shown in Fig. 12.1.

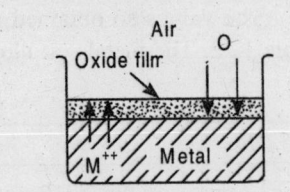

Fig. 12.1. Direct chemical corrosion.

The metal ions have been found to possess much more mobility. It is due to this reaction that the diffusion of metal ions is more rapid than oxygen ions. Once the surface of a metal is covered with oxide layer, the growth of layer continues and forms a thick layer of oxide. Both the chemical reactions of oxidation of metal may be expressed as

$$M \longrightarrow M^{++} + 2e^- \qquad \ldots \text{(Metal ion)}$$

$$2e^- + \frac{1}{2}O_2 \longrightarrow O^{--} \qquad \ldots \text{(Oxygen ion)}$$

It will be interesting to know that the corrosion of ferrous metals such as wrought iron, cast iron, steel and alloy steels takes place through direct corrosion and is commonly known as rusting. The rusting of iron or steel is caused by the continuous action of oxygen, carbon dioxide and moisture. It converts the metal into hydrated ferric oxide. The complete chemical reaction for rusting of iron as follow:

$$Fe + O + 2CO_2 \longrightarrow Fe(HCO_3)_2 \qquad \ldots (i)$$
(Ferrous bicarbonate)

$$2Fe(HCO_3)_2 + 2O \longrightarrow 2Fe(OH)CO_3 + 2CO_2 + H_2O \qquad \ldots (ii)$$
(Ferric carbonate)

$$Fe(OH)CO_3 + H_2O \longrightarrow Fe(OH)_3 + CO_2 \qquad \ldots (iii)$$
(Hydrated ferric oxide)

The first equation shows the reaction of iron with oxygen and carbon dioxide. It converts iron into a soluble ferrous bicarbonate. The second equation indicates the reaction of ferrous bicarbonate with oxygen. In this reaction, the iron is oxidised to a basic ferric carbonate. In the third equation, the ferric carbonate reacts with the moisture thereby producing a hydrated ferric oxide along with the liberation of carbon dioxide gas. The rusting of iron may be slowed down by immersing iron in lime water solution.

---

* The firist chemical reaction (*i.e.*, metal ion) takes place at the junction of metal and oxide film. The chemical reaction indicates that the metal (M) dissociates into metal ions (M^{++}) and electrons ($2e^-$). The second chemical reaction (*i.e.*, oxygen ion) takes place at the junction of air and oxide film. The chemical reaction indicates that the combinations of elecrrons ($2e^-$) with oxygen gas ($\frac{1}{2}O_2$) from the air results in the formation of oxygen ions (O^{--}).

Corrosion of Metals

The other dry gases, which cause corrosion effect are sulphur dioxide, carbon dioxide, chlorine, hydrogen sulphide etc. In addition to this, the dry corrosion may also take place due to the chemical action of flowing liquid metal at high temperatures on the solid metal or alloy. Such a dry corrosion is known as liquid-metal corrosion. This type of corrosion occurs in devices used for nuclear power. In this case, the corrosion reaction involves either a dissolution of solid metal by a liquid-metal or internal penetration of the liquid-metal into the solid metal. This causes the weakening of the solid metal.

The direct chemical corrosion of a metal surface may also take place in the presence of liquids (instead of gases). The liquids may be acidic or alkaline. This classification can be differentiated clearly by knowing the pH value of the liquid under consideration. The liquids with pH value between 0 and 7 are acidic while those with pH value between 7 and 14 are alkaline. Thus liquids with pH value below 6 cause corrosion by chemical reactions similar to those of acids. Similarly, the liquids with pH value more than 7 cause corrosion by chemical reactions similar to those of alkalies.

12.6 ELECTROCHEMICAL CORROSION (WET CORROSION)

The type of corrosion, which involves the flow of electric current between two dissimilar metals, is known as electrochemical corrosion. It takes place at or near the room temperature, because of the reaction of metals which takes place with water or aqueous solution of acids and bases. This phenomenon may be explained with the help of an electrochemical cell as shown in Fig. 12.2.

An electrochemical cell, in its simplest form, consists of a vessel containing electrolyte (liquid), two dissimilar electrodes (known as anode and cathode) and a metallic wire connecting the two electrodes as shown in the figure.

In this cell, the two principal reactions take place one at the cathode and other at anode. The reactions taking place at the anode (known as anodic reaction) are always [1]oxidation reactions. These reactions always tend to destroy the anode metal by causing it to dissolve in the electrolyte. And the reactions taking place at the cathode (known as cathodic reaction) are always [2]reductions reactions. These reactions, usually, do not affect the cathode metal, because most of the metals cannot be reduced further. The electrons, which are produced by the anodic reaction flow through the metal, are used up in the cathodic reaction.

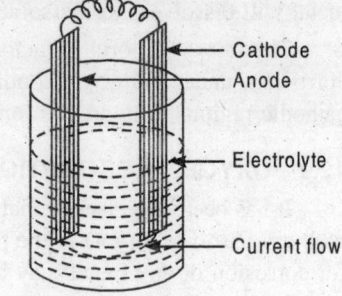

Fig. 12.2. Electrochemical cell.

12.7 MECHANICS OF CORROSION

We have already discussed in the last article that the phenomenon of electrochemical corrosion of metals is a result of anodic and cathodic reactions, which take place in an electrochemical cell. The anodic reaction is always associated with the dissolution of the metal in the electrolyte and formation of the corresponding ions. The cathodic reaction, which usually does not affect the metal, may involve the following two different processes depending upon the nature of corrosive environment:

1. Hydrogen evolution corrosion, and
2. Oxygen absorption corrosion reaction.

---

[1] The oxidation reaction leads to the liberation of electrons from the substance with oxygen.
[2] The reduction reaction leads to the gain of electrons by the metal.

12.8 MECHANISM OF ELECTROCHEMICAL CORROSION

This reaction, usually, takes place in the presence of acids, alkali or salt of a chemically strong metal. Bot the anodic and cathodic [1]chemical reactions for hydrogen evolution corrosion of a metal represented as:

$$M \rightleftarrows M^{++} + 2e^-$$ .... (Anodic)

$$2H^+ + 2e^+ \longrightarrow \uparrow H_2$$ .... (Cathodic)

The above mentioned two chemical reactions may also be represented by the a [2]combined chemical reaction (by adding both the reactions) as:

$$M + 2H^+ \rightleftarrows M^{++} + H_2$$

It will be interesting to know that all the metals above hydrogen in the electrochemical series (given in Table 12.1 on page 236) have a tendency to dissolve in acid solution. The hydrogen evolution corrosion reaction, in its simplest (form, consists of anode and cathode in contact with electrolyte (acid solution) as shown in Fig. 12.3. The anodes are shown as larger areas, whereas cathodes as smaller ones. Since there is an electric potential difference between the anode and cathode, an electric current flows from anode to cathode.

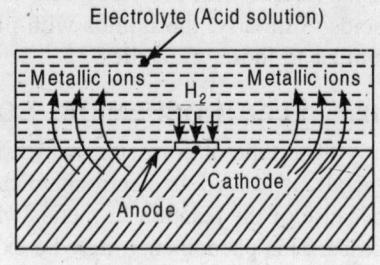

Fig. 12.3. Hydrogen evolution corrosion reaction.

The hydrogen ions, which are separated from the electrolyte, are deposited on the cathode during the process and evolve hydrogen gas. In order to maintain the flow of electric current, the metal will dissolve into the solution.

The most important factors, which influence the course and rate of electrochemical corrosion, under the given conditions, are electric potential difference between the anodic and cathodic regions to hydrogen ion concentration in the solution and temperature etc.

12.9 OXYGEN ABSORPTION CORROSION REACTION

It has been experienced that a metal surface is attacked by aerated solution (*i.e.*, salt solution with air dissolved into it) in the presence of oxygen. The anodic and cathodic [3]chemical reactions for corrosion of any metal may be represented as

$$M \rightleftarrows M^{++} + 2e^-$$ .... (Anodic)

$$2e^- + \frac{1}{2} O_2 + H_2O = 2OH^-$$ .... (Cathodic)

---

[1] The anodic chemical reaction indicates the dissociation of metal (M) into metal ions (M^{++}) and electrons ($2e^-$). The cathodic chemical reaction indicates the combination of hydrogen ions ($2H^+$) and the electrons (e^-) which leads to the formation of hydrgogen (H_2). The electrons obtained in the first reaction are absorbed in the second reaction.

[2] The combined chemical reaction indicates the reaction of metal (M) with hydrogen ions (H^+). It results in the displacement of hydrgoen ions from the solution by metallic ions (M^{++}) along with the formation of hydrgen gas.

[3] The anodic chemical reaction indicates the dissociation of metal (M) into metal ions (M^{++}) and electrons ($2e^-$). The cathodic chemical reaction indicates the interception of electrons ($2e^-$) by oxygen $\left(\frac{1}{2}O_2\right)$ in the presence of water (H_2O)-forming hydroxide ions ($2OH^-$).

Corrosion of Metals

In an oxygen absorption corrosion reaction, the anodic areas on the metal surface are formed due to the presence of crack in oxide film coating of the metal, whereas the entire surface of the coated metal represents the cathodic area as shown in Fig. 12.4.

In the figure, we see that as the anodic area is very small, as compared to the cathodic area, therefore the total corrosion current will be concentrated at a very small area. As a result of this, severe corrosion takes place. Moreover, as the sodium chloride solution is always present in the form of sodium (Na^+) and chloride (Cl^-) ions in the electrolyte, the product obtained at the cathode is sodium hydroxide and at the anode is metal chloride.

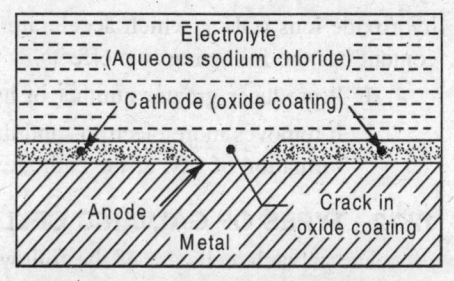

Fig. 12.4. Oxygen absorption corrosion.

Both these products diffuse from the neighbouring areas, of their respective electrodes, which results in the corrosion of metals.

12.10 PASSIVITY

It is a phenomenon in which a metal or an alloy exhibits a much higher corrosion resistance than expected from its position in the electrochemical or galvanic series. The passivity of a metal is the result of formation of a highly protective, but very thin and quite invisible film on the surface of a metal or an alloy, which makes it more noble (or inactive). This film is insoluble, non-porous and of a self-healing nature. It means that when a film is broken it will repair, itself, on the exposure to oxidizing conditions. Titanium, aluminium, chromium and a variety of stainless steels (containing chromium) are the commonly known metals and alloys, which are considered to be passive.

12.11 GALVANIC CELL

A galvanic cell, in its simplest form, consists of two dissimilar metal electrodes, which allow the current to flow through the wire, when the two electrodes are connected. It is an electrochemical type of corrosion, which involves, transfer of electrons from one electrode to another.

If an electric contact is made between any two dissimilar electrodes, the greater potential at anode will permit the electrons to flow from anode to cathode as shown in Fig. 12.5.

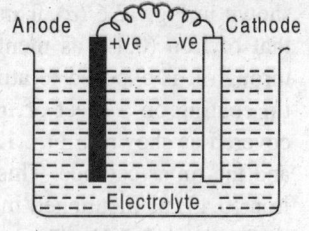

Fig. 12.5. Galvanic cell.

The introduction of surplus electrons at the cathode upsets the equilibrium condition, which dissociates the water into hydrogen ions (H^+) and hydroxide ions (OH^-). The chemical reaction may be represented as:

$$H_2O \rightleftharpoons H^+ + OH^-$$

As a result of dissociation of water, hydrogen gas is formed at the cathode due to the interception of electrons by the hydrogen ions. This chemical reaction may be represented as:

$$2H^+ + 2e^- \longrightarrow H_2 \uparrow$$

The following factors affect the rate of corrosion in a galvanic cell:
1. Difference in electrode potential between anode and cathode.
2. Acidity of electrolyte.
3. Presence of oxygen content.

It has been experienced that if the difference in electrode potential is higher, the rate of corrosion will be more. Similar increased acidity of electrolyte produces greater number of hydroxide ions (OH⁻), which accelerate the corrosion rate at the anode. An increase in oxygen content has the following two effects.

1. It produces greater number of hydroxide ions (OH⁻).
2. It removes more electrons and therefore accelerates corrosion at the anode alongwith the formation of rust at the cathode.

12.12 TYPES OF GALVANIC CELLS

The galvanic cells are of the following three types:

1. Composition cells. 2. Stress cells. 3. Concentration cells.

Now we shall discuss all the abovementioned types of galvanic cells one by one in the following pages.

12.13 COMPOSITION CELL

A composition cell may be established between any two dissimilar metal electrodes. In each case, the metal of higher electrode potential acts as the anode. Examples of composition cells are galvanised steel, sheet, tin, plate etc. The zinc coating on steel protects the iron even if the surface is not completely covered.

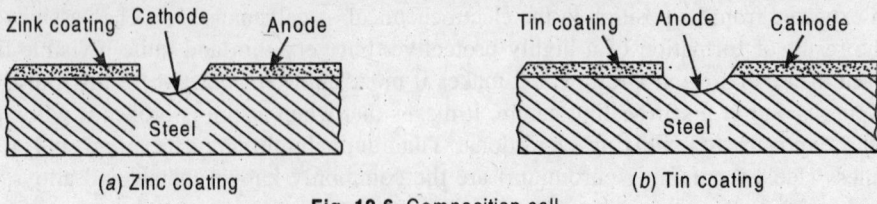

(a) Zinc coating

(b) Tin coating

Fig. 12.6. Composition cell.

The zinc coating serves as anode and iron, in the presence of crack, serves as cathode as shown in Fig. 12.6 (a). This happens because the electrode potential of zinc (0.76) is higher than that of iron (0.44) as mentioned in Table 12.1. It is thus obvious that iron is protected even through it is exposed to atmosphere, where the zinc coating is scrapped off. On the other hand, tin coating on a sheet of iron provides protection so long as the surface of iron is completely covered as shown in Fig. 12.6 (b). As soon as the coating is broken, the iron becomes the anode and the tin as cathode. This happens as the electrode potential of tin (0.14) is lower than that of iron (0.44) as mentioned in Table 12.1. Other examples of composition cells are:

1. Steel screws in brass or marine hardware.
2. Steel shaft in bronze bearings.
3. Lead-tin solder around the wire.

12.14 STRESS CELLS

These cells are established at over-stressed points in materials. The effect becomes evident after a metal has been cold worked as shown in Fig. 12.7 (a) and (b). This happens due to the

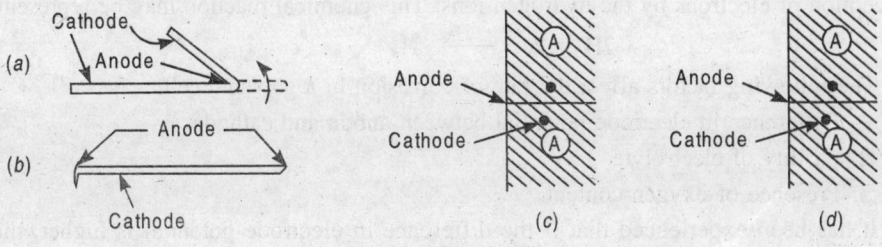

Fig. 12.7. Stress cells.

reason that during cold working of metals, internal stresses are piled up at certain points due to the dislocation of metal grains. It results in the creation of over-stressed regions within the metal. These over-stressed regions have a tendency to form a crack. As a result of this, anodes and cathodes are developed.

It has been observed that the stress cells are also formed at the grain boundaries (*i.e.*, junction points of the same or different metals as shown in Fig. 12.7 (*c*) and (*d*). This happens due to the fact that the atoms at the grain boundaries have a different electrode potential than the atoms within the grains. As a result of this, anodes and cathodes are developed.

12.15 CONCENTRATION CELLS

A concentration cell is established between a pair of same metal electrodes, when they are exposed to an electrolyte of varying concentration. The concentration cell, in its simplest form, consists of a vessel divided into two parts with a passage between them. The electrolyte can pass from one side of the vessel to the other side.

The metal electrode (say copper) in portion (*A*) is immersed in a dilute solution, whereas the metal electrode (*i.e.*, copper) in portion (*B*) is immersed in a concentrated solution. When the two electrodes are connected through a voltmeter (*V*), the cell is formed as shown in Fig. 12.8.

It may be noted that the metal electrode in the dilute electrolyte (side *A*) acts as an anode. And the metal electrode in the concentrated electrolyte (side *B*) acts as cathode. It will be interesting to know that the electrode in the left side becomes more anodic with greater concentration of copper ions (Cu^{++}) in the right side.

The [1]chemical reaction taking place in the left side (anode) of the cell may be represented as:

$$Cu \longrightarrow Cu + 2e^-$$

and the [2]chemical reaction taking place in the right side (cathode) of the cell may be represented as:

$$Cu^{++} + 2e^- \longrightarrow Cu$$

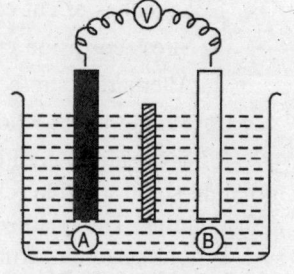

Fig. 12.8. Concentration cell.

As a result of the above two chemical reactions, the electrons will continue to flow from anode to the cathode (*i.e.*, from electrode in side *A* to that in side *B*). It is due to this fact, that the anode undergoes further corrosion to produce additional electrons.

12.16 TYPES OF CORROSIONS

Though there are innumerable types of corrosion, yet the following are important from the subject point of view:

1. Uniform corrosion
2. Pitting corrosion
3. Intergranular corrosion
4. Stress corrosion
5. Season corrosion
6. Crevice corrosion
7. Fatigue corrosion
8. Erosion corrosion
9. Atmospheric corrosion
10. Underground corrosion
11. Fretting corrosion
12. Selective corrosion.

Now we shall discuss all the abovementioned types of corrosion in the following pages.

---

[1] The copper metal (Cu dissociates into copper ions (Cu^{++}) and electrons ($2e^-$).

[2] The combination of copper ions (Cu^{++}) with electrons ($2e^-$) leads to the formation of copper.

12.17 UNIFORM CORROSION

It is a type of corrosion which occurs where metal or alloy is completely homogeneous (*i.e.*, of the same nature both chemically as well as mechanically). As a result of this, the galvanic cells are established between any two points (on the metal surface) in the presence of acids or alkalies.

It may be noted that it is a type of corrosion (as the name indicates) in which the rate of corrosion is [1]uniform.

12.18 PITTING CORROSION

It is a type of corrosion, which takes place at a particular point like a pin hole, thus forming a cavity on the metal surface. The pin hole penetrates deep into the metal and the corrosion product comes out as shown in Fig. 12.9.

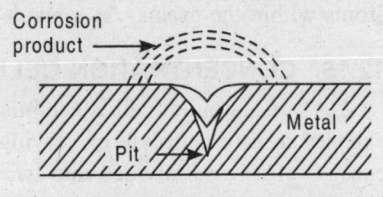

Fig. 12.9. Pitting corrosion.

It has been observed that the pitting corrosion occurs more commonly in tubes, pipes and vessels due to breakdown or cracking of the protective film on the metal surface. This occurs due to some mechanical factors during the turbulent flow of solution over the metal surface. The various mechanical factors involved are:

1. Surface roughness or non-uniform finish.
2. Scratches or cut edges.
3. Stress corrosion cracking.
4. Alternating stresses.
5. Sliding under load.

It has been observed that the stainless steel and aluminium produces a characteristic pitting in halide solutions (like chloride or bromide solutions). Oxygen concentration cells may also initiate pitting. The [2]plating out a noble metal (a metal, which does not react with other substances such as gold, platinum) from a salt solution may also result in pitting due to the formation of local galvanic cells.

Strictly speaking, the pitting corrosion results in the break-down of a metal at a particular location. This gives rise to the formation of small anodic and large cathodic areas. In the proper environment, it produces a high corrosion current, and thus the rate of corrosion is rapid.

The pitting corrosion does not affect mechanical properties except raising the fatigue strength of a metal.

12.19 INTERGRANULAR CORROSION

It is a type of corrosion, which occurs along grain boundaries of a metal sensitive to corrosive attack. The intergranular corrosion, generally, occurs it the grain boundaries contain material which shows a solution potential more anodic than that the grain centre in the particular corroding medium. This potential difference may be either due to the difference in grain orientations or [3]precipitation in grain boundaries.

It has been observed that the intergranular corrosion begins commonly at the surface and then rapidly processes inwards and causes some damage in the internal structure of the metal.

---

[1] If the surface is not uniform, the corrosion does not remain uniform. But it becomes concentrated at that place. This leads to pitting corrosion.

[2] It is a process of depositing a thin layer of some metal by chemical process. It is also known as electroplating.

[3] It is a process by which any metal in solution is made to separate as a solid and settle at the bottom of the containing vessel.

(a) Grain boundary corrosion

(b) Intergranular corrosion

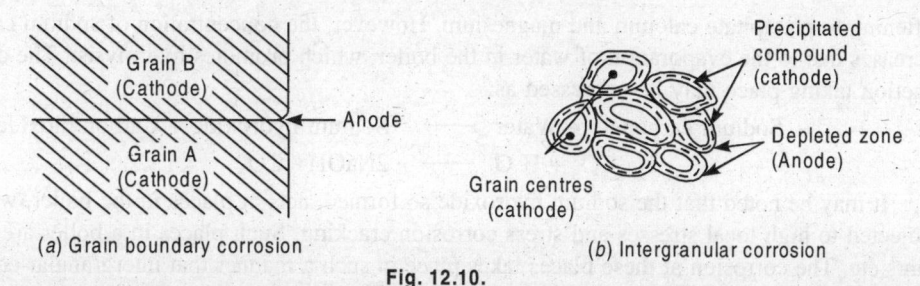

Fig. 12.10.

Fig. 12.10 (a) shows a grain boundary corrosion. Anodes and cathodes are developed due to the difference of potential between the atoms at the boundaries and those at grain centres. It results in the formation of galvanic cells, which cause corrosion. Fig. 14.10 (b) illustrates the corrosion by the precipitation of certain compounds at grain boundaries. This leaves the solid solution adjacent to grain boundary impoverished (i.e., depleted) in one constituent. The impoverished solution acts as an anode with respect to the centre of grain and to the precipitated compound. As a result of this, it may be attacked by the corrosion medium.

The intergranular corrosion may also occur in non-stabilised austenitic stainless steels and copper-aluminium alloys. The corrosion resistance may be restored by proper heat treatment and rapid quenching to prevent * heterogeneous precipitation that occurs due to slow cooling.

12.20 STRESS CORROSION (STRESS CORROSION CRACKING)

It is a type of corrosion, which occurs in internally stressed engineering components used in corrosive environments. The stress corrosion failures involve high residual stresses, which may result from precipitation and phase transformation, unequal cooling, cold working and welding.

Fig. 12.11 shows a very simple example of nail in which, the cold worked area (head and tail) serves as the anode and the strain-free area as cathode. It has been observed that the magnitude of stress, required to cause cracking, depends upon nature of the corrosive environment, micro-structure and geometry of the metal specimen. It has also been observed that many failures, due to stress corrosion, involve high residual stresses approaching yields stress of the metal. But in some cases, the stresses below the yield stress may also produce failure, It may be noted that pure metals are relatively immune to stress corrosion cracking.

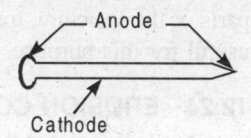

Fig. 12.11. Stress corrosion.

The prevention of stress corrosion cracking is generally done by the removal of tensile stresses or removal of corrosive medium.

12.21 SEASON CORROSION

It is a type of corrosion, which occurs in brass especially in the presence of moisture and traces of ammonia. It has been observed that pure copper is not susceptible to stress corrosion. But small amounts of alloying elements such as zinc, aluminium, silicon, antimony, arsenic and phosphorus result in marked sensitivity to intergranular attack. The susceptibility to cracking in the case of brass increases with the increased zinc content. Zinc and copper are known to be very active electrochemically in ammonia solutions to form stable complex ions. This causes the dissolution of metal initiating fissures (i.e., splitting of a metal). The fissures, in the presence of the high tensile stress, result in the formation of crack. It has been observed that caustic embrittlement of steel exposed to solution containing sodium hydroxide is another well known example of season cracking. In this case, a slight amount of sodium hydroxide is added in water

---

* It is a process of precipitation in which the solid metal product consists of different parts.

softening to precipitate calcium and magnesium. However, the concentration of sodium carbonate increases due to the evaporation of water in the boiler, which promotes hydrolysis,. The chemical reaction taking place may be expressed as:

$$\text{Sodium carbonate} + \text{Water} \longrightarrow \text{Sodium hydroxide} + \text{Carbon dioxide}$$

or

$$Na_2CO_3 + H_2O \longrightarrow 2NaOH + CO_2$$

It may be noted that the sodium hydroxide so formed, acts at places in the boiler, which are subjected to high local stresses and stress corrosion cracking. Such places in a boiler are rivetted joints etc. The corrosion at these places takes place in such a manner that intergranular cracks are formed running from one rivet to another. This phenomenon is known as caustic embrittlement.

12.22 CREVICE CORROSION

It is a type of corrosion, which occurs in cracks or crevices formed between mating surface metal assemblies and takes the form of pitting or etched patches. The mating surfaces may be of similar metal, a metal and a non-metal or dissimilar metals. The crevice corrosion may also occur under scale and surface deposits, under lose fitting washers and gaskets.

12.23 FATIGUE CORROSION

It is a type of corrosion, which occurs in components, which are subjected to cyclic stresses. It generally takes place by the propagation of transcrystalline cracks, which occur on nearly all materials and almost any corrosive liquid under the effect of cyclic stresses. The influence of corrosion, on fatigue strength, is expressed by a parameter called damage ratio. The damage ratio is the ratio of corrosion fatigue strength to the normal fatigue strength. The ratio for salt water as a corroding medium is about 0.2 for carbon steels, 0.4 for aluminium alloys and 1.0 for copper.

The corrosion fatigue may be prevented by avoiding presence of crevices between adjacent parts of the structure, treatment of the corroding medium etc. Sometimes nitriding of steels is very useful for this purpose.

12.24 EROSION CORROSION

It is a type of corrosion, which is caused by the combined effect of erosion by the turbulent flow of gases or liquids and the rubbing of solids over a metal surface. The erosion corrosion occurs due to the breakdown of a protective film at the place of impingement and its subsequent inability to repair itself under the abrasing conditions. It results in the formation of galvanic cells at such areas and leads to corrosion.

The erosion corrosion occurs more commonly in piping agitators, condenser tubes and in such vessels in which stream of liquid or gases emerge from an opening and strikes the side wall at very high velocities.

12.25 ATMOSPHERIC CORROSION

It is a type of corrosion, which occurs either due the oxide-film formation or due to the film breakdown on the metal surface. The factors, which greatly affect the atmospheric corrosion, are the humidity, presence of impurities in the atmosphere, nature of corrosion products and the presence of suspended particles in the atmosphere. The atmospheric corrosion is very severe in ferrous metals.

12.26 UNDERGROUND CORROSION

It is a type of corrosion, which occurs in the pipes buried under the earth's surface. The underground corrosion depends upon the corrosive nature of the soil and may cause serious damage to materials, unless some protective measures are taken. The corrosive nature of a soil depends upon a number of factors such as acidity, degree of aeration, electrical conductivity,

Corrosion of Metals

moisture and salt content, presence of bacteria and microorgnism and soil texture. The corrosion takes place due to the formation of galvanic or concentration cells between different metals. It also occurs due to the pipelines passing through the different soils on their way across the land.

12.27 FRETTING CORROSION

It is a type of corrosion, which occurs in situations where there is slight relative movement of contacting surfaces due to the action of an alternating load. The fretting corrosion occurs, more frequently, in bolted joints and other fitted assemblies. Surface contact, at high spots, results in localized plastic flow and cold working. The welds subsequently rupture and loose metal particles are formed, These particles oxidise in the presence of oxygen and may become more abrasive. If protective films are broken, the bare metal is exposed. This leads to more oxidation. This effect produces localized pitting, which may become the source of fatigue cracks.

12.28 SELECTIVE CORROSION

It is a type of corrosion, which occurs in alloys in which one component is removed selectively, leaving behind the other, with the help of a corrosive medium. The selective corrosion is generally found in marine boilers. The marine boilers use condenser tubes made of brass with a high content of zinc. The condenser tubes use the sea water which is, usually, a weak acidic solution. During the process of heating, both zinc and copper pass into the sea water together. But it has been found that copper (being less active than zinc) is redeposited subsequently on the base metal surface in the form of a porous spongy layer. On the other hand, zinc is removed continuously from the alloy during this attack. This process of losing zinc from the condenser tubes is called dezincification.

12.29 PREVENTION AND CONTROL OF CORROSION

The following methods are, generally, adopted to prevent or control of corrosion of metals:

1. Suitable design and fabrication procedure.
2. Use of inhibitors.
3. Modification of the corrosive environment.
4. Use of protective coating (metallic and non-metallic).
5. Use of cathodic protection.
6. Alloying of metals.
7. Heat treatment of metals.

Now we shall discuss all the abovementioned methods of prevention and control of corrosion one by one.

12.30 SUITABLE DESIGN AND FABRICATION PROCEDURE

The corrosion may be prevented (or minimised) by selecting a suitable design and fabrication procedure for a particular shape of the component. The selection of material for a component should be such that:

1. The use of dissimilar metal contacts be prevented. This helps in preventing galvanic corrosion. However, if the use of dissimilar metals is essential, the metals selected should have their electrode potential as close as possible to each other.
2. The design procedure should also avoid the presence of cracks.
3. Sharp corners and recesses should also be avoided, as they give rise to the formation of stagnant areas and accumulation of solids.
4. Welded joints should be preferred over the rivetted joints to prevent stress corrosion

12.31 USE OF INHIBITORS

An inhibitor is a substance which is added in a small quantity in the electrolyte, to reduce the rate of corrosion. The inhibitors may be organic or inorganic. But they should be able to dissolve in the corroding medium. Moreover, they should be able to form a protective layer, of some kind, either at anodic or cathodic areas. It has been observed that chromates and phosphates are anodic inhibitors, whereas magnesium and calcium salts are cathodic inhibitors. The anodic inhibitors are used to prevent corrosion in radiators, steam boilers and others containers.

12.32 MODIFICATION OF CORROSIVE ENVIRONMENT

The rate of corrosion may be greatly reduced by small changes in the corroding environment. e.g., changes in composition, nature and temperature. Since the rate of corrosion is, usually, an exponential function of the temperature, therefore a small decrease in temperature, causes an appreciable decrease in the rate of corrosion. The removal of dissolved gases also help in reducing the rate of corrosion.

12.33 USE OF PROTECTIVE COATING

Sometimes, a protective coating is applied on the base metal in order to prevent or reduce the rate of corrosion. The protective coatings, may be broadly classified into the following two categories:

1. Metallic coatings, and 2. Non-metallic coatings.

12.34 METALLIC COATINGS

These are the coatings of metals, which are applied on the base metal. Though there are many types of metallic coatings, yet the following are important from the subject point of view:

1. Electroplating. 2. Dipping 3. Spraying. 4. Cladding. 5. Cementation.

12.35 ELECTROPLATING

It is a process of depositing a very thin layer of metal coating, on the base metal by passing a direct current through an electrolyte solution containing some salt of the coating metal. Now a days, electroplating (or sometimes known as electrodepositing) is one of the best methods for the commercial production of a metallic coating.

In this process, the component of base metal is made to act as a cathode whereas the coating metal as an anode in a solution containing some salt of the coating metal *i.e.*, electrolyte as shown in Fig. 12.12 Now direct current is passed for a known time to obtain the coating of desired thickness. The commonly used metals, which are used for protective coating, are copper, nickel, silver, gold, chromium, cadmium and tungsten etc.

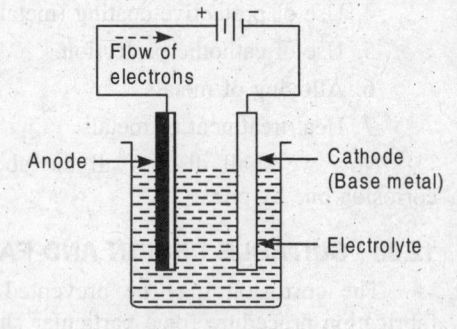

Fig. 12.12. Electroplating.

It will be interesting to know that the electroplating is the reverse of corrosion. In electroplating, some metal is deposited from the solution, whereas in corrosion, some metal is dissolved into the solution. It may be noted that the corrosion takes place at anode and electroplating at cathode.

Corrosion of Metals

12.36 DIPPING

In this process, the component, to be coated, is cleaned and then dipped in a bath of molten metal. The component is then taken out from the bath and then finished properly. The most common processes of dipping are:

1. *Galvanising*. It is a process of providing a thin layer of zinc coating on iron and steel components by dipping them in a bath of molten zinc. The galvanising improves the resistance against corrosion due to atmosphere and water.
2. *Tinning*. It is a process of providing a very thin layer of tin coating on steel parts, by dipping them in a bath of molten tin. The tinning is generally done on the metal sheets, which are used to make containers for storing oil, ghee or some chemicals.

12.37 SPRAYING

It is a process of providing a thin coating by depositing an atomised metal on the surface of a metal. The most common examples of spraying are:

1. *Wire gun method*. In this method, the coating metal is melted by oxyacetylene flame. Now compressed air is used to spray the coating metal uniformly over the surface.
2. *Powder method*. In this method, the coating metal is first powdered and then sucked from the chamber. The powered metal gets heated as it passes through the flame of blow pipe.

It has been observed that the method of spraying has many advantages overall the other method of coating. The spraying offers greater working speed. A large surface area of irregular objects can also be given a uniform coat more easily as compared to other methods. The spraying has been successfully used for applying coatings of aluminium, brass, copper, zinc, tin etc.

12.38 CLADDING

It is a process of providing a comparatively thicker layer of coating on the metal surface. The cladding is, usually, done by hot rolling. The main objective cladding is to produce a corrosion resistant surface. For example, alcaled (aluminium alloys) are used for cladding where a high corrosion resistance is required. It is also used to make bimetal strips for temperature-controlled devices.

12.39 CEMENTATION

It is a process of providing a thin layer of powdered metal coating by heating the metal parts. The temperature, to which the parts are heated, is always more than the melting temperature of the coating metal. The most common processes of cementation are:

1. *Sheradizing*. It is a process of providing of zinc on steel parts. It improves the resistance against atmospheric corrosion.
2. *Chromizing*. It is a process of providing a thin coating of chromium on steel parts. The chromizing greatly improves the resistance against oxidation. It is also used for steam turbine buckets.
3. *Calorizing*. It is a process of providing a thin coating of aluminium on steel parts. The calorizing improves the resistance to oxidation at high temperatures. It is very suitable for treating steel parts for furnaces, oil refineries, driers and kilns etc.

12.40 NON-METALLIC COATINGS

These are the coatings of non-metals, which are applied on the base metal. Though there are many types of non-metallic coatings, yet the following are important from the subject point of view:

1. Paints and lacquering. 2. Plastic coatings, 3. Vitreous coatings. 4. Oxide coatings, 5. Chemical dip coatings.

Now we shall discuss all the abovementioned types of non-metallic coatings one by one.

12.41 PAINTS AND LACQUERING

In this process, a thin coating of paint or a lacquer is proved on a metal surface, to avoid its contact with the corrosive environment. The paint is a mixture of pigments a drying oil and a solvent thinner. We know that whenever a thin coating of paint is applied on any surface, the thinner evaporates from the painted surface and the drying oil gets oxidised, thus forming a dry pigmented film.

The lacquer is a solution of resins and plasticizers with or without any pigments. When a thin coat of lacquer is applied, it dries out quickly, thus leaving a very fined coating, which does not allow the surface to come in contact with the corrosive environment. It will be interesting to know that the evaporation of solvents forms a film from its non-volatile constituents.

12.42 PLASTIC COATINGS

In this process, the metal part to be coated is cleaned and dipped in a bath containing hot liquid plastic (usually ethyl cellulose). After some time, the metal part is taken out of the bath. On cooling, the metal part gets a thick adherent film of the plastic. It will be interesting to know that the plastic coatings are essential for steel parts like tools, saws and other components to protect them against rusting.

12.43 VITREOUS COATINGS

In this process, a coating of vitreous enamel powder is applied to any surface. The powder is fused on the surface by heating. It will be interesting to know that the vitreous enamels are glass like materials, having wide-range of composition. The common example of vitreous enamelling is known as glazing, which is provided as a protective coating on crockery, porcelain, stoneware, ceramics etc. The vitreous coating is generally provided to make parent boy non-absorbant of moisture or chemical reaction proof.

12.44 OXIDE COATINGS

In this process, an oxide coating is provided on the metallic surface by immersing it in an electrolyte both of suitable composition. The oxide coating, formed on the metal surface, acts as an anode. That is why, this process is also termed as anodizing. It will be interesting to know that oxide coatings are provided on door and bathroom fittings of zinc and magnesium. They have good resistance against corrosion.

12.45 CHEMICAL-DIP COATINGS

In this process, the metal, to be coated, is cleaned and dipped in a chemical solution in the same way as the plastic coating. Sometimes, a chemical solution is also sprayed on the metal surface. The chemical solution is generally, prepared by dissolving a metal having higher resistance against corrosion in some suitable chemical. It will be interesting to know that the chemical solution reacts with the metal surface and produces an adherent coating of the metal compound. Following chemical dip coatings are important from the subject point of view:

1.7 *Chromatic coatings.* In this process, the metal to be coated, is dipped in the solution of sodium dichromate and concentrated nitric acid. The metal parts are kept in solution for about one minute at room temperature and then taken out. After coating, the surface provides adequate resistance to corrosion. The chromate coatings are generally provided for protection of zinc, aluminium, magnesium parts etc.

2. *Phosphate coatings.* In this process, the metal to be coated is dipped in a hot solution of manganese phosphate. The metal parts are kept tin the solution for several minutes. After coating, the surface provides adequate resistance to coating. The chromate coatings are generally provided for the protection of iron, steel, zinc, aluminium, tin parts etc.

12.46 ALLOYING OF METALS

It has been observed that pure metals possess greater resistance against corrosion. But they possess inadequate mechanical properties. Moreover, the cost of production to obtain a pure metal is very high. It well be interesting to know that the corrosion resistance of some metals may be increased by alloying them with suitable constituents. The alloying constituents are capable of thoroughly mixing up with the parent metal. The resulting alloys have adequate mechanical properties and corrosion resistance.

12.47 CATHODIC PROTECTION

We have already discussed that in an electrochemical type of corrosion, a current flows from anode to cathode. Thus a metal, which acts as cathode, gains some weight (or it is protected) and the metal which acts as anode loses some weight (i.e., its corrosion takes place). It is thus obvious, that the corrosion may be controlled by making the whole surface as cathodic with respect to some other metal, which acts as anode. This technique is known as cathode protection. In this technique, the following methods are important to protect the metal against corrosion.

1. Sacrificial anode method. 2. External voltage method.

12.48 ANODIC OR SACRIFICIAL ANODE METHOD

In this method, the metal surface to be protected is made cathode by means of another metal, which acts as anode. This is done by selecting the anode metal, which has a higher electrode potential than that of the metal to be protected. The metal, which acts as anode, is known as sacrificial anode. It has been observed that this method may be adopted both for ferrous and non-ferrous metals.

Fig. 12.13. (*a*) shows a mild steel pipe buried in the earth's surface. In this case, a magnesium plate is connected to the pipe, in order to prevent the corrosion of pipe. The magnesium plate (with electrode potential 2.40) acts as an anode and the buried pipe (with electrode potential 0.44) as cathode. Fig. 12.13 (*b*) shows a mild steel tank filled with water. We know that the outer surface of the tanks is generally painted and thus is free from corrosion. But the inner surface, in the absence of water, is likely to be corroded. In this case, a magnesium plate is also connected to the tank in order to prevent the corrosion of the tank. The magnesium plate acts as an anode and the water tank as cathode in the same way as that of steel pipe.

(*a*) Burried pipe   (*b*) Tank filled with water

Fig. 12.13. Sacrificial anode method.

It will be interesting to know that in the abovementioned cases, the corrosion is prevented at the expense of another metal, which serves as anode. The anode metals, after their corrosion, are replaced.

12.49 EXTERNAL VOLTAGE METHOD

In this method, a small external D.C. voltage is connected in such a way that the metal to be protected is made to act as a cathode. Fig. 12.14 shows the protection of a mild steel pipe burried in the earth's surface. The corrosion of this pipe may be prevented by connecting the positive terminal of the battery to the earth and the negative terminal to the pipe as shown in the figure.

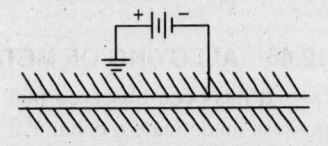

Fig. 12.14. External voltage method.

It will be interesting to know that the D.C. voltage provides the electrons to make the pipe cathode. The method is very useful for the protection of ferrous structures such as steam boilers, condensers, oil and water storage tanks. etc.

A.M.I.E. (I) EXAMINATION QUESTIONS

1. Explain the effect of pH value on corrosion. *(Winter, 1993)*
2. Write a short note on season cracking. *(Winter, 1993)*
3. (a) Briefly explain the major types of corrosion of metallic materials.
   (b) Describe the techniques generally used to control corrosion of metals. *(Summer, 1994)*
4. (a) Define corrosion of metals and briefly explain the mechanism of corrosion.
   (b) Discuss galvanic cell corrosion.
   (c) Discuss the method of preventing corrosion. *(Winter, 1994)*
5. Explain the corrosion mechanism. What are the different methods of preventing corrosion? Explain with the aid of sketches wherever necessary. *(Summer, 1996)*
6. What is corrosion? Explain the different mechanisms of corrosion. *(Summer, 1997)*
7. Name the type of corrosion, which may take place even at room temperature. *(Summer, 1997)*
8. (a) Discuss the electrochemical phenomenon of corrosion.
   (b) Describe (i) cathodic protection (ii) anodic protection in this connection.
   (c) Explain the phenomenon of sensitization in stainless steel. *(Summer, 2000)*
9. What alloying element is most important to make steel corrosion resistant? *(Summer, 2000)*

MULTIPLE CHOICE QUESTIONS

1. Which of the following is not true :
   (a) Corrosion is a process of destruction of a metal through an unwanted chemical attack by its environment
   (b) Corrosion is the process used to produce metal from its ore
   (c) Rusting of an iron is a good example of corrosion
   (d) All are true.
2. Fatigue corrosion occurs in components
   (a) along grain boundaries of a metal
   (b) due to moisture and traces of ammonia
   (c) due to cracks formed between mating surfaces
   (d) which are subjected to cyclic stresses

ANSWERS

1. (a)   2. (d)

13
Ferrous and Non-Ferrous Alloys

1. Introduction. 2. Ferrous Alloys. 3. Cast Iron, 4. Effect of Impurities on Cast Iron. 5. Types of cast iron, 6. Steel. 7. Steels Designated on the basis of Mechanical properties. 8. Steels Designated on the basis of chemical Composition. 9. Plain Carbon Steel. 10 Effects of Impurities on Steel. 11. Alloy Steel. 12. Tool and Die Steels. 13. Special Steel. 14. High Speed Steel. 15. Stainless Steel. 16. Heat Resisting Steel. 17. Free Cutting Steels. 18. Spring Steels. 19. Non-ferrous Metals. 20. Aluminium. 21. Aluminium Alloys. 22. Copper. 23. Copper Alloys. 24. Brass. 25. Bronze. 26. Gun metal. 27. Lead. 28. Lead-base Alloys. 29. Tin. 30. Tin-base Alloys (Babbit Metal). 31. Zinc. 32. Zinc-base Alloys. 33. Nickel. 34. Nickel-base Alloys. 35. Magnesium. 36. Cadmium. 37. Vanadium. 38. Antimony. 39. Bearing Metals. 40. Zirconium Alloys. 41. Shape-memory Alloys. 42. Biomaterials 43. Materials for Hip joint Replacement.

13.1 INTRODUCTION

The engineering metals play an important role in an industry, because the process of all manufacturing starts with the raw materials. The materials, mainly used in actual practice may be broadly divided into the following two groups:

1. *Ferrous metals*. The metals, which contain iron as their main constituent, are called ferrous metals. The various ferrous metals used in industry are pig iron, cast iron, wrought iron and steel. In this chapter, we shall discuss only about cast iron and steel.

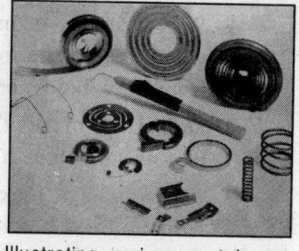

Illustrating various metal parts like springs and clips. These are characteristics of wide range of applications of metals.

2. *Non-ferrous metals*. The metals, which contain a metal. The iron as their main constituent, are called non-ferrous metals. Other than various non-ferrous metals used in industry are aluminium, copper, zinc, lead, brass, tin etc.

We shall now discuss, the important ferrous and non-ferrous metal in the following articles. It may be noted that we shall discuss in detail the mechanical properties and uses of the ferrous and non-ferrous metals. The production or extraction of these metals is beyond the scope of this book.

13.2 FERROUS ALLOYS

We have already discussed in the last article that ferrous metals are those which contain iron as their main constituent. As a matter fact, the ferrous metals are extensively used in engineering industry due to the following characteristics:

1. Each of fabrication processes (casting, rolling, welding and machining),
2. Resistance to corrosion,
3. Magnetic properties, and
4. Weight.

The important ferrous metals used in engineering industry are pig iron, cast iron, wrought iron and steel. However, we shall discuss only about cast iron and steel, which are important from the subject point of view.

13.3 CAST IRON

The cast iron is an eutectic alloy of iron and carbon. Thus it has relatively low melting point (about 1200°C). This is advantageous because it can be easily melted, requires less fuel and more easily operated in furnaces. Moreover, the molten metal easily fills intricate moulds completely. These characteristics lead to an inexpensive material and versatility in product design.

The carbon content in cast Iron varies from 2% to 4.5%. It also contains small amounts of silicon, sulphur, manganese, and phosphorus. The cast iron is a brittle material, therefore it cannot be used in those parts which are subjected to shocks. The properties of cast iron, which makes it a valuable material for engineering purposes are its low cost, good casting characteristics, high compressive strength, wear resistance, and excellent machinability. The compressive strength of cast iron is much greater than its tensile strength. Following are the *values of ultimate strengths of cast iron:

$$\text{Tensile} = 100 \text{ to } 200 \text{ MPa}$$
$$\text{Compressive} = 400 \text{ to } 1000 \text{ MPa}$$
$$\text{Shear} = 120 \text{ MPa}$$

13.4 EFFECT OF IMPURITIES ON CAST IRON

We have discussed that the cast iron contains small percentages of silicon, sulphur, manganese and phosphorus. The effect of these impurities on cast iron are as follows:

1. *Silicon.* It may be present in cast iron upto 3%. It provides the formation of free graphite, which makes the iron soft and easily machinable. It also produces sound castings free from blow holes, because of its high affinity for oxygen.

2. *Sulphur.* It makes the cast iron hard and brittle. Since too much sulphur gives unsound casting, therefore it should be kept well below 0.1% for most foundry purposes.

3. *Manganese.* It makes the cast iron white and hard. It is often kept below of 0.75%. It helps to exert a controlling influence over the harmful effect of sulphur.

4. *Phosphorus.* It aids fusibility and fluidity in cast iron, but induces brittleness. It is rarely allowed to exceed 1%. Phosphoric irons are useful for casting of intricate design and for many light engineering castings when cheapness is essential.

13.5 TYPES OF CAST IRONS

Though there are many types of cast irons, yet the following are important from the subject point of view.

1. *Gray cast iron.* It is an ordinary commercial cast iron having the following approximate compositions:

$$\text{Carbon} = 3\% \text{ to } 3.5\%, \text{ Silicon} = 1\% \text{ to } 2.75\%$$
$$\text{Manganese} = 0.4\% \text{ to } 1\%; \text{ Phosphorus} = 0.15 \text{ to } 1\%$$
$$\text{Sulphur} = 0.02\% \text{ to } 0.15\% \text{ and the remaining is iron.}$$

The gray colour is due to the fact the carbon is present in the form of free graphite. It has a low tensile strength, high compressive strength and no ductility. It can be easily machined. A very good property of gray cast iron is that the free graphite in its structure

---

* MPa means megapascal. One MPa = 1×10^6 N/m^2

acts as a lubricant. Due to this reason, it is very suitable for those parts where sliding action is desired. The gray iron casting are widely used for machine tool bodies, automobile cylinder blocks, pipes and pipe fittings and agricultural implements.

2. *White cast iron.* It is a particular variety of cast iron, which, shows a white fracture. It has the following approximate compositions:

$$\text{Carbon} = 2\% \text{ to } 2.3\%; \text{ Silicon} = 0.85\% \text{ to } 1.2\%$$
$$\text{Manganese} = 0.1\% \text{ to } 0.4\%; \text{ Phosphorus} = 0.05\% \text{ to } 0.2\%$$
$$\text{Sulphur} = 0.12\% \text{ to } 0.35\% \text{ and the remaining is iron.}$$

The white colour is due to the fact that the carbon is in the form of carbide (known as cementite) which is the hardest constituent of iron. This cementite is caused by quick cooling of molten iron. The white cast iron has a high tensile strength and a low compressive strength. Since it is hard, therefore it cannot be machined. It is used for inferior casting and in places where hard coating is required as in the outer surface of car wheels. The white cast iron is also used as a raw material in the production of malleable cast iron and wrought iron.

3. *Chilled cast iron.* It is a white cast iron produced by quick cooling of molten iron. The quick cooling is generally called chilling. The iron so produced is known as chilled iron. All castings are chilled at their outer skin by contact of the molten iron with the cool sand in the mould. But in most castings, this hardness penetrates to a very small depth (less than 1 mm). Sometimes, a casting is chilled intentionally and sometimes chilled becomes accidently to a considerable depth. Intentional chilling is carried out by putting inserts of iron or steel (chills) into the mould. When the molten metal comes in contact with the chill, its heat is rapidly conducted away and the hard surface is formed. Chills are used on the faces of a casting, which are required to be hard to withstand wear and friction.

The process of chilling is used in the casting of rolls for crushing grains and jaw crusher plates. The running surface of railcarriage wheels is also chilled.

4. *Molted cast iron.* It is a product in between gray and white cast iron in composition, colour and general properties. It is obtained in castings where certain wearing surfaces have been chilled.

5. *Malleable cast iron.* The malleable cast iron is obtained from white cast iron by a suitable heat treatment process (*i.e.*, annealing). The annealing process separates the combined carbon of the white cast iron into nodules of free graphite.

The malleable cast iron is ductile and may be bent without breaking or fracturing the section. Its tensile strength is usually higher than that of gray cast iron and has excellent matching qualities. It is used for making machine parts for which the steel forgings would be expensive and in which the metal should have a fair degree of accuracy, *e.g.*, hubs of wagon wheels, small fittings for railway rolling stock, brake supports, parts of agricultural machinery, pipe fittings, door hinges, locks etc.

6. *Nodular cast iron.* It is also known as ductile cast iron spheroidal graphite (i.e., S.G.) cast iron or high strength cast iron. The nodular cast iron is produced by adding magnesium to the molten cast iron. The magnesium converts the graphite of cast iron from flake form to spheroidal or nodular form. This way, the mechanical properties are considerably improved.

The nodular cast iron behaves like steel. It is usually used for pressure-resisting castings, hydraulic cylinder heads, rolls for rolling mill and centrifugally cast products.

7. *Alloy cast iron.* The cast irons discussed above are called plain cast irons. The alloy cast iron is produced by adding alloying elements like nickel, chromium, molybdenum,

copper, silicon and manganese. These alloying elements give more strength and result in improvement of properties. The alloy cast iron has special properties like increased strength, high wear resistance, corrosion resistance or heat resistance. The alloy cast irons are extensively used for automobile parts like cylinders, pistons, piston rings, crank cases, brake drums, parts of crushing and grinding machinery etc.

13.6 STEEL

It is an alloy of iron and carbon with carbon content upto a maximum of 1.4%. The carbon occurs in the form of iron carbide (Fe_3C), because of its ability to increase the hardness and strength of the steel. Other elements *e.g.*, silicon, sulphur, phosphorus, and manganese are also present to a greater or lesser amount to impart certain desired properties to it. Most of the steel produced now-a-days is plain carbon steel. A plain carbon steel is defined as a steel which has its properties mainly due to its carbon content and does not contain more than 0.5% of silicon and 1.5% of manganese. The plain carbon steels varying from 0.06% carbon to 1.4% carbon are divided into the following four types depending upon the carbon content:

1. Dead mild steel = 0.06%–0.12% carbon.
2. Low carbon or mild steel = 0.10%–0.25% carbon.
3. Medium carbon = 0.25%–0.55% carbon.
4. High carbon steel = 0.55–1.4% carbon.

According to Indian Standard [IS : 1762 (Part-I) – 1974], a new system of designating the steel is recommended. According to this standard, steels are designated on the following two basis:

(*a*) On the basis of mechanical properties and

(*b*) On the basis of chemical composition.

We shall now discuss, in detail, the designation of steel on the above two basis, in the following pages.

Illustrating a basic oxygen furnace used in the production of steel

Illustrating a close up of the rocker assembly on a mountain bike

13.7 STEELS DESIGNATED ON THE BASIS OF MECHANICAL PROPERTIES

These steels are carbon and low alloy steels where the main criterion in the selection and inspection of steel is the tensile strength or yield stress. According to Indian Standard [IS : 1570 (Part-I) – 1978], these steels are designated by a symbol 'Fe' or 'FeE' depending on whether the

Ferrous and Non-Ferrous Alloys

steel has been specified on the basis of minimum tensile strength or yield strength followed by the figure indicating minimum tensile strength or yield stress in N/mm². For example, 'Fe-290' means a steel having minimum tensile strength of 290 N/mm² and 'Fe-220' means a steel having yield strength of 220 N/mm².

Note: According to IS : 1570–1961, these steels were designated by the symbol 'St' followed by a figure indicating the tensile strength in kgf/mm².

13.8 STEELS DESIGNATED ON THE BASIS OF CHEMICAL COMPOSITION

According to Old Indian Standard (IS : 1570–1961) the carbon steels were designated by the alphabet 'C followed by numerals which indicate the average percentage of carbon in it. For example, C-40 means a plain carbon steel containing 0.35% to 0.45% carbon (0.40% average). Now, according to New Indian Standard [IS : 1570 (Part-II)–1970, the carbon steels, are designated in the following order:

1. Figure indicating 100 times the average percentage of carbon content.
2. Letter C, and
3. Figure indicating 10 times the average percentage of manganese content. The figure after multiplying shall be rounded off to the nearest integer.

For example, 20 C 8 (old designation (C-20) means a carbon steel containing 0.15% to 0.25% (0.20% on an average) carbon and 0.60 to 0.90% (0.75% rounded off to 0.8% on an average manganese.

13.9 PLAIN CARBON STEELS

We have already discussed in last article that steels which have its properties mainly due to its carbon content and do not contain more than 0.5% silicon and 1.5% manganese are called plain carbon steels. These steels are strong, tough, ductile and used in expensive materials. They can be cast, worked, machined and heat treated to a wide range of properties. Unfortunately, plain carbon steel has poor atmospheric corrosion resistance. But it can be protected easily by painting, enamelling or galvanizing.

The properties of plain carbon steels depend upon the presence of carbon content. The hardness and strength increases with an increase in carbon content. These properties increase due to the presence of hard and brittle cementite. The ductility and toughness decreases with an increase in the carbon content. Table 13.1 shows some of the important applications of plains steels.

Table 13.1 Applications of plain carbon steels

	Types of steels	*Uses*
1.	Low-carbon steels or mild steels	Chain links, nails, rivets, ship hulls, car bodies, bridges, cams, light duty gears etc.
2.	Medium carbon steels	Axles, connecting rods, gears, wheels for trains and rails etc.
3.	High carbon steels	Clutch plates, razor blades scissors, knives, files, punches, dies etc.

13.10 EFFECTS OF IMPURITIES ON STEEL

The following are important effects of impurities like silicon, sulphur, the manganese and phosphorus on steel:

1. *Silicon.* The amount of silicon in the finished steel usually ranges from 0.05% to 0.30%. Silicon is added in low carbon steels to prevent them from becoming porous. It removes the gases and oxides, prevent blow holes and thereby makes the steel tougher and harder.
2. *Sulphur.* It occurs in steel either as iron sulphide or manganese sulphide. Iron sulphide, because of its melting point produces red shortness, whereas manganese sulphide does not effect so much. Therefore manganese sulphide is less objectionable in steel than iron sulphide.
3. *Manganese.* It serves as a valuable deoxidising and purifying agent in steel. Manganese also combines with sulphur and thereby decreases the harmful effects of this element in the steel. When decreases the harmful effects of this element in the steel. When used in ordinary low carbon steels, manganese makes the metal ductile and of good bending qualities. In high speed steels, it is used to toughen the metal and to increase its critical temperature.
4. *Phosphorus.* It makes the steel brittle. It also produces cold shortness in steel. In low carbon steels, it raises the yield point and improves the resistance to atmospheric corrosion. The sum of carbon and phosphorus, usually, does not exceed 0.25%.

13.11 ALLOY STEEL

We have already discussed in the previous articles about plain carbon steels. These steels contain carbon together with small amounts of manganese and silicon. A steel in which elements other than carbon are added in sufficient quantity, in order to obtain special properties, is known as alloy steel. The alloying of steel is generally done to increase its strength, hardness, toughness, resistance to abrasion and wear and to improve electrical and magnetic properties. The various alloying elements are nickel, chromium, molybdenum, cobalt, vanadium, manganese, silicon and tungsten. The effects of these alloying elements are discussed below:

1. *Nickel.* It is one of the most important alloying elements. Steel sheets contain 2% to 5% nickel and 0.1% to 0.5% carbon. In this range, nickel improves tensile strength, raises elastic limit, imparts hardness, toughness and reduces rust formation. It is largely used for boiler plates automobile engine parts, large forgings, crankshafts, connecting rods etc. When nickel is added to steel in appreciable proportions (about 25%) it results in higher strength steels with improved shock and fatigue resistance. It makes the steel resistant to corrosion and heat. It is used in the manufacture of boiler tubes, valves for gas engines, pump barrels, sparking plugs for petrol engines, liners and pump parts etc. A nickel steel alloy containing about 36% nickel and 0.5% carbon is known as invar. It can be rolled, forged, turned and drawn. It has nearly zero coefficient of expansion. So it is widely used for making pendulums of clocks, precision measuring instruments etc.
2. *Chromium.* Addition of chromium to steel increases its strength, hardness and corrosion resistance. A chrome steel containing 0.5 to 2% chromium is used for balls, rollers and races for bearings, dies, rolls for rolling mills, permanent magnets, etc.

    A steel containing 3.25% nickel, 1.5% chromium and 0.25% carbon is known as nickel chrome steel. The combination of toughening effect of nickel and the hardening effect of chromium produces a steel of high tensile strength with great resistance to shock. It is extensively used for motor car crank shafts, axles and gears requiring great strength and hardness.
3. *Vanadium.* It is added in low and medium carbon steels in order to increase their yield and tensile strength properties. A very useful effect of vanadium is the improvement it imparts to the harden ability of steel. In constructional steels, it is added to an extent of 0.25% while for tool steels and other special steels, increased percentage vanadium is

Ferrous and Non-Ferrous Alloys

used. In combination with chromium (*i.e.* chrome-vanadium steel containing about 0.5% to 1.5% chromium, 0.15% to 0.3% vanadium and 0.13% to 1.1% carbon), it produces a marked effect on the properties of steel and makes the steel extremely tough and strong. These steels are largely used for making spring steels, high speed tools, steels crank shafts, locomotives and wagon axles, chassis and other parts of automobiles.

4. *Tungsten.* The addition of tungsten raises the critical temperature of steel and hence it is used for increasing strength of alloyed steels at high temperatures. It imparts cutting hardness and abrasion resistance properties to steel. When added to the extent or 5% to 6%, it gives the steel good magnetic properties. Thus it is commonly used for magnets, in electrical instruments etc.

The tungsten is, usually, used in conjunction with other elements. Steel containing 18% tungsten, 4% chromium, 1% vanadium and 0.7% carbon is called tool steel or high speed steel. Since the tool made with this steel the ability to maintain its sharp cutting edge even at elevated temperatures, therefore it is used for making high speed cutting tools such as cutters, drills, dies, broaches, reamers etc.

5. *Manganese.* It is added to steel in order to reduce the formation of iron sulphide by combining with sulphur. It is, usually, added in the form of ferro-manganese or silico-manganese. It makes the steel hard, tough and wear resisting. The mangnese alloy steels, containing over 1.5% manganese with a carbon range of 0.40% to 0.55% are widely used for gears, axles, shafts, and other parts where high strength combined with fair ductility is required.

A steel containing manganese varying from 10% to 14% and carbon from 1% to 1.3% form an alloy steel, which is extensively hard and tough and has a high resistance to abrasion. It is largely used for mining, rock crushing and railways equipment.

6. *Silicon.* It increases the strength and hardness of steel without lowering its ductility. Silicon steels containing from 1% to 2% silicon and 0.1% to 0.4%% carbon have good magnetic permeability and high electrical resistance. It can withstand impact and fatigue even at elevated temperatures. These steels are principally used for generator and transformers in the form of laminated cores.

7. *Cobalt.* It is added to high speed steel from 1% to 12%, to give red hardness by retention of hard carbides at high temperatures. It ends to decarburise steel during heat treatment. It increases hardness and strength. But too much of cobalt it decreases impact resistance of steel. It also increases residual mangnetism and coereive magnetic force in steel for magnets.

8. *Molybdenum.* A small quantity (0.15% to 0.30%) of molybdenum is generally used with chromium and manganese (0.5% to 0.8%) to make molybdenum steel. These steels possess extra tensile strength and are used for aeroplane and automobile parts. It can replace tungsten in high speed steels.

13.12 TOOL AND DIE STEELS

These are the steels used in making tools and dies which are required for cutting, shaping, forming and blanking of materials. These steels should have high hardness, greater abrasion or wear resistance, greater toughness, high impact strength, high thermal conductivity, low coefficient of friction etc. The tool and dies steels are of the following types:

1. *Plain carbon steels.* These steels contain carbon from 0.0% to 1.4% and are hardened either by oil or water quenching. The important advantage of these steels is that they low cost, good machinability and high impact resistance, These steels are used for keys, sledge hammers, stamping dies, spanners, cold heading and cold drawing dies, twist drills, general wood working and leather cutting tools etc.

2. *Low-alloy tool steels*. These steels contain alloying elements like vanadium chromium, tungsten and silicon. The presence of alloying elements refines the structure and increases the toughness and impact resistance. These steels have a greater ability to retain the sharpness and serviceability of the cutting edge than plain carbon steels. The low-alloy tool steels are extensively used for heavy duty pneumatic tools, pavement breakers etc.

3. *High speed steels*. These steels are used for cutting metals at a much higher speed than ordinary carbon tool steel. The important high speed steels are 18-4-1 high speed steel, molybdenum high speed steel and super high speed steel. For details please refer to Art. 13.14.

4. *High carbon high chromium steels*. These steels are much cheaper than high speed steels, but have greater importance than high speed steels. These are widely used for various types of dies like those used for drawing, coining, blanking, forming and thread rolling. They are also used for reamers gauges, rolls for forming and bending sheet metal etc.

13.13 SPECIAL STEEL

Steels manufactured for special purposes, such as high speed steel, stainless steel, heat resisting steel, free cutting steel and spring steel, are called special steels. We shall now discuss these types of steels in the following pages.

13.14 HIGH SPEED STEEL

These steels are used for cutting metals, at a much higher cutting speed than ordinary carbon tool steel. The carbon steel cutting tools do not retain their sharp cutting edges under heavier loads and higher speeds. This is due to the fact that at high speeds, sufficient heat may be developed during the cutting operation and causes the temperature of the cutting edge of the tool to reach a red hot. This temperature would soften the carbon tool steel and thus the tool will not work efficiently for a longer period. The high speed steels have the valuable property of retaining their hardness even when heated to red hot. Most of the high speed steels contain tungsten as the chief alloying element. But other alloying elements like cobalt, chromium, vanadium etc., may be present in some proportions. Following are the different types of high speed steels.

1. *18-4-1 High speed steel*. This steel contains 18% tungsten, 4% chromium and 1% vanadium. It is considered to be one of the best of all purpose tool steels. It is widely used for drills, lathes, planer and shaper tools, milling cutters, reamers, threading dies, punches etc.

2. *Molybdenum high speed steel*. This steel contains 6% tungsten, 6% molybdenum, 4% chromium and 2% vanadium. It has an excellent toughness and cutting ability. The molybdenum high speed steels are better and cheaper than other types of steels. It is particularly used for drilling and tapping operations.

3. *Super high speed steel*. This steel is also called cobalt high speed steel, because cobalt is added from 2% to 4% in order to increase the cutting efficiency especially at high temperatures. This steel contains 20% tungsten, 4% chromium, 2% vanadium and 12% cobalt. Since the cost of this steel is more, therefore it is principally used for heavy cutting operations, which impose high pressures and temperatures on the tool.

13.15 STAINLESS STEEL

It is a steel, which when correctly heat-treated and finished, resists oxidation and corrosive attack from corrosive media. Following are the different types of stainless steels:

1. *Ferritic stainless steels*. These steels contain a greater amount of nickel (from 15% to 20%) and about 0.1% carbon. They have great strength, toughness and extremely good resistance to corrosion. These steels can be welded, forged, rolled and machined. They

are used in the manufacture of vats and pipes in the chemical and food plant equipments to resist nitric acid corrosion. Thy are also used in highly stressed fittings of engines and machines, in the manufacture of bars, sheets, strips, wires etc.

2. *Martenisic stainless steels.* These steels contain 11% to 14% chromium and about 0.35% carbon. These steels can be hardened by suitable heat treatment and have satisfactory corrosion resistance qualities. These can be welded and machined. Such steels are used for making steam valves, turbine blades, shafts, scissors, knives, gears, ball bearings, springs, cutlery, surgical and dental instruments and other purposes where hard edges are required.

3. *Austentic stainless steels.* These steels contain 18% chromium and 8% nickel (commonly referred to as 18/8 steel). These steels have greatest resistance to corrosion and good tensile strength. These are very tough and can be welded, forged or rolled, but they offer great difficulty in machining. They are used in the manufacture of pump sets, rail road car frames, screw nuts and bolts etc. They are also used in chemical plants, appliances, storage and transport tanks for chemical industries, utensils and cutlery.

13.16 HEAT RESISTING STEEL

A steel which can resist the creep and oxidation at high temperatures and retain sufficient strength is called heat resisting steel. A number of heat-resisting steels have been developed, which are discussed below:

1. *Low alloy steels.* These steels contain 0.5% molybdenum. The main applications of these steels are for superheater tubes and pipes in steam plants, where service temperatures are in the range of 400° to 500°C.

2. *Valve steels.* The chromium-silicon steels such as silchrome (0.4% C, 8% Cr, 3.5% Si) and volmax (0.5% C, 8% Cr, 3.5% Si, 0.5% Mo) are used for automobile valves. They possess good resistance to scaling at dull red heat, although their strength at elevated temperatures is relatively low. For aeroplane engines and marine diesel engine valves 13/13/3, nickel-chromium tungsten valve steel is usually used.

3. *Platinum chromium steel.* The plain chromium steels consists of
   (a) Martensitic chromium steels with 12% to 13% Cr, and
   (b) Ferritic chromium steels with 18% to 30% Cr.
   These steels are very good for oxidation resistance at high temperatures as compared to their strength, which is not high at such conditions, Maximum operating temperature for martensitic steels is about 750°C, whereas for ferritic steels it is about 1000°C–1150°C.

4. *Austenitic chromium-nickel steels.* These steels have good mechanical properties at high temperatures with good scaling resistance. These alloys contain a minimum of 18% chromium and 8% nickel stabilised with titanium. Other carbide forming elements such as molybdenum or tungsten may also be added in order to improve creep strength. Such alloys are suitable for use upto 1100°C and are used for gas turbine discs and blades.

13.17 FREE CUTTING STEELS

These steels are used where rapid machining is the prime requirement. They have higher sulphur content than other carbon steels. In general, the carbon content of such steels vary from 0.08% to 0.45% and sulphur upto 0.3%. The machinability of this steel improves with the increase in sulphur content within reasonable limits. But it is always at the sacrifice of some of its cold working, welding and forging qualities. Now-a-days lead is used from 0.05% to 0.2% instead of sulphur, because lead also greatly improves the machinability of steel without the loss of toughness.

13.18 SPRING STEELS

The most suitable materials for springs are those which can store the maximum amount of work or energy in a given weight or volume of spring material, without permanent deformation. These steels should have a high elastic limit as well as high deflection value. For aircraft and automobile purposes, the spring steel should possess maximum strength against fatigue effects and shocks. The steels most commonly used for making springs are as Follows:

1. *High carbon steels*. These steels contain 0.6% to 1.1% carbon, 0.2% to 0.5% silicon and 0.6% to 1% manganese. These steels are heated to 780°C-850°C according to the composition and quenched in oil or water. It is then tempered at 200°C–500°C to suit the particular application. These steels are used for laminated springs for locomotives, carriages, wagons for heavy road vehicles. The higher carbon content oil hardening steels are used for volute, spiral and conical springs and for certain types of petrol engine inlet valve springs.

2. *Chrome-vanadium steels*. These are high quality spring steels and contains 0.45% to 0.55% carbon, 0.9% to 1.2% chromium, 0.15% to 0.20% vanadium, 0.3% to 0.5% silicon and 0.5% to 0.8% manganese. These steels have high elastic limit, resistance to fatigue and impact stresses. Moreover, it can be machined without difficulty and given a smooth surface free from tool marks. It is hardened by oil quenching at 850°C–870°C and tempered for vehicle and other spring purposes at 470°C–510°C. It is used for motor car laminated and coil springs for suspension purposes, automobile and aircraft engine valve springs.

3. *Silicon manganese steel*. These steels contain 1.8% to 2% silicon, 0.5% to 0.6% carbon and 0.8% to 1% manganese. These steels have high fatigue strength, resistance and toughness. It is hardened by quenching in oil at 850°C–900°C and then tempered at 475°C–525°C. It is the usual standard quality modern spring material and is much used for many engineering purposes.

13.19 NON-FERROUS METALS

We have already discussed that the non-ferrous metals are those which contain metal other than iron as their chief constituent. The non-ferrous metals are usually employed in industry due to the following characteristics:

1. Ease of fabrication (casting, rolling, forging, welding and machining).
2. Resistance to corrosion,
3. Electrical and thermal conductivity, and
4. Weight.

The various non-ferrous metals used in engineering practice are aluminium, copper, lead, tin zinc, nickel etc. and their alloys. We shall now discuss these non-ferrous metals and their alloys, in detail, in the following pages.

13.20 ALUMINIUM

It is a white metal produced by electrical process from the oxide (alumina), which is prepared from a clayey mineral called bauxite. It is a light weight metal having specific gravity 2.7 and melting point 660°C.

In its pure state, the metal would be weak and soft for most purposes. But when mixed with small amounts of other alloys, it becomes hard and rigid. So, it may be blanked, formed, drawn turned, cast, forged and die cast. Its good electrical conductivity is an important property and is widely used for overhead cables. The high resistance to corrosion and its non toxicity makes it a useful metal for cooking utensils under ordinary conditions. It is extensively used in aircraft and automobile components where saving of weight is an advantage.

Ferrous and Non-Ferrous Alloys

13.21 ALUMINIUM ALLOYS

The aluminium may be alloyed with one or more other elements like copper, magnesium, manganese, silicon and nickel. The addition of small quantities of alloying elements converts the soft and weak metal into hard and strong metal, while still retaining its light weight. The main aluminium alloys are:

1. Duralumin
2. Y-alloy
3. Mangalium
4. Hindalium

These alloys are discussed as below:

1. *Duralumin*. It is an important and interesting wrought alloy. Its composition is as follows:

    Copper = 3% to 4.5%; Manganese = 0.4% to 0.7%
    Magnesium = 0.4% to 0.7% and rest is aluminium.

    The alloy possesses maximum strength after heat treatment and age hardening. After working, if the metal is allowed to age for 3 or 4 days, it will be hardened. This phenomenon is known as age hardening.

    It is widely used in wrought conditions for forging, stamping, bars, sheets, tubes and rivets. It can be worked in hot conditions at a temperature of 500°C. However, after forging and annealing, it can also be cold worked. Due to its high strength and light weight, this alloy may be used in automobile and aircraft components.

2. *Y-alloy*. It is also called copper-aluminium alloy. The addition of copper to pure aluminium creases its strength and machinability. The composition of this alloy is as follows:

    Copper = 3.5% to 4.5%; Manganese =1.2% to 1.7%
    Nickel = 1.8% to 2.3%, Silicon, Magnesium, Iron = 0.6% each and the rest is aluminium.

    This alloy is heat treated and age hardened like duralumin. The ageing process is carried out at room temperature for about five days.

    It is mainly used for casting purposes. But it can also be used for forged components like duralumin. Since Y-alloy, has better strength (than duralumin) at high temperatures, therefore it is mainly used in aircraft engines, for cylinder heads and pistons.

3. *Magnalium*. It is made by melting the aluminium with 2% to 10% magnesium in a vacuum and then cooling it in a vacuum or under a pressure of 100 to 200 atmospheres. It also contains about 1.75% copper. Due to its light weight and good mechanical properties, it is mainly used for aircraft and automobile components.

4. *Hindalium*. It is an alloy of aluminium and magnesium with a small quantity of chromium. It is the trade name of aluminium alloy produced by Hindustan Aluminium Corporation Ltd., Renukoot (U.P). It is produced as a rolled product in 16 gauge, mainly for anodized utensil manufacture.

13.22 COPPER

It is one of the most widely used non-ferrous metals in industry. It is a soft, malleable and ductile material with a reddish-brown appearance. Its specific gravity is 8.9 and melting point is 1083°C. It is a good conductor of electricity. It is largely used in making electrical cables and wires for electrical machinery and appliances in electrotyping, electroplating in making coins and household utensils.

It may be casted, forged, rolled and drawn into wires. It is non-corrosive under ordinary conditions and resists weather very effectively. Copper in the form of tubes is widely used in mechanical engineering. It is also used for making ammunitions. It is used for making useful alloys with tin, zinc, nickel and aluminium.

13.23 COPPER ALLOYS

The copper alloys are broadly classified into the following two groups:
1. Copper-zinc alloys (Brasses), in which zinc is the principal alloying metal, and
2. Copper-tin alloys (Bronzes), in which tin is the principal alloying metal.

13.24 BRASS

The most widely used copper zinc alloy is brass. There are various types of brasses, depending upon the proportion of copper and zinc. This is fundamentally a binary alloy of copper with zinc each 50%. By adding small quantities of other elements, the properties of brass may be greatly changed. For example, the addition of lead (1% to 2%) improves the machining quality of brass. It has a greater strength than that of copper, but has a lower thermal and electrical conductivity. Brasses are very resistant to atmospheric corrosion and can be easily fabricated by processes like spinning and Y-chromium. The Table 13.2 shows the composition of various types of brasses according to Indian Standards.

Table 13.2

I.S.I. designation	Composition in percentages	Uses
Cartridge brass	Copper = 70 Zinc = 30	It is a cold working brass used for cold rolled sheets, wire drawing, deep drawing pressing and tube manufacture.
Yellow brass (Muntz metal)	Copper = 60 Zinc = 40	It is suitable for hot working by rolling, extrusion and stamping.
Leaded brass	Copper = 62.5 Zinc = 36 Lead = 1.5	These are used for plates, tubes etc.
Admiralty brass	Copper = 70 Zinc = 29 Tin = 1	
Naval brass	Copper = 59 Zinc = 40 Tin = 1	It is used for marine castings.
Nickel brass (German silver)	Copper = 60.45 Zinc = 35.20 Nickel = 5.35	It is used for valves, plumbing fittings, automobile fittings, type writer parts and musical instruments.

13.25 BRONZE

The alloys of copper and tin are, usually, termed as bronzes. The useful range of composition is 75 to 95% copper and 5 to 25% tin. The metal is comparatively hard, resists surface wear and can be shaped or rolled into wires, rods and sheets very easily. In corrosion

Ferrous and Non-Ferrous Alloys

resistant properties, bronzes are superior to brasses. Some of the common types of bronzes are as follows:

1. *Phosphor bronze.* When bronze contains Phosphorus, it is called phosphor bronze. Phosphorus increases the strength, ductility and soundness of castings. The alloy possesses good wearing qualities and high elasticity. The metal is resistant to salt water corrosion. The composition of the metal varies according to whether it is to be forged wrought or made into castings. A common type of phosphor bronze has the following composition according to Indian Standards:

   Copper = 87% to 90%
   Tin = 9% to 10%
   Phosphorus = 0.1% to 0.3%

   It is used for bearings, worm wheels, gears, nuts for machine lead screws, pump parts, linings and for many other purposes. It is also suitable for making springs.

2. *Silicon bronze.* It contains 96% copper, 3% silicon and 1% manganese or zinc. It has good general corrosion resistance of copper combined with higher strength. It can be casted, rolled, stamped, forged and pressed either hot or cold and it can be welded by all the usual methods.

   It is widely used for boilers, tanks, stoves or where high strength and good corrosion resistance is required.

3. *Beryllium bronze.* It is a copper base alloy containing about 97.75% copper and 2.25% beryllium. It has high yield point, high fatigue limit and excellent cold and hot corrosion resistance. It is particularly by suitable material for springs, heavy duty electrical switches, cams and bushes. Since the wear resistance of beryllium copper is five times that of phosphor bronze, therefore it may be used as a bearing metal in place of phosphor bronze. It has a film forming and soft lubricating property, which makes it more suitable as bearing metal.

4. *Manganese bronze.* It is an alloy of copper, zinc and little percentage of manganese. The usual composition of this bronze is as follows :

   Copper = 60%
   Zinc = 35%
   Manganese = 5%

   This metal is highly resistant to corrosion. It is harder and stronger than phosphor bronze. It is generally used for bushes, plungers, feed pumps, rods etc. Worm gears are frequently made from this bronze.

5. *Aluminium bronze.* It is an alloy of copper and aluminium. The aluminium bronze with 6 to 8% aluminium has valuable cold working properties. Thy are most suitable for making components exposed to severe corrosion conditions. When iron is added to these bronzes, the mechanical properties are improved by refining the grain size and improving the ductility.

   The aluminium bronzes are widely used for making gears, propellers condenser bolts, pump components, tubes, air pumps, slide valves and bushes, etc. Cams and rollers are also made from this alloy. A 6% aluminium alloy has a fine gold colour, which is used for imitation jewellery and decorative purposes.

13.26 GUN METAL

It is an alloy of copper, tin and zinc. It usually contain 88% copper, 10% tin and 2% zinc. This metal is also known as admirality gun metal. The zinc is added to clean the metal and to increase its fluidity.

It is not suitable for working in the cold state. But it may be forged when at about 600°C. The metal is very strong and resistant to corrosion by water and atmosphere. Originally, it was made for casting guns. It is extensively used for casting boiler fittings, bushes, bearings, glands etc.

13.27 LEAD

It is a bluish grey metal having specific gravity 11.36 and melting point 326°C, It is soft and can be cut even with a knife. It has no tenacity. It is extensively used for making solders, as a lining for acid tanks, cisterns, water pipes and as coating for electrical cables.

13.28 LEAD BASE ALLOYS

The lead base alloys are employed where a cheap and corrosion resistant material is required. An alloy containing 83% lead, 15% antimony, 1.5% tin and 0.5% copper is used for large bearing subjected to light service.

13.29 TIN

It is a brightly shining white metal. It is soft, malleable and ductile. It can be rolled into very thin sheets. It is used for making important alloys, fine solder, as a protective coating for iron and steel sheets and for making tin foil used as moisture proof packing.

13.30 TIN BASE ALLOYS (BABBIT METAL)

A tin base alloy containing 88% tin, 8% antimony and 4% copper is called babbit metal. It is a soft material with a low coefficient of friction and has little strength. It is the most common bearing metal used with cast iron boxes where the bearings are subjected to high pressure and load. An alloy of 60% tin and 40% lead is a common solder alloys. But alloys containing 85% tin and 15% lead are commonly used solder alloys.

13.31 ZINC

The zinc is a bluish white metal which in pure state has bright smooth crystals at its fracture. Its specific gravity is 7.1. And its melting point is 420°C. It boils at 940°C and can easily be distilled. It is not very malleable and ductile at ordinary temperatures. But it can be readily worked and rolled into thin sheets or drown into wires by heating it to 100°C–150°C. At about 200°C, it becomes so brittle that it may be powdered. Its tensile strength is 19 to 25 MPa. It offers high resistance to atmospheric corrosion.

It is used for covering steel sheets to form galvanised iron due to its high resistance to atmospheric corrosion. The covering is done by dipping the sheets into the molten metal after initial fluxing. The galvanised wire, nails etc. are also made by this process. When rolled into sheets, zinc is used for roof covering and for providing a damp proof non-corrosive lining to containers etc. The other important uses of zinc are in the manufacture of brasses and in the production of zinc base die casting alloys. The oxide of zinc is used as pigment in paints.

13.32 ZINC BASE ALLOYS

The most of the die castings are produced from zinc base alloys. These alloys can be casted easily with a good finish at fairly low temperatures. They have also considerable strength and low cost. The usual alloying elements for zinc are aluminium, copper and magnesium and they are all held in close limits. The composition of two standard die casting zinc alloys are as follows :

1. Aluminium 4.1%, copper 0.1%, magnesium 0.04% and the rest is zinc.
2. Aluminium 4.1%, copper 1%, magnesium 0.04% and the rest is zinc.

The aluminium improves mechanical properties and also reduces the tendency of zinc to dissolve iron. Copper increases the tensile strength, hardness and ductility. Magnesium has the beneficial effect of making the castings permanently stable. These alloys are widely used in the automotive industry and for other high production markets such as washing machines, oil burners, refrigerators, radios, phonographs, television sets, business machines etc.

13.33 NICKEL

It is a silvery white metal capable of taking a high polish. Its specific gravity is 8.85 and melting point 1452°C. It is almost as hard as soft steel. When it contains a small amount of carbon, it is quite malleable. It can be satisfactorily rolled with as little as 0.005 per cent of carbon. It is somewhat less ductile than soft steel. But small amount of magnesium improves the ductility considerably. It resists the attacks of most of the acids. But it dissolves readily in nitric acid.

It is used an alloying metal in some types of steels and cast irons. It is extensively used as a coating for other metals such as steel, copper, brass etc., for both decorative and corrosion protection purposes.

13.34 NICKEL BASE ALLOYS

The nickel base alloys are widely used in engineering and industry on account of their high mechanical properties, corrosion resistance etc. The most important nickel base alloys are discussed below:

1. *Monel metal.* It is an important alloy of nickel and copper. It contains 68% nickel, 29% copper and 3% other constituents like iron, manganese, silicon and carbon. Its specific gravity is 8.87 and melting point 1360°C. It resembles with nickel in appearance and is strong, ductile and tough. It is superior to brass and bronze in corrosion resisting properties. It is used for making propellers, pump fittings, condenser tubes, steam turbine blades, sea water exposed parts, tanks and chemical and food handling plants.

2. *Iconel.* It consists of 80% nickel, 14% chromium and the rest is iron. Its specific gravity is 8.55 and melting point 1395°C. This alloy has excellent mechanical properties at ordinary and elevated temperatures. It can be casted, rolled and cold drawn. It is used for making springs, which have to withstand high temperatures and are exposed to corrosive action. It is also used for exhaust manifolds of aircraft engines.

3. *Nichrome.* It consists of 65% nickel, 15% chromium and 20% iron. It has a high heat and oxidation resistance. It is used in making electrical resistance wire for electric furnaces and heating elements.

4. *Nimonic.* It consists of 80% nickel and 20% chromium. It has high strength and ability to operate under intermittent heating and cooling conditions. It is widely used in gas turbine engines.

13.35 MAGNESIUM

It is the lightest metal used as an engineering material. The tensile strength of cast metal is 90 MPa which is the same as that of ordinary cast aluminium. It is preferable when reduction of weight is important, because of its low density of 1.74. The tensile strength of rolled and annealed magnesium is 175 MPa which is equal to that of a good quality of grey cast iron. It is harder than aluminium. It can be readily machined. It takes high polish under the buffing wheel.

It readily forms alloys with most metals (not with iron and chromium). The metals with which it is mostly alloyed are aluminium, zinc, cadmium, and copper. It is employed in the form of sheets, wires, rods, tubes etc. The magnesium ribbon is now chiefly used for degasification of radio tubes. In the powdered form, mixed with an oxidising agent such as barium nitrate or potassium chlorate, it is employed in the manufacture of flash light powder and in military pyrotechnique for the production of rockets, signals and flares.

13.36 CADMIUM

It is a white metal with a bluish tinge, capable of taking a high polish. Its specific gravity is 8.65 and melting point 321°C. It is slightly harder than tin but is softer than zinc. It is malleable and ductile and can be readily rolled and hammered into foils and drawn into wire. At 80°C, it becomes so brittle that it can be pulverised under the hammer.

It is chiefly used in antifriction alloys for bearings. It is also used as a rust proof coating for iron and steel. Generally bolts, nuts and other small parts employed in automobile manufacture, refrigerator trimmings, locks and wire products are plated with it.

13.37 VANADIUM

It is silvery white in colour. Its specific gravity is 5.68 and melting point 1710°C. It is harder than quartz. But it is sufficiently malleable and tough, so that when heated to a suitable temperatures, it may be rolled and hammered into rods and drawn into wires. The chief use of vanadium is in manufacture of alloy steels. Non-ferrous alloys of vanadium are those of copper and aluminium, from which excellent casting may be made.

13.38 ANTIMONY

It is silvery white, hard, highly crystalline and so brittle that it may be readily powdered under the hammer. Its specific gravity is 6.62 and melting point 630°C. It is mostly used as an alloying elements with most of the heavy metals. The metals, with which antimony is most commonly alloyed are lead, tin and copper.

13.39 BEARING METALS

The following are widely used bearing metals:
1. Copper-base alloys;
2. Lead-base alloys,
3. Tin-base alloys and
4. Cadmium-base alloys.

The copper-base alloys are the most important bearing alloys. These alloys are harder and stronger than the white metals (lead base and tin base alloys) and one used for bearings subjected to heavy pressures. These include brasses and bronzes which are discussed in Art. 13.23 and 13.24. The lead base and tin base alloys are discussed in Art 13.27 and 13.29. The cadmium base alloys contain 95% cadmium and 5% silver. It is used for medium loaded bearings subjected to high temperature.

The selection of a particular type of bearing metal depends upon the conditions under which it is to be used. It involves factors relating to bearing pressures, rubbing speeds, temperatures, lubrication etc. A bearing material should have the following properties :
1. It should have low coefficient of friction.
2. It should have good wearing qualities.
3. It should have ability to withstand bearing pressures.
4. It should have ability of operate satisfactorily with suitable lubrication means at the maximum rubbing speeds.
5. It should have a sufficient melting point.
6. It should have high thermal conductivity.
7. It should have good casting qualities.
8. It should have minimum shrinkage after casting.
9. It should have non-corrosive properties.
10. It should be economical in cost.

Ferrous and Non-Ferrous Alloys

13.40 ZIRCONIUM ALLOYS

Although zirconium is relatively abundant in the earth's crust, yet it was not until recent times that commercial refining techniques were developed. Zirconium and its alloys are ductile. They have other mechanical characteristics that are comparable to those of titanium alloys and the austenitic stainless steels. However the major characteristic of these alloys is their resistance to corrosion in a host of corrosive environment including superheated water.

Moreover, zirconium is transparent to thermal neutrons. Because of this, zirconium alloys have been used as cladding for uranium fuel in water – cooled nuclear reactors. Because of their lower cost, the zirconium alloys are the "materials of choice" for heat exchangers, reactor vessels, and piping systems for the chemical processing and nuclear industries. They are also used in scaling devices for vacuum tubes, in industries for making military weapons. Refer to the picture of a heat exchanger in a petro-chemical plant shown in Fig. 13.1.

Fig. 13.1.

13.41 SHAPE-MEMORY ALLOYS

The shape-memory alloys provide the engineer a means of restoring a bent metal wire to some trained alternate shape. These alloys have applications ranging from frames for optical glasses to repair parts for the human body. For example fine wire made of equal parts of nickel and titanium can be woven into cylindrical shape for various applications. One such application is vascular stents to reinforce blood vessels in the human body. The stent is crushed and inserted through a cannula into the proper location in the blood vessel.

13.42 BIOMATERIALS

A biomaterial is any material, natural or man made, that comprises whole or part of a living structure or biomedical device which performs, augments or replaces a natural function. For the past one decade a significant work has been done in the field of biomaterials and biochemicals. Biomechanics is a field of engineering which involves the structure and function of biological systems using the methods of mechanics.

13.43 MATERIALS FOR HIP JOINT REPLACEMENT

Some of the most interesting development in the application of materials has been in the field of medicine. For example the artificial hip replacement is one of the most successful applications. Before we discuss about the hip replacement device let us understand more about the hip joint.

The hip joint is located at bone junctions where loads may be transmitted from bone to bone by muscular action. This is normally accompanied by some relative motion of the component bones. As a matter of fact, the bone tissue is a complex natural composite material. It consists of

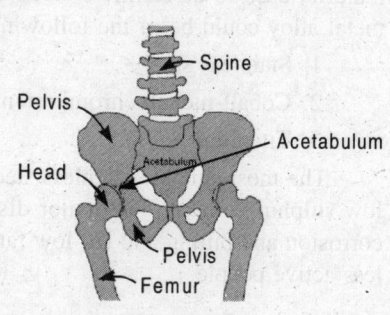

Fig. 13.2.

soft and *strong* protein *collagen* and brittle *apatite*. The density of this material is around 1.6 to 1.7 g/cm³. It is an isotropic material, *i.e.* its properties differ in longitudinal and transverse directions. The articulating or connecting surface of each joint is coated with cartilage. The cartilage consists of body fluids that lubricate and provide an interface having a low coefficient of friction. This interface helps to produce bone-sliding movement.

Fig. 13.2 shows a human hip joint along with some adjacent skeletal components. As seen from this figure, the hip joint occurs between the pelvis and the upper leg (or thigh) bone. The upper leg (or thigh bone) is called *femur*. A relatively large range of rotary motion is permitted at the hip by a *ball-and-socket* type of joint. The top of femur terminates in a ball-shaped head that fits into a cup-like cavity (called acetabulum) within the pelvis.

The hip joint is susceptible to fracture. The fracture normally occurs at the narrow region just below the head. In the fractured situation, the hip may become diseased. Usually in such a case, small lumps of bone form on the rubbing surfaces of the joint. This causes a pain as the head rotates in the acetabulum.

Damaged and diseased hip joints have been successfully replaced with artificial hip joints. The total hip replacement surgery involves the removal of head, the upper portion of femur and some of the bone marrow at the top of the remaining femur segment.

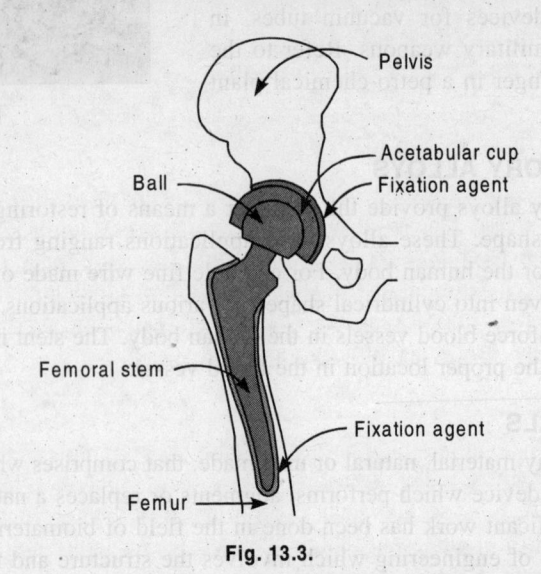

Fig. 13.3.

Fig. 13.3 shows a schematic diagram of the artificial total hip replacement : Early artificial hip designs called for both, the femoral stem and ball to be of the same materials. Those days this material used to be stainless steel. Currently the femoral stem is made from a metal alloy. The metal alloy could be of the following three possible types:

1. Stainless steel
2. Cobalt-nickel-chromium-molybdenum and
3. Titanium

The most suitable stainless steel has been found to be 316L. This stainless steel has a very low sulphur content. The major disadvantages of this alloy are (*i*) its susceptibility to crevice corrosion and pitting and (*ii*) low fatigue strength. Normally 316L alloy is implanted in older and less active people.

Ferrous and Non-Ferrous Alloys

Many cobalt–chromium–molybdenum and cobalt-nickel-chromium-molybdenum alloys have been employed for artificial hip prosthesis. The MP35N alloy has been found to be most suitable in this group. The MP35N alloy has a composition of 35 wt % cobalt, 35 wt % nickel, 20 wt % chromium and 10 wt % molybdenum. This alloy is formed by hot forging and hence has high tensile and yield strengths. Moreover the corrosion and fatigue strength of MP35N alloy are excellent titanium alloy is probably the most biocompatible material for the artificial hip joint replacement device the titanium alloy used for human hip replacement consists of 90 wt % titanium, 6 wt % aluminium and 4 wt % vanadium.

Fig. 13.4 shows a photograph of two artificial total hip replacement designs. There have been some further improvements in the materials for the hip replacement device recently. The engineers have suggested. The use of ceramic material (aluminium oxide) for the ball component instead of the metal alloy. Aluminium oxide is harder and more wear resistant.

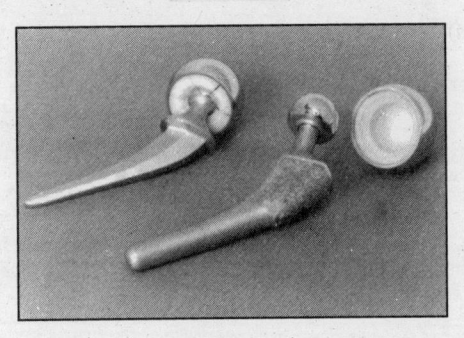

Fig. 13.4.

As mentioned earlier, the artificial hip joint device has a stem, a ball and an acetabular cup. We have discussed already the materials for the femoral stem and a ball. Let us talk about the material for the acetabular cup. The cups are made from one of the biocompatible alloys or aluminium oxide. However these days the use of ultrahigh-molecular weight polyethylene ($\approx 4 \times 10^6$ atomic mass units) is more common. This material has been found to be virtually *inert* in the human body environment. It has excellent wear resistance characteristics. Moreover it has very low coefficient of friction when in contact with the materials for the ball component of the socket.

A.M.I.E. (I) EXAMINATION QUESTIONS

1. Distinguish between white and grey cast-irons. *(Winter, 1993)*
2. Explain the effects of various alloying elements added to carbon steels. *(Winter, 1994)*
3. Distinguish between high carbon steel and alloy tool steel. *(Winter, 1994)*
4. Why is cast iron used as a material for machine tool beds ? *(Summer, 1995)*
5. (a) What is an alloy steel ? How are alloy steels classified ? List four important alloying elements added to steel and their function.
   (b) List properties required for a material to withstand high temperature. Discuss the use of alloy steels as heat resisting materials.
   (Winter, 1996)
6. What is duralumin ? Give its composition and one application. *(Winter, 1996)*
7. What is meant by the term high carbon steel ? Give its application. *(Winter, 1996)*
8. Give the composition of high speed steel ? Where it is used ? *(Winter, 1996)*
9. (a) Why is alloying done ? What are the effects of chromium and nickel an alloying elements or properties of steel ?
   (b) Enumerate the composition and uses of any two types of iron-ferrous alloys. *(Summer, 1997)*

10. What is maximum solubility of carbon in iron ? (*Summer, 1997*)
11. What is 18-4-1 high speed steel ? (*Summer, 1997*)

MULTIPLE CHOICE QUESTIONS

1. The Y-alloy is an alloy of
   - (*a*) copper and zinc
   - (*b*) copper and nickel
   - (*c*) copper–aluminium alloy
   - (*d*) steel and nicke
2. Which of the following material is more suitable for hip joint replacement
   - (*a*) titanium
   - (*b*) high-speed steel
   - (*c*) zirconium alloy
   - (*d*) brass

ANSWERS

1. (*c*)   2. (*a*)

14

Organic Materials

1. Introduction 2. Organic Compounds. 3. Types of Organic Compounds 4. Saturated Organic Compounds. 5. Unsaturated Organic Compounds. 6. Aromatic Organic Compounds. 7. Polymers, 8. Addition to Polymers. 9. Types of Polymerizations. 10. Addition Polymerization. 11. Copolymerization. 12. Mechanism of Addition Polymerization. 13. Condensation Polymerization. 14 Strengthening Mechanism of Polymers. 15. Mechanical Behaviour of Polymers. 16. Shape of the Polymer. 17. Degree of Polymerization. 18. Temperature of Polymer. 19. Crystalallinity in Polymers. 20. Polymer Crystals. 21. Electrical Behaviour of Polymers. 22. Plastics (Resins). 23. Types of Synthetic Plastics. 24. Thermoplastics (Thermoplastic Resins). 25. Thermosetting Plastics. (Thermosetting Resins). 26. Comparison Between Thermoplastics and Thermosetting Plastics. 27. Rubber. 28. Types of Rubbers. 29. Natural Rubbers. 30. Synthetic Rubbers (Elastomers). 31. Vulcanization 32. Wood. 33. Fibre Polymers. 34. Advanced Polymeric Materials. 35. Ultrahigh Molecular Weight Polyethylene. 36. Liquid crystal polymers. 37. Thermoplastic Elastomers.

14.1 INTRODUCTION

The materials, which are derivatives of carbon, chemically combined with hydrogen, oxygen or any other non-metallic substance, are known as organic materials. The organic materials may be natural or synthetic (*i.e.*, manufactured artificially). The natural organic materials include wood, cotton, natural rubber, coal, petroleum and food products etc. The synthetic organic materials include synthetic rubber, plastics, lubricants, soap oils, synthetic fibres etc. It has been observed that the organic materials, consists of very large molecules. In natural organic materials, the molecules are readymade. But in synthetic materials, the molecules are built up by joining simpler molecules by some chemical reactions.

These days, with the development of Organic Chemistry, many new organic materials are manufactured and extensively used in various industries. The properties of these materials may be properly understood in terms of molecular structure, which will be discussed in the following pages. The research and development units of large industrial houses are doing lot of research for the improvement of various engineering properties of these organic materials.

14.2 ORGANIC COMPOUNDS

All the compounds, which essentially contain carbon atoms along with other non-metal atoms such as hydrogen, nitrogen, oxygen, chloride etc. in each molecule, are known as organic compounds. It will be interesting to know that the characteristics of the organic compounds play a very important role in the various engineering properties of the organic materials.

14.3 TYPES OF ORGANIC COMPOUNDS

Though there are a number of organic compounds, yet the following are important from the subject point of view:

1. Saturated organic compounds.
2. Unsaturated organic compounds.
3. Aromatic organic compounds.

14.4 SATURATED ORGANIC COMPOUNDS

Sometimes, the compounds in the carbon atoms, present in each molecule, are joined by single covalent bonds. And the remaining bond is completed by the atoms of carbon and hydrogen. All the carbon atoms may be either in a single chain or side branching.

The simplest member of the saturated organic compound series is methane (CH_4). Each molecule of the methane contains one carbon atom joined to four hydrogen atom. All these atoms are arranged in such a manner that they form a regular tetragon. Though the saturated organic compounds are three dimensional structures, yet it is convenient to represent them by two dimensional structural formula as shown in Fig. 14.1.

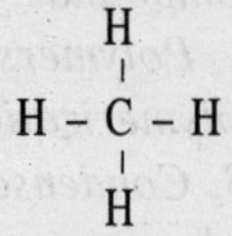

Fig. 14.1. Methane.

Similarly, other two members of the series are ethane (C_2H_6) and propane (C_3H_8) which may be represented as shown in Fig. 14.2. (*a*) and (*b*) respectively.

Fig. 14.2.

It may be noted that each member, in the series, has been derived from the previous one by replacing a hydrogen atom with a *methyl (— CH_3) group.

If this process is repeated, each time, to a carbon atom at the end of the chain, we get a homologous series known as alkane series with general formula as $C_n H_{2n+2}$. In a displaced form, this may be written as shown in Fig. 14.3

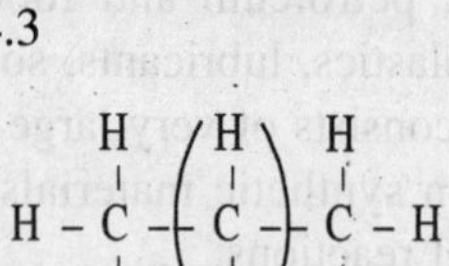

Fig. 14.3. General formula for alkane series.

Where '*n*' is the number of carbon atoms in the compound. The names of the next few numbers (from propane) in the alkanes series are;

1. butane (C_4H_{10})
2. pentane (C_5H_{12})
3. hexane (C_6H_{14})
4. heptane (C_7H_{16})
5. octane (C_8H_{18})
6. nonane (C_9H_{20})

It will be interesting to know that the substitution of methyl group for a hydrogen atoms need not be carried out at the end of the chain. Thus the butane (C_4H_{10}) may be derived from the propane (C_3H_8) in either of the following two ways:

---

* The methyl group has been represented as (—CH_3). As a matter of fact, the actual representation methyl is CH_3. The dash indicates an incomplete covalent bond.

Organic Materials

1. By replacing a hydrogen atom at the end of the carbon atom as shown in Fig. 14.4 (a). It is known as normal butane.
2. By replacing hydrogen atom on the central carbon atom by the methyl group (—CH$_3$) as shown in Fig. 14.4 (b). It is known as isobutane.

```
    H  H  H  H                    H   H   H
    |  |  |  |                    |   |   |
H - C -C -C -C - H            H - C - C - C - H
    |  |  |  |                    |   |   |
    H  H  H  H                    H   H   H
                                      |
                                  H - C - H
                                      |
                                      H
    (a) Normal butane             (b) Isobutane
                    Fig. 14.4.
```

It may be noted that the normal butane is considered to be more stable than the isobutane. This is due to the fact that the normal-butane molecules have a single chain of atoms, whereas the isobutane has side branching atoms as shown in the figure. Moreover, the isobutane has, somewhat, lower melting and boiling points than those of normal butane. It has been observed that the higher members of alkane series (i.e., members containing more than 8 carbon atoms in the molecule) contain more *isomers. It may be noted that the lowest members of the alkane series (upto pentane (C$_5$H$_{12}$) are gases at room temperature. The medium ones (i.e., between C$_6$H$_{14}$ and C$_{16}$H$_{34}$) are liquids. The highest ones (i.e., C$_{17}$A$_{36}$) and above) are solids.

It has been observed that the alkanes do not react easily with other compounds. Their only notable reactions are with oxygen and chlorine. They also react with oxygen to form carbon dioxide and water. Moreover, they react with chlorine to form a mixture of different chloride compounds. For example, when methane reacts with chlorine, a mixture of methyl chloride (CH$_3$Cl), methylene chloride (CH$_2$Cl$_2$), chloroform (CHCl$_3$) and carbon tetrachloride (CCl$_4$) is formed. A wide range of compounds is formed by replacing hydrogen atoms of radicals (i.e., group of atoms).

14.5 UNSATURATED ORGANIC COMPOUNDS

The compounds in which any two carbon atoms, present in a molecule, are joined by double covalent bonds and the remaining carbon bonds completed by hydrogen atoms are known as unsaturated organic compounds. The simplest member of this series is ethylene (C$_2$H$_4$). The two dimensional formula for ethylene is given in Fig. 14.5 (a).

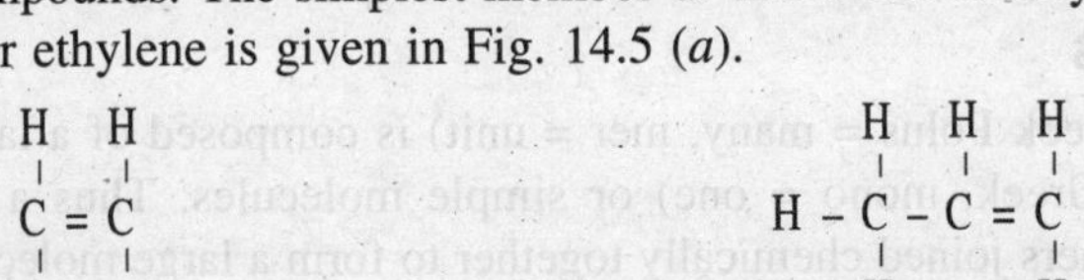

```
    H   H                         H   H   H
    |   |                         |   |   |
    C = C                     H - C - C = C
    |   |                         |   |
    H   H                         H   H
   (a) Ethylene                (b) Propylene
                 Fig. 14.5.
```

The next member of this series is propylene or propene (C$_3$H$_6$). Its two dimensional formula is given in Fig. 14.5 (b). A little consideration will show that the next members in this series may be derived by replacing hydrogen atom with methyl group (—CH$_3$). The series so obtained is known as alkene series with general formula as C$_n$H$_{2n}$, where 'n' is the number of carbon atoms in the compound. The names of the next few members in the alkene series are:

1. butene (C$_4$H$_8$)   2. pentene (C$_5$H$_{10}$)   3. hexene (C$_6$H$_{12}$)

---

*   Isomers are the different compounds with same molecular formula. Because of the difference in structural formulae, the properties of isomers are also different.

The double bonds in the unsaturated organic compounds are relatively weak and revert back easily to single bonds by taking part in additive reactions.

14.6 AROMATIC ORGANIC COMPOUNDS

The organic compounds, which are derived from benzene (C_6H_6) by replacing the hydrogen atom with other groups of atoms are known as aromatic organic compounds. The structure of benzene is in the form of a ring in which the carbon atoms are arranged in a regular hexagon with the hydrogen atoms attached to each carbon atom outside the ring as shown in Fig. 14.6 (a) and (b).

Fig. 14.6. Structure of benzene.

It may be noted that a single line between two carbon atoms indicates single covalent bond the double line indicates double covalent bond is the benzene structure. The difference between the two structures is in the position of double bond in the ring. In structural formulae, the general practice for the benzene ring is that it is, usually, drawn as a simple hexagon, the carbon and hydrogen atoms are not shown. Any groups substituted for the hydrogen atoms are indicated adjacent to the appropriate corner of the hexagon.

Some examples of the aromatic compounds are shown in Fig. 14.7 (a) and (e).

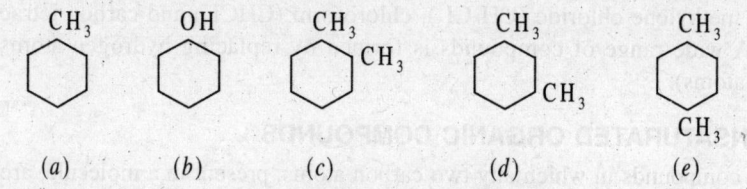

Fig. 14.7. Aromatic compounds.

14.7 POLYMERS

A polymer (Greek Polus = many, mer = unit) is composed of a large number of repetitive called monomers (Greek, mono = one) or simple molecules. Thus a polymer is made up of thousands of monomers joined chemically together to form a large molecule. It has been observed that each molecule of a polymer is either a long chain or a network of repetitive units or monomers. The following terms regarding polymers are important from the subject point of view:

1. *Polymerization*: It is the process of forming a polymer.
2. *Degree of polymerization.* It is the number of repetitive units (or mers) present in one molecule of a polymer.

   Mathematically, degree of polymerization

   $$= \frac{\text{Molecular weight of a polymer}}{\text{Molecular weight of a single monomer}}$$

3. *Linear polymer.* It is a polymer, which is obtained by simply adding the monomers together to form long chains Refer to Fig. 14.8 (a).
4. *Copolymer.* It is a polymer, which is obtained by adding different types of monomers.

Organic Materials 277

5. *Branched polymer.* It is a polymer which is obtained by connecting side branches to the main one. Refer to Fig. 14.8(b).
6. *Cross-linked polymer.* It is a polymer which is obtained by connecting the long chains through a covalent bond. Refer to Fig. 14.8. (c).
7. *Network polymer.* It is a polymers having 3-dimensional network. Refer to Fig. 14.9.
8. *Thermoplastics.* These are linear polymers, whose plasticity increases with the rise in temperature.
9. *Thermosets.* These are the polymers, whose plasticity does not change with the rise in temperature.

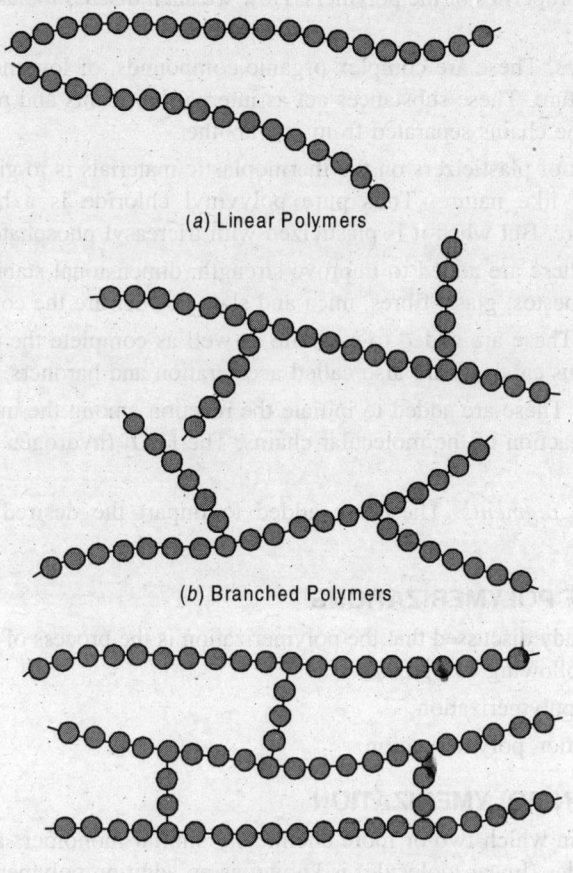

(a) Linear Polymers

(b) Branched Polymers

(c) Cross linked polymers

Fig. 14.8.

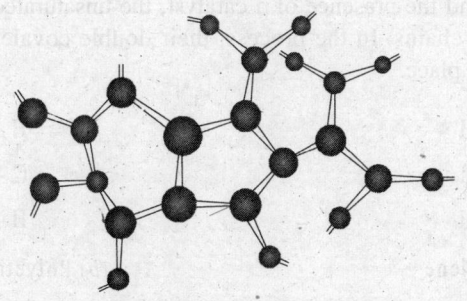

Fig. 14.9. Network Polymers.

14.8 ADDITION TO POLYMERS

In most of the cases, it becomes very essential to add some extra materials into the monomers before or during the process of polymerization in order to impart certain desired properties to the polymers. The various substances, which are usually added into the monomers are:

1. Plasticizers. 2. Fillers. 3. Catalysts. 4. Initiators.
5. Different types of dyes and pigments.

It will be interesting to know that all the above mentioned additions are not added into the monomers at one time. But one or more of these additions are added, in different proportions, to obtain the desired properties of the polymers. Now we shall discuss the usefulness of the additions one by one.

1. *Plasticizers*. These are complex organic compounds, of low molecular weight, and are oily in nature. These substances act as internal lubricants and prevent crystallisation by keeping the chains separated from one another.

    The effect of plasticizers on the thermoplastic materials is to give them a more flexible or rubber like nature. Thus pure polyvinyl chloride is a hard substance at room temperature. But when it is plasticized with tricreasyl phosphate, it becomes flexible.

2. *Fillers*. These are added to improve strength, dimensional stability and heat resistance. Wood, asbestos, glass, fibres, mica and slate powder are the common filler materials.

3. *Catalyst*. These are added to expedite as well as complete the polymerization reaction. The various catalysts are also called acceleration and hardners.

4. *Initiators*. These are added to initiate the reaction among the monomers and to stabilise the end reaction of the molecular chains. The H_2O_2 (hydrogen per oxide) is a common initiator.

5. *Dyes and pigments*. These are added to impart the desired colour to the finished polymers.

14.9 TYPES OF POLYMERIZATIONS

We have already discussed that the polymerization is the process of forming a polymer. This process is of the following two types :

1. Addition polymerization
2. Condensation polymerization

14.10 ADDITION POLYMERIZATION

The process in which two or more chemically similar monomers are polymerized to form long chain molecules (linear molecule) is known as an addition polymerization. It takes place in unsaturated organic compounds. These compounds are, relatively, unstable as compared to the saturated organic compounds. It has been observed that under suitable conditions such as high pressure, temperature and the presence of a catalyst, the unsaturated organic compounds react in a manner to form long chains. In the process, their double covalent bond is broken and single bonds are formed in its place.

$$\begin{array}{c} H \quad H \\ | \quad \; | \\ C = C \\ | \quad \; | \\ H \quad H \end{array} \qquad \begin{array}{c} H \quad H \quad H \quad H \\ | \quad \; | \quad \; | \quad \; | \\ -C - C - C - C - \\ | \quad \; | \quad \; | \quad \; | \\ H \quad H \quad H \quad H \end{array}$$

(a) Ethylene (b) Polyethylence (Polythene)

Fig. 14.10. Addition polymerization.

Organic Materials

Fig. 14.10 (a) shows an ethylene monomer, in which carbon atoms are joined by a double covalent bond. Now if one more ethylene monomer is added to this monomer, then the double covalent bonds between their carbon atoms are broken and single covalent bonds are formed between the carbon atoms. The new molecule formed is known as polyethylene (or simply polythene) as shown in Fig. 14.8 (b). Similarly, more ethylene monomers may also be added to molecule. It results in the formation of a long chain molecule.

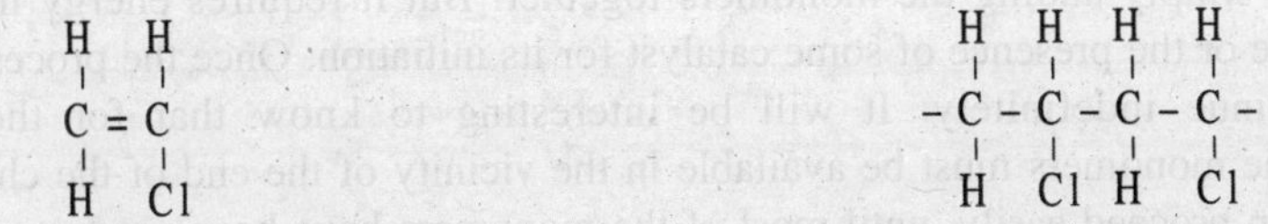

```
   H   H                        H   H   H   H
   |   |                        |   |   |   |
   C = C                      - C - C - C - C -
   |   |                        |   |   |   |
   H   Cl                       H   Cl  H   Cl

  (a) Venyl chloride           (b) Polyvinyl chloride (p.v.c.)
```

Fig. 14.11. Addition polymerization.

Similarly Fig. 14.11 (a) shows vinyl chloride monomer, in which carbon atoms are joined by a double covalent bond. Now if one more vinyl chloride monomer is added to this monomer, then the double covalent bond between the two carbon atoms is broken and single covalent bond is formed between the two carbon atoms. The new molecule formed is known as polyvinyl chloride (or simple p.v.c.)] as shown in Fig. 14.11 (b). Similarly, one more vinyl chloride monomers may also be added to the molecule. It also results in the formation of a long chain molecule.

Note: Some of other important monomers and polymers formed are propylene and polypropylene: strene and polystrene; methyl methacrylate and polymethyl methacrylate; tetrafluro ethylene and polytetrafluoro ethylene etc.

14.11 COPOLYMERIZATION

It is a particular type of addition polymerization in which two or more chemically different monomers are polymerised to form a long chain molecules.

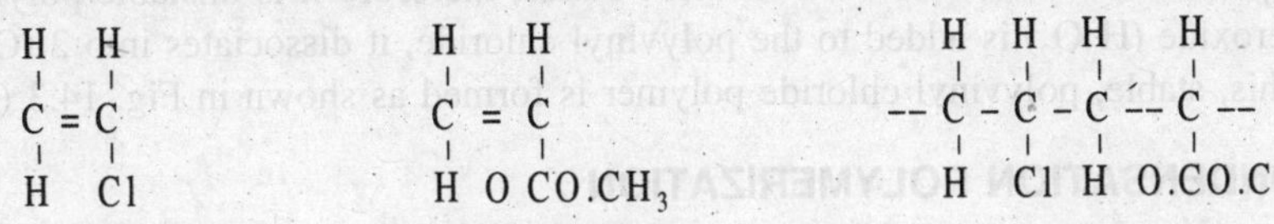

```
   H   H               H   H                  H   H   H   H
   |   |               |   |                  |   |   |   |
   C = C               C = C                -- C - C - C - C --
   |   |               |   |                  |   |   |   |
   H   Cl              H   O CO.CH₃           H   Cl  H   O.CO.CH₃

  (a) Venyl chloride  (b) Vinyl acetate       (c) Copolymer
```

Fig. 14.12. Copolymerization.

Fig. 14.12 (a) shows a vinyl chloride monomer, in which carbon atoms are joined by a double covalent bond. Fig. 14.12 (b) shows a vinyl acetate monomer in which the carbon atoms are also joined by a double covalent bond. Now if these two monomers are polymerized, then double covalent bond between their carbon atoms are broken. And single covalent bond is formed between the carbon atoms as shown in Fig. 14.12 (c). The newly formed molecule is known as copolymers vinyl chloride and vinyl acetate. Similarly, more such monomers may also be added to the molecule to form a long chain copolymer.

It will be interesting to know that copolymers have different properties than either of their constituents. The copolymers can be made in a number of different geometric arrangements, e.g., regular, random and blocks depending upon the sequence of different monomers. The process of copolymerization is applied extensively in the field of manufacturing synthetic rubbers.

14.12 MECHANISM OF ADDITION POLYMERIZATION

We have already discussed in the last two articles the process of addition polymerization. It has been observed that the long chain of the molecules are linked to each other by Van der Waals forces. Since these forces are weak, therefore the melting points of the long chain molecules are, relatively, low as compared to those of ceramics and metals.

It has been observed that the process of addition polymerization does not take place, automatically, by simply adding the monomers together. But it requires energy in the form of heat, light pressure or the presence of some catalyst for its initiation. Once the process is initiated, it does not continue indefinitely. It will be interesting to know that for the process of polymerization, the monomers must be available in the vicinity of the end of the chain. Thus the polymerization can proceed easily, until most of the monomers have been used up. But after that unpolymerized monomers must also diffuse to the regions where they can polymerize with other monomers.

It has been observed that the polymers, which grow simply from the monomers are unstable. It is due to the reason that the carbon atoms, lying at both ends of the long chain, do not have four covalent bonds. In order to provide stability to the polymers, some terminal radicals (*i.e.*, group of atoms) are necessary. The commonly used terminal radical is hydrogen peroxide (H_2O_2). It provides stability to the long chain by dissociating into 2(OH) ions. It may be noted that each (OH) ion makes its link on either end of the chain.

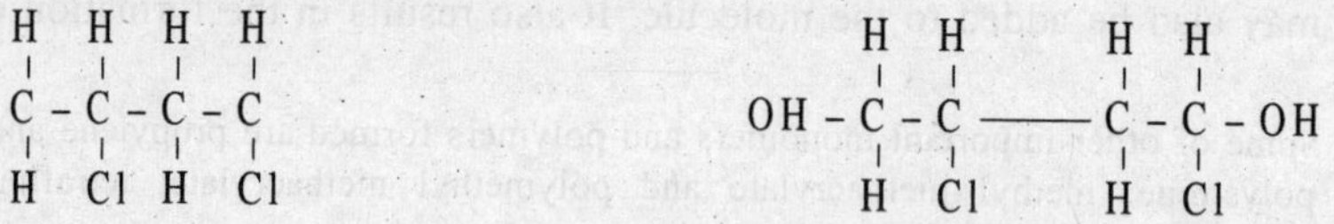

```
    H  H  H  H                          H  H            H  H
    |  |  |  |                          |  |            |  |
    C - C - C - C                OH - C - C ———— C - C - OH
    |  |  |  |                          |  |            |  |
    H  Cl H  Cl                         H  Cl           H  Cl

  (a) Unstable polyvinyl              (b) Stable polyvinyl chloride
        chloride
```

Fig. 14.13. Mechanism of addition polymerization.

Fig. 14.13 (*a*) shows a polyvinyl chloride polymer. Since the carbon atoms, lying at both the ends of the polymer, do not have four covalent bonds, therefore it is unstable polymer. Now if hydrogen peroxide (H_2O_2) is added to the polyvinyl chloride, it dissociates into 2 (OH) ions. As a result of this, stable, polyvinyl chloride polymer is formed as shown in Fig. 14.3 (*b*).

14.13 CONDENSATION POLYMERIZATION

The process in which two or more chemically different monomers are polymerized to form a cross-link or linear polymer, along with a by product such as water or ammonia, is known as condensation polymerization. It takes place in unsaturated organic compounds and requires suitable conditions such as high pressure, temperature and presence of a catalyst in the same way as that of addition polymerization. Thus methyl alcohol (CH_3OH) and acetic acid (CH_3COOH) condenses to form an ester with water as a by-product.

Fig. 14.14 (*a*) shows two separate monomers of methyl alcohol and acetic acid. Now if these two monomers are polymerized, they will form a long chain polymer known as ester as shown in Fig. 14.14 (*b*). It may be noted that in this case, the byproduct is water (H_2O).

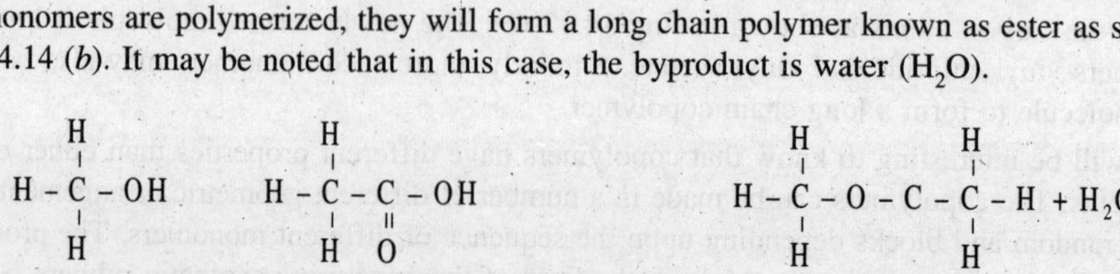

```
      H              H                          H            H
      |              |                          |            |
  H - C - OH     H - C - C - OH             H - C - O - C - C - H  + H₂O
      |              |  ||                      |            |
      H              H  O                       H            H

  (a) Monomers of methyl alcohol                           (b) Ester
       and acetic acid
```

Fig. 14.14.

Organic Materials

In addition to the chain polymers, three dimensional network polymers can also be formed with polyfunctional molecules that can link to three or more other molecules in polymerization. Such polymers are known as cross-linked polymers. One example of cross-linked polymers is in the reaction of one monomer of formaldehyde (CH_2O) and two monomers of phenol (C_6H_5OH). In this case the polymer is urea formaldehyde and the byproduct is water.

Note: Another important example of condensation polymerization is that one monomer of formaldehyde and two monomers of urea polymerize together to form urea fomaldehyde along with water as a byproduct. Similarly, one monomer of formaldehyde and two monomers of malamine polymerize together to form malamine formaldehyde along with water as byproduct.

14.14 STRENGTHENING MECHANISM OF POLYMERS

Sometimes it is required to produce polymers, which are rigid and resistant to corrosive chemicals and temperature. Though there are many strengthening mechanisms of polymers, yet the followings are important from the subject point of view :

1. *Crystallisation.* This mechanism makes the polymers to attain regularity in their atomic or molecular arrangements. It depends upon the orientation of the chains.

2. *Cross-linking.* This mechanism ties the chains of molecules together by covalent bonds. It depends upon chemical reactions rather than chain orientations. The cross linking is strongly affected by temperature, *i.e.*, it gets accelerated by the increase in temperature and is not reversible.

3. *Chain-stiffening.* This mechanism can be used to produce stiff chains through a number of methods. One of the methods is to 'hang' the bulky groups of atoms on the chain to restrict bending. Certain polymers are intrinsically stiff. In such polymers, the chains cannot bend but can twist on its bonds. Similarly, some monomers, which have ring shaped structure, are inherently rigid.

14.15 MECHANICAL BEHAVIOUR OF POLYMERS

Though there are a number of factors which affect the mechanical behaviour of polymers, yet the following are important from the subject point of view:

1. Shape of the polymer.
2. Degree of polymerization.
3. Temperature of polymer.
4. Crystallinity in polymer.

14.16 SHAPE OF THE POLYMER

As a matter of fact, the shape of a polymer affects, to a great extent, the resistance to its plastic deformation. It has been observed that the unsymmetrical polymers have greater resistance to the plastic deformation than that of symmetrical polymers. It is due to the fact, that the chain molecules, in unsymmetrical polymers, are held together by greater forces of attraction than those of symmetrical polymers. Moreover, it is difficult for the chain molecules to slip over one another. For example, polyvinyl chloride polymer is an unsymmetrical polymer and the polyethylene is a symmetrical polymer. Therefore polyvinyl chloride has a greater resistance to the plastic deformation than that of polyethylene.

14.17 DEGREE OF POLYMERIZATION

We have already discussed in Art. 14.7 that the number of repetitive units (or mers) present in one molecule of a polymer is known as degree of polymerization and is abbreviated as D.P. It has been observed that the substance, in which the value of D.P is less than 20 are liquids at room temperature. Similarly, the substances in which the value of D.P. is less than 100 are greasy at room temperature. And the substances in which the D.P. is more than 1000 are solids at room temperature.

14.18 TEMPERATURE OF POLYMER

It has been observed that at very low temperature, the polymers are hard, rigid and glass like materials. They give limited elastic deformation and fracture in a brittle manner. But as the temperature of polymer is increased, above the room temperature, it gradually deforms under the applied load. Moreover, when the load is removed, it recovers back to its original shape and size, after a certain time. This type of behaviour in a polymer, is known as visco-electric behaviour. It exists below a certain temperature known as glass transition temperature.

When the temperature of the polymer is further increased, it undergoes an irrecoverable plastic deformation under the applied load. But it may be noted that the polymer is still in a solid form at these temperatures. This type of behaviour exists below a certain temperature known as flow temperature or melting temperature of the polymer. Above this temperature, the polymers are present in its liquid form. For example, polymers such as polyethylene, polyvinyl chloride etc., are well above their glass transition temperature even at room temperatures. On the other hand, the polymers such as polystrene, perspex etc. are well below their glass transition temperature even at room temperature. The polymers, which exhibit this type of behaviour with the increase in temperature, are known as thermoplastics or thermoplastic resins.

It will be interesting to know that the above discussion holds good only for the linear polymers and not for cross-linked polymers. It has been observed that the cross-linked polymers do not deform under the applied load due to the increase in temperature. Or in other words, cross-linked polymers do not change their shape with the increase in temperature. It is thus obvious, that they remain hard and fracture, in a brittle manner, even at high temperatures. Such polymers are known as thermosetting plastics or thermosetting resins.

14.19 CRYSTALLINITY IN POLYMERS

As a matter of fact, the polymer are rarely crystalline in structure, But some polymers have grains of about 10 nanometre (10^{-9}m) in size, which have an ordered three dimensional structure. This type of structure is possible only in cases where the polymer chain has a regular structure.

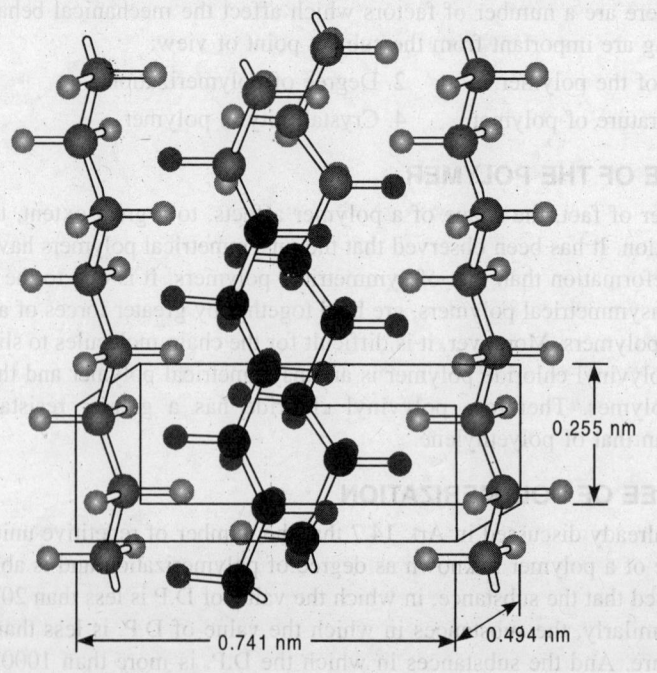

Fig. 14.15. Atomic arrangement in polymer crystals.

Organic Materials

Fig. 14.16 shows the unit cell for polyethylene and its relationship to the molecular chain structure. As seen from this figure, since the polymer structure involves molecules instead of atoms or ions, the atomic arrangement is more complex.

It has been observed that polyethylene and polytetrafluoro ethylene (teflon) have a crystalline structure. But in vinyl polymers (such as vinyl chloride and vinyl acetate) the side groups may be randomly arranged on one or other side of the chain.

In these polymers, the crystallinity is possible only when all the side groups are arranged either on one side or alternatively on both sides of the chain. It has been observed that in copolymers, units of two or more monomers do not necessarily occur in a regular sequence, so that these are more likely to be non-crystalline.

It has been observed that the behaviour of partially crystalline polymers is intermediate between those for crystalline and amorphous polymers. Whenever flexibility in a material is required, the crystallinity is undesirable. When a polymer comprising linear chain is elongated, there is an initial and almost linear elastic region. This is followed by yielding and elongation, which may be several hundred percent of its original length. Then the elongation becomes constant at higher values of loads. The above results are indicated on a load-extension curve as shown in Fig. 14.16. The polymers, which possess this type of behaviour, are known as elastomers.

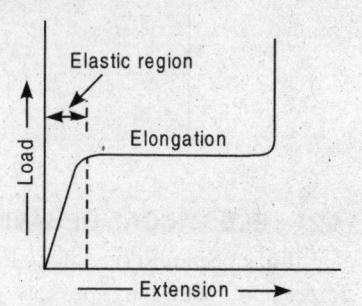

Fig. 14.16. Crystallinity in polymer.

It has been observed that during the extension or drawing of a polymer, the molecular chain gets oriented in the direction of its drawing. This alignment gives them a crystalline structure.

14.20 POLYMER CRYSTALS

Fig. 14.17 shows the model that has been proposed to describe the spatial arrangement of molecular chains in polymer crystals. This model is called the fringed-micelle model. According to this model, a semicrystalline polymer consists of small crystalline regions called crystallites or micelles. Each such region has a precise alignment which is embedded within the amorphous matrix composed of randomly orient molecules.

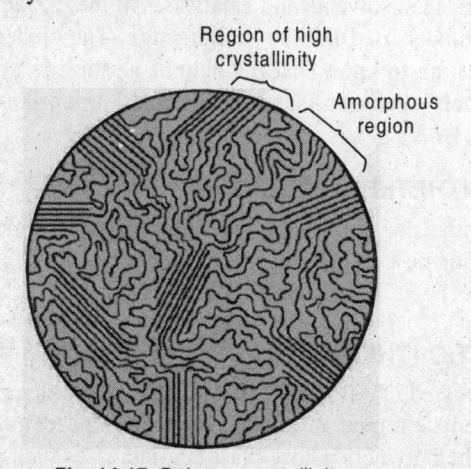

Fig. 14.17. Polymer crystallinity.

More recent studies on polymer single crystal grown from dilute solutions have indicated that these crystals are regularly shaped, thin platelets. A typical size of these platelets are approximately 10-20 nm thick and about 10 μm long. These platelets will form a multilayered structure as shown in Fig. 14.18. It is proposed that the molecule chains within each platelet fold back and forth on them selves with fold occurring at the faces. Such a structure is known as chain-folded model.

Fig. 14.18. Microstructure of Polyethylene.

14.21 ELECTRICAL BEHAVIOUR OF POLYMERS

It has been observed that almost all the polymers are poor conductors of electricity at room temperature. But they possess good insulation resistance and high dielectric constant (or permittivity). That is why the polymers are extensively used as insulating as well as dielectric materials. They are also applied on wires and cables and unpolymerized or partially polymerized (in liquid form) to obtain a uniform thin coating.

14.22 PLASTICS (RESINS)

A plastic may be defined as an organic polymer, which can be moulded into any desired shape and size with the help of heat, pressure or both. The plastic, in its liquid form, is known as resin. These days, the plastics have found greatest utility because of their low specific gravities, ease of fabrication, resistance to solvent action and low thermal as well as electrical conductivities.

The plastics may contain a number of constituents including binders, fillers, dyes or pigments, plasticisers, lubricants, solvents and catalysts. All the plastics are, usually, classified on the basis of the nature of binder used in their manufacture. The binders may be natural synthetic polymers. It will be interesting to know that the natural plastics do not find much importance in actual practice. It is the only synthetic plastics, which are important from the subject point of view.

14.23 TYPES OF SYNTHETIC PLASTICS

All the synthetic plastics may be broadly classified in to the following two types depending upon the nature of the synthetic polymers used as binders :

    1. Thermoplastics.   2. Thermosetting plastics.

14.24 THERMOPLASTIC (THERMOPLASTIC RESINS)

We have already discussed in Art., 14.17 that the thermoplastic resin are the polymers, which are viscoelastic at intermediate temperatures and fluids at high temperatures. Or in other words, thermoplastics resins are the polymers whose plasticity increases with the increase in temperature. Some important examples of such polymers are polyethylene, polystrene, vinyl polymers, amides, acrylics etc. It will be interesting to know that most of thermoplastics are formed by addition polymerization. We know that as the addition polymerization produces chain molecules or linear

Organic Materials

molecules, therefore thermoplastics can be mechanically deformed and softened at high temperatures. Moreover, they can be easily moulded or extruded due to the absence of cross links. On cooling, they regain their original low temperature properties. But they retain the shape and size into which they were moulded. A brief description of some of the important thermoplastics is given below:

1. *Polyethylene (Polythenes)*. These polymers are produced by polymerizing ethylene molecules. They have excellent electrical insulation properties and are widely used as insulating coatings for electric wire, films, sheets, pipes, bottles, buckets etc.

2. *Polystrenes*. These polymers are produced by polymerizing styrene. They have excellent dielectric properties and high resistance to chemicals. They are widely used as refrigerator door liners, hot drink cups, radio and television cabinets, food containers etc.

3. *Polyvinyls*. These polymers are produced by polymerizing vinyl compounds. They have good electrical insulation resistance, low flammability and toughness. They are widely used to manufacture rain coats, handbags, lead wire insulation, vinyl flooring etc.

4. *Acrylics*. These polymers are produced by the chemical reaction of esters of acrylic acid and methacrylic acid. The most important member of this group is polymethyl methacrylate, which is commonly known as perpex. These acrylics have excellent light transmitting power, ease of fabrication and resistance to moisture. They are widely used as light covers, lamp shades lenses, sign boards, plastic jewellery etc.

5. *Polyamides (Nylons)*. These polymers are produced by polycondensation of diabasic acids and hexamethylene diamine. They have good mechanical and dielectric strength. They are widely used in high frequency insulators, gasoline filter bowl, components in automatic lubrication systems etc.

6. *Polytetrafluoro ethylenes (Teflon)*. These polymers are produced by polymerising tetrafluoro ethylene molecules. The have excellent dielectric properties and resistance to heat. They are widely used in packaging delicate equipments, packaging drugs, fabrication of machine components, lead wire insulation etc.

7. *Cellulosics*. The cellulose is a natural polymer formed during growth of its plant. Its useful derivatives are esters and ethers. The cellulose esters are compounds such as cellulose nitrates and acetates. And cellulose ethers are compounds such as ethyl cellulose. They have excellent mechanical properties and electrical insulation properties. They are widely used in toys, flashlight cases, helmets, trays, plastic films, synthetic fibres etc.

14.25 THERMOSETTING PLASTICS (THERMOSETTING RESINS)

We have already discussed in Art. 14.18 that the thermosetting resins are the polymers, which do not deform on mechanical stressing. Some important examples of such polymers are polyester, phenolics, urea formaldehyde, epoxides etc. It will be interesting to know that the thermosetting plastics are formed by condensation polymerization. We know that as the condensation polymerization produces crosslinked molecules, therefore that thermosetting plastics cannot be softened, once they are moulded, even at high temperatures. Or in other words, their plasticity does not change with the temperature. As a result of this, they cannot be remoulded into any new shape. It has been observed that on raising the temperature upto the point, where crosslinks are broken, and irreversible reaction takes place. At this high temperature, all the useful properties, of the material, are destroyed. Such a reaction is known as degradation depolymerization of the polymers. A brief description of some of the important thermosetting plastics is given below:

1. *Polyesters.* These polymers are produced by the polycondensation of polycarboxylic acids and polyhydric alcohols. They have excellent dielectric properties and surface hardness. They are widely used in paper, mat, cloth, insulation for wire and cable etc.
2. *Phenolic.* These polymers are produced by the polycondensation of phenol and formaldehyde. They have good dielectric properties and surface hardness. They are widely used in making electric iron handles, socket boxes, fan-motor housings, switch covers, etc.
3. *Urea formaldehyde.* These polymers are produced by the chemical combination of urea and formaldehyde. They have good bonding quality, mechanical properties and dielectic properties. They are widely used in making instrument dials, electric mixer housings, cosmetic boxes, distributor heads etc.
4. *Melamine formaldehyde.* These polymers are produced by copolymerization of melamines and formaldehydes. They have excellent tensile strength and flame resistance. They are widely used for plastic crockery, automobile parts etc.
5. *Epoxides.* These polymers are produced by the condensation of epichlorohydrin and dioxydiphenyl propane. The most important member of this group of polymers is araldite. They have excellent adhesive properties, resistance to chemicals, toughness and electrical insulation properties. They are widely used for bonding the materials together such as wood, metal, porcelain etc. They are also used in the manufacture of high voltage insulating materials, laminates, varnish etc.

14.26 COMPARISON BETWEEN THERMOPLASTICS AND THERMOSETTING PLASTICS

The following table gives the comparison between thermoplastics and thermosetting plastics:

S. No.	*Thermoplastics*	*Thermosetting plastics*
1.	These polymers are composed of chain molecules.	These polymers are composed of cross-linked molecules.
2.	They are produced by addition polymerization.	They are produced by the condensation polymerization.
3.	These polymers can be mechanically deformed and softened at high temperatures.	These polymers cannot be mechanically deformed or softened at high temperature.
4.	Their plasticity increases with the increase in temperature.	Their plasticity does not increase with the increase in temperature.
5.	They can be easily moulded and remoulded into any shape.	They can not be remoulded into any new shape.

14.27 RUBBER

The rubber may be defined as an organic polymer, which elongates on stretching and regains its original shape after the removal of the stress. Today, the rubber is considered to be one of the most important materials in the world. A major portion of the rubber is consumed in the field of manufacture of tyre and tubes for the vehicles.

The rubbers may contain a large number of constituents such as sulphur, accelerators, (i.e., catalysts like lime, magnesia, litharge etc.) antioxidants (negative catalysts like complex organic compounds), reinforcing agents (like carbon black, zinc oxide etc.), colouring agents like zinc sulphide, lead chromate etc.) and plasticizers like vegetable oils, stearic acid etc.). These

Organic Materials

constituents are added to the polymers to obtain certain desirable properties in them. These properties include elasticity, tensile strength, resistance against chemical reagents, resistance to abrasion, low electrical and thermal conductivity etc.

The structure of a rubber consists of long chain molecules which are interlocked with one another. It has been observed that when a tensile stress is applied to a rubber molecule, the chain gets stretched and a considerable elongation takes place. On removing the stress, the thermal agitation will return the chain to the interlocked form, thus causing the rubber to return to its original length.

14.28 TYPES OF RUBBERS

All the rubbers may be broadly classified into the following two types depending upon the nature of their origin :

   1. Natural rubber.   2. Synthetic rubber or elastomers.

14.29 NATURAL RUBBER

It is an elastic material present in the latex (sap or fluid) of certain plants such as have a brasiliensis and guayule shrub etc. It has been observed that the natural rubber is a polymer of isoprene with the structure as shown in Fig 14.19.

It will be interesting to know that the latex is treated in two ways to obtain the rubber. In the first method, the latex is coagulated with organic acid to produce crude rubber. It may be noted that the crude rubber does not possess the desirable properties. Therefore its properties are improved by addition of compounding materials such as sulphur, accelerators, colouring agents etc. It is then passed through the calendering machine to produce sheets of desired thickness. In the second method, latex itself is mixed with the compounding materials and then precipitated directly from the solution in the final shape to be used. The natural rubber is of the following three types:

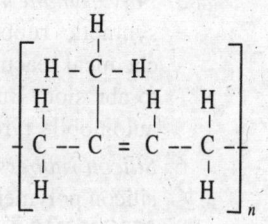

Fig. 14.19. Isoprene.

   1. Chlorinated rubber.   2. Rubber hydrochloride.
   3. Cyclised rubber.

The chlorinated rubber is widely used in the production of protective coatings and adhesives. The rubber hydrochloride is used in wrapping and packaging of the delicate equipments. And the cyclised rubber is used in combination with paraffine wax (solid alkanes) to manufacture papers.

14.30 SYNTHETIC RUBBERS (ELASTOMERS)

It is one of the most interesting and useful development in the rubber industry in these days. The synthetic rubbers are manufactured from raw materials such as coke, limestone, petroleum, natural gas, salt, alcohol, sulphur, ammonia, coal tar etc. Strictly speaking, the synthetic rubbers or elastomers are not exactly a structure like that of natural rubber. But they are rubber like materials, which have may properties like those of natural rubber. Moreover, some of the synthetic rubbers resemble with the natural rubbers in their chemical structure also.

It will be interesting to know, that the processing of synthetic rubbers involved approximately the same steps as that of crude rubber. Moreover, some of the properties of the synthetic rubbers are better than those of natural rubbers. For example, some synthetic rubbers are more resistant to sunlight than the natural rubbers. Similarly, some synthetic rubbers have greater solvent resistance and others have greater elasticity than that of natural rubber. A brief description of some of the important synthetic rubber is given below:

1. *Polychloroprenes (Neoprenes).* These rubbers are produced by polymerizing chloroprene molecules. These are closely related to the natural rubber. But they have higher resistance to oils, greases, aging, high temperature etc. as compared to that of natural rubber. They are widely used for parts such as oil seals, gaskets, low voltage insulations, tank linings, adhesives etc.

2. *Butyl rubbers.* These rubbers are produced by polymerizing isobutylene together with small amount of isoprene or butadiene. They have great impermeability to gases, high resistance to aging and oxidizing agents etc. They are widely used for tyres, inner tubes, adhesives, tank linings, wire and cable insulation etc.

3. *Butadiene rubbers.* These rubbers are produced by the copolymerization of butadience and styrene. They have high resistance to abrasion and weathering as compared to that of natural rubbers. These rubbers are widely used for automobile tyres, belts, shoes soles, flooring, electric wire insulation etc.

4. *Nitrite rubbers (Buna-N).* These rubbers are produced by copolymerization of crylonitrite and butadiene. They have excellent oil, grease and solvent resistance. They also possess good resistance to abrasion and aging. These rubber are widely used for tank linings, conveyor belts etc.

5. *Polyurethane rubbers.* As a matter of fact, polyurethane is a name given to a group of synthetic rubbers containing urethane linkage. These rubbers are produced by the chemical reaction of polyester with dissocynates. They have high strength and resistance to abrasion. But they are susceptible to acid and alkali attack. They are widely used for automobile tyres, expanded flexible foams, conveyor belts etc.

6. *Silicon rubbers.* These rubbers are produced by condensation polymerization of dimethyl silicon polymers. They are the most stable rubbers, over a wide temperature ranging from 55°C to 315°C. But they possess poor physical properties. Moreover, they are costly and mechanically weak. But they have excellent electrical properties. These rubbers are widely used or wire and cable insulation, coatings, packaging, tubing, gaskets etc.

14.31 VULCANIZATION

The term 'vulcanization' may be defined as the process of heating the rubber compound with sulphur. It has been observed that the sulphur combines chemically, by the double bonds in the

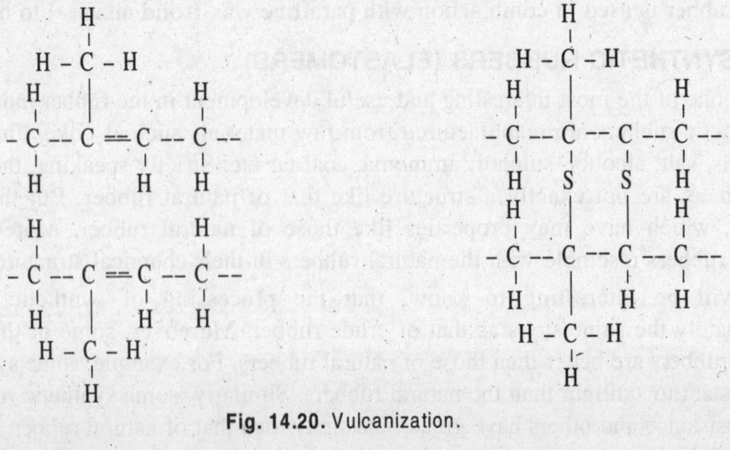

Fig. 14.20. Vulcanization.

rubber molecules causing major changes in its properties. The vulcanization is essential to provide certain desirable properties such as elasticity and strength to the rubber. It will be interesting to

Organic Materials

know that after vulcanization, the rubber becomes less sensitive to changes of temperature. It acquires increased elasticity and tensile strength. It is more durable when exposed to weather and more resistant to the chemical reagents. The vulcanization in the case of polyisoprene rubber (natural rubber) is given in Fig. 14.20 (a) and (b).

It may be noted that the vulcanization with sulphur breaks the double bonds between the two carbon atoms and the sulphur atoms bridge the gaps between the two chains, *i.e.*, they establish cross-links between the chains. With higher sulphur contents, more cross linkage takes place. But it restricts the range of elastic stretching.

14.32 WOOD

It is a natural organic material, which can be cut or pressed into suitable shapes and size by mechanical or chemical means. The primary structural (or microconstituents) of wood are cellulose and an amorphous carbohydrate material lignin. The cellulose is a long chain polymer, which has the structure as shown in Fig. 14.21 (a). Here 'n' is the number of molecules per chain. It value may be as high as three to four thousand molecules per chain.

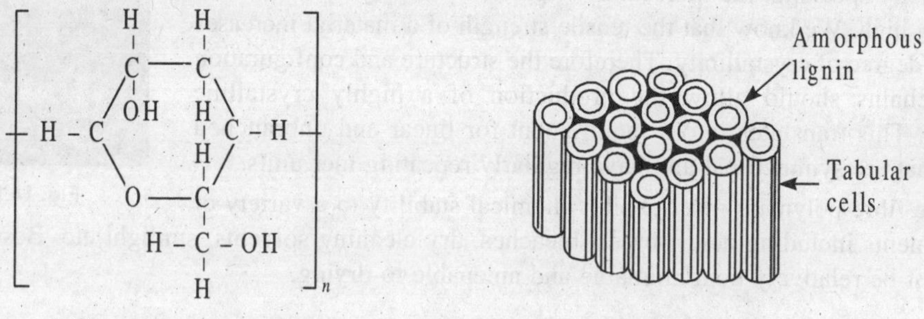

(a) Structure of wood   (b) Tabular cell

Fig. 14.21. Wood.

The presence of hydroxyl (OH groups), in the cellulose structure, makes it highly polarised. This gives rise to the formation of secondary bonds between the chains. As a result of this, the cellulose possesses a crystalline structure. The crystals of the cellulose take the form of tabular cells as shown in Fig. 14.21. (b). These crystals bonded together by amorphous lignin. The amorphous lignin also provides a passage for moisture and sap to flow during the growth wood (*i.e.*, tree or plant). It has been observed that the growth of wood is most rapid in spring giving thin and open cells. Whereas in summer, the growth is slower producing denser and stronger cells. The annual cycle of growth produces the rings and crystalline structure of wood.

It will be interesting to know that the wood possesses greater strength in the longitudinal direction than that in the transverse direction. It is due to the fact the cellulose chains run parallel to the crystals, which are directed in the longitudinal direction.

The strong polar nature of cellulose chains provides the strong affinity of wood for moisture, which causes swelling. When the wooden articles are exposed to air, for a sufficient length of time, they lose moisture more easily from the surface perpendicular to the crystals than from those parallel to the crystals. This tends to cause warping in the wooden articles. In larger pieces, the moisture is lost more easily from the surfaces than from the inside. As a result of this, the surface shrinking takes place which leads to cracking. These difficulties can be overcome by laminating (*i.e.*, preparing thin sheets) which are easily dried and then fixed together with adhesives impregnating wood with plastics and high temperature compression and binders etc.

It has been observed that the wood, on heating in the air, decomposes before it softens. But by heating in steam, it is prevented from drying out, softened and can be easily shaped.

14.33 FIBRE POLYMERS

The fibre polymers are capable of being drawn into long filaments having at least 100 : 1 length-to-diameter ratio. Most commercial polymer fibres are utilized in textile industry, being woven or knit into cloth or fabric. They are also utilized in composite materials (refer to Fig. 14.22.)

To be useful as a textile material, a fibre polymer must have a host of rather restrictive physical and chemical properties, while in use, fibres are subjected to a veriety of mechanical deformations. The deformation may be due to stretching, twisting, shearing and abrasion. Hence the fibres must have high tensile strength, a high modulus of elasticity and abrasion resistance. These properties depend upon the chemistry of the polymer chains and the fibre drawing process.

Strictly speaking, the molecular weight of fibre polymer should be relatively high. We know that the tensile strength of a material increases with the degree of crystallinity. Therefore the structure and configuration of the chains should allow the production of a highly crystalline polymer. This translates into a requirement for linear and unbranched chains that are symmetrical and have regularly repeating mer units.

Fig. 14.22.

The fibre polymers must exhibit chemical stability to a variety of environments including acids, bases, bleaches, dry cleaning solvents, sunlight etc. Besides this they must be relatively nonflammable and amenable to drying.

14.34 ADVANCED POLYMERIC MATERIALS

Since past few decades, a number of polymers having unique and desirable combinations of properties have been developed. Many polymers have found applications in either new technologies or have replaced other materials.

Followings are some of the advanced polymeric materials which are important from the subject point of view.

1. Ultrahigh molecular weight polyethylene
2. Liquid crystal polymers
3. Thermoelastic elastomers.

Now we shall discuss these polymers one by one in the following pages.

14.35 ULTRAHIGH MOLECULAR WEIGHT POLYETHYLENE

This advanced polymer is a linear polyethylene. It has an extremely high molecular weight (in the range of 1×10^6 to 4×10^6 atomic mass units). In the fibre form, this polymer is called spectra. The ultrahigh molecular weight polyethylene has some of the following outstanding characteristics:

1. Extremely high impact resistance
2. Excellent resistance to wear and abrasion
3. Very low coefficient of friction
4. Self lubricating and nonstick surface

This material has been found to have a relatively low melting temperature. Because of this characteristic, its mechanical properties diminish rapidly with the increase in temperature.

Organic Materials

Applications. Although ultrahigh molecular weight polyethylene has many applications, yet the following are important from the subject point of view:

1. Bullet proof vests
2. Composite military helmets
3. Fishing line
4. Sky bottom surfaces
5. Golf ball cores
6. Bowling alley
7. Ice skating rink surfaces
8. Biomedical prosthesis
9. Blood filters
10. Marking pen nibs
11. Bulk material handling equipment
12. Bushings
13. Pump impellers
14. Valve gaskets

14.36 LIQUID CRYSTAL POLYMERS

These polymers are a group of chemical complex and structurally distinct materials that have unique properties and are employed in diverse applications. The detailed discussion of the chemistry of liquid crystal polymers is beyond the scope of this book.

The liquid crystal polymers are composed of extended, rod shaped and rigid molecules. These materials do not fall within any conventional liquid, amorphous, crystalline or semicrystalline category. In fact, the liquid crystal polymers are considered as a new state of matter, *i.e.* the liquid crystalline state. In this state the material is neither crystalline nor liquid. In the melt or liquid condition, whereas other polymers are randomly oriented, the liquid crystal polymers can become aligned in high ordered configurations. As solids, this ordered molecular alignment remains. Besides this, the molecules in the solid condition form in domain structures having characteristic intermolecular spacings.

Based on orientation and positional ordering, the liquid crystal polymers may be classified into the following three types:

1. Smectic
2. Nematic
3. Cholesteric.

Applications. The liquid crystal polymers are widely used in liquid crystal displays, digital watches, notebook computers, high definition LCD television.

Some of the nematic type of liquid crystal polymers are rigid solids at room temperature. These materials are widely used in electronic industry as interconnect devices, relay and capacitor housings, brackets etc. They are also used by the medical industry in components which need to be repeatedly sterlized. The nematic type liquid crystal polymers are also used in photo copiers and fibre-optic components.

The liquid crystal polymers have excellent thermal stability, stiff and strong, high impact strength, chemical inertness, extremely low shrinkage and warpage during molding, excellent dimensional repeatability and low heat of fusion.

14.37 THERMOPLASTIC ELASTOMERS

The thermoplastic elastomers are a type of polymeric material that exhibits elastomeric (or rubbery) behaviour at ambient conditions yet is thermoplastic in nature. Most elastomers are categorised as thermosets because they become cross linked during vulcanisation.

Of the several varieties of thermoplastic elastomers, one of the widely used material is block copolymer. It consists of block segments of a hard and rigid thermoplastic mer (styrene and butadiene or isoprene). These two block type alternate positions — for a common molecule, hard polymerized segments are located a chain ends as shown in Fig. 14.23 (*a*) & (*b*). On the other hand the soft central region consists of polymerized butadiene or isoprene units.

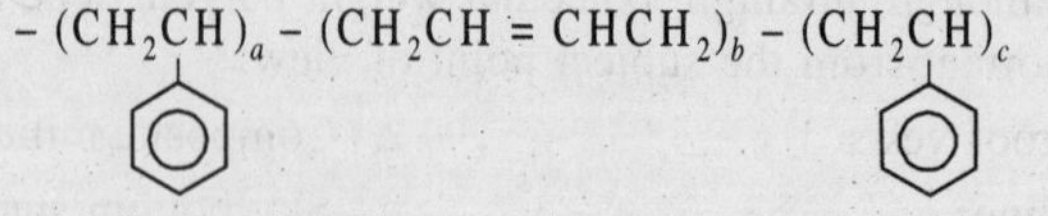

$$-(CH_2CH)_a - (CH_2CH = CHCH_2)_b - (CH_2CH)_c$$
$$\bigcirc\bigcirc$$

(a) Styrene-butadiene-styrene thermoplastic elastomer

$$(CH_2CH)_a - (CH_2C = CHCH_2)_b - (CH_2CH)_c$$
$$\bigcirc|\bigcirc$$
$$CH_3$$

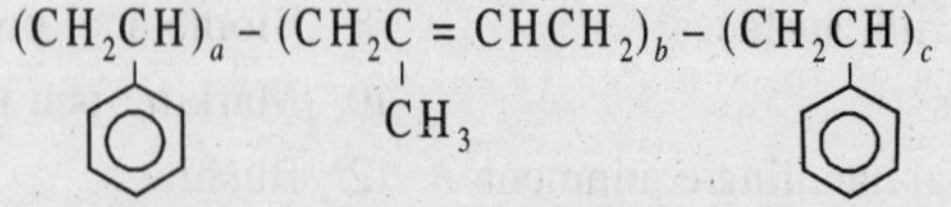

(b) Styrene-isoprene-styrene thermoplastic elastomer

Fig. 14.23.

It has been observed that at ambient temperatures, the soft, amorphous, central segment (*i.e.* butadiene or isoprene) impart the rubbery, elastomeric behaviour to the material. For temperature below the T_m of the hard component (styrene), hard chain-end segments from numerous adjacent chains aggregate together to form rigid domain regions. These domains are "physical crosslinks" that act as anchor points so as to restrict soft segment motions. They function as the same way as "chemical cross links" for the thermoset elastomers.

The major advantage of thermoplastic elastomers over the thermoset elastomers is that they may be processed by conventional thermoplastic techniques such as blow, moulding, injection moulding etc. Moreover since melting-solidification process is reversible and repeatable, thermoplastic elastomer parts may be reformed into other shapes. In other words, they are recyclable.

Applications. Thermoplastic elastomers have replaced thermoset elastomers in a veriety of applications. Some of these include:

1. automotive exterior trim (bumpers, fascia etc.)
2. automatic underhood components (electrical insulation and connectors, gaskets etc.)
3. shoe soles and heels
4. sporting goods (e.g. bladders for footballs and soccer balls)
5. medical barrier films and protective coatings
6. as components in scalants, caulking and adhesives.

A.M.I.E. (I) EXAMINATION QUESTIONS

1. Write the formula for natural rubber. *(Summer, 1993)*
2. What do you understand by polymerization ? What is the difference between addition polymerization and condensation polymerization ? *(Winter, 1993)*
3. (a) Discuss briefly the different polymerization mechanisms.
   (b) What are the general properties of polymeric materials ? *(Summer, 1994)*
4. (a) What feature is necessary in a monomer for addition polymerization to be possible ? Is this same feature a necessity for condensation Polymerization ?
   (b) Describe the difference between thermoplastic and thermosetting polymer in terms of ?
      (i) Applied stress
      (ii) Increased temperature
      (iii) Atomic structure.
   (c) Describe briefly how each of the following modifies the strength of a polymer:
      (i) Increased degrees of polymerization
      (ii) Increased branching

Organic Materials

      (*iii*) Increased cross-linking

      (*iv*) Increased crystallinity.   *(Summer, 1995)*

5. What is meant by the polymerization ? What are the two broad classifications of polymers. Discuss them.   *(Summer, 1996)*
6. Define copolymerization.   *(Summer, 1996)*
7. Define the term 'polymerization' Explain the different types of polymers with examples and applications.   *(Winter, 1996)*
8. Give two examples of thermoplastics and thermosetting plastics.   *(Winter, 1996)*
9. What is polymerization ? With the help of suitable examples, compare and contrast the process of addition polymerization and condensation polymerization.   *(Summer, 1997)*
10. What is the effect of the use of plasticizer or thermoplastic materials ?   *(Summer, 1997)*
11. Name a thermoplastic and thermosetting polymers. State the characteristics in each case.

      (Winter, 2000)

MULTIPLE CHOICE QUESTIONS

1. A polymerization is a process in which
   - (*a*) two or more chemically different monomers are polymerized to form long chain molecules
   - (*b*) two or more chemically same monomers are polymerized to form long chain molecules
   - (*c*) two or more chemically same monomers to form a cross-linked polymer
   - (*d*) two or more chemically different monomers to form a cross-linked polymer
2. Thermoplastic is a polymer which
   - (*a*) is visco elastic at intermediate temperatures and fluid at high temperature
   - (*b*) do not deform on mechanical stressing
   - (*c*) is produced by condensation polymerization
   - (*d*) cannot be remoulded into any new shape.

ANSWERS

1. (*b*)   2. (*a*)

15
Composite Materials and Ceramics

1. Introduction 2. Types of Composite Materials. 3. Agglomerated Materials. 4. Important Terms in Agglomerated Materials. 5. Particle size. 6. Packing Factor. 7. Density and porosity. 8. Change in Volume in Agglomerated Materials. 9. Important Agglomerated Materials 10. Cements. 11. Setting and Hardening of Cement. 12. Cement Concrete. 13. Laminated Materials (or Laminates). 14. Important Laminated Materials. 15. Plywood. 16. Tufnol. 17. Reinforced Materials. 18. Important Reinforced Materials. 19. Reinforced Cement Concrete. 20. Nylon Reinforced Rubber. 21. Glass-fibre Reinforced Plastics. 22. Advanced Composite Materials. 23. Ceramics. 24. Classification of Ceramics on the Basis of Application. 25. Glasses 26. Clay Products. 27. Refractory Materials. 28. Properties of Refractory Materials. 29. Dimensional Stability of Refractory Materials. 30. Abrasives 31. Advanced Ceramics. 32. Optical Fibres. 33. Microelectromechanical Systems, (MEMS). 34. Ceramic Ball Bearings. 35. Piezoelectric ceramics 36. Other Advanced Ceramic Materials.

15.1 INTRODUCTION

Sometimes two or more materials are combined together to produce a new material, which possesses much superior properties than any one of the constituent materials. Such a material is known as composite material. The common example of a natural composite material is wood, which consists of long cellulose fibre held together by amorphous lignin. Some of the artificial (or synthetic) composite materials are cement concrete, glass reinforced plastic plywood etc.

15.2 TYPES OF COMPOSITE MATERIALS

The composite materials, which are important from the subject point of view, are of he following three types:

1. Agglomerated materials
2. Laminated materials
3. Reinforced materials.

Now we shall discuss all the above mentioned types of composite materials one by one.

15.3 AGGLOMERATED MATERIALS

The materials, in which the particles are condensed together to form an integral mass, are known as agglomerated materials. A common example of agglomerated materials is cement concrete, which is formed by mixing coarse aggregate, fine aggregate, cement and water in different proportions. Other useful agglomerated material is grinding wheel (abrasive material) which is formed by mixing asphalt, stone and resin of different sizes. Some other useful agglomerated materials are cemented carbide tools and ceramics. The cemented carbide tools are manufactured by agglomerating small particles of hard tungsten carbide with alloys of cobalt and nickel. These alloys act as a binder between the tungsten carbide particles. The cermets are produced by combining a ceramic material and metal through various processes. In this process,

Composite Materials and Ceramics

the mixture of ceramic material and metal is taken in the form of powder, which is mixed, pressed and finally sintered to produce a composite material. In cermets, the metal provides high toughness and thermal shock resistance. The ceramic materials provide high refractoriness and creep resistance. They have high abrasion resistance, excellent chemical stability and machinability etc.

15.4 IMPORTANT TERMS IN AGGLOMERATED MATERIALS

Following are the important terms, which are frequently used in the study of agglomerated materials:

1. Particle size   2. Packing factor   3. Density and porosity.

15.5 PARTICLE SIZE

It has been observed that most of the particles used in agglomerated materials are not perfect spheres. As a result of this, it is, usually, difficult to measure the exact particle size of an agglomerated material. In most of the cases, the variation in spherical shape is quite common. In order to have the particle size of he various constituent materials, the particles are made to pass through standard size screens, which have a fixed number of openings per square centimetre.

It may be noted from the practical point of view that it is not sufficient to depend upon the average size of an aggregate, as the aggregate mixture will be composed of a varying range of screen sizes.

15.6 PACKING FACTOR

The term 'packing factor' may be defined as the ratio of true volume of the total constituent materials to the bulk volume of the agglomerated materials. It has been observed that whenever uniformity sized material is packed into a large mass, a considerable amount of porosity exists within the body, whose value depends upon the size and shape of the packing material. Mathematically packing factor,

$$P.F. = \frac{\text{True volume}}{\text{Bulk volume}}$$

where true volume is the total actually occupied by all the particles and the bulk volume is the total volume of the agglomerated material.

Mathematically bulk-volume

$$= \text{True volume} + \text{Total pore volume}$$

$$\therefore P.F. = \frac{\text{Bulk volume} - \text{Total pore volume}}{\text{Bulk volume}}$$

$$= 1 - \frac{\text{Total pore volume}}{\text{Bulk volume}}$$

$$= 1 - \text{porosity}$$

It will be interesting to know that, from practical point of view, it is always desirable to increase the packing factor of an agglomerated material. It is achieved by the following two ways.

1. By using the non-spherical particles with matching surfaces.
2. By mixing particles of different sizes.

15.7 DENSITY AND POROSITY

We have already discussed in the last article that a considerable amount of porosity exists within the body of an agglomerated material. It has been observed that this porosity is due to open or closed pores. It will be interesting to know that the open pores are the vacant spaces, between the packed particles, which are interconnected with one another. And the closed pores are the vacant spaces, between the packed particles, which are not interconnected with one another. The following relations are used to obtain density and porosity of an agglomerated material:

$$\text{True density} = \frac{\text{Mass}}{\text{True volume}}$$

$$= \frac{\text{Mass}}{\text{Bulk volume} - \text{Total pore volume}}$$

$$\text{Apparent density} = \frac{\text{Mass}}{\text{Bulk volume} - \text{Open pore volume}}$$

$$\text{Bulk density} = \frac{\text{Mass}}{\text{Bulk volume}}$$

$$\text{True porosity} = \frac{\text{Mass}}{\text{True volume} + \text{Total pore volume}}$$

$$\text{Apparent porosity} = \frac{\text{Open pore volume}}{\text{Bulk volume}}$$

15.8 CHANGE OF VOLUME IN AGGLOMERATED MATERIALS

It has been observed that the volume of an agglomerated material changes, during the manufacturing processes, whose value depends upon a number of factors. As a result of this, an allowance is always provided in agglomerated materials for this change in volume. It has been observed that the volume of a brick reduces during drying and burning processes due to excessive water contained by the clay, which is removed. On the removal of water, the clay particles come closer to each other and thus the volume reduces. In the firing process, the reduction in volume is due to sintering of particles also.

In agglomerated materials, sintering is necessary to attain porosity and resistance against mechanical wear. The porosity, in certain machine parts, is very essential, e.g., in self lubricating bearings and insulation materials. The presence of pores in such bearings permits lubricating oil to reach the bearing surface. Similarly, in heat or sound insulating materials the presence of pores reduces the transmission of heat or sound etc.

15.9 IMPORTANT AGGLOMERATED MATERIALS

Though there are a number of agglomerated materials these days, yet the following are important from the subject point of view:

1. Cements. 2. Cement concrete.

15.10 CEMENTS

The cement is one of the agglomerated materials, which is one of the most commonly used material of construction work these days. It is manufactured by a partial fusion of a mixture containing clay, limestone and oxides of silicon, aluminium, iron and often magnesium. This mixture is heated to a temperature of about 1450°C. At this temperature, the material sinters and partially fuses into balls known as clinkers. The clinkers are then cooled and grounded into a very fine powder and some gypsum is also added to it. The resulting product is cement, commonly known as portland cement. The main constituents of cement are tricalcium silicate (C_3S), dicalcium silicate (C_2S), tricalcium aluminate (C_3A) and tetra allumino-ferrite (C_4Af).

Composite Materials and Ceramics

15.11 SETTING AND HARDENING OF CEMENT

We know that water is always added to the cement, before it is used in the construction work. It has been observed that the water reacts chemically with the surface of cement grains and produces a gelatinous material called 'gel'. It seres as a cementing material between the grains. With the passage of time, hardening of this material takes place due to loss of water. The following chemical equations indicate the process in which the constituents of cement react with water. The products, so obtained, are called hydrated products.

$$C_3A + 6H_2O \longrightarrow C_3A.6H_2O \qquad ...(i)$$

$$C_4Af + 7H_2O \longrightarrow C_3A.6H_2O - Cf.H_2O \qquad ...(ii)$$

$$C_3S + xH_2O \longrightarrow C_2S.xH_2O + Ca(OH)_2 \qquad ...(iii)$$

$$C_2A + xH_2O \longrightarrow C_2A.xH_2O \qquad ...(iv)$$

It will be interesting to know that the equations (i) and (ii) indicate the starting of hardening process of cement. The resulting hydrated products coat the coarse aggregate particles and seal them off so that they do not absorb water. And equations (iii) and (iv) indicate the main subsequent reactions for the hardening of the cement.

All the above mentioned reactions are known as exothermic as they liberate energy in the form of heat. This heat is known as heat of hydration. The rate at which the setting and hardening takes place, depends upon the amount of heat of hydration. It has been observed that the tricalcium silicate (C_3A) requires about 30 days to attain 70 per cent of its ultimate strength. Whereas dicalcium silicate (C_2S) requires more than 6 months to attain 2/3 of its ultimate strength. It is thus obvious that strength of cement concrete goes on increasing with the time.

15.12 CEMENT CONCRETE

It is a mixture of cement, sand, brick ballast of stone ballast and water in different proportions. On hardening, the mixture forms a stone like material known as cement concrete. It has been observed that the cement and water form a paste, which binds the aggregates to a permanent mass after hardening. The sand, which is usually known as fine aggregate, is required to fill the voids present between the ballast. The brick or stone ballast is commonly known as coarse aggregate.

The purpose for the addition of water, to the mixture, is to produce hydrates and make it more workable. The strength of cement concrete depends upon the water-cement ratio. It has been observed that if the water-cement ratio is too low, the trapped air in the mixture can not be removed. This produces porosity in the concrete mixture which decreases its strength. On the other hand, if the water-cement ratio is too high, the bonding in the mixture is imperfect. Thus optimum water-cement ratio is required to produce optimum strength. Sometimes, some chemicals are also added to the cement concrete to reduce porosity at the optimum water-cement ratio and to avoid failure of the structure.

It will be interesting to know that the cement concrete produces excellent resistance to compressive stress, shear stress, and abrasion. The limitations of cement concrete are low tensile strength shrinkage and expansion etc. The above limitations are removed by laying the cement concrete with steel to from reinforced cement concrete.

15.13 LAMINATED MATERIALS (OR LAMINATES)

The materials, which are produced by bonding two or more layers of different materials completely to each other, are known as laminated materials or laminates. The materials, constituting a laminated material, may be metallic or non-metallic depending upon the type of application.

The common examples of laminated materials, are plywood, tufnol, sunmica, linolium etc. In laminated materials the top most layer provides the desired appearance and workability, while the lower layer contributes to its strength. These days, a variety of techniques are available to manufacture laminated composite materials. The most common techniques are roll bending, coextrusion, explosive welding and brazing.

15.14 IMPORTANT LAMINATED MATERIALS

Though there are a number of laminated materials these days, yet the following are important from the subject point of view :
1. Plywood
2. Tufnol.

15.15 PLYWOOD

It is produced by binding together an odd number of thin layers (called plies) of fine wood by a resin under pressure. These plies are laid in such a way that alternate layers have their grain structure at right angles to each other. The number of plies, usually, range from 3 to 7. But in special cases, the number of plies may be more. The plywood has the advantage over the ordinary wood, that it is less liable to warp. Moreover, it is equally strong in all directions.

15.16 TUFNOL

It is produced by combining a laminated material having layers of woven textiles with a thermosetting resin. In the composite material, the resin provides the necessary appearance and rigidity, while the woven textile provides the necessary strength.

15.17 REINFORCED MATERIALS

The materials, which are produced by combining some suitable material to provide additional strength which does not exist in a single material, are known as reinforced materials. The common example or reinforced material is reinforced cement concrete (R.C.C.). Glass-fibre reinforced plastic etc.

15.18 IMPORTANT REINFORCED MATERIALS

Though there are a number of reinforced materials these days, yet the following are important from the subject point of view:
1. Reinforced cement concrete.
2. Nylon reinforced rubber.
3. Glass-fibre reinforced plastic.

15.19 REINFORCED CEMENT CONCRETE

It is produced by placing the steel rods (*i.e.*, reinforcement bars) in the cement concrete mixture. The resulting product, which provides excellent resistance to both the compressive and tensile stresses, is known as reinforced cement concrete (R.C.C.). It will be interesting to know that the reinforcement, being strong in tension is provided in the tensile zone. And the cement concrete, being strong in compression, is provided in the compression zone. Sometimes, the reinforcement is provided in both the compression and tensile zones. This is generally done when the maximum moment is very high.

These days, torsteel (*i.e.*, twisted steel rods) is used as reinforcement in all types of building. And thick steel wires are used as high tensile steel in long span bridges.

Composite Materials and Ceramics

15.20 NYLON REINFORCED RUBBER

It is commonly used in the manufacture of automobile and cycle tyres. In this composite material, the nylon thread (sometimes steel wires, glass wires etc.) provides the necessary strength and the rubber provides the necessary surface of the tyres.

15.21 GLASS-FIBRE REINFORCED PLASTIC

It is produced by combining glass fibre with plastic. In this composite material, the glass-fibre provides the necessary strength and the plastic reduces brittleness. The commonly used plastics are polyester resin, phenolics, silicons etc. The fibres can be employed either in the form of continuous lengths, staples or whiskers.

A complete material with desired yield strength and modulus of elasticity can be obtained by selecting the maximum number of fibres per unit volume. This will allow each fibre to takes its full share of load. The fibre-reinforced composites are generally anisotropic (i.e., the material has different properties in different directions). Moreover, the maximum strength of the composite is in the direction of alignment of fibres.

The following three conditions are essential to prepare a glass-fibre reinforced plastic material :

1. The coefficient of expansion of the fibre should match closely to that of the plastic material.
2. The fibre and the plastic material should be chemically compatible with each other. This would avoid any undesirable reaction taking place between them.
3. The fibre should be stable at room temperature. And it should retain its strength at high temperatures.

Example 15.1 *Determine the volume ratio of aluminium and boron in aluminium-boron composite which can have the same Young's modulus equal to that of iron. The Young's modulus of aluminium, iron and boron are 71, 210 and 440 GN/m² respectively. (A.M.I.E. Summer, 1993)*

Solution. Given: $Y_{Al} = 71 GN/m^2$; $Y_{Fe} = 210 GN/m^2$ and $Y_B = 440 GN/m^2$.

Let V_{Al} = Volume fractions of Aluminium in Al-B composite

and V_B = Volume fraction of Boron in Al-B composite

and we know that volume of Al-B composite

$$V_{Al} + V_B = 1 \qquad \qquad ....(i)$$

and Young's Modulus of iron (Y_{Fe}),

$$210 = V_{Al} \cdot Y_{Al} + V_B \cdot Y_B = V_{Al} \times 71 = V_B \times 440 \qquad \qquad ....(ii)$$

$\therefore \qquad 71 V_{Al} + 440 V_B = 210$

Solving equation (i) and (ii),

$V_{Al} = 0.623$ and $V_B = 0.377$ **Ans.**

15.22 ADVANCED COMPOSITE MATERIALS

It will be interesting to know that recreational equipment is heavily dependent on material technology. For example snow boards shown in Fig. 15.1 (*a*) is fabricated from advanced composite materials. The snow board is a free riding, turn-tip board with a cap and a full wrap around edge. It is stiff and torsionally rigid so one can rail them at high speed, launch and land the hugest airs. Fig. 15.1 (*b*) shows the sectional view of the snow board. The base of a snow board is usually made of compressed carbon. The other layers are made of fibre glass ABS plastic, poleurethane plastic and hardened steel.

A mountain bike shown in Fig. 15.2 is another piece of recreational equipment which makes use of advanced composite materials. The mountain bike is actually an integration of composite

(a) Fig. 15.1. (b)

materials and a number of other structural materials like metals, elastomers (*i.e.* rubber) etc. Thus it is, in fact, a composite system. Such bikes have a weight less than 8 kg and still meet the tough requirements of the sports.

Fig. 15.2.

The integration of ceramic, metallic, plastic and semiconductor materials is a necessary requirement to the fabrication of the microelectronics package shown in Fig. 15.3. It is a composite system and its function is to provide interface between the central integrated circuit (IC) device and the other items like printed circuit board, connector pins etc.

Fig 15.3.

Various composite materials are used in aircraft structures such as the Boeing 777 or Airbus 380 because of their strength and weight savings. Composites offer resistance to fatigue, corrosion and impact damage. The materials used in aircraft structures may be broadly classified into two categories:

1. Metals — aluminium alloy, titanium alloy and alloy steel
2. Composites — fibre glass, graphite and toughened graphite.

Fig. 15.4 shows the breakup of the materials used in aircraft structures. As seen from this diagram, in Boeing 747 structure, 81 percent of the entire structure makes use of aluminium, 13 percent steel, 4 percent titanium and only 1 percent, composites and 1 percent other materials. On the other hand, the Boeing 777 structure is composed of 70 percent aluminium, 11 percent steel, 11 percent composites, 7 percent titanium and 1 percent other materials.

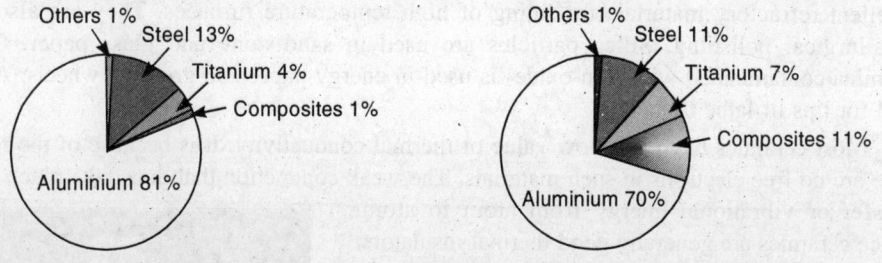

Materials used in Boeing 747 structure   Materials used in Boeing 777 structure

Fig. 15.4.

15.23 CERAMICS

The term "ceramics" is used to denote those products which are made from inorganic materials and have non-metallic properties. Ionically, bonded magnesia (MgO) and covalently bonded silicon carbide (SiC) are the simple examples of ceramics. The traditional ceramic materials are stone, brick, concrete, clay, glass, vitreous enamel and refractories. Majority of these are composed of silicates. New ceramic materials includes oxides, carbides, borides and other similar compounds which are being developed. Such materials have properties of special advantage in particular applications such as high temperature work, solid-state electronics and in nuclear reactors, Table 15.1 gives the values of some mechanical and physical properties of some ceramics.

Table 15.1

S. No.	Ceramic type	Melting Point (°C)	Young's Modulus in GPa	Coeffi. of Thermal Expansion	Thermal Conductivity (k), W/mk	Tensile strength MPa	Thermal Shock resistance $R = k\sigma\rho/\alpha E$
1.	Magnesia (MgO)	2800	210-310	α, 10^{-6} X 13.5	36-45	100	$Wm^{-1}, 10^3$ X 1.2
2.	Sintered alumina (Al_2O_3)	2040	350-380	7-9	12-32	400-500	3.4
3.	Vitreous silica (SiO_2)	1710	70	0.55	1-2	80-160	3.2
4.	Hot pressed Silicon carbide (SiC)	decomposes at 2300	350-470	4.5	100	350-800	3.1
5.	Hotpressed Silicon nitride (Si_3N_4)	above 1900	150-320	2.9	10-16	500-900	13.4

From the above table, we find that above ceramics have high melting points. It is because of the strong primary bonding (ionic or covalent) between the constituent atoms. Materials such as magnesia and alumina, which have melting points of 2800°C and 2040°C respectively make excellent refractory materials for lining of high temperature furnaces. They are also used for metallurgical polishing. Silica particles are used in sand-stone and glass paper. Corundum (alumina contaminated with iron-oxide) is used in energy paper and grinding wheels. Alumina is used for tips of lathe tools etc.

Most ceramics have very low value of thermal conductivity. It is because of the reason that there are no free electrons in such materials. The weak conduction that does take place is only by transfer or vibrational energy from atom to atom. Hence ceramics are generally good thermal insulators. Alumina is used for spark plug insulators. These articles have to withstand rapid fluctuations of temperature and pressure, the maximum being about 850°C and 6 MPa and voltage upto 12000 V. They also maintain gas-tight joints with the metal conductor and base.

In general, ceramics are very durable, being chemically resistant to most acids, alkalies and organic solvents. Moreover they are not affected by oxygen.

Fig. 15.5 shows some common ceramics for traditional engineering applications. These miscellaneous parts with characteristic resistance to damage by high temperatures and corrosive environments are used in a variety of furnaces and chemical processing systems.

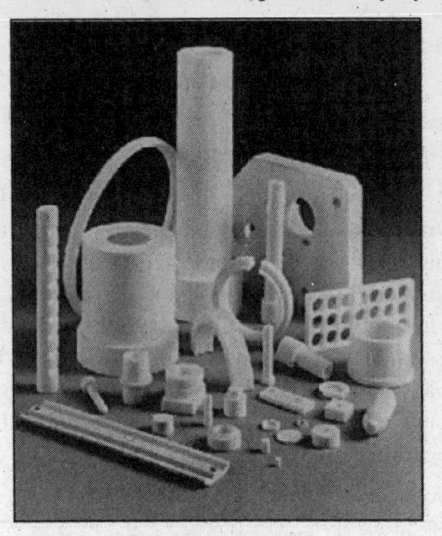

Fig. 15.5.

15.24 CLASSIFICATION OF CERAMICS ON THE BASIS OF APPLICATION

Fig. 15.6 shows the classification of ceramic materials on the basis of application. As seen from this diagram, the ceramic materials can be classified into the following six groups:

1. Glasses
2. Clay products
3. Refractories
4. Abrasives
5. Cements
6. Advanced materials.

Ceramic Materials

Glasses | Clayproducts | Refractories | Abrasives | Cements | Advanced ceramics

Glasses | Glasses Ceramics | Structural clayproducts | Whitewares | Silica | Fireclay | Basic | Special

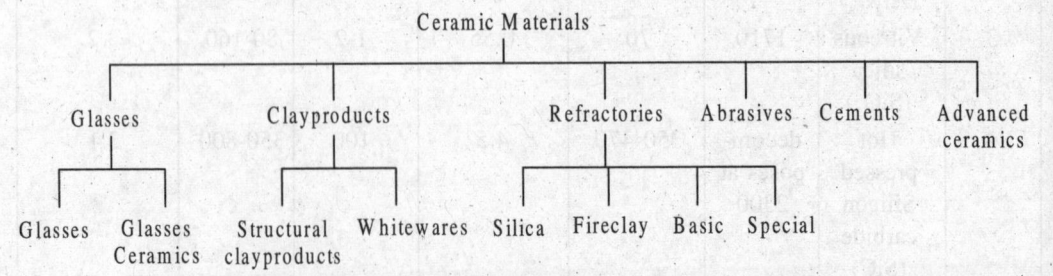

Fig. 15.6.

Now we shall discuss these materials one by one in the following pages. Please note that the cements have been discussed already under the composite materials section.

Composite Materials and Ceramics

15.25 GLASSES

This is the most familiar group of ceramic materials. The typical applications of glasses include containers, windows, lenses and fibre glass. The glasses are non crystalline silicates containing other oxides such as CaO, Na$_2$O, K$_2$O and Al$_2$O$_3$. Table 15.2 shows some common commercial glasses, their composition, characteristics and applications.

Table 15.2

Glass type	\multicolumn{6}{c}{Composition (wt%)}	Characteristics and applications
	S$_i$O$_2$	Na$_2$O	CaO	Al$_2$O$_3$	B$_2$O$_3$	Others	
Fused silica	99.5	–	–	–	–	–	High melting temperature, shock resistant.
Vycor	96	–	–	–	4	–	Chemically resistant, used in laboratory equipment
Borosilicate (PYREX)	81	3.5	–	2.5	13	–	Chemically resistant, used in ovenware
Containers	74	16	–	5	1	4MgO	Low melting temperature, easily worked, durable
fibreglass	55	–	16	15	10	4MgO	Easily drawn into fibres-glass-resin composites
Optical flint	54	1	–	–	–	37PbO	High density and high index of refraction used in optical lenses.
Glass-ceramic	43.5	14	–	30	5.5	6.5 TiO$_2$ 0.5 As$_2$O$_3$	easy fabrication, strong, used in ovenware

15.26 CLAY PRODUCTS

Clay is one of the most widely used ceramic raw material. It is an inexpensive ingredient and is found naturally in great abundance. The clay can be used very easily to form products. When mixed in the proper proportions, clay and water form a plastic mass that is very flexible to shaping. The formed piece is dried to remove some of the moisture. Later it is fired at elevated temperature to improve its mechanical strength.

Broadly speaking, the clay products may be classified into the following two categories:

1. *Structural Clay products*: These include boiling bricks, tiles and sewer pipes.
2. *White wares*: These include porcelain, pottery, table ware, china and sanitary ware.

15.27 REFRACTORY MATERIALS

The ceramic materials capable of withstanding high temperatures without appreciable deformation under service conditions are called refractory materials. The ceramic materials such as magnesia (with melting point of 288°C) and alumina (with melting point of 2040°C) are excellent choice as a refractory material. Such materials are used for lining of high temperature furnaces, boilers, crucibles, converters and supports for hot wires etc. Some common refractory materials are silica, magnesite, dolomite, silicon carbide, zircon and graphite.

15.28 PROPERTIES OF REFRACTORY MATERIALS

Following are some of the basic properties of refractory materials which are important from the subject point of view :

1. These materials do not fuse or soften at the temperature at which it is used.
2. These materials have excellent resistance against thermal shock and abrasion. Thermal shock refers to the sudden changes in temperature.
3. These materials do not crumble or crack under the prevailing pressure when in use at high temperatures.
4. These materials have low coefficient of thermal conductivity.
5. These materials have high resistance to corrosion and therefore do not react chemically with the molten metal or slag in the furnace.
6. These materials are impermeable to gases and liquids.
7. These materials have extremely low values of electrical conductivity.

15.29 DIMENSIONAL STABILITY OF REFRACTORY MATERIALS

The resistance of a refractory material to any volume changes, which may occur due to change in temperature is called dimensional stability. The change in temperature may be during heated or prolonged exposure to high temperatures. Dimensional changes in a refractory material could be reversible or irreversible. The reversible changes are related directly to the coefficient of thermal expansion. While irreversible changes occur due to phase transformations and result in either contraction or expansion of the material permanently. The contraction is caused by the formation of liquid in increasing amounts from relatively low melting constituents present within the material when subjected to high temperature over a long period. The liquid fills the pores gradually causing shrinkage of the material. This behaviour is exhibited by the fire clay bricks.

Permanent contraction can also be caused by polymorphic transformation. A magnesite brick, on heating undergoes contraction from an amorphous magnesium oxide (with density $3.05 \times 10^3 \text{ kg/m}^3$) to the crystalline form magnesium oxide (with density $3.54 \times 10^3 \text{ kg/m}^3$). On the other hand, silica bricks undergo permanent expansion. It is due to the transformation of quartz in silica to tridymite and cristoballite at higher temperatures.

15.30 ABRASIVES

These ceramic materials are used to wear, grind or cutaway other materials. Hence these materials has a very high hardness or wear resistance and toughness. Diamonds both natural and synthetic have the characteristics necessary for an abrasive material but they are expensive. The commonly used abrasives are silicon carbide, tungsten carbide, aluminium oxide (called corundum) and silica sand.

15.31 ADVANCED CERAMICS

Since the last 5 decades, there has been a tremendous progress in the development of ceramics. Most of the materials in this category have been developed because of their electrical, magnetic and optical properties. Although the advanced ceramics are utilized in many applications yet the following are important from the subject point of view:

1. in optical fibre communication systems
2. in micro-electro-mechanical systems (MEMS)
3. as ball bearings
4. in application that exploit piezoelectric behaviour of a number of ceramic materials.
5. Others.

Now we shall discuss these applications one by one in the following pages.

Composite Materials and Ceramics 305

15.32 OPTICAL FIBRES

Optical fibre is a critical component in modern optical communication systems. It is made of extremely high purity silica. Very advanced and sopishiticated processing techniques have been developed that meet the rigid restrictions required for application.

A light in the glass fibre can pass with great efficiency because of total internal reflection and no losses due to refraction to the surrounding environment. As a practical matter, commercial optical fibres are not bare glass. The glass is a core embedded in a cladding that is, inturn, covered by a coating as shown in Fig. 15.7.

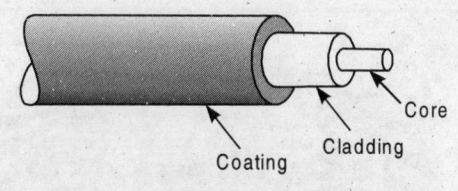

Fig. 15.7.

The light signals pass through the glass core. The cladding is lower index of refraction glass providing the phenomenon of total internal reflection. The coating protects the core and cladding from environmental damage.

The core is made from a high purity silica glass ranging in diameter from 5 μm to 100 μm (1 μm = 10^{-6} m). Fig. 15.8 shows three common configurations of an optical fibre.

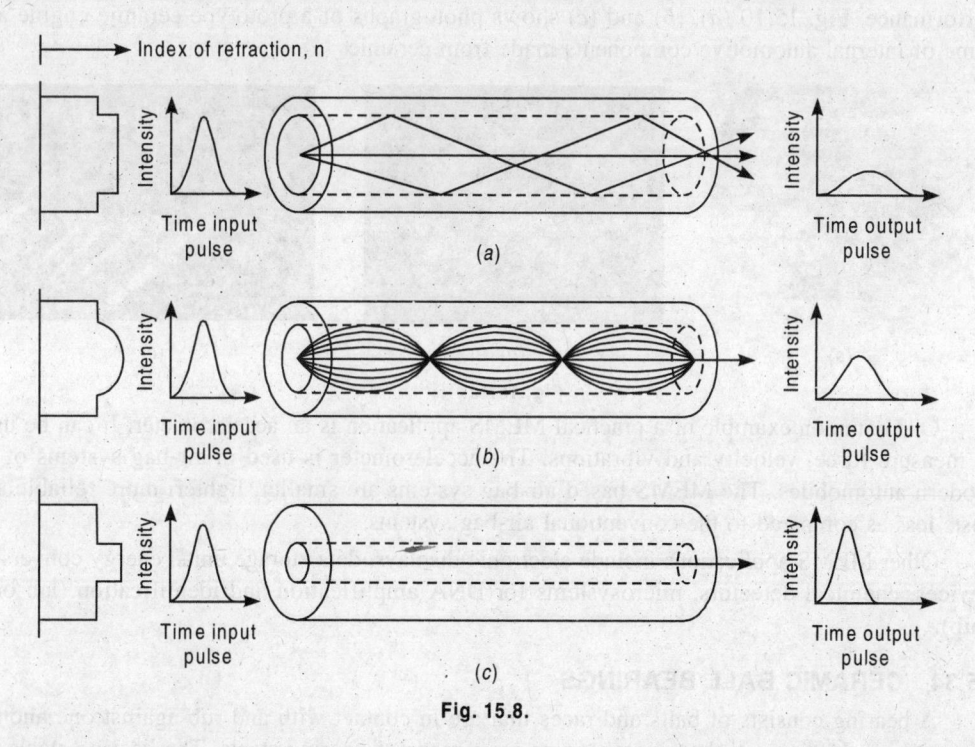

Fig. 15.8.

Fig. 15.8 (a) shows a step-index configuration, Fig. 15.8 (b), graded-index and Fig. 15.8 (c), single-mode optical fibre design. Notice the shape of input and output signals, as well as the transmission of optical signal within the optical fibre.

15.33 MICROELECTROMECHANICAL SYSTEMS (MEMS)

These are miniature "intelligent" systems consisting of several mechanical devices that are integrated with large number of electrical elements on a silicon substrate. The mechanical devices are microsensors, micromotors microgears, micropumps, microvalves etc. Fig. 15.9 shows a ceramic turbine in a millimeter range for a very small engine. This device is only 4 mm in diameter. The material used for the fabrication of this nano turbine is Sintered Silicon Carbide-silicon-nitride alloy. Notice that silicon has a limitation of low fracture toughness, a low softening temperature and is highly active to the presence of water and oxygen. Hence silicon is not a suitable material for MEMS.

Fig. 15.9.

The ceramic engine can offer better fuel economy, efficiency, weight savings and performance. Fig. 15.10 (a), (b) and (c) shows photographs of a prototype ceramic engine and some of internal automotive components made from ceramics

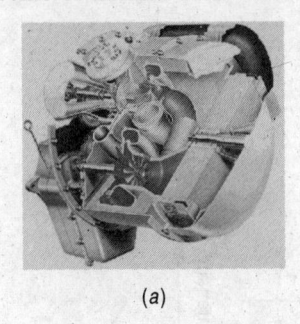

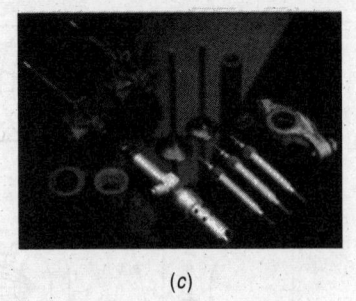

(a)   (b)   (c)

Fig. 15.10.

One common example of a practical MEMS application is an accelerometer. It can be used to measure force, velocity and vibrations. The accelerometer is used in air-bag systems of all modern automobiles. The MEMS based air-bag systems are smaller, lighter, more reliable and costs less as compared to the conventional air-bag systems.

Other MEMS applications include electronic displays, data storage units, energy conversion devices, chemical detectors, microsystems for DNA amplification and identification (lab on a chip).

15.34 CERAMIC BALL BEARINGS

A bearing consists of balls and races that are in contact with and rub against one another when in use. In the past, these components were made of bearing steels. The bearing steels are very hard, extremely corrosion resistant and may be polished to give a very smooth surface finish.

These days, silicon nitride (Si_3N_4) balls are replacing steel balls in most of the applications. It may be noted that races are still made of steel because its tensile strength is superior to that of silicon nitride. This combination of ceramic balls and steel races is known as a hybrid bearing.

Composite Materials and Ceramics

Following are some of the main advantages of hybrid bearings over the conventional steel bearings:

1. *Weight less.* Because of this the hybrid bearings operate at higher speed (20% to 40% higher).
2. *Lower levels of noise and vibration.* This is because of the fact that silicon nitride balls are more rigid and experience lower deformation when in use.
3. *Greater lifetime.* The hybrid bearings have 3 to 5 times greater life time. It is due to the higher hardness of silicon nitride than that of steel (about 30% high).
4. *Lower wear rate.* The coefficient of friction of silicon nitride is 30% of that of the steel. This leads to an increase in grease life.
5. *Higher corrosion resistance.* Because of this advantage, hybrid bearings are used in more corrosive environment and at higher operating temperatures.

15.35 PIEZOELECTRIC CERAMICS

As a matter of fact, a few ceramic materials exhibit the phenomenon of piezoelectricity. Whenever a piezoelectric ceramic is subjected to an external force, an electric voltage is produced across its two ends. If the direction of the external force is reversed, (*i.e.* from tension to compression), the polarity of the voltage is also reversed. Refer to Fig. 15.11 below.

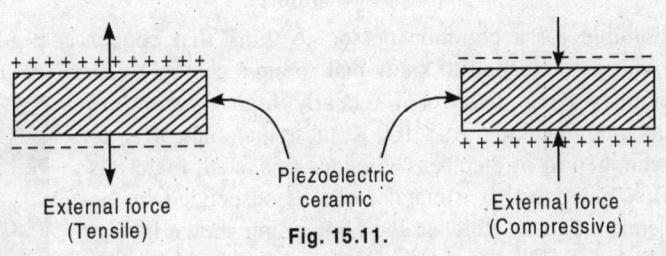

Fig. 15.11.

The piezoelectric ceramic also exhibits inverse piezoelectric effect, *i.e.* if an electric voltage is applied across the two ends of such a material, the mechanical strain is produced in it. Some of the commonly used piezoelectric ceramics are barium titanate, lead zirconate-titanate (PZT), lead titanate and sodium-potassium niobate. The piezoelectric ceramic materials are utilized as transducers. Such transducers are used in SONAR (sound navigation and ranging system). They help in detecting underwater objects (such as submarines). This is achieved by using an ultrasonic transmitting and receiving system. A piezoelectric crystal is caused to oscillate by an electrical signal. This produces high-frequency (> 20 kHz) mechanical vibrations that are transmitted through the water. Upon encountering an object, these signals are reflected back. Another piezoelectric ceramic crystal receives this reflected vibration signals and convert them back into an electrical signal. By measuring the time elapsed between the transmitting and receiving signals, distance from the ultrasonic source and reflecting body is determined.

These days piezoelectric ceramics are employed in many applications. Some of the important ones are :

1. *Automotive* – wheel balances, seat belt buzzers, treadwear indicators, keyless door entry and air-bag sensors.
2. *Computer* – microactuators for hard disks and notebook transformers
3. *Commercial/consumer* – ink-jet printing heads, strain gauges, ultrasonic welders and smoke detectors.
4. *Medical* – insulin pumps, ultrasonic therapy and ultrasonic cataract-removal devices.

15.36 OTHER ADVANCED CERAMIC MATERIALS

It will be interesting to know that quartz envelopes make light bulbs and other lamps possible. Some of the lamp applications are shown in Fig. 15.12 (a). Fig. 15.12 (b) shows a Quartz tubes fabricated from beach sand. The sand is produced into a quartz ingot. A rather large ingot used to produced furnace quartzware is also shown in Fig. 15.12 (c). General Electric Company of America produces quartz products in great quantity.

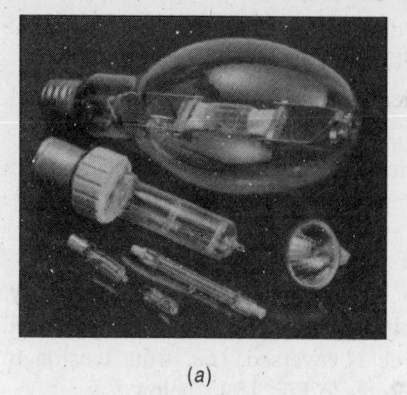

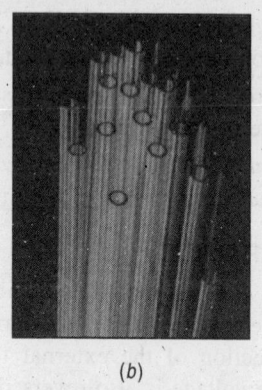

(a)   (b)   (c)

Fig. 15.12.

It is quite possible for a common person to think that copper is a good conductor of electricity. But it will be interesting to know that ceramic can be a better conductor of electricity than copper. Scientists have discovered recently high temperature superconducting ceramic materials. At ≤ 100 K, these materials offer no resistance to the conduction of electrons. Besides this, such materials reject magnetic flux lines so that a magnet can be suspended in the space above the semiconductor. Refer to the photograph shown in Fig. 15.13. In Japan, a high-speed levitated train is being developed based on this principle called "Meissner effect".

Graphites are known to be refractory, light-weight and corrosion resistance materials. Such properties are essential for many applications, e.g. dies for continuous casting, rocket nozzles and heat exchangers for the chemical industry. However, the relatively poor resistance of graphites to wear and oxidation limits there use. The addition of titanium carbide (TiC) coatings greatly extends the use of graphites. The TiC coatings possess excellent resistance to wear, oxidation and corrosion as well as having other desirable properties. Fig. 15.14 shows some of the TiC coated quartz parts.

Fig. 15.13.

Fig. 15.14.

Composite Materials and Ceramics

A.M.I.E. (I) EXAMINATION QUESTIONS

1. (a) What is a composite material ?
   (b) What are the ingradient of various FRP ? *(Summer, 1993)*
2. What are refractory materials ? Describe the use of such materials as insulating materials.
   (Winter, 1993)
3. What are composite materials ? Explain with examples. *(Summer, 1994)*
4. Discuss the factors affecting the dimensional stability of refractory materials. *(Winter, 1994)*
5. What are refractory materials ? State their basic properties and used. *(Summer, 1996)*
6. Explain the term 'Refractory'. Discuss the important properties of refractories. Give examples and applications. *(Winter, 1996)*
7. Distinguish between Fire-clay and sillimanite refractories. Discuss the importance of pure oxide refractories for modern furnaces. *(Winter, 2000)*

MULTIPLE CHOICE QUESTIONS

1. Ceramic material is a material made of a metal alloy
   (a) a metal alloy
   (b) an organic material
   (c) an inorganic material
   (d) cement and concrete
2. Optical fibres are made from
   (a) clay
   (b) non-ferrous metals
   (c) silica
   (d) refractory materials

ANSWERS

1. (c)   2. (c)

16

Semiconductors

1. Introduction. 2. Bonding of Semiconductors. 3. Classification of Semiconductors. 4. Intrinsic Semiconductors. 5. Extrinsic Semiconductors. 6. N-type semiconductor. 7. P-type Semiconductors. 8. Expression for Conductivity of a semiconductor. 9. P-N Junction. 10. Application of Voltage Across a P-N Junction. 11. Flow of Current in Forward Biased P-N Junction. 12. Volt-Ampere (V-I) Characteristics of a P-N Junction. 13. V-J Characteristic of a Forward Biased P-N Junction. 14. V-I Characteristic of a Reverse Biased P-N Junction. 15. Semiconducting Materials. 16. Semiconductor Devices. 17. Semiconductor Diodes. 18. Zener Diode. 19. Transistors. 20. Types of Transistors. 21. N-P-N Transistors. 22. P-N-P Transistors. 23. Silicon Controlled Rectifier (SCR). 24. Thermistors. 25. Varistors. 26. Integrated Circuits. 27. Photo-cells. 28. Laser.

16.1 INTRODUCTION

We have already discussed in Art. 5.13 that the materials, whose electrical conductivity lies in between those of conductors and insulators, are known as semiconductors. We have also discussed in Art. 5.18 the energy band diagram of a semiconductor. It has been observed, from the energy band diagram, that the semiconductors have almost an empty conduction band and almost filled valence band. These bands are separated by a small energy gap (of the order of 1 to 2 eV).

It will be interesting to know that at O K, there is no electron in the conduction band of a material, whereas the valence band is completely filled. However, if the temperature is increased, some of the covalent bonds break. As a result of this, some electrons, jump from the valence band into the conduction band. This results in the flow of current through the semiconductor. This shows that the electrical conductivity of a semiconductor increases with the increase in temperature. The common examples of semiconductors are germanium, silicon, indium, antimonide, gallium, arsenide etc. These materials are widely used in the fabrication of electronic devices such as diodes, transistors, silicon controlled rectifiers, photocells etc.

16.2 BONDING OF SEMICONDUCTORS

It has been observed that the semiconductors like germanium and silicon have a crystalline structure. Both these materials, are tetravalent (*i.e.,* their atoms have four valence electrons). In order to acquire a stable electronic configuration, each atom shares its four electrons with four neighbouring atoms and forms a covalent bond. The cross-section of germanium crystal lattice is shown in Fig. 16.1

In this figure, the circles represent the atom cores. These cores consists of nuclei and the inner '28' electrons. A pair of line represents a covalent bond and a dot represents the valence electrons.

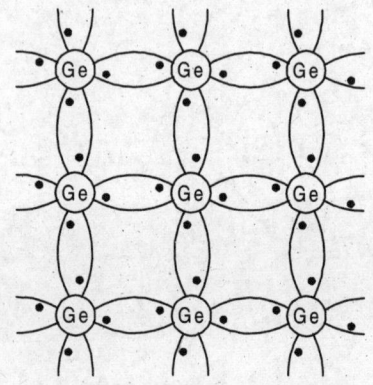

Fig. 16.1. Bonding in semiconductors.

Semiconductors

16.3 CLASSIFICATION OF SEMICONDUCTORS

The semiconductors may be broadly classified into the following two types:

1. Intrinsic semiconductors.   2. Extrinsic semiconductors.

16.4 INTRINSIC SEMICONDUCTORS

A semiconductor, which is made of its extremely pure form, is known as an intrinsic semiconductor. The examples of such semiconductors are pure germanium and silicon.

We have already discussed in Art. 16.1 that in semiconductor, there exists a small energy gap between the valence and conduction bands as shown in Fig. 16.2. It has been observed that the value of this gap is 0.72 eV for germanium and 1.1 eV for silicon, A little consideration will show, that this energy gap is so small that even at room temperature, some electrons jump from valence band to conduction band. However, for each electron liberated from the valence to the conduction band, a vacant site is created in the valence band. This vacant site is called 'hole'. This hole behaves like a positively charged particle. It has been observed that whenever an electric field is applied at room temperature, the electrons move in a direction opposite to the direction of the electric field, while the holes moves in the same direction as that of the electric field. Thus, in a semiconductor, the current consists of two components. The first component, of current, is due to the movement of electrons and second due to the movement of holes in the opposite direction.

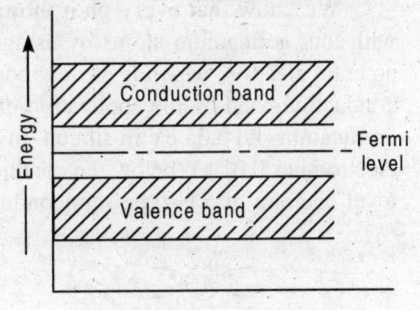

Fig. 16.2. Energy band diagram of intrinsic semiconductor.

In an intrinsic semiconductor, the covalent bonds break up due to the thermal agitation of electrons between them. As a result of this, new electron-hole pairs are produced continuously. But at the same time, other electron-hole pairs disappear due to the *recombination.

It will be interesting to know that at any given temperature, the number of electrons present in the conduction band is equal to the number of holes present in the valence band. This there exists an equilibrium between the concentration of holes and electrons. The energy level, which corresponds to this equilibrium position is known as fermi level. This is an intrinsic semiconductors, fermi level lies in the middle of energy gap as shown in Fig. 16.2

16.5 EXTRINSIC SEMICONDUCTORS

An intrinsic semiconductor, having extremely small amount (about I part in 10^8 parts) of suitable impurity is known as extrinsic semiconductor. The process of adding impurity is known as doping and the impurity as doping agent. Usually, the impurities are either of pentavalent atoms (*i.e.,* atoms containing five valence electrons) or of trivalent atoms (*i.e.,* atoms containing three valence electrons). The examples of pentavalent impurities are arsenic, antimony, phosphorus etc. And the examples of trivalent impurities are gallium, indium, boron etc. It has been observed that the addition of an impurity increases the conductivity by several powers of ten.

It will be interesting to know that the pentavalent impurity is known as donor atom, as it donates one electron to the conduction band of the intrinsic semiconductor. On the other hand, a trivalent impurities is known as acceptor atom, as it accepts an electron from the intrinsic

---

* It is the process by which conduction band electrons fall into the holes lying in the valence band. As a result of this, the holes in the valence band disappear, and electrons which fall into the holes become valence electrons.

semiconductor. The extrinsic semiconductors may be further sub-divided into the following two types depending upon the type of impurity of the doping agent used:

    1. N-type semiconductors   2. P-type semiconductors.

16.6 N-TYPE SEMICONDUCTORS

An intrinsic semiconductor, which is doped by a pentavalent impurity like arsenic, antimony, phosphorus etc. is known as a N-type semiconductor. The cross-section of a N-type semiconductor crystal lattice, with phosphorus as impurity in germanium crystal lattice is shown in Fig. 16.3 (a).

We know that every phosphorus atom has five valence electrons. It forms covalent bond with four germanium atoms by using its for valence electrons only. The fifth valence electron is an extra electron, which is loosely bonded to the nucleus of atom. This electron can jump into the conduction band by the application of very little energy. The amount of this energy is 0.01 eV in germanium and 0.05 eV in silicon. In other words, the energy level (called donor level) of the fifth electron lies 0.01 eV below the conduction band for germanium of 0.05 eV for silicon. The energy level diagram of a N-type semiconductor is shown in Fig. 16.3 (b).

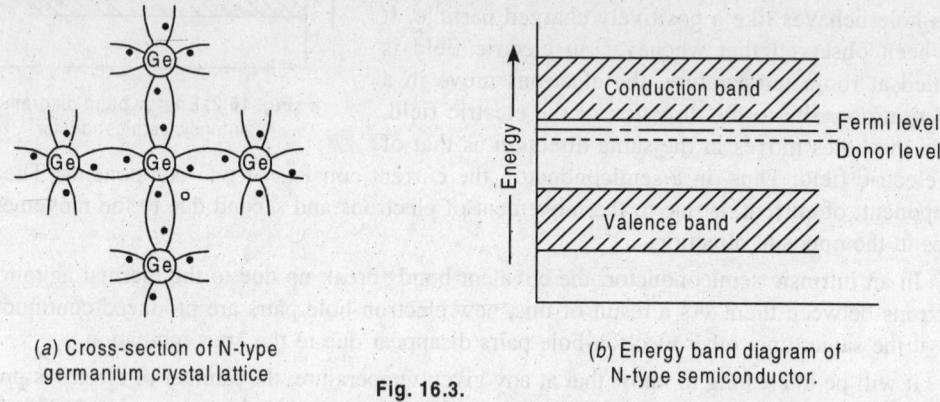

(a) Cross-section of N-type germanium crystal lattice

(b) Energy band diagram of N-type semiconductor.

Fig. 16.3.

It will be interesting to know from the above mentioned energy level diagram, that the fermi level has shifted upwards from the centre. It is because of the fact, that the addition of phosphorus atoms greatly increases the number of electrons in the conduction band. It has been observed that the addition of phosphorus decreases the number of holes below that which would be available in the intrinsic semiconductor. This is because of the reason that larger number of present electrons increase the rate of recombination of electrons with holes.

As a matter of fact, when the pentavalent atom gives away its fifth valence electron (as mentioned above) it becomes a positively charged ion. This ion is bonded by four covalent bonds in the germanium crystal lattice. As a result of this, the pentavalent ion does not contribute to the conduction of electricity. Thus in N-type semiconductors, the current flows due to the movement of electrons and holes. But a major part of the current flows due to the movement of electrons only. Hence in N-type semiconductors, electrons are known as majority carriers and holes as minority carriers.

16.7 P-TYPE SEMICONDUCTORS

An intrinsic semiconductor, which is doped by a trivalent impurity like gallium, indium aluminium etc., is known as a P-type semiconductor. The cross-section of P-type semiconductor crystal lattice with aluminium as impurity in germanium crystal lattice as shown in Fig. 16.4 (a).

We know that every aluminium atom has three valence electrons. It forms covalent bonds with for germanium atoms. But one bond remains incomplete. This gives rise to a hole in the

(a) Cross-section of P-type germanium crystal lattice

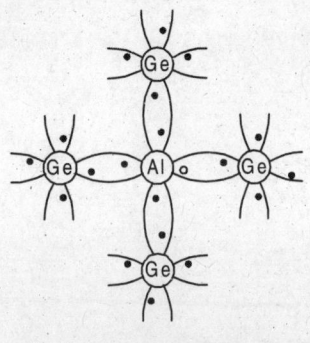

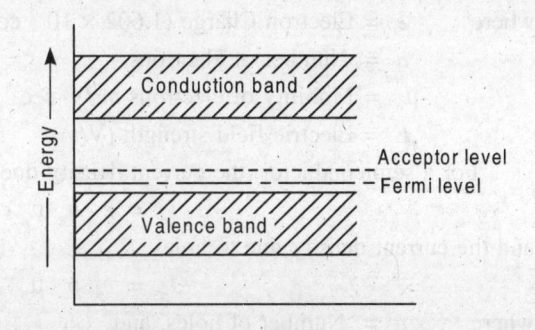

(b) Energy band diagram of a P-type semiconductor.

Fig. 16.4.

crystal lattice. Now the aluminium atom seeks its surrounding atoms to acquire the fourth electron. It does so by taking an advantage of thermal motion that brings one electron from the surrounding atoms. Thus an electron, which is in a favourable position, is captured by the aluminium atom. After capturing the electrons, the aluminium atom establishes covalent bond with the four nearest germanium atoms and becomes an immobile ion. The energy involved in capturing an electric is very small. It has been observed that the amount of energy is 0.01 eV in germanium and 0.05 eV in silicon. In other words, the energy level (called acceptor level) of the fourth electron lies 0.01 eV above the valency band for germanium and 0.05 eV above the valence band for silicon. The energy level diagram of a P-type semiconductor is shown in Fig. 16.4 (b).

It will be interesting to know from the above mentioned energy level diagram, that the fermi level has shifted downwards from the centre. It is because of the fact that the addition of aluminium atom greatly increases the number of holes in the valence band. It has been observed that the addition of aluminium decreases the number of electrons below that which would be available in the intrinsic semiconductor. This is because of the reason that larger number of present holes increases to rate of recombination of holes with the electrons.

As a matter of fact, when the trivalent atom accepts an electron (as mentioned above) it becomes a negatively charged ion. The ion is bonded by four covalent bonds in the germanium crystal lattice. As a result of this, the trivalent ion cannot contribute to the conduction of electricity. Thus in P-type semiconductors, the current flow due to the movement of electrons and holes. But a major part of the current flows due to the movement of holes only. Hence in P-type semiconductors, the holes are known as majority carriers and the electrons as minority carriers.

16.8 EXPRESSION FOR CONDUCTIVITY OF A SEMICONDUCTOR

There is one fundamental difference between all the metals and semiconductors. The difference is that in metals. the current flows due to charge carriers (i.e., electrons) of one sign only, whereas in semiconductors, the current flows due to charge carriers (i.e., electrons and holes) of opposite signs. The magnitude of charge on the electrons and holes is equal, but it is negative for electrons and positive for holes. When an electric field is applied to a semiconductor, the holes move in the direction of the field whereas the electrons move in the opposite direction as shown in Fig. 16.5. The total current in a semiconductor is the sum of hole current and electron current.

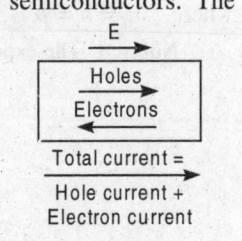

Fig. 16.5. Total current in a semiconductor.

We know that current density, which results due to the movement of electrons in a metal as derived in chapter 7 is given by the relation.

$$J = e \cdot n \cdot \mu_n \cdot E \ A/m^2$$

where e = Electron Charge (1.602×10^{-9} coulomb)

n = Number of electrons,

μ_n = Mobility of electrons m^2/V–sec

E = Electric field strength (V/m)

For a semiconductor, the current density due to electron

$$J_n = e \cdot n \cdot \mu_n \cdot E$$

and the current density due to holes.

$$J_p = e \cdot p \cdot \mu_p \cdot E$$

where p = Number of holes, and

μ_p = mobility of holes.

∴ Total current density in a semiconductor,

$$J = J_n + J_p = e \cdot n \cdot \mu_n \cdot E + e \cdot p \cdot \mu_p \cdot E$$
$$= e \cdot (n \cdot \mu_n + p \cdot \mu_p) E \qquad ...(i)$$

If we substitute σ (called conductivity) for $e(n \cdot \mu_n + P \cdot \mu_p)$ in the above equation, then

$$J = \sigma \cdot E \qquad ...(ii)$$

The equation (*ii*) may be recognised as another form of Ohms Law, which states that conduction current density, in a semiconductor, is proportional to the applied electric field. Thus conductivity of a semiconductor,

$$\sigma = e(n \cdot \mu_n + p \cdot \mu_p)$$

The above equation is a general expression for the conductivity of a P-type or N-type semiconductor. A more simple form for the conductivity expression may be obtained as follows:

If the number of holes (*p*) in a semiconductor is very large as compared to the number of electrons (*n*) as in a P-type semiconductor, then the total current in a semiconductor is mainly due to the holes only. Thus if $P \gg n$, then conductivity of a semiconductor is given by the relation.

$$\sigma = e \cdot p \cdot \mu_p \qquad \text{... (For P-type semiconductor)}$$

If the number of electrons (*n*) in a semiconductor is very large as compared to number of holes (*p*) as in a N-type semiconductor, then total current in a semiconductor is mainly due to the electrons only. Thus if $p \gg n$, then conductivity of a semiconductor is given by the relation

$$\sigma = e \cdot n \cdot \mu_n \qquad \text{... (For N-type semiconductor)}$$

We know that in an intrinsic semiconductor, the number of holes and electrons are equal. Therefore conductivity for an intrinsic semiconductor,

$$\sigma = e \cdot n_i (\mu_n + \mu_p)$$

where $n_i = n = p$ is the intrinsic concentration.

Notes: 1. The expression for resistivity of a semiconductor may be written as follows:

$$\rho = \frac{1}{\sigma} = \frac{1}{e(n\mu_n + p\mu_p)} \qquad \text{.... (for any semiconductor)}$$

$$= \frac{1}{e \cdot p \cdot \mu_p} \qquad \text{.... (for P-type semiconductor)}$$

$$= \frac{1}{e \cdot n \cdot \mu_n} \qquad \text{... (for N-type semiconductor)}$$

Semiconductors

2. For intrinsic semiconductor the resistivity,

$$\rho_i = \frac{1}{\rho_i} = \frac{1}{e \cdot n_i \cdot (\mu_n + \mu_p)} \quad \text{... (for intrinsic semiconductor)}$$

Example 16.1 *Determine the concentration of conduction electrons per cm^3 in pure silicon, if its conductivity is 5×10^{-4} Ω/m, electron mobility is $0.14 m^2/V$-sec and hole mobility is 0.05 m^2/V-sec.*

Solution. Given : $\sigma_i = 5 \times 10^{-4}$ Ω/m : $\mu_n = 0.14$ m^2/V-sec and $\mu_p = 0.05$ m^2/V-sec.

Let n_i = concentration of conduction electrons per cm^3 in pure silicon.

We know that conductivity (σ_i),

$$5 \times 10^{-4} = n_i \cdot e(\mu_n + \mu_p) = n_i \times 1.602 \times 10^{-19} \times (0.14 + 0.05)$$
$$= 0.304 \times 10^{-9} \, n_i$$

$$\therefore \quad n_i = \frac{5 \times 10^{-4}}{0.304 \times 10^{-19}} = 1.64 \times 10^{16}/m^3$$

$$= 1.64 \times 10^{22}/cm^3 \textbf{ Ans.}$$

Example 16.2 *The electron and hole mobilities in In-Sb are 6 and $0.2 m^2/volt$-sec respectively. At room temperature (300 K) the resistivity of In Sb in 2×10^{-4} ohm-m. Assuming that the material is intrinsic, determine its intrinsic carrier density at 300 K.*

Solution : Given : $\mu_n = 6 m^2/$ volt-sec ; $\mu_p = 0.2$ $m^2/$ volt-sec and $\rho_i = 2 \times 10^{-4}$ ohm-m.

Let n_i = Intrinsic carrier density (or concentration).

We know that resistivity of an intrinsic semiconductor (ρ_i).

$$2 \times 10^{-4} = \frac{1}{e \cdot n_i (\mu_n + \mu_p)} = \frac{1}{1.602 \times 10^{-19} \times n_i (6 + 0.2)}$$

$$n_i = \frac{1}{1.602 \times 10^{-19} \times (6 + 0.2) \times (2 \times 10^{-4})}$$
$$= 5.03 \times 10^{21} \, /m^3 \textbf{ Ans.}$$

Example 16.3 *How much N-type impurity is necessary to explain the conductivity of 10^{-12} mhos/m in diamond. If the mobility is 0.18 m^2/V-sec for electrons. Neglect the intrinsic conductivity. Take $e = 1.6 \times 10^{-19}$ C.*

Solution. Given : $\sigma = 10^{-12}$ mhos/m and $\mu_n = 0.18$ m^2/V-sec and $e = 1.6 \times 10^{-19}$ C.

Let N = Number of atoms required to produce 'n' electrons in diamond.

We know that conductivity of N-type material (σ),

$$10^{-12} = e \cdot n \cdot \mu_n = 1.6 \times 10^{-19} \times n \times 0.18 = 0.288 \times 10^{-19}$$

$$\therefore \quad n = \frac{10^{-12}}{0.288 \times 10^{-19}} = 3.47 \times 10^{7}/m^3$$

Now assuming every atom of an N-type impurity produces one electron in the diamond, then number of impurity atoms (N) will be equal to the number of conduction electrons (n) *i.e.*,

$$N = n = 3.47 \times 10^{7} /m^3 \textbf{ Ans.}$$

Example 16.4. *An N-type semiconductor has a resistivity of 20×10^{-2} ohm-m. The mobility of electrons through a separate experiment was found to be 100×10^{-4} m^2/V-sec. Find the number of electron carriers per m^3.*

Solution. Given: $\rho = 20 \times 10^{-2}$ ohm-m and $\mu_n = 100 \times 10^{-4}\ m^2 \cdot V^{-1}\ S^{-1}$

Let n = Number of electron carriers m^3

We know that for an N-type of semiconductor the resistivity/(ρ),

$$20 \times 10^{-2} = \frac{1}{e \cdot n \cdot \mu_n} = \frac{1}{1.602 \times 10^{19} \times n \times 100 \times 10^{-4}}$$

$$\therefore \quad n = \frac{1}{1.602 \times 10^{-19} \times n \times 100 \times 10^{-4} \times 20 \times 10^{-2}}$$

$$= 3.13 \times 10^{21}\ /m^3\ \textbf{Ans.}$$

Example 16.5 *A rod of P-type germanium, 10 mm long and 1 mm diameter, has a resistance of 100 ohm. What is the concentration of impurity in this rod ? Given mobility of holes in germanium. = 0.19 m^2 / V-sec.*

Solution. Given: $l = 10$ mm $= 10 \times 10^{-13}$ m ; $d = 1$ mm $= 1 \times 10^{-3}$ m ; $R = 100\ \Omega$ and $\mu_p = 0.19\ m^2$ V-sec.

Let N = concentration of impurity in the rod.

We know that area of cross-section of the rod,

$$A = \frac{\pi}{4}(d)^2 \times (1 \times 10^{-3})^2 = 7.854 \times 10^{-7}\ m^2$$

and resistivity,

$$\rho = \frac{R.A}{l} = \frac{100 \times 7.854 \times 10^{-7}}{10 \times 10^{-3}} = 7.854 \times 10^{-3}\ \Omega\text{-m}$$

We also know that resistivity of a P-type germanium, (ρ),

$$7.854 \times 10^{-3} = \frac{1}{e \cdot n \cdot \mu_p} = \frac{1}{1.602 \times 10^{-19} \times n \times 0.19}$$

$$\therefore \quad n = \frac{1}{1.602 \times 10^{-19} \times 0.19 \times 7.854 \times 10^{-3}}$$

$$= 4.18 \times 10^{21}\ /m^3$$

Assuming every impurity atom produces one electron in the germanium, and neglecting intrinsic concentration of germanium, the concentration of impurity,

$$N = n = 4.18 \times 10^{21}/m^3\ \textbf{Ans.}$$

Example 16.6 *Determine the intrinsic carrier density of pure silicon. Its resistivity at room temperature is 3000 ohm/m. The mobility of electrons and holes in silicon at room temperature 0.14 and 0.05 m^2/V-sec. Take Electron charge, (e)=1.602 $\times\ 10^{-19}$ C. (A.M.I.E, Summer, 1993)*

Solution. Given : $\rho = 3000\ \Omega/m$: $\mu_n = 0.14\ m^2/V$-sec ; $\mu_p = 0.05\ m^2$ /V-sec and $e = q = 1.60 \times 10^{19}$ C.

Let n_i = Intrinsic carrier density.

We know that conductivity,

$$\sigma = \frac{1}{\rho} = \frac{1}{3000} = 3.33 \times 10^{-4}\ \Omega^{-1}\ m$$

We also know that conductivity, (σ)

$$3.33 \times 10^{-4} = q \cdot n_i (\mu_i + \mu_\rho) = 1.602 \times 10^{-19} \times n_i \times (0.14 + 0.05)$$
$$= 0.304 \times 10^{-19} n$$
$$\therefore \quad n_i = \frac{3.33 \times 10^{-4}}{0.304 \times 10^{-19}} = 10.95 \times 10^{15} /m^3 \textbf{ Ans.}$$

Example 16.7. *The energy gap of Si is 1.1 eV. Its electron and hole mobilities at room temperature are 0.48 and 0.013 $m^2V^{-1}s^{-1}$ respectively. Evaluate its conductivity.*

(Anna Univ., April 2002)

Solution. Given: $\mu_n = 0.48$ m^2 V^{-1} s^{-1}; $\mu_p = 0.013$ m^2 V^{-1} s^{-1}.

Let us assume the intrinsic concentration for silicon.
$$n_i = 5.021 \times 10^{15} \text{ m}^{-3}$$

We know that conductivity,
$$\sigma = e \cdot n_i (\mu_n + \mu_p)$$
$$= (1.602 \times 10^{-19}) \times (5.021 \times 10^{15}) \times (0.48 + 0.013)$$
$$= 3.966 \times 10^{-4} \ \Omega^{-1}\text{-m}^{-1} \textbf{ Ans.}$$

16.9 P-N JUNCTION

When a single piece of semiconductor is doped in such a manner that its one half portion is P-type and the other half is N-type, the plane dividing the two halves is known as P-N junction as shown in Fig. 16.6.

We have already discussed in the last article that in a P-type semiconductor, holes are majority carriers. And in N-type semiconductors, the electrons are majority carriers. We know that like charges repel each other and unlike charges attract each other.

A little consideration will show that as the charges in P-type and N-type semiconductors are unlike, therefore the electrons cross over the junction from N-type to P-type regions. It has been observed that these electrons have a short life span because of the fact that a large number of holes are present around them. The electrons recombine with the holes and become valence electrons. As the electron moves away from its parent atom, it creates a positive charge on its parent atom in the crystal lattice. After this charge, the parent atom also becomes positive ion or donor atom. Similarly, the holes also cross over the junction from P-region to N-region and they recombine themselves with the electrons there. The movement of holes, in the P-region creates a negative charge on its parent atom. After this charge, the parent atom becomes a negative ion or acceptor atom. As the electrons and holes continue to cross over the junction, the positive and negative ions form a thin layer on either side of the P-N junction as shown in Fig. 16.7 (a). This thin layer is also known as depletion layer.

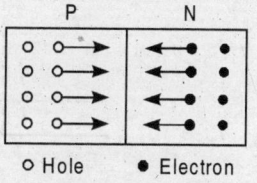

Fig. 16.6. P-N junction.

In this figure, the circles with negative sign in the depletion layer represent acceptor ions and the circles with positive sign represent donor ions. It may be noted that there are no free or mobile charge carriers in the depletion layer. But it contains only immobile positive and negative ions. It has been observed that an electric potential is developed across the P-N junction due to charge separation on its either side. This electric potential is known as barrier pentential and is represented by the symbol V_B as shown in Fig. 16.7 (b). The

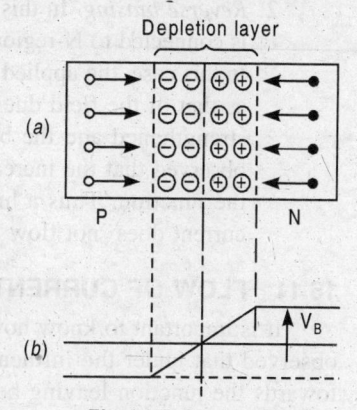

Fig. 16.7. Barrier potential.

value of carrier potential at room temperature is about 0.3 V for germanium and 0.7 V for silicon. It has been observed that there is no further diffusion of electrons or holes across the junction, due to the barrier potential, unless the voltage or potential difference is applied across the P-N junction.

16.10 APPLICATION OF VOLTAGE ACROSS A P-N JUNCTION

The voltage or potential difference across a P-N junction may be applied in any of the following two methods:

    1. Forward biasing.   2. Reverse biasing.

    1. *Forward biasing.* In this method, the positive terminal of the voltage source (*i.e.,* battery is connected to P-region and negative terminal to N-region as shown in Fig. 16.8 (*a*). The applied potential difference establishes an electric field, which acts against the field due to the barrier potential. As a result of this, the resultant field is weakened and the barrier height (V_B) decrease as shown in the figure. As the value of barrier potential

(a) Forward biasing   (b) Reverse biasing

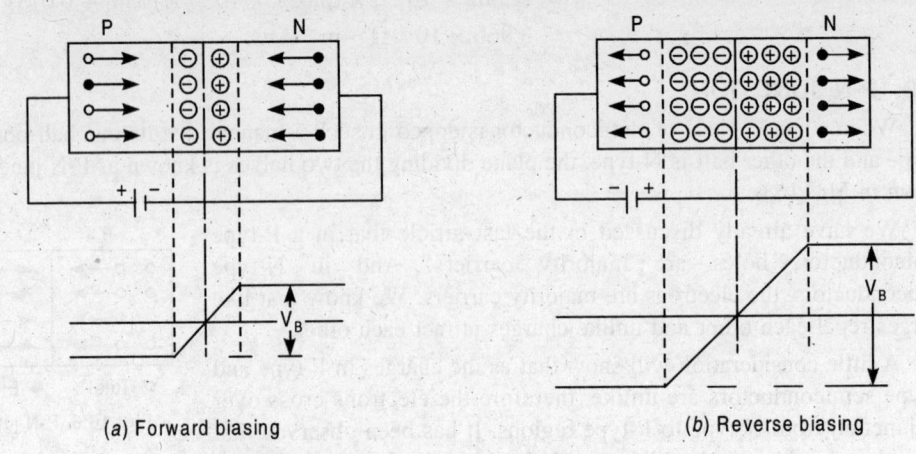

Fig. 16.8. Application of a voltage across a P-N junction.

is very small (0.3 V to 0.7 V), therefore a small forward voltage is sufficient to completely eliminate the barrier. Once the potential barrier is eliminated, by the forward voltage, the junction resistance becomes almost zero. Now a very low resistance path is established for the entire circuit and the current starts flowing in it. This current is known as forward current whose magnitude depends upon the applied voltage.

    2. *Reverse biasing.* In this method, the positive terminal of the voltage source (*i.e.,* battery) is connected to N-region and the negative terminal to P-region as shown in Fig. 16.8 (*b*). In this case, the applied reverse voltages establishes an electric field in the same direction as that of the field due to barrier potential. As the same direction as that of the field is strengthened and the barrier height (V_B) increases as shown in the figure. It has been observed that the increased barrier potential prevents the flow of charge carriers across the junction. Thus a high resistance path is established, for the entire circuit, and the current does not flow through the junction.

16.11 FLOW OF CURRENT IN FORWARD BIASED P-N JUNCTION

It is important to know how the current flows in a forward biased P-N junction. It has been observed that under the influence of forward voltage, the free electrons in the N-region move towards the junction leaving behind positively charged atoms. The positions, vacated by these electrons, are occupied by electrons, which arrive from the negative terminal of the battery. As the

Semiconductors

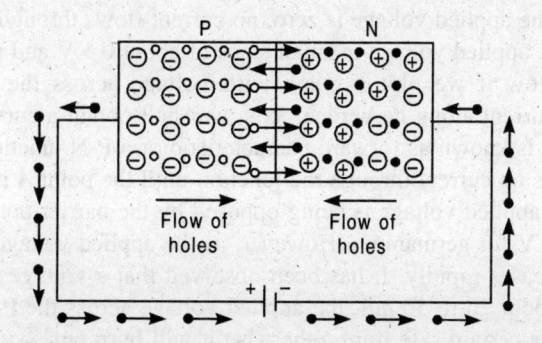

Fig. 16.9. Current flow in a forward biased P-N terminal.

free electrons cross the junction, they recombine themselves with holes and become valence electrons. It has been observed that these valence electrons move through the P-region towards left end of the crystal and then enter the positive terminal of the battery as shown in Fig. 16.9.

Similarly, the holes move from the P-region towards P-N junction because they are repelled by the positive terminal of the voltage source. As the holes cross the junction, they recombine with free electrons and become valence electrons. Therefore in a forward biased P-N junction, the current flows due to the movement of electrons from N-region to P-region and the movement of holes from P-region to N-region.

16.12 VOLT-AMPERE (V-I) CHARACTERISTICS OF A P-N JUNCTION

The Volt-Ampere (V-I) characteristics of a P-N junction is a curve between the voltage across the current through it. Usually, the voltage is plotted along the horizontal axis and the current along the vertical axis. The V-I characteristics of a P-N junction may be discussed under the following two heads :

1. V-I characteristic of a forward biased P-N junction.
2. V-I characteristic of a reverse biased P-N junction.

16.13 V-I CHARACTERISTIC OF A FORWARD BIASED P-N JUNCTION

Fig. 16.10 (*a*) shows the circuit arrangement for obtaining the V-I characteristic of a forward biased P-N junction. In this circuit, the P-N junction is connected to a d.c. battery (*B*), through a potentiometer (*L*) and a resistance (*R*). The potentiometer (*L*) helps in varying the voltage applied across the P-N junction, while the resistance (*R*) limits the current through it. A voltmeter (*V*) and a milliameter (*mA*) is connected to measure the current and voltage respectively through the P-N junction.

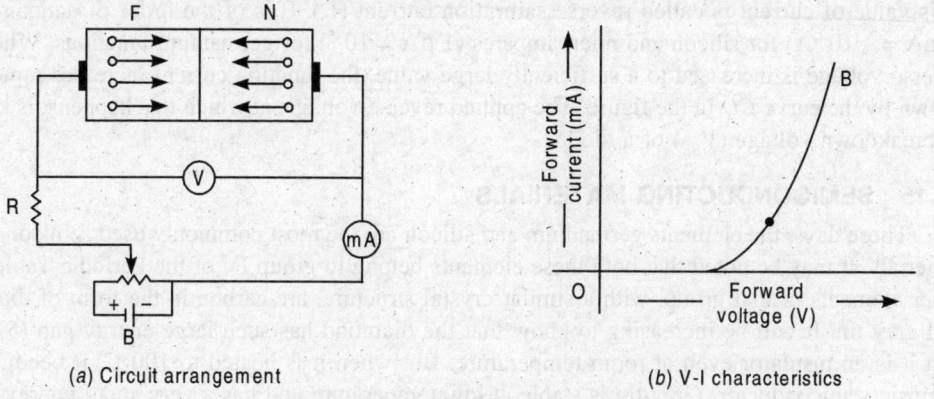

(*a*) Circuit arrangement (*b*) V-I characteristics

Fig. 16.10. Forward biased P-N junction.

Initially, when the applied voltage is zero, no current flows through the P-N junction. Let us gradually increase the applied voltage in small steps of about 0.1 V and record and corresponding values of current. Now if we plot a graph with voltage across the P-N junction along the horizontal axis and current along he vertical axis, we shall obtain a curve OAB as shown in Fig. 16.10 (b). This curve is known as forward characteristic of a P-N junction. It may be noted from this curve that there is no current through the junction until the point A is reached. It is due to the fact that the external applied voltage is being opposed by the barrier potential whose value is 0.7 V for silicon and 0.3 V for germanium. However, as the applied voltage is further increased, the junction current increases rapidly. It has been observed that a voltage of about 1 V produces a junction current of about 20 to 50 mA, the applied voltage across the P-N junction should never be increased beyond a certain safe limit, otherwise it will burn out.

16.14 V-I CHARACTERISTICS OF A REVERSE BIASED P-N JUNCTION

The circuit arrangement for obtaining the V-I characteristic of a reverse biased P-N junction is as shown in Fig. 16.11 (a). This circuit is similar to that shown in Fig. 16.10 (a) except two changes namely the battery terminals are reversed and the miliameter is replaced by a microammeter. Initially, when the applied voltage is zero, no current flows through the junction. The applied reverse voltage is increased above zero in suitable steps and the values of junction current are recorded at each step. Now if we plot a graph with reverse voltage along the horizontal axis and the junction current along the vertical axis, we shall obtain a curve OCD as shown is Fig. 16.11 (b). It is known as reverse characteristic of a P-N junction.

(a) Circuit arrangement (b) V-I characteristics

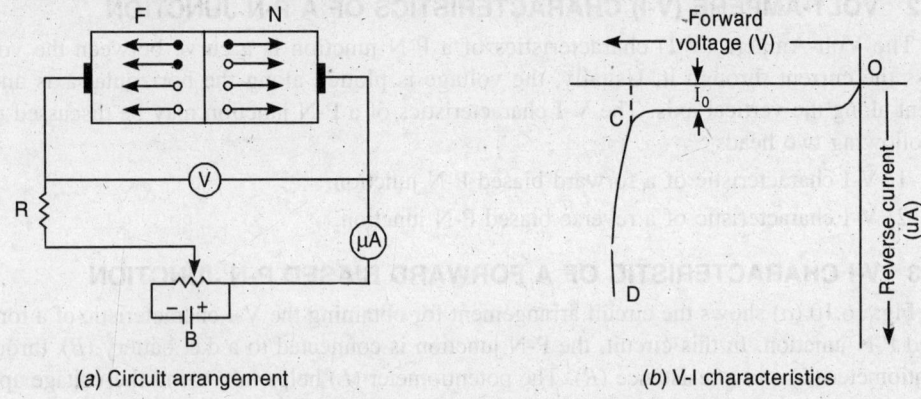

Fig. 16.11. Reverse biased P-N junction.

It may be noted from this characteristic curve that when the applied reverse voltage is below the breakdown voltage (V_{BR}) the current through the P-N junction is small and remains constant. This value of current is called reverse saturation current (I_0). It is of the order of nanoamperes (1 nA = -10^{-9}A) for silicon and microamperes (1 μA = 10^{-6}) for germanium junctions. When the reverse voltage is increased to a sufficiently large value, the junction current increases rapidly as shown by the curve CD in the figure. The applied reverse voltage, at which this happens, is known as breakdown voltage (V_{BR}) of a diode.

16.15 SEMICONDUCTING MATERIALS

These days, the elements germanium and silicon are the most commonly used semiconductor materials. It may be noted that both these elements belong to group IV of the Periodic Table. The other elements of that group, with a similar crystal structure, are carbon in the form of diamond and grey tin. It will be increasing to know that the diamond has such large energy gap (5.3 eV) that it is an insulator even at room temperature. But when it is heated to 100°C, it becomes an intrinsic semiconductor. Grey tin is stable at low temperature and has a very small energy gap.

Semiconductors

The elemental semiconductors such as a germanium and silicon consists of only one kind of atom. In addition to these, a large number of semiconductor crystals made of two or more different atoms are possible. Such compound semiconductors are important for devices, because they extend the range of properties available from the elemental semiconductors. A brief list of compounds, which are important and widely used in semiconductor devices alongwith their chemical symbols, forbidden gap and area of applications is given in Table 16.1

Table 16.1

S.No.	Element	Chemical symbol	Forbidden gap in eV	Applications
1.	Indium Antimonide	InSb	0.18	Infrared detectors
2.	Lead telluride	PbTe	0.33	–do–
3.	Lead sulphide	PbS	0.37	–do–
4.	Germanium	Ge	0.72	P-N junction devices
5.	Silicon	Si	1.1	–do–
6.	Gallium arsenide	GaAs	1.34	Tunnel diodes
7.	Cadmium telluride	CdTe	1.45	Photo-cells
8.	Cadmium sulphide	Cds	2.45	–do–

The manner is which the semiconducting properties arise in compounds can be understood from the indium antimonide, which contains indium atoms. They have three valence electrons and the antimony atoms have five valence electrons. It is thus obvious, that when they form a solid with 1 : 1 ratio of indium and antimony atoms, the average number of electrons per atom is four. This results in crystal properties, which are quite similar to those of germanium and silicon. It will be interesting to know that other semiconductors also form their compounds in the same way as that of indium antimonide.

16.16 SEMICONDUCTOR DEVICES

Though there are innumerable semiconductor devices, which are in use today, yet the following are important from the subject point of view:

1. Semiconductor diodes.
2. Zener diodes.
3. Transistors.
4. Silicon controlled rectifiers.
5. Thermistors.
6. Varistors.
7. Integrated circuits.
8. Photo-cells.
9. Lasers.

Now we shall discuss all the above mentioned semiconductor devices in the following pages.

16.17 SEMICONDUCTOR DIODES

We have already discussed P-N junction in Art. 16.9. As a matter of fact, the P-N junction is also known as semiconductor diode. The P-region of this device is known as anode, whereas the N-region as cathode.

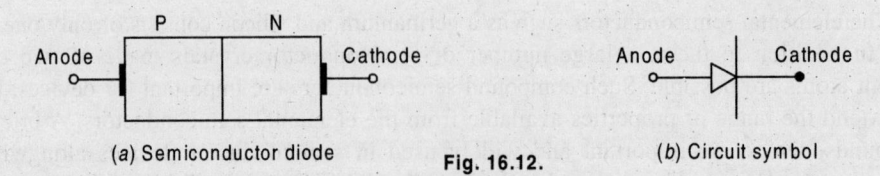

Fig. 16.12.
(a) Semiconductor diode
(b) Circuit symbol

A semiconductor diode, alongwith its circuit symbol, is shown in Fig. 16.12. The arrowhead, in the circuit symbol, indicates the direction of flow of the conventional current. The working of a semiconducting diode can be understood from the P-N junction. A semiconductor diode is widely used to convert an alternating current into direct current, to detect information in communication circuits, in logic circuits of computers etc.

16.18 ZENER DIODES

We have already discussed in Art. 16.14 that when the reverse bias on a P-N junction is increased to a critical voltage, its junction breaks down. At this stage, the P-N junction offers a very low resistance and the reverse saturation current rises very sharply. The critical voltage, at which the breakdown occurs, is known as breakdown voltage. It depends upon the amount of doping of P and N-regions. It has been observed that if the diode is heavily doped, the depletion layer will be thin. As a result of this, the breakdown of the junction will occur at a lower reverse voltage. On the other hand, if the diode is lightly doped, the breakdown of the junction occurs at a higher reverse voltage. When an ordinary semiconductor diode is doped in such a manner that it has a sharp breakdown zener diode. The zener diodes are widely used in voltage stabilizers, peak clippers etc.

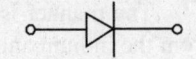

Fig. 16.13. Zener diode.

16.19 TRANSISTORS

A transistor is a device which consists of two P-N junctions J_1 and J_2 formed by sandwitching either P-type or N-type semiconductor between a pair of opposite types as shown in Fig. 16.14 (a) and (b).

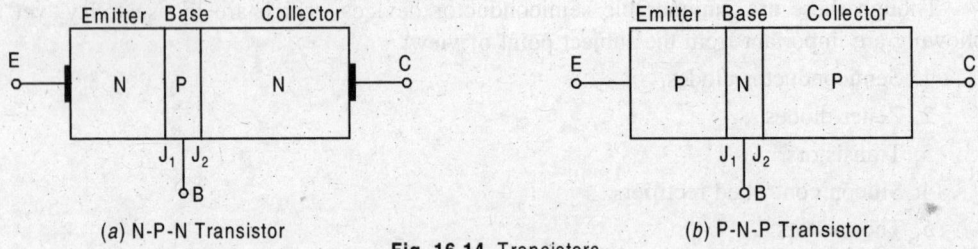

(a) N-P-N Transistor
(b) P-N-P Transistor

Fig. 16.14. Transistors.

A transistor has, essentially, three regions known as emitter, base and collector. All these three regions are provided with terminals, which are labelled as E (for emitter), B (for base) and C (for collector) respectively. A brief description of the above regions is discussed below:

1. *Emitter.* It is a region situated in one side of the transistor, which supplies charge carriers (*i.e.*, electrons for holes) to the other two regions.

2. *Base.* It is the middle region that forms P-N junctions in the transistor. The base of a transistor is very thin as compared to the emitter.

3. *Collector.* It is a region situated in the other side of transistor (*i.e.*, the side opposite to the emitter), which collects charge carriers (*i.e.*, electrons or holes). The collector of a transistor is always larger than the emitter and base of a transistor.

Semiconductors

As a matter of fact, the transistor has two P-N junctions J_1 and J_2 as shown in Fig. 16. 14 (a) and (b). The junction J_1 is a junction between emitter and base regions. Thus it is known as emitter-base junction. Similarly, the junction J_2 is a junction between collector and base regions. Thus it is known as collector-base junction. In the normal working of a transistor, the junction J_1 is forward biased, whereas the junction J_2 is reversed biased. These days, the transistors are widely used in amplifiers, oscillators, voltage stabilizing circuits etc.

16.20 TYPES OF TRANSISTORS

The transistors are of the following two types depending upon the arrangement and types of sandwitch:

1. N-P-N Transistor   2. P-N-P Transistors

16.21 N-P-N TRANSISTORS

An N-P-N transistor, as the name indicates, is composed of two N-type semiconductors separated by a very thin layer of P-type semiconductor as shown in Fig. 16.15 (a).

Now let us consider an N-P-N transistors, whose emitter b-base junction (J_1) is forward biased by the voltage source V_{EB} and the collector-base junction (J_2) is reverse biased by the voltage source V_{CB} as shown in Fig. 16.15 (b).

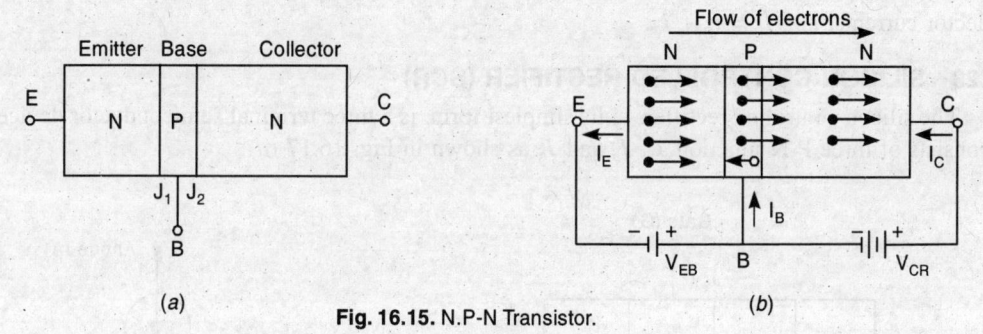

Fig. 16.15. N.P-N Transistor.

We know that the forward bias of the emitter-base junction causes the electrons in the N-type emitter to flow forwards the emitter-base junction. This constitutes the emitter current I_E as shown in the figure. Since the barrier potential at the emitter-base junction is reduced due to forward bias, the electrons cross over the emitter-base junction and enter the P-type base region. Now the electrons in the base region tend to recombine with the holes. Since the base is very thin, therefore only about 5 per cent of the emitter electrons combine with the holes in the base region and are lost as charge carriers. This constitutes the base current (I_B) as shown in the figure. It has been observed that most of the electrons cross over the base-region and enter the N-type collector region, where they are readily swept by the positive collector voltage V_{CB}. This constitutes the collector current (I_C) as shown in the figure. It is thus obvious that emitter current is equal to the sum of base and collector currents, i.e., $I_E = I_B + I_C$.

16.22 P-N-P TRANSISTORS

A P-N-P transistor, as the name indicates, is composed of two P-type semiconductors separated by a very thin layer of N-type semiconductor as shown in Fig. 16.16. (a).

Now let us consider P-N-P transistor, whole emitter base junction (J_1) is forward biased by the voltage source V_{EB} and the collector-base junction (J_2) is reversed by the voltage source V_{CB} shown in Fig. 16.16 (b). We know that the forward base of the emitter base junction causes the holes in the P-type emitter to flow towards the emitter-base junction. This constitutes the emitter current I_E as shown in the figure. Since the barrier potential at the emitter-base junction is reduced,

                    Emitter   Base   Collector

```
E ○──┤ P │ N │  P  ├──○ C
         J₁ J₂
            ○
            B
```

(a)

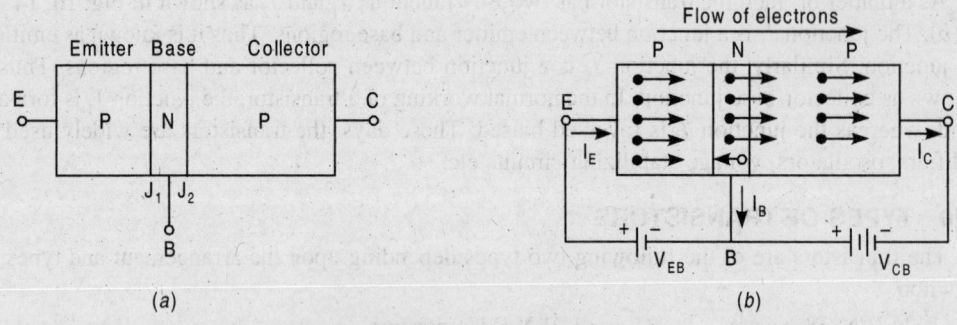

```
          Flow of electrons
          P      N      P
E ○→  ●→  │●→    │●→  ○ C
    Iₑ    │      │      I_C
          │  ↓I_B
       +├─┤-   +├─┤-
         V_EB  B  V_CB
```

(b)

Fig. 16.16. P.N-P. Transistor.

due to forward bias, the holes cross over the emitter-base junction and enter into the N-type base junction. Now the holes in the base region tend to recombine with the electrons. Since the base is very thin, therefore only about 5 percent of the holes combine with the electrons and are lost as charge carriers. This constitutes the base current I_B as shown in the figure. It has been observed that most of the holes cross over the base region and enter the collector region, where they are readily swept by the negative collector voltage V_{CB}. This constitutes the collector current (I_C) as shown in the figure. It is thus obvious that emitter currents is equal to the sum of the base and collector currents, i.e., $I_E = I_B + I_C$.

16.23 SILICON CONTROLLED RECTIFIER (SCR)

The silicon controlled rectifier, in its simplest form, is a three terminal semiconductor device. It consists of three P-N junction J_1, J_2 and J_3 as shown in Fig. 16.17 (a).

```
              Gate (G)
               ○
         J₁   J₂   J₃
Anode (A) ○─┤ P │ N │ P │ N ├─○ Cathode (C)
```

(a) Silicon controlled rectifier

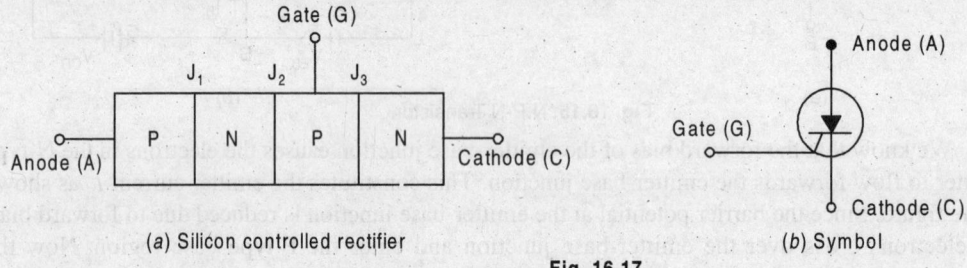

```
          ● Anode (A)
          │
Gate (G) ─┤▷│
          │
          ○ Cathode (C)
```

(b) Symbol

Fig. 16.17.

As a matter of fact, the silicon controlled rectifier is essentially, a combination of an ordinary P-N junction and a N-P-N transistor. Or in other words, P-N junction and a N-P-N transistor are combined together in one unit to form P-N-P-N device. The silicon controlled rectifier has three terminals. One of the terminals is taken from the outer P-type semiconductor and is known as anode (A). This second terminal is taken from the outer N-type semiconductor and is known as cathode (C). The third terminal is taken from the inner P-type semiconductor (or base of the N-P-N transistor) and is known as gate (G) as shown in Fig. 16.11 (b). During the normal working of a silicon controlled rectifier, the anode is held at high positive potential and the gate at small positive potential with respect to the cathode.

Fig. 16.17 (b) shows the circuit symbol of a silicon controlled rectifier. It is widely used in switching and control circuits such as speed control of D.C shunt motors, power control overlight detectors etc.

Semiconductors

16.24 THERMISTORS

The semiconductor devices, which are used for the temperature measurements, are known as thermistors. These devices are based on the fact that connectivity of a semiconductor increases with an increase in temperature. It will be interesting to know that for the same materials, the conductivity variation is so large, that it is possible to record temperatures even as small as 10^{-6} K. The thermistors are, usually, made from sintered oxides such as manganese oxide containing dissolved lithium ions, magnetic or germanium etc.

These days, the thermistors are widely used for temperature measurement in electronic circuits, as sensing elements in pressure gauges and flow meters etc.

16.25 VARISTORS

The semiconductor devices, with non-linear voltage-current (or V-I) characteristics are known as varistors. The varistors are usually, made from silicon carbide. This material has mathematical voltage-current relationship as:

$$I = k\, V^\alpha$$

where I = Current through the devices,

k = A constant of proportionality,

V = Voltage across the devices, and

α = Another constant, whose value depends upon the composition of the material. Usually, its value lies between 4 and 6.

These days, the varistors are widely used in the electronic equipments, which require protection against over voltages and lighting surges. They are also used for many other purposes such as voltage stabilization, arc suppression, motor speed control etc. in the low current and high frequency circuits.

16.26 INTEGRATED CIRCUITS

An integrated circuit (abbreviated as IC) is an electronic circuit, in which many devices such as transistors, diodes, resistors, capacitors etc., are fabricated on a single small silicon chip. It is different from a discrete circuit, which is built by connecting separate devices. The integrated circuits offer a number of advantages over the discrete circuits. Some of the important advantages are extremely small physical size, very small weight, reduced cost, extremely high reliability, low power consumption and easy replacement. However, they suffer from some limitations such as non-fabrication of coils and inductors functioning at fairly low voltages (3 to 30 volts) and difficult in producing IC's for handling large amount of power.

Though there are many types of integrated circuits, yet the most common types is the monolithic type. The monolithic integrated circuits are fabricated bipolar, unipolar or metal-oxide semiconductor technique. The metal-oxide semiconductors are slower than bipolar IC's. But they consume less power and need less space.

16.27 PHOTO-CELLS

These are the semiconductor devices, which convert light energy into electrical energy. The semiconductor materials, which are used in the fabrication of photo-cells are selenium, cadmium sulphide, lead sulphate etc. These devices have already been discussed in the last chapter. These days the photo-cells are widely used in television cameras, reproduction of sound in cinematography, fire alarms etc.

16.28 LASER

The word 'laser' is an acronym for Light Amplification by Simulated Emission. It is a source of highly directional, monochromatic, coherent light. As such, it has revolutionized some long-standing applied problems and has created some new fields of basic and applied optics. The light from a laser, depending on the type, can be continuous beam of low or medium power; or it can be a short burst of intense light delivering millions of watts.

The lasers are of many types such as Ruby laser, CO_2 laser (carbon dioxide laser), He-Ne laser and semiconductor laser. But the semiconductor laser has become very popular with the advancement of semiconductor device technology these days. The semiconductor laser consists of a heavily doped P-N junction made from the gallium-arsenide (GaAs) or the gallium-arsenide-phosphide (GaAsP) materials.

When a P-N junction is forward biased, a large number of electrons flow across the junction from the N-side to P-side. This increase the concentration of electrons to great extent in the conduction band of the P-region much above the equilibrium value. Similarly, a large number of holes flow across the junction from the P-side to N-side. This increases the concentration of holes in the valence band of the N-region much above the equilibrium value. When the electrons recombine with the holes into N-and P-regions, they emit a coherent monochromatic beam of light called laser beam. The wavelength of radiation for a GaAs laser is 8400 Å, which lies in the visible region. On the other hand, the wavelength of radiation for a GaAsP laser lies in the infrared region.

The lasers are widely used in variety of applications such as communication engineering, fabrication of electronic devices, high accuracy distance measurements, three-dimensional lensless photography (called holography), in metal working as machine tools, for cutting and drilling extremely small holes and in bio-engineering as sophisticated surgery tools.

A.M.I.E. (I) EXAMINATION QUESTIONS

1. Distinguish between intrinsic and extrinsic semiconductors. *(Winter, 1994)*
2. (a) Which elements are to be used for doping silicon to make it a P-type semiconductor ?
   (b) Define drift mobility of electrons of electrons in semiconducting materials.
3. Distinguish between P-type and N-type semiconductors. *(Summer, 1995)*
4. (a) Discuss the behaviour of a metal, a semiconductor and an insulator under the influence of an electric field.
   (b) Briefly discuss the factors affecting the electrical resistance of materials. *(Winter, 1996)*
5. What is meant by an 'N-type semiconductor ? Give example. *(Winter, 1996)*
6. Explain the factor affecting the electrical resistance of engineering materials. *(Summer, 1997)*
7. How are resistivity and conductivity related ? *(Summer, 1997)*
8. Explain why the Fermi level in an N-type semiconductor falls with increasing temperature, approaching the middle of the energy gap. *(AMIE., Summer, 1999)*
9. (a) What is the nature of bonding in semiconductor materials? What is meant by semiconductor device and what are the typical components of a semiconductor device.
   (b) Is it possible for compound semiconductor to exhibit intrinsic behaviour. Explain in brief ? *(AMIE., Summer, 2000)*
10. Describe N-type and P-type semiconductors with structural reasoning. Give an example of each. *(AMIE., Winter, 2000)*

MULTIPLE CHOICE QUESTIONS

1. The process of adding impurities to a pure semiconductor is called
   (a) omissing   (b) doping   (c) diffusing   (d) refining
2. The pentavalent impurities like antimony and phosphorus, added to intrinsic semiconductors are called
   (a) acceptor or P-type impurities   (b) donor or P-type impurities
   (c) acceptor or N-type impurities   (d) donor or N-type impurities
3. In a N-Type semiconductor, the Ferm's level
   (a) is lower than the centre of energy gap   (b) is at the centre of energy gap
   (c) is higher than the centre of energy gap   (d) does not exist
4. The mobility of electrons in a material is expressed in units of
   (a) V/S   (b) m^2/V-S   (c) m^2/S   (d) J/K

ANSWERS

1. (b)   2. (d)   3. (c)   4. (b)

17
Insulating Materials

1. Introduction. 2. Electric Field. 3. Electric Field Strength. 4. Electric Flux Density. 5. Permittivity. 6. Polarization. 7. Dielectric Polarization. 8. Polarization. 8. Polarization Mechanisms. 9. Total Polarization of a Dielectric Material. 10. Capacitor. 11. Dielectric Properties. 12 Dielectric Constant. 13. Effect of Frequency on Dielectric Constant. 14. Dielectric Loss. 15. Dielectric Strength. 16. Important Points for Selection of a Dielectric Material. 17. Ferroelectric Materials. 18. Hysteresis Curve. 19. Electrostriction. 20. Piezoelectricity. 21. Important Insulating Materials. 22. Glass. 23. Mica. 24. Ceramics. (or Refractories). 25. Asbestos. 26. Resins. 27. Rubber.

17.1 INTRODUCTION

We have already discussed in Art. 5.12 that the materials, in which the electrical conduction cannot occur, are known as insulating materials. These materials are also known as dielectric materials or dielectrics. It will be interesting to know that when these materials are used to prevent the flow of electricity through them, on the application of potential difference, they are called insulators. But when they are used to store electricity and release the same under controlled conditions, they are known as dielectrics.

We have also discussed in Art. 5.17 the energy band diagram of insulating materials as shown in Fig. 17.1 We know that the valence band of these materials is separated from the conduction band by a wide energy gap which is of the order of 3 eV or more. This energy gap is so wide that at ordinary temperature, the electrons cannot jump from valence band to the conduction band. As a result of this, no electrical conduction takes place. However, if the temperature is increased, some of the covalent bonds may break. As a result of this some electrons jump from valence band to the conduction band. This results in the flow of small current through the insulating material. This shows that the insulating materials have very large insulating resistance at ordinary temperatures. This resistance decreases with the increase in temperature. The common examples of such materials are rubber, bakelite, mica, glass etc. These materials are used as insulators in electrical machines and as dielectrics in the fabrication of capacitors.

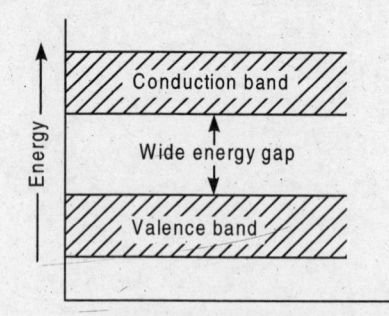

Fig. 17.1. Energy band diagram of insulators.

Table 17.1 shows the electrical resistivity of some polymers and ceramics used as insulators.

Insulating Materials

Table 17.1

No.	Material	Resistivity (Ohm-m)
1.	*Polymers*	
	Polyethylene	10^{13}
	Polytetrafluoroethylene	10^{16}
	Polystrene	10^{15} to 10^{16}
	Epoxy	10^{10} to 10^{15}
2.	*Ceramics*	
	Alumina	10^{12}
	Silica glass	10^{15}
	Boron nitride	10^{11}

As seen from this table, all the polymers and ceramics have the resistivity greater than 10^{11} ohm-m.

17.2 ELECTRIC FIELD

We have already discussed in Art. 2.2 that every substance consists of a large number of atoms or molecules. Each atom, in turn, consists of protons (positively charged particles), electrons (negatively charged particles), etc. In its neutral state, an atom contains equal number of protons and electrons. But when two or more substances are combined together, some of the electrons transfer from one substance to another. During this process, one of the substance loses its electrons and the other gains them. The substance which loses its electrons, is said to be positively charged. And the substance, which gains electrons, is said to be negatively charged. The total excess or deficiency of the electrons, in a substance, is known as negative or positive charge respectively. This charge is generally designated by the letter 'q' and is measured in terms of units called coulomb (C). It has been observed that if two charges of opposite polarity are placed closer to each other, they experience a force of attraction between them. Similarly, if the charges are of the same polarity, they experience a force of repulsion between them.

The region around the charge, in which it exerts a force on the other charge, is called electric field. The electric field is assumed to consist of lines of force. It has been observed that these lines of forces emanate from the positive charge and end on the negative charge as shown in Fig. 17.2.

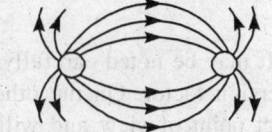

Fig. 17.2. Distribution of electric field.

The direction of the electric field at any point is determined by drawing a tangent to the line of force at that point.

17.3 ELECTRIC FIELD STRENGTH

The term 'electric field strength' at any point may be defined as the force experienced by a unit positive charge placed at that point. It is generally designated by the letter 'E' and is measured in terms of units known as newton per coulomb of charge (N/C) or volts per meter (V/m).

Let
$\quad q$ = Magnitude of the charge in Coulombs, and
$\quad F$ = Force experienced by that charge in newtons.

Then the electric field strength,

$$E = \frac{F}{q} \text{ in N/C or V/m}$$

17.4 ELECTRIC FLUX DENSITY

The total number of lines of force emanated from the positive charge is known as electric flux. It is generally represented by the symbol (ψ) and is measured in terms of units called Coulombs. In other words, the electric flux is numerically equal to the charge and has the same units as that of charge.

The term 'electric flux density' at any point may be defined as the flux (ψ) passing normally through the unit cross-section at the point. It is generally designated by the letter 'D' and is measured in terms of the units called coulomb per square meter. Sometimes, electric flux density is also known as electric displacement.

Now let ψ = Total flux in coulombs, and
A = Cross-sectional area of section in square metres.

Then the electric flux density,

$$D = \frac{\psi}{A} \text{ in C/m}^2$$

17.5 PERMITTIVITY

The term 'permittivity' may be defined as the ratio of electric displacement (D) in a dielectric medium to the applied electric field strength (E). It is designated by the symbol ε (epsilon) and expressed in farad/metre. Thus permittivity,

$$\varepsilon = \frac{D}{E} \text{ in F/m}$$

The permittivity indicates the degree to which the medium can resist the flow of electric charge and is always greater than unity. Sometimes permittivity is referred to as an absolute permittivity of the dielectric medium. In order to measure the absolute permittivity, it is convenient to express it by the relation,

$$\varepsilon = \varepsilon_0 \cdot \varepsilon_r$$

where ε_0 = Absolute permitting of the free space or vacuum.
Its value is 8.854×10^{-12} F/m.

ε_r = Relative permittivity or dielectric constant of a medium.

It may be noted carefully that ε_0 has no physical meaning except that, it is a fundamental conversion factor. On the other hand, dielectric constant is an important parameter from the subject point of view and will be discussed in detail in this chapter. In order to measure the dielectric constant of different dielectric materials, the vacuum or free space is considered as a reference medium. Therefore it is allotted a dielectric constant, $\varepsilon_r = 1$. Hence absolute permittivity of the free space or vacuum,

$$\varepsilon_r = \varepsilon_0 \times 1 = \varepsilon_0 = 8.854 \times 10^{-12} \text{ F/m}$$

Therefore dielectric constant of dielectric medium, can now be expressed as the ratio of absolute permittivity of a dielectric medium to the absolute permittivity of free space, *i.e.*,

$$\varepsilon_r = \frac{\varepsilon}{\varepsilon_o}$$

The dielectric constant is a dimensionless quantity, because it is the ratio of the two permittivities. Table 17.1 shows dielectric constants of some of the important dielectric materials or insulating materials.

Insulating Materials

Table 17.1 Dielectric constant of some materials

Material	ε_r	Material	ε_r
Air or Vacuum	1	Paper (Waxed)	3—5
Glass	3.7—10	Polysterene	2.6
Mica	5—7.5	Bakelite	4.5—5.5
Ceramics	4—10	Porcelain	5—6

17.6 POLARIZATION

The charge of equal magnitude, but of opposite polarity, separated by the some distance is called dipole. The product of magnitude of charge and the distance between them is called dipole moment. It is generally designated by the symbol (p) and is measured in terms of unit called coulomb metre (C-m).

Let q = Magnitude of each charge, and
d = Distance between the charges.

Then the dipole moment,

$$p = q \cdot d \text{ in Coulomb metre.}$$

The dipole moment per unit volume is known as polarization. It is generally designated by letter (P) and is measured in terms of units called coulomb-metre per cubic metre.

Let p = Dipole moment in (C-m), and
V = Volume of dielectric in cubic metres.

Then the polarization,

$$P = \frac{q}{V} \text{ in C-m}/m^3 \text{ or } C/m^2$$

The dipole moment per unit volume of the material is the sum of all individual dipole moments within that volume.

17.7 DIELECTRIC POLARIZATION

When an electric field is applied to dielectric material, its positive charge undergoes a small displacement in the direction of the field and the negative charge in the opposite direction. The displacements are very small as the charges are bound and not free. Its net effect is equivalent to a series of dielectric dipoles oriented in the direction of field. Thus electric charges appear on the surface of a dielectric material as shown in the Fig. 17.3. This phenomenon is called dielectric polarization. It has been observed that in a dielectric material, the polarization increases linearly with the applied electric field.

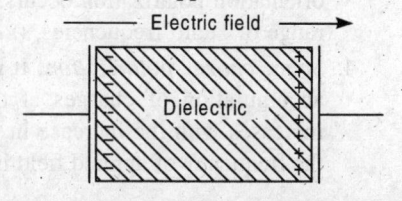

Fig. 17.3. Charges on the dielectric surface.

For a dielectric material, the expression for electric displacement may be obtained as follows:
We know that electric displacement,

$$D = \varepsilon \cdot E = \varepsilon_0 \cdot \varepsilon_r \cdot E$$

Adding and subtracting the term $\varepsilon_0 \cdot E$ on the R.H.S. of the above expression,

$$D = \varepsilon_0 \cdot \varepsilon_r \cdot E + \varepsilon_0 \cdot E - \varepsilon_0 \cdot E$$
$$= \varepsilon_0 \cdot E + \varepsilon_0 \cdot E (\varepsilon_r - 1)$$
$$= \varepsilon_0 \cdot E + P$$

where P is the polarization of the dielectric medium. The above equation indicates that electric displacement in a dielectric medium consists of two components: (*i*) the electric displacement in the absence of dielectric medium (*i.e.*, $\varepsilon_0 \cdot E$) and (*ii*) the electric displacement due to the polarization of the dielectric medium (*i.e.*, $\varepsilon_0 \cdot E (\varepsilon_r - 1)$. It is evident from the above discussion that the polarization measures the additional electric displacement arising from the presence of dielectric as compared to the free space. The polarization has the same units as that of electric displacement (*i.e.*, C/m^2).

It has been observed that in neutral atoms and symmetrical molecules, the centres of gravity of positive and negative charges coincide. The application of an electric field causes some displacement of these charges. This leads to the creation of dipoles and polarization. However, in unsymmetrical molecules, the permanent dipoles are created even in the absence of the electric field. In these molecules, the applied electric field tends to orient these dipoles parallel to the direction of the field. Such molecules are called polar molecules. The molecules in which diploes are created on the application of electric field, are called non-polar molecules.

17.8 POLARIZATION MECHANISMS

The Polarization occurs due to several mechanisms. But the following four are important from the subject point of view:

1. *Electronic polarization*. This polarization is the result of displacement of the positively charged nucleus and the negatively charged electrons of an atom in the opposite directions on the application of an electric field. It may occur in dielectric, even if the frequency of the applied electric field is 10^{13} Hz. Such frequencies lie in the ultraviolet part of the electromagnetic spectrum and are referred to as optical frequencies. The electronic polarization is independent of the temperature.

2. *Ionic Polarization*. This polarization is the result of displacement of the positively and negatively charged ions in the ionic bonded solid in opposite directions on the application of an electric field. The ionic polarization occurs, when the frequency of the applied electric field is below the optical range of frequencies. It is also independent of temperature.

3. *Orientational polarization*. This polarization is the result or alignment of permanent electric dipoles in polar molecules with the applied electric field. The orientational polarization decreases with the increase in temperature. It is because of the fact that thermal energy tends to randomize the alignment of permanent dipole moment. The orientation polarization occurs, when the frequency of the applied electric field is in the range of audio frequencies, (*i.e.*, about 20 kHz).

4. *Space charge polarization*. It is also called interfacial polarization. It occurs due to the accumulation of charges at the electrode or interfaces in a multiphase material. It increases with the increase in temperature. The space charge polarization occurs, when the frequency of applied field is in the range of power frequencies (*i.e.*, about 50-60 Hz.).

17.9 TOTAL POLARIZATION OF A DIELECTRIC MATERIAL

The total polarization of a dielectric material is the sum or contribution of the electronic, ionic orientational and space charge polarizations. Thus, $P_{total} = P_e + P_i + P_o + P_s$

where $P_e + P_i + P_o$ and P_s are electronic, ionic, orientational and space charge polarizations respectively.

17.10 CAPACITOR

It is an arrangement to store electricity, when connected to a voltage source. It consists of two metal plates separated by a dielectric as shown in Fig. 17.4. It has been observed that when the capacitor is connected to a voltage source, one of the plates acquire positive charge and the other negative as shown in Fig. 17.4.

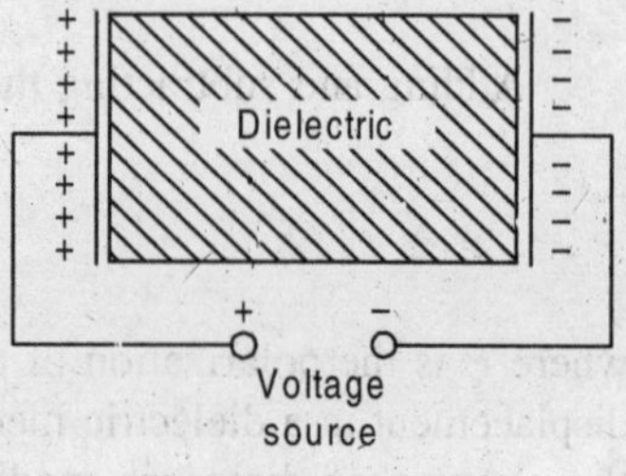

Fig. 17.4. Capacitor.

Insulating Materials

The property of a capacitor to store electricity (or electric charge,) when its plates are at different potentials is known as its capacitance or capacity. It is generally designated by the letter 'C'. The capacitance of a capacitor is defined as the amount of charge required to create a potential difference of 1 volt between the plates.

Suppose we give Q coulombs of charge to one of the two plates of a capacitor and if a potential difference of V volts is established between the plates, then its capacitance is given by the relation:

$$C = \frac{Q}{V}$$

The capacitance is measured in terms of units called farad (F). A farad is a much larger unit of capacitance. Therefore, we use a much smaller unit called microfarad (μF) or picofarad (pF).

The value of capacitance for a parallel plate capacitor is given by the relation:

$$C = \frac{\varepsilon_0 \cdot \varepsilon_r \cdot A}{d}$$

Where ε_0 = Absolute permittivity of the free space,
ε_r = Relative permittivity of the dielectric,
A = Area of each plate, and
d = Distance between the two plates.

Example 17.1. *Two capacitors are made, one using glass plate ($\varepsilon_r = 6.0$) of thickness 0.25 mm and the other using plastic field ($\varepsilon_r = 2.6$) of thickness 0.1 mm, between the metal electrodes, which one holds greater charge ?*

Solution. Given: For capacitor with glass plate, $\varepsilon_r = 6$ and thickness, $d = 0.25$ mm and for capacitor with plastic film, $\varepsilon_r = 2.6$ and thickness, $d = 0.1$ mm.

We know that capacitance of a capacitor,

$$C \propto \frac{\varepsilon_r \cdot A}{d}$$

∴ For a capacitor with glass plate,

$$C_1 \propto \frac{6.A}{0.25} = 24A \qquad \ldots(i)$$

and for a capacitor with plastic film,

$$C_2 \propto \frac{2.6A}{0.1} = 26\,A \qquad \ldots(ii)$$

Assuming that both the capacitors have metal electrodes of same cross-section (*i.e.*, same A), we find from equation (*i*) and (*ii*) that C_1 is its capacitance (*i.e.*, $Q \propto C$) therefore capacitor with plastic film will hold greater charge.

17.11 DIELECTRIC PROPERTIES

Though there are many properties of dielectrics, yet the following are important from subject point of view:

1. Dielectric constant. 2. Dielectric losses. 3. Dielectric strength.

17.12 DIELECTRIC CONSTANT

We have already discussed in the last articles about the dielectric constant. It is the ratio of absolute permittivity (ε) of a dielectric to the absolute permittivity (ε_0) of the free space. It is an important property of dielectric, because it determines the capacity of a dielectric to develop charges on its surface due to polarization. Thus a good dielectric should have a high value of dielectric constant.

For static electric fields, the dielectric constant (ε_0) is a real number. However, for time varying (*i.e.*, alternating) fields, dielectric constant is no consists of two components– a real component and an imaginary component. Thus for time varying fields, the complex dielectric constant,

$$\varepsilon_r = \varepsilon_r' - j\varepsilon_r''$$

where ε_r' = Real component of dielectric constant, and
ε_r'' = Imaginary component of dielectric constant.

It has been found that the absorption of energy by a dielectric (called dielectric loss) is proportional to imaginary part of a complex dielectric constant. Dielectric constant depends upon the frequency of the applied electric field. It decreases with the increase in frequency. Dielectric constant also depends upon temperature. It decreases in dielectrics in which orientational polarization is predominant whereas it increases in dielectrics in which space charge polarization is predominant. These topics are discussed in the next articles.

17.13 EFFECT OF FREQUENCY ON DIELECTRIC CONSTANT

It has been observed that when all the four polarizations (*i.e.*, electronic, ionic, orientational and space charge polarization occur in a dielectric material, the dielectric constant will decrease with increase in frequency of the applied electric field. Table 17.3 shows dielectric constants of some important dielectric materials at a frequency of 50 Hz and 1 MHz.

Table 17.2 Properties of Some Dielectric Materials

Material	Dielectric Constant (ε_r) 50 Hz	1 MHz	Dissipation factor Tan δ (1 MHz)	Dielectric strength KV/mm
Fused Silica	4	3.8	0.0001	10
Soda-lime glass	7	7	0.005	10
Mica	8	5	0.0005	100
Polyethylene	2.3	2.3	0.0004	4
Polyvinyl chloride	7	3.4	0.05	2
Nylon 66	4	3.5	0.02	15
Bakelite	4.5	4.5	0.028	15
Vulcanized rubber	4	2.7	0.003	25
Transformer oil	5	2.5	0.0001	10
Porcelain	6	6	0.02	5

It may be noted from the table shown above that dielectric constants of some of the materials like soda-lime glass, polyethylene, bakelite and porcelain do not decrease with the increase in frequency from 50 Hz to 1 MHz

17.14 DIELECTRIC LOSS

It has been observed that when an alternating voltage is applied across a capacitor, having free space or vacuum as dielectric, the polarization of the dielectric is in phase with the voltage. In such a case, the resulting current leads the applied voltage by an angle of 90° as shown in Fig. 17.5 (*a*). The vectors *I* and *V* represent the R.M.S values of current and voltage respectively.

Insulating Materials

A phase angle of 90° indicates that there is an electrical energy in the free space during the charging of the capacitor.

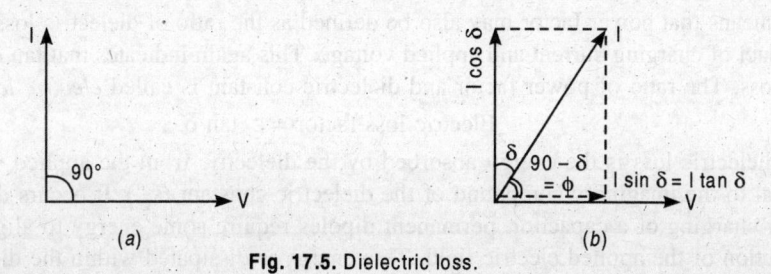

Fig. 17.5. Dielectric loss.

Now consider a capacitor with a real dielectric inside it. When the alternating voltage is applied to such a capacitor, the polarization is no longer in phase with the applied voltage. In such a case, the resulting current leads the applied voltage by an angle, $\phi = 90° - \delta$ as shown in Fig 17.5 (b), where δ is known as loss angle. The current can be factorized into the following two components:

1. A component of current leads the applied voltage by an angle equal to 90°. This component is denoted by $I \cos \delta$ and is called imaginary component of current. It is similar to a current in an ideal capacitor.
2. Another component of current is in phase or parallel to the applied voltage. This component is denoted by $I \sin \delta$ and is called real component of current. This component results in a power loss within the dielectric.

If δ is small then,

$$\sin \delta \approx \delta \approx \tan \delta$$

The quantity $\tan \delta$ is a measure of power loss. It is commonly known as loss tangent, dissipation factor or power factor. It may be defined as the ratio of imaginary component of dielectric constant (ε_r'') to the real component (ε_r'). Thus power factor,

$$\tan \delta = \frac{\varepsilon_r''}{\varepsilon_r'}$$

It is evident from the above relation that if there are no losses then $\varepsilon_r'' = 0$. This means that $\delta = 0$ and current in a capacitor leads the applied voltage by 90°. Hence $\tan \delta$ is a measure of power loss. Table 17.2 shows the values of $\tan \delta$ for some of the important dielectrics at a frequency of 1 MHz.

The power loss in a dielectric is known as dielectric power loss or simply dielectric loss. It is given by the relation,

$$P = V \cdot I \tan \delta$$

where V = Applied voltage in volts. ... (i)

I = Charging Current of a capacitor in amperes. Its value is given by the relation V/x_E where $X_E = 1/2 \pi f \cdot C$ is the capacitor reactance in ohms, C is the capacitance of capacitor and its value is $\varepsilon_0 \varepsilon_r A/d$ and f is the frequency of the alternating voltage.

$\tan \delta$ = Power factor or dissipation factor of a dielectric.

The equation (i) is a very useful relation to determine the dielectric loss for any applied voltage. This equation implies that a capacitor with dielectric loss may be represented by an equivalent circuit, which consists of a pure capacitance and a parallel resistance. Also from equation (i), we find that

$$\tan \delta = \frac{P}{V \cdot l}$$

This means that power factor may also be defined as the ratio of dielectric loss in a material to the product of charging current and applied voltage. This again indicates that tan δ is a measure of power loss. The ratio of power factor and dielectric constant is called *electric loss factor*.

$$\text{Electric loss factor} = \varepsilon_r \tan \delta$$

The dielectric loss is the energy absorbed by the dielectric from the applied voltage and is proportional to the imaginary compound of the dielectric constant (ε_r''). It occurs due to the fact that during charging of a capacitor, permanent dipoles require some energy to align themselves in the direction of the applied electric field. This energy is dissipated within the dielectric in the form of heat.

The dielectric loss is a function of the frequency of the applied field. Its value is very high at certain frequencies lying in the audio and radio range as shown by the points A, B and C in Fig. 17.6 (b). The increased dielectric loss at such frequencies is because of a resonance. The resonance occurs at these frequencies, when the period of the voltage is in the same range as the relaxation time of the polarization process.

17.15 DIELECTRIC STRENGTH

It has been observed that whenever the electric field strength applied to a dielectric exceeds a critical value, a large current flows through it. As a result of this, the dielectric loses its insulating properties. This phenomenon is known as dielectric break down. The electric field strength, at which this occurs, is known as dielectric strength.

Though there are several mechanisms by which this break down takes place, yet the following two are important from the subject point of view:

1. *Avalanche breakdown.* In this mechanism, a few electrons are liberated from the points of imperfections in the dielectric material. These electrons acquire high velocity and energy due to the greater value of applied electric field strength. It has been observed that these electrons collide with the valence electrons and transfer their own energy to the valence electrons. The valence electrons, by acquiring this energy, jump from the valence band to the conduction band. This process gets multiplied as more and more valence electrons jump to the conduction band. As a result of this, a large current flows through the dielectric and breakdown is said to occur.

2. *Thermal breakdown.* In this mechanism, the heat generated by the ionic current increases due to the greater value of the applied electric field strength. It has been observed that when the heat generated is greater than the dissipation, the temperature inside the dielectric increases. This increases the electric conductivity in the dielectric and breakdown is said to occur.

It will be interesting to know that the dielectric strength of a material is not constant. But it varies with the thickness of the material. As the thickness of the dielectric increases, its dielectric strength also increases. The dielectric strength of a material is measured in volts per unit thickness. Thus, if V is the potential difference in volts across the dielectric and d is the thickness of the dielectric, in metres then dielectric strength

$$E = \frac{V}{d} \text{ (Volts/ metre)}$$

Usually, volt is a smaller unit for potential difference, therefore a much bigger unit called kilovolt is used to denote the potential difference. Similarly, the metre is a bigger unit for the dielectric thickness, therefore a much smaller unit called millimetre is used to denote the dielectric strength.

Insulating Materials

The factors which affect dielectric strength of a material are moisture, temperature, ageing etc. Any increase in the values of these factors reduce the dielectric strength. The dielectric strength of some important materials is shown in Table 17.2. It may be noted that mica has the highest dielectric strength and hence it is used a large number of applications.

17.16 IMPORTANT POINTS FOR THE SELECTION OF A DIELECTRIC MATERIALS

Following are some of the important points for the selection of a dielectric material:

1. It should have a high value of dielectric constant. An electric constant determines the capacity of a dielectric to develop charges on its surface, due to polarization. Therefore higher the value of dielectric constant, higher is the capacity of a dielectric.
2. It should have a low dissipation factor (or power factor). Therefore lower the dissipation factor, smaller is the dielectric loss.
3. It should have a sufficiently high dielectric strength. Therefore higher the dielectric strength, greater is the voltage per unit thickness, which it can withstand before breakdown.
4. It should have a high insulation resistance. Therefore higher the insulation resistance, lower is the leakage current through the dielectric.

17.17 FERROELECTRIC MATERIALS

For the dielectric materials discussed so far, the polarization is a linear function of the applied field. However, there are a number of substances for which polarization is not a unique function of applied field. But it depends upon the history of the substance. Such materials are Rochelle salt, Barium Titanate ($BaTiO_3$), Potassium Niobate ($KNbO_3$), Lead Titanate ($PbTiO_2$), Potassium Dihydrogen Phosphate (KH_2PO_4) etc. The dielectric constants of such materials are much larger than those of ordinary dielectrics. For example, dielectric constant of Barium Titanate is more than 2000. This value is much higher as compared to a dielectric constants of ordinary dielectrics which are usually below 10. These materials exhibit hysteresis effect similar to those observed in ferromagnetic materials (discussed in the next chapter). Therefore as an analogy, these materials are known as ferroelectric materials. In a ferromagnetic material, the electric dipoles in a ferroelectric material are all aligned in the same direction even in the absence of an electric field.

17.18 HYSTERESIS CURVE

It has been observed that if a ferroelectric material, initially non-polarized, is subjected to an electric field, the polarization varies as shown in Fig. 17.6. In this figure, the polarization is plotted along the vertical axis and the electric field strength along the horizontal axis. It will be interesting to know that when the electric field strength is gradually increased from zero, the polarization increases rapidly as shown by the curve OA. If the electric field strength is further increased, the rate of increase of the polarization increases until a saturation is reached as shown by the curve AB in the figure. The polarization corresponding to the saturation is called saturation polarization (or spontaneous polarization) and is designated by P_s.

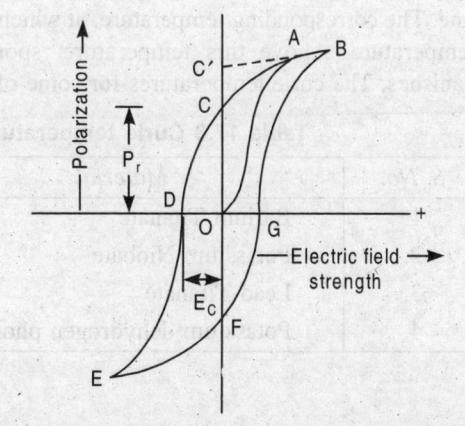

Fig. 17.6. Hysteresis curve for a ferroelectric material.

Now if the electric field strength is reversed to zero, the polarization decreases along the curve *BAC*. And it does not become Zero. The value of polarization (represented by *OC*), which is still left in the material, is known as remnant polarization. It is designated by the letter P_r. In order to make the polarization equal to zero, a certain value of electric field strength is applied in the reversed direction. This value of electric field strength is known as coercive field and is designated by the letter E_c. If the field strength is further increased in the negative direction, the reverse polarization increases rapidly until the saturation is reached at *E*. If the electric field strength is again reversed to make it positive, the polarization will follow the curve *EFGB* as shown in the figure. It may be noted from the figure that the curve *ABCDEFGB* is a closed one. It may also be noted from the figure that polarization always lags behind the electric field strength. This phenomenon of a ferroelectric material is known as dielectric hysteresis and the closed curve is the hysteresis loop.

The existence of a hysteresis loop, in a ferroelectric material, indicates the presence of spontaneous polarization, *i.e.,* polarization which exists when the applied electric field is zero and which requires a certain field in the reverse direction to change its polarity. The spontaneous polarization is indicated by the curve *OC* in the figure. The point *C* may be obtained by producing the *AB* backwards till it meets the vertical axis through the origin. The spontaneous polarization is due to the alignment of electric dipoles in the same direction even in the absence of an electric field. This induces a large polarization in a material.

The hysteresis loop may be explained as discussed below. At room temperature, a ferroelectric material does not exhibit any polarization in the absence of electric field. This can be explained easily if we visualize the presence of small domains* in the ferroelectric material. These domains are polarized to saturation. But these domains are oriented in such a direction that the total polarization in the absence of an external field is zero. The domains are separated by intermediate regions, called domain walls, in which there is gradual transition of the orientation of polarization. When the electric field is applied, the domains, for which the direction of the polarization is parallel to the field, grow at the expense of other domains. The process is shown by the curve *OA* in the figure. As the applied field is increased further, the material reduces to a single domain. As a result of this, the polarization reaches its maximum value. This process is shown by the curve *AB* in the figure.

It has been observed that the hysteresis loop of a ferroelectric material changes its shape as the temperature is increased. Moreover with the increase in temperature, the height of the hysteresis loop decreases slightly, but the width decreases considerably, until it becomes a straight line. The corresponding temperature, at which the loop reduces to a straight line, is known as curie temperature. Above this temperature, spontaneous polarization, of a ferroelectric material vanishes. The curie temperatures for some of the ferroelectric material are shown in Table 17.3.

Table 17.3 Curie temperature of some ferroelectric materials

S. No.	Material	Curie temperature (°C)
1.	Barium Titanate	120
2.	Potassium Niobate	439
3.	Lead Titanate	490
4.	Potassium dehydrogen phosphase	−150

---

* These are areas of territories in which electric dipoles are aligned in the same direction. However, the direction of polarization of adjacent domains need not be parallel.

Insulating Materials

17.19 ELECTROSTRICTION

We have already discussed in our previous discussions that an applied electric field induces dipole moments in atoms or ions and generally displaces ions relative to each other. The effect of this is to produce changes in the dimensions of a specimen. The dimension of a specimen can also be changed by applying mechanical stresses. But in general, such charges do not produce dipole moment. In other words, in most dielectric materials, the polarization produces a mechanical distortion. But a mechanical distortion does not produce polarization. This electromechanical effect is called electrostriction and is present in all dielectric materials. In pure electrostrictive materials, the mechanical deformation produced by a polarization, in a given direction, is the same as that produced by a polarization in the opposite direction. The principle of electrostriction is used in transducers to convert electrical energy into mechanical energy.

17.20 PIEZOELECTRICITY

The term 'piezoelectricity' may be defined as the ability of a material to develop charge on its surface when it is mechanically stress. This effect is observed in those dielectric materials which have permanent dipoles in them. These dipoles get polarized, when the material is stressed and hence develops a charge on its surface. Similarly, when a dielectric material possessing permanent dipoles, is subjected to an electric field, it aligns itself in the direction of the field and produces a mechanical strain. This process is known as inverse piezoelectric effect. This effect is, generally, observed in all ferroelectric materials.

Some of the important piezoelectric materials are rochelle salt, barium titanate, quartz, lead zinconate titanate etc. These materials are used in transducers, which are used to convert electrical energy into mechanical energy and vice versa. They are also employed in microphones, phonograph pickups, strain gauges etc.

17.21 IMPORTANT INSULATING MATERIALS

As a matter of fact, the insulating materials may be solids, liquids or even gases. But here we shall discuss the solid insulating materials only. Though there are a number of solid insulating materials, yet the following are important from the subject point of view.

17.22 GLASS

It is an inorganic insulating material, which comprises of complex system of oxides. Silica (SiO_2) is the most essential constituent of many commercial glasses. It is fused with alkali (like potash, soda etc.) and some base (like lime, lead oxide etc.). The silica glass (having 100 per cent SiO_2 is the best insulating material. Its dielectric constant varies between 3.7 and 10, loss tangent between 0.003 and 0.01 and dielectric strength between 2.5 and 50 kV/mm.

These days, the glass is widely used as an insulating material to form an envelop for electrical bulbs, electrical valves, mercury switches, X-ray tubes etc. It is also used as a dielectric material in capacitors.

17.23 MICA

It is an inorganic mineral compound of silicates of aluminium, soda potash and magnesia. It is crystalline in nature and can be easily split into very thin flat sheets. The two important types of mica are muscovite and phlogopite. The muscovite mica is known to be one of the best insulating material. The mica has a good dielectric and mechanical strength. Its dielectric constant varies between 5 and 7.5, loss tangent between 0.0003 and 0.015 and dielectric strength between 700 and 10000 kV/mm.

The mica is widely used as an insulator for commutator segment separator in electrical machines, switchgears, armature winding, electrical heating devices like electric irons, hot plates, toasters etc. It is also used as a dielectric material for high frequency applications.

17.24 CERAMICS OR REFRACTORIES

These are generally non-metallic inorganic compounds such as silicates, aluminates, oxides, carbide, borides, nitrides and hydroxides. These materials are brittle in nature and may be crystalline or amorphous. The ceramics used as dielectrics may be broadly described as porcelains, alumina ceramics, titanates etc. The ceramics have excellent dielectric and mechanical properties. The dielectric constant of the commonly used ceramics varies between 4 and 10.

The ceramics are widely used as insulators for switches, plug holders, telephone thermocouples, cathode heaters, vacuum type ceramic metal seals etc. These are also used as dielectric material in capacitors. These capacitors may be operated at high temperatures and be moulded into any shape and size.

17.25 ASBESTOS

It is an inorganic material which is used to designate a group of naturally occurring fibre material. The asbestos generally consists of magnesium silicate composition. The most commonly occurring mineral serpentine. The asbestos has good dielectric and mechanical properties.

It is widely used as an insulator in the form of paper, tape, cloth and board. It is also used in dielectric applications where it is impregnated with a liquid or solid. It is also used to manufacture panel boards, arcing barriers, insulating tubes and cylinders used in the construction of air cooled transformers.

17.26 RESINS

These are organic polymers and may be natural or synthetic. The natural resins are derived from plant, animal sources etc. The resins have very little applications in engineering industry. The synthetic resins are produced artificially. The commonly used synthetic resins are polyethylene, polystrene, polyvinyl chloride, acrylic resins, teflon, nylon etc. The resins have good dielectric and mechanical properties. The dielectric constant of resins varies between 2 and 4.5, the loss tangent between 0.0002 to 0.04 and dielectric strength is quite high.

The resins are widely used in engineering industry as dielectric and insulating materials. They are also used as an insulating material in radio, television, power and submarine cables. They are also used as a dielectric material in D.C. and high frequency capacitors.

17.27 RUBBER

These are also organic polymers and may be natural or synthetic. The natural rubber is obtained from rubber tree hevea brasiliensis. The natural rubber has a limited applications, because of its resistance at high temperatures. The synthetic rubbers are produced artificially by copolymerisation of isobutylene and isoprene. These rubbers have good electrical and thermal properties. The dielectric constant of rubber varies between 2.5 and 5, and loss tangent between 0.01 and 0.03.

The rubbers are widely used as an insulating material for electrical wires, cables, ropes, coatings, transformers, motor winding etc.

Insulating Materials

341

A.M.I.E. (I) EXAMINATION QUESTIONS

1. Write a short note on the use of dielectric materials. *(Summer, 1993)*
2. What are refractory materials ? Describe the use of such materials as insulating materials. *(Winter, 1993)*
3. What is an insulator ? What is the reason that a material behaves as an insulator. *(Winter, 1993)*
4. What is meant by dielectrics. Answer in two sentences. *(Winter, 1994)*
5. Describe with examples, the three principal categories of dielectrics. *(Summer, 1995)*
6. Explain 'dielectric strength' and dielectric losses' in materials. *(Summer, 1996)*
7. (a) What are dielectric ? Discuss their important properties and give applications.
   (b) Discuss ferroelectric behaviour and piezoelectric behaviour with examples. Where are these materials used ? *(Winter, 1996)*
8. Explain the term 'refractory'. Discuss the important properties of refractories. Give examples and applications. *(Winter, 1996)*
9. Give two examples of dielectric materials. *(Winter, 1996)*

MULTIPLE CHOICE QUESTIONS

1. An insulator is a material having resistivity approximately in the range:
   (a) 10^3 to 10^{17} Ω-m
   (b) 10^7 to 10^9 Ω-m
   (c) 10^{-3} to 10^3 Ω-m
   (d) none of the above

2. The value of capacitance for a parallel plate capacitor is
   (a) directly proportional to d
   (b) inversely proportional to A
   (c) inversely proportional to the permeability
   (d) in inversely proportioal to d

3. The power loss in a dielectric is given by
   (a) $P = VI \cos \delta$
   (b) $P = VI \sin \delta$
   (c) $P = VI \tan \delta$
   (d) $P = \dfrac{V}{I} \tan \delta$

4. Piezoelectricity is an ability of a material
   (a) to develop charge on its surface when it is mechanically stressed
   (b) to produce deformation when it is polarized
   (c) due to which it leads to a high value of dielectric constant
   (d) none of these

ANSWERS

1. (b)   2. (d)   3. (a)   4. (a)

18
Magnetic Materials

1. Introduction. 2. Magnetic Field. 3. Magnetic Moment 4. Origins of Magnetic Moment. 5. Magnetic Field Strength. 6. Magnetic Flux Density. 7. Magnetic Permeability. 8. Magnetisation. 9. Magnetic Susceptibility. 10. Classification of Magnetism in Magnetic Materials. 11. Diamagnetism. 12. Susceptibility of Diamagnetic Materials. 13. B-H Curve for Diamagnetic Materials. 14. Paramagnetism. 15. Susceptibility of Paramagnetic Materials. 16. B-H Curve for paramagnetic Material. 17. Ferromagnetism. 18. The Domain structure. 19. Susceptibility of Ferromagnetic Materials. 20. B-H Curve for Ferromagnetic Materials. 21. Magnetic Hysteresis. 22. Hysteresis and Eddy Current Loss 23. Ferrimagnetism. 24. Ferrites. 25. High Frequency Applications of Ferrites. 26. Classification of Magnetic Materials. 27. Hard Magnetic Materials. 28. Soft Magnetic Materials.

18.1 INTRODUCTION

The materials, which can be magnetised, are known as magnetic materials. These days, the magnetic materials are widely used in electrical machines, computers, meters, transducers, television tubes etc. In order to understand the magnetic behaviour of different materials, a thorough knowledge of such materials is very essential. The properties of magnetic materials are analogous to those of dielectric materials. In this chapter we shall discuss the various types of magnetic materials, in terms of magnetic properties of their atoms and interaction among them.

18.2 MAGNETIC FIELD

We know that if magnet is suspended freely to swing in its horizontal plane as shown in fig. 18.1 (a). Its end points *i.e.,* North pole and South pole (briefly written as N-pole respectively) always take up particular directions. These directions always point towards the earth's North and South poles respectively.

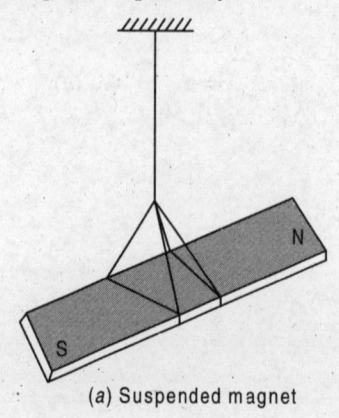

(a) Suspended magnet   Fig. 18.1.   (b) Magnetic field

Magnetic Materials

It has been observed that the two poles of a magnet cannot be separated, even it is cut into two or more parts. Thus each magnet is said to constitute a dipole system, which consists of one North and the other South pole. The region around the magnet, in which it exerts its influence, is called magnetic field. The magnetic field is assumed to consist of lines of magnetic force. These lines emanate from the N-pole, pass through the surrounding medium and then re-enter the S-pole. It has been observed that these lines form closed loops through the magnet as shown in Fig. 18.1 (*b*). The direction of the magnetic field, at any point, is determined by drawing a tangent to the line of force at that point.

18.3 MAGNETIC MOMENT

We have already discussed in the last article that the magnetic field is assumed to consist of lines of the magnetic forces. The distance between the two poles is known as magnetic length. Now consider a magnetic field.

Let m = Pole strength of any pole in webers (*Wb*), and
l = Magnetic length in metres,

∴ Magnetic moment
$= m \times l$ in Wb-m

18.4 ORIGINS OF MAGNETIC MOMENT

In fact, the macroscopic magnetic properties of materials are a consequence of magnetic moments associated with individual electrons. Each electron in an atom has magnetic moments that originate from two sources. The first source is related to the orbital motion of electron around the nucleus. Being a moving charge, an electron may be considered to be a small current loop, generating a very small magnetic field. This contributes to a magnetic moment along its axis of rotation as shown in Fig. 18.2 (*a*).

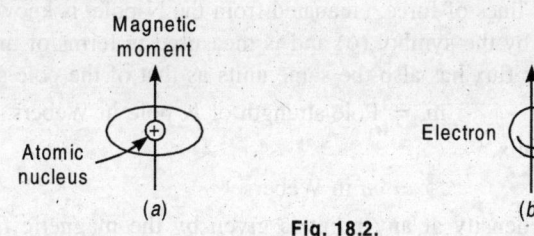

Fig. 18.2.

We know that each electron may also be thought of as spinning around an axis. The second source of magnetic moment originates from this electron spin. The direction of the magnetic moment due to electron spin is directed along the spin axis as shown in Fig. 18.2 (*b*). It may be carefully noted that spin magnetic moments may be only in an "up" direction or in an antiparallel "down" direction.

Thus each electron in an atom may be thought of as being a small magnet having permanent orbital and spin magnetic moments.

As a matter of fact, the most fundamental magnetic moment is the "Bohr magneton". It is represented by the symbol, μ_B. Its magnitude is equal to 9.27×10^{-24} A.m². For each electron in an atom, the spin magnetic moment is $\pm \mu_B$. The plus sign is considered for spin up, and negative sign for spin down. Moreover, the orbital magnetic moment contribution is equal to $m_l \mu_B$ where m_l is the magnetic quantum number of the electron.

It has been observed experimentally that in each individual atom, orbital moments of some electron pairs cancel each other. This fact also holds true for the spin moments. For example, the spin moment of an electron with spin up will cancel that of one with spin down. Then the net magnetic moment for an atom is just the sum of the magnetic moments of each of the constituent electrons, including both orbital and spin contributions and of course taking into account moment cancellation.

For an atom with completely filled electron shells or subshells, when all the electrons are considered, there is a total cancellation of both orbital and spin moments. Thus materials composed of atoms having completely filled electron shells are not capable of being permanently magnetised. This category includes the inert gases (like He, Ne, Kr, Ar, Xn), as well as some ionic material. All the other materials composed of atoms having incompletely filled electron shells are capable of having permanently magnetised. The magnetism in materials is of three main types: (a) diamagnetism, (b) paramagnetism and (c) ferromagnetism. The antiferromagnetism and ferromagnetism are considered to be subclasses of ferromagnetism. All these types of magnetism are discussed later in this chapter.

18.5 MAGNETIC FIELD STRENGTH

The magnetic field strength at any point is measured by the force experienced by a North pole of one weber placed at that point. It is, generally, designated by the letter 'H' and its units are expressed in terms on newton per weber (N//Wb) or ampere turns per metre (AT/m).

Now let m = Pole strength of N-pole,

F = Force experienced by N-pole when placed at some point in the magnetic field.

∴ Magnetic strength,

$$H = \frac{F}{m} \text{ in N/Wb or AT/m}$$

18.6 MAGNETIC FLUX DENSITY

The total number of lines of force, emanated from the N-pole, is known as magnetic flux. It is generally, represented by the symbol (ϕ) and is measured in terms of units called Webers. In other words, the magnetic flux has also the same units as that of the pole strength.

Let m = Pole strength of N-pole in Webers

Now the magnetic flux,

$$\phi = m \text{ in Webers}$$

The magnetic flux density at any point is given by the magnetic flux passing normally through the unit cross-section of that point. It is, generally, designated by the letter (B) and is measured in terms of Weber per square metre or tesla. (Remember 1 tesla = 1 Weber /m²)

Now let ϕ = Total magnetic flux, and

A = Area of cross section.

Then the magnetic flux density,

$$= \frac{\phi}{A} \text{ in Wb/m}^2 \text{ or tesla}$$

18.7 MAGNETIC PERMEABILITY

Consider an unmagnetised bar of magnetic material placed in a uniform magnetic field as shown in Fig. 18.3 (a).

It has been observed that the bar gets magnetised by induction and develops a polarity. After magnetisation, the magnetic lines in the bar emanate from N-pole, pass through the outer region and then re-enter the S-pole as shown in Fig. 18.3 (a). These lines form a closed loop within the magnet by passing from S-pole to N-pole. It will be interesting to know, that the lines of the

Magnetic Materials

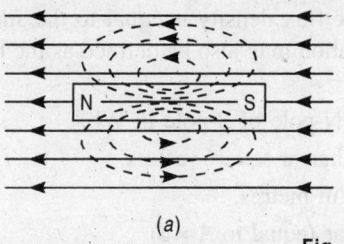

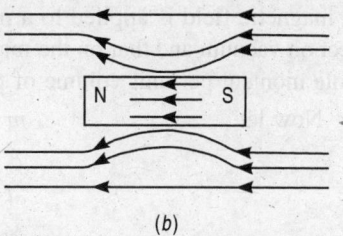

Fig. 18.3. Magnetic permeability.

magnetised bar oppose the lines of the original field outside the magnet and favour inside the magnet. The resultant field is shown in Fig. 18.3 (b). As a result of this, the magnetic field strength (H) is increased inside the magnet bar and decreased outside it. Similarly, the magnetic flux density (B) becomes high inside the magnetic bar and low outside it. Thus we find the flux density (B) is directly proportional to the magnetic field strength (H). Mathematically flux density,

$$B \propto H$$
$$= \mu \cdot H$$

where μ is a constant of proportionality and is known as permeability or absolute permeability of the medium. It the flux density is established in air or vacuum or in a non-magnetic material, by keeping it in the same magnetic field strength (H), then the above equation may be written as,

$$B_0 = \mu_0 \cdot H$$

where B_0 is the flux density in air or vacuum and μ_0 is the absolute permeability of air or vacuum. In S.I. system of units the value of μ_0 is $4\pi \times 10^{-7}$ henry/metre.

The ratio μ/μ_0 is known as the relative permeability of the medium. It is designated by the symbol μ_r. Mathematically relative permeability.

$$\mu_r = \frac{\mu}{\mu_0} = \frac{\text{Absolute permeability of the medium}}{\text{Absolute permeability of air}}$$

Since $B = \mu \cdot H$ and $B_0 = \mu_0 \cdot H$, therefore for the same value of magnetic field strength (H), the relative permeability of a medium may be defined as the ratio of magnetic flux density (B) established in a medium to the magnetic flux density (B_0) established in air. Therefore relative permeability,

$$\therefore \quad \mu_r = \frac{\mu}{\mu_0} = \frac{B}{B_0}$$

It may be noted that the relative permeability of air a non-magnetic material is unity. But for certain nickel iron alloys, its value may be as high as 10^5.

18.8 MAGNETIZATION

The term 'magnetization' may be defined as the process of converting a non-magnetic bar into a magnetic bar. This term is analogous to the polarization in dielectric materials. We have already discussed in the last article that the flux density

$$B = \mu \cdot H = \mu_0 \cdot \mu_r H \quad \quad \ldots (\because \mu = \mu_0 \cdot \mu_r)$$
$$= \mu_0 \cdot \mu_r \cdot H + \mu_0 \cdot H - \mu_0 \cdot H$$
$$\quad \quad \ldots \text{(Adding and subtracting } \mu_0 \cdot H)$$
$$= \mu_0 \cdot H + \mu_0 (\mu_r - 1) H$$
$$= \mu_0 \cdot H + \mu_0 \cdot M$$

where M is equal to $(\mu_r - 1) H$ and is known as magnetization. It is expressed in terms of coulomb/metre/second (C/m sec) or ampere/metre (A/m). From the above equation, we find that if a magnetic field is applied to a material, the magnetic flux density is equal to the sum of the effect on vacuum and that on the material. The magnetization may also be defined as the magnetic dipole moment per unit volume of the bar.

Now let
- m = Pole strength of N-pole of S-pole in Wb,
- A = Pole area of the bar in square metres,
- l = Magnetic length in metres,
- V = Volume of the bar (equal to $A \times l$)

∴ Magnetic dipole moment or magnetic moment
$$= M \times l$$

and magnetization $= \dfrac{\text{Magnetic moment}}{\text{Volume of the bar}} = \dfrac{m \times l}{A \times l} = \dfrac{m}{A}$

18.9 MAGNETIC SUSCEPTIBILITY

The term 'magnetic susceptibility' may be defined as the ratio of magnetization to the magnetic field strength. It is, generally, designated by the symbol χ_m (known as chi-m)

Let
- M = Magnetization, and
- H = Magnetic field strength.

We have already discussed in Art. 18.7 that the magnetization
$$M = (\mu_r - 1) H$$

∴ Magnetic susceptibility,
$$\chi_m = \dfrac{M}{H} = \dfrac{(\mu_r - 1) H}{H} = \mu_r - 1$$

From the above equation, we find that the magnetic susceptibility is a dimensionless quantity as the term (μ_r) is a pure number. It may be noted that M (magnetization) and H (magnetic field strength) have same units.

As a matter of fact, magnetic susceptibility is an important property of a magnetic material. The sign and magnitude of magnetic susceptibility are used to determine the nature of the magnetic materials.

Table 18.1 shows the values of magnetic susceptibility (χ_m) and relative permeability (μ_r) of silver, copper, aluminium, manganese and iron.

Table 18.1

No.	Material	Susceptibility (χ_m)	Relative Permeability (μ_r)
1.	Silver	-2×10^{-5}	0.99998
2.	Copper	-0.8×10^{-5}	0.99999
3.	Aluminium	2×10^{-5}	1.00002
4.	Manganese	8.3×10^{-4}	1.00083
5.	Iron	5×10^3	5×10^3

As seen from this table, silver and copper have negative values of susceptibility whereas aluminium, manganese and iron have positive values of susceptibility.

Example 18.1. *A paramagnetic material has a magnetic field intensity of 10^4 A/m. If the susceptibility of the material at room temperature is 3.7×10^{-3}, calculate the magnetization and flux density in the material.* (Anna Univ., April, 2002)

Solution. Given: $H = 10^4$ A/m; $\chi_m = 3.7 \times 10^{-3}$

Let M = The magnetization and
B = Flux density of the material

We know that the relative permeability,
$$\mu_r = 1 + \chi_m = 1 + 3.7 \times 10^{-3}$$
and the magnetic flux density,
$$B = \mu_0 \mu_r H = (4\pi \times 10^{-7}) \times (1 + 3.7 \times 10^{-3}) \times 10^4$$
$$= 12.6 \times 10^{-3} \text{ Wb/m}^2 \textbf{ Ans.}$$

We also know that magnetisation,
$$M = (\mu_r - 1) H = x_m H$$
$$= (3.7 \times 10^{-3}) \times 10^4 = 37 \text{ A/m } \textbf{Ans.}$$

18.10 CLASSIFICATION OF MAGNETISM IN MAGNETIC MATERIALS

It has been observed that the motion of electrons around the nucleus, as well as about their own axes, creates magnetic moments in the atoms of a material. These magnetic moments produce magnetic field in an atom, which affect the nature of magnetism in the materials. This magnetism may be classified into the following three types:

1. Diamagnetism 2. Paramagnetism 3. Ferromagnetism.

Now we shall discuss all the above three types of magnetism in the following pages.

18.11 DIAMAGNETISM

The term 'diamagnetism' may be defined as a magnetism, in which a material gets weakly magnetised in the direction opposite to that of the applied field. The diamagnetism is originated due to the motion of the electrons in circular and elliptical orbits around their nuclei. Each electron moving in an orbit, is equivalent to an electron current flowing in a closed loop. This current produces a magnetic field in a direction at right angles to the plane of the orbit. According to the law of electromagnetism, this magnetic field induces a magnetic moment in the atom in a direction opposite to it. We know that material contains a large number of electrons and the orbits of these electrons are randomly oriented in space. Therefore the magnetic moments of all such electrons are also randomly oriented. As a result of this, the magnetic moment of all the electrons gets cancelled. Thus the net magnetism in the material is zero.

It has been observed that when an external field is applied to a material, its electrons experience a force which changes their angular momentum. It also changes the orientation of their magnetic moments in the atoms. As a result of this, some of the magnetic moments change their orientation with respect to the direction of the field, while others remain unchanged. The net effect of this phenomenon is that the magnetic moments in an atom do not cancel each other completely. But some magnetism is still left in the atom. This magnetism, which is in the opposite direction to that of the external field, is known as diamagnetism.

It will be interesting to know that diamagnetism is present in all the material since the motion of electrons is a universal phenomenon. But the materials, in which the diamagnetism is present to a large extent are copper, gold, germanium, silicon, etc.

18.12 SUSCEPTIBILITY OF DIAMAGNETIC MATERIALS

It has be observed that the susceptibility (χ_m) of a diamagnetic material is of the order of -10^{-5}. The negative sign indicates that the material is magnetized in a direction opposite to that of the applied field. As a matter of fact, the direction of the applied field is taken as positive.

It has also been observed that the susceptibility of a diamagnetic material is independent of the change in temperature.

18.13 B-H CURVE FOR DIAMAGNETIC MATERIALS

We have already obtained in Art. 18.9 and equation for the magnetic susceptibility, i.e.,

$$\chi_m = \mu_r - 1$$

or $\quad \mu_r = 1 + \chi_m$

Since the value of χ_m for a diamagnetic material is negative, and very small as compared to unity, therefore the value of relative permeability (μ_r) is slightly less than unity. We have also obtained the following equation for flux density in Art. 18.8

$$B = \mu_0 \cdot \mu_r \cdot H$$

Now if we substitute the value μ_r equal to unity is the above equation, we find that

$$B = \mu_0 \cdot H$$

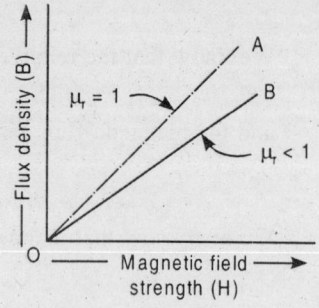

Fig. 18.4. B-H curve for diamagnetic material.

If we plot a graph with flux density (B) along vertical axis and magnetic field strength (H) along horizontal axis, we shall obtain a curve OA as shown in Fig. 18.3. Such a curve is known as B-H curve. This curve is for air. Now if we plot a similar curve, for diamagnetic material by substituting the actual value of μ_r, we shall obtain a curve OB as shown in the figure. Since the value of μ_r is slightly less than unity, therefore the slope of B-H curve for a diamagnetic material, is slightly less than that of air as shown in the figure.

18.14 PARAMAGNETISM

The term 'paramagnetism' may be defined as a type of magnetism in which the material gets weakly magnetized in the same direction as that of the applied field. The materials, which exhibit paramagnetism are alkali metals, transition metals and rare earth elements, (i.e., lanthanides and actinides).

The origin of paramagnetism arises due to the presence of permanent magnetic moments in the atoms or molecules. The permanent magnetic moments in single atom is created due to:

1. Motion of an electron in circular and elliptical orbits around the nucleus (called orbital normal motion).
2. Motion of an electron about its own axis (called spin motion).

It has been observed that the magnetic moment, created due to the orbital motion of electrons, disappear due to the effect of electric field of the neighbouring charges. But *magnetic moments due to spin motion of electrons remain unaffected by this field. It will be interesting to know that the magnetic moments are composed of two groups. One of the groups contains electrons with clockwise spin and the other electrons with anticlockwise spin. In the absence of external field, these magnetic moments are randomly distributed. Or in other words, they have no mutual interaction among them.

As a matter of fact, when the external field is applied, the magnetic moments tend to line up in the direction of field. If there is no opposing force, all the magnetic moments line up, and material acquires a large magnetization. But the thermal agitation of the atoms opposes such a tendency of the magnetic moments, and keeps them at random. This results in a partial alignment of the magnetic moments in the direction of the external field and the material is weakly magnetized.

---

* The magnetic moment of an electron is taken as one unit and is known as one Bohr magneton. Its value is 9.273×10^{-24} A/m^2.

Magnetic Materials

18.15 SUSCEPTIBILITY OF PARAMAGNETIC MATERIALS

It has been observed that the susceptibility (χ_m) of a paramagnetic material is positive and of the order 10^{-3}. The positive value of the susceptibility indicates that the material is magnetized in the same direction as that of the applied field. It has also been found that the susceptibility of paramagnetic material varies inversely with the change in temperature.

Mathematically, susceptibility of a paramagnetic material is given by the relation:

$$\chi_m \propto \frac{1}{T}$$

$$= \frac{C}{T}$$

where T = Absolute temperature in degrees Kelvin, and

C = A constant of proportionality known as Curie constant.

The above expression is known as Curie's Law. This law holds good only for paramagnetic gases and dilute solutions containing magnetic atoms or ions. It has been observed that the solids obey a modified Curie's Law, which is known as Curie-Weiss's Law

This law is given by the relation,

$$\chi_m = \frac{C}{T - \theta}$$

where C = Curie's constant,

T = Absolute temperature in degrees Kelvin, and

θ = Another constant known as paramagnetic curie temperature.

The above relation holds good only for those values of T_1, which are considerably higher than θ. It has been observed that for most of the paramagnetic materials, the value of θ is very low and sometimes, even, negative. It will be interesting to know that the susceptibility of alkali metals and some alkaline earth metals is independent of temperature.

18.16 B-H CURVE FOR A PARAMAGNETIC MATERIALS

We have already obtained in Art. 18.9, an expression for the magnetic susceptibility, i.e.,

$$\chi_m = \mu_r - 1$$
$$\mu_r = 1 + \chi_m$$

Since the value of χ_m of a paramagnetic material is positive and very small as compared to unity, therefore the value of relative permeability (μ_r) is slightly greater than unity.

Now if we proceed in the same way as discussed in Art. 18.12 and plot a graph with flux density (B) along vertical axis and magnetic field strength (H) along horizontal axis, we shall obtain curve as shown in Fig. 18.5. The curve OA represents the B-H curve for air and the curve OB represents the B-H curve for a paramagnetic material. Since the value of μ_r is slightly greater than unity, the slope of the B-H curve for a paramagnetic material is slightly greater than that of air as shown in the figure.

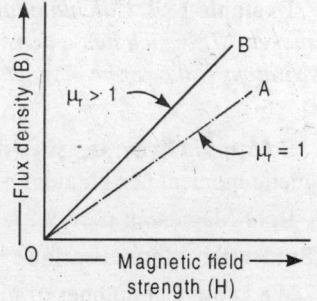

Fig. 18.5. B-H curve for paramagnetic material.

18.17 FERROMAGNETISM

The term 'ferromagnetism' may be defined as a type of magnetism in which a material gets magnetized to a very large extent in the presence of external field. The direction, in which the material gets magnetized is the same as that of the external field.

The origin of ferromagnetism arises due to the presence of permanent magnetic moments in the atoms or molecules of the material. It has been observed that when the external field is applied, the magnetic moments line up in the same direction as that of the external field. This alignment takes place in the same way as in paramagnetism. We have already discussed that in paramagnetism, the magnetic moments are unable to align themselves in the direction of the field, due to thermal agitation between the atoms. But in ferromagnetism, there is no such force, which may oppose the alignment of magnetic moments. Thus the material is magnetized to a very large extent. Some of the materials, which possess ferromagnetism are iron, cobalt, nickel etc. The relative permeability (μ_r) of these materials is very high i.e., to the order of 10^5.

The maximum possible magnetisation, M_s of a ferromagnetic material represents the magnetisation that results when all the magnetic dipoles in a solid piece are mutually alligned with the external field. The corresponding saturation flux density is B_S. The value of the saturation magnetisation is given by the equation:

$$M_S = \text{Net magnetic moment for each atoms} \times \text{No. of atoms present}$$

For each of iron, cobalt and nickel, the net magnetic moments per atom are 2.22, 1.72 and 0.60 Bohr magnetions respectively. Thus the saturation magnetisation for nickel is given by,

$$M_S = 0.60 \, \mu_B \times \text{No. of atoms present.} \qquad \ldots (i)$$

Now, recall the discussion from chapter 3, the number of atoms present in a material,

$$n = \frac{\rho N_A}{A} \text{ atoms/m}^3$$

where
ρ = density of a material (kg/cm³)
N_A = Avogadro's number (= 6.023×10^{23} atoms/mol)
A = Atomic weight of the material.

∴ The equation for saturation magnetisation in nickel can be rewritten as,

$$M_S = 0.60 \, \mu_B \, n \, (\text{A.m}^2) \qquad \ldots (ii)$$

Note: If the density (ρ) is substituted in g/cm³, then the number of atoms (n) will also be in atoms/cm³. So in that case, in order to use the value of 'n' in equation (ii) the number of atoms "n" must be converted from atoms/cm³ to atoms/m³.

Example 18.2. *Calculate (a) the saturation magnetisation and (b) the saturation flux density for nickel (Ni), which has a density of 8.9 g/cm³. Atomic weight of Ni = 58.71, magnetic moment per Ni atom = 0.6. Bohr magneton: $\mu_B = 9.27 \times 10^{-24}$ A m² and $\mu_0 = 4\pi \times 10^{-7}$ H/m.*

(AMIE., Summer, 2000)

Solution. Given: density (d) = 8.9 gm/cm³ = 8900 kg/m³, Atomic weight = 58.71 g/mol; Magnetic moment per Ni atom = 0.6

Bohr Magneton; $\mu_B = 9.27 \times 10^{-24}$ A.m²; $\mu_0 = 4\pi \times 10^{-7}$ H/m.

(a) Saturation Magnetisation

We know that number of nickel atoms/cm³,

$$n = \frac{\rho N_A}{A} = \frac{8.9 \times (6.023 \times 10^{23})}{58.71}$$

$$= 9.13 \times 10^{22} \text{ atoms/cm}^3$$

$$= 9.13 \times 10^{28} \text{ atoms/m}^3$$

and the saturation magnetisation,

$$M_S = 0.60 \, \mu_B \, n$$

$$= 0.60 \times (9.27 \times 10^{-24}) \times (9.13 \times 10^{28})$$

$$= 5.1 \times 10^5 \text{ A/m } \textbf{Ans.}$$

(b) Saturation flux density

We know that saturation flux density,

$$B_S = \mu_0 M_S$$
$$= (4\pi \times 10^{-7}) \times (5.1 \times 10^5)$$
$$= 0.64 \text{ tesla } \textbf{Ans.}$$

Example 18.3. *The saturation magnetic induction of Ni is 0.65 Wb/m². If the density of Ni is 8906 kg/m³ and its atomic weight is 58.7, calculate the magnetic moment of the Ni atom in Bohr magneton.* (*Anna Univ., Dec. 2001*)

Solution. Given: Saturation magnetic induction, $B_S = 0.65$ Wb/m²; $\rho_{Ni} = 8906$ kg/m³ = 8.906 g/cm³; Atomic weight, $A = 58.7$.

Let m = magnetic moment per atom of Ni and
$m\mu_B$ = magnetic moment in Bohr magneton
M_S = saturation magnetisation

We know that number of nickel atoms/cm³.

$$n = \frac{\rho N_A}{A} = \frac{8.9 \times (6.023 \times 10^{23})}{58.71}$$
$$= 9.13 \times 10^{22} \text{ atoms/cm}^3$$
$$= 9.13 \times 10^{28} \text{ atoms/m}^3$$

and saturation magnetic induction of nickel (B_S)

$$0.65 = \mu_0 M_s = 4\pi \times 10^{-7} \times M_S$$

$$\therefore M_S = \frac{0.65}{4\pi \times 10^{-7}} = 5.17 \times 10^5 \text{ A/m}$$

We also know that saturation magnetisation, (M_S),

$$5.17 \times 10^5 = m \mu_B n$$
$$= (m\mu_B) \times 9.13 \times 10^{28}$$

$$\therefore m \mu_B = \frac{5.17 \times 10^5}{9.13 \times 10^{28}} = 5.67 \times 10^{-24} \text{ A.m}^2 \textbf{ Ans.}$$

18.18 THE DOMAIN STRUCTURE

The term ferromagnetism was explained nicely by Wiess. He suggested that the group of atoms are organised into tiny bounded regions called *domains*. In each individual domain the magnetic moments of the atoms are aligned in the same direction. Thus the domain is magnetically saturated and behaves like a magnet with its own magnetic moment and axis. In an

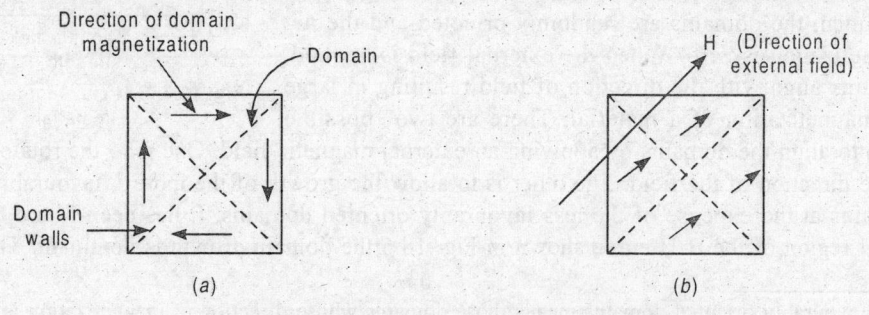

Fig. 18.6. Domain structure in a ferromagnetic material.

unmagnetized sample, the domains are randomly oriented as shown in Fig. 18.5 (*a*), so that the magnetization of the specimen as a whole is zero. The boundaries separating the domains are called domain walls. The domain walls are analogous to the grain boundaries in a polycrystalline material. However, the domain walls are thicker than the grain boundaries. The domain walls can also exist in a grain. Like grain growth, the domain size can also grow due to the movement of domain walls.

When a magnetic field is applied externally to a ferromagnetic material, the domain align themselves with the field as shown in Fig. 18.5 (*b*). This results in a large net magnetization of the material.

18.19 SUSCEPTIBILITY OF FERROMAGNETIC MATERIALS

It has been observed that the susceptibility of a ferromagnetic material is very high but below a certain temperature (T_c) known as ferromagnetic curie temperature or simply curie point. If the temperature is increased above the curie point, the behaviour of a ferromagnetic material becomes similar to that of a paramagnetic material. It is because, of the fact that the increase in temperature reduces the magnetic strength of the material due to the increased vibrations of atoms and molecules.

Table 18.2 shows the Curie temperature for iron, cobalt and nickel. As seen from this table, the curie temperature for iron is 770°C, for cobalt, 1131°C and nickel, 358°C.

Table 18.2

No.	Material	Curie temperature (°C)
1	Iron	770
2	Cobalt	1131
3	Nickel	358

18.20 B-H CURVE FOR FERROMAGNETIC MATERIALS.

The B-H curve for a ferromagnetic material may be plotted in the same way as discussed for diamagnetic and paramagnetic materials as shown in Fig. 18.6.

From the B-H curve for a ferromagnetic material, we find that the curve is not a straight line. This indicates that the relation between flux density (*B*) and magnetic field strength (*H*) in a ferromagnetic material is not linear.

The characteristic shape of the B-H curve may be explained by considering the domain behaviour in a ferromagnetic material. We know that in an unmagnetised specimen, the domains are randomly oriented and the net magnetization is zero. When the external field is applied, domains align with the direction of field resulting in large net magnetization of a material. There are two possible

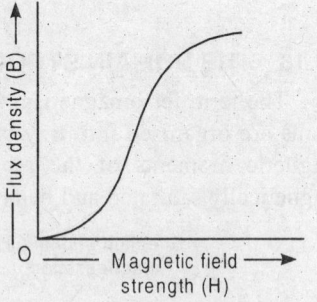

Fig. 18.7. B-H curve for ferromagnetic material.

ways to align the domains by applying an external magnetic field. One is to the rotate a domain in the direction of the field. The other is to allow the growth of the more * favourably oriented domains at the expense of the less favourably oriented domains. It has been found that in the initial region of the B-H curve shown in Fig. 18.6 the domain growth is dominant. The domain

---

* Favourably oriented domains means those domains whose direction of magnetization is parallel or nearly parallel to the direction of the external field.

Magnetic Materials

growth occurs as shown in a sequence of Fig. 18.8 (*a*), (*b*) and (*c*). Fig. 18.8 (*a*) shows an unmagnetized specimen in which domains are randomly aligned.

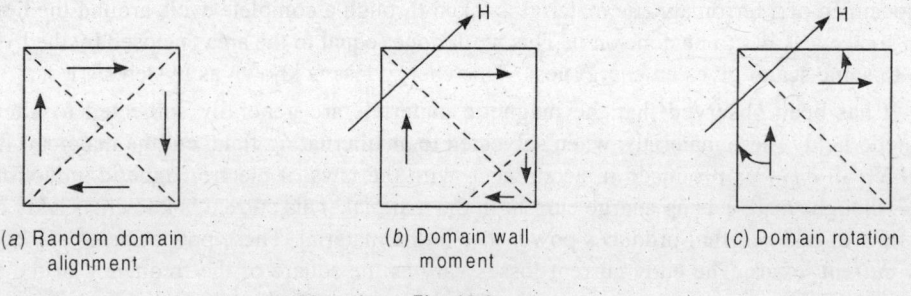

(a) Random domain alignment
(b) Domain wall moment
(c) Domain rotation

Fig. 18.8.

When a small magnetic field is applied, the domains with magnetization direction parallel or nearly parallel to the field, grow at the expense of others as shown in Fig. 18.7 (*b*). The domain growth occurs due to the movement of domain walls away from the minimum energy state. As the magnetic field is increased to a large value (*i.e.*, near saturation) further domain growth becomes impossible. Therefore, most favourably oriented and fully grown domains tend to rotate so as to be in complete alignment with the field direction as shown in Fig. 18.7 (*c*). It may be noted that the energy required to rotate and entire domain is much more than that required to move the domain walls during growth. It is due to this fact that slope of the *B-H* curve decreases on approaching saturation, as shown by the curve *AB* in Fig. 18.6

18.21 MAGNETIC HYSTERESIS

The term magnetic hysteresis may be defined as a phenomenon of variation of flux density (*B*) with the change in magnetic field strength (*H*) in a ferromagnetic material. It may be best understood from the following discussion:

Consider an unmagnetized specimen of ferromagnetic material subjected to an external magnetic field. We know that when the magnetic field strength is gradually increased form zero, the flux density increases rapidly as shown by the curve *OA* in Fig. 18.9. It has been observed that if the electric field strength is further increased, the rate of increase in flux density is reduced, until a saturation is reached as shown by the curve *AB* in the figure.

Now if the magnetic field strength is reversed to zero, then the flux density decreases along the curve *BAC*. But it does not become zero. The value of flux density (represented by (*OC*), which is still left in the material, is known as remnant flux density or residual magnetism. It designated by the symbol (B_r). In order to make the flux density to zero, a certain amount of magnetic field strength (represented by *OD*) is applied in the reverse direction. This value of magnetic field strength is known as coercive field and is designated by the symbol (H_c). It has been observed that if the field strength is further increased in the negative direction, the flux density increases rapidly along the curve *DE*, until a negative saturation is reached at *E*. Now if the magnetic field strength is again reversed to make it positive, the flux density will follow the curve *EFGB* as shown in the figure.

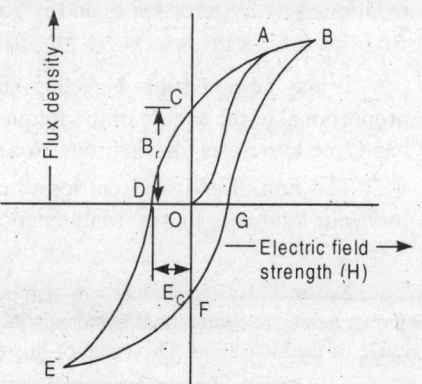

Fig. 18.9. Magnetic hysteresis.

It may be noted from the figure that the curve *ABCDEFGBA* is a closed one. Moreover, the flux density always lags behind the magnetic field strength. This phenomenon is known as magnetic hysteresis and the closed curve as hysteresis loop. It will be interesting to know, that if the specimen of a ferromagnetic material is taken through a complete cycle around the hysteresis loop, some work has to be done on it. This work done (equal to the area enclosed by the hysteresis loop to some scale) gives an energy loss. This energy loss is known as hysteresis loss.

It has been observed that the magnetic materials are generally subjected to alternating magnetic field. These materials, when subjected to an alternating field, cut the magnetic flux due to which an *e.m.f.* is produced in accordance with the laws of electromagnetic induction. This *e.m.f.*, though small, sets up a large current in the material. This current is known as eddy current. The flow of eddy current produces power loss in the material. These power losses are known as eddy current losses. The eddy current losses vary as the square of the frequency of the applied alternating field.

18.22 HYSTERESIS AND EDDY CURRENT LOSS

It has been experimentally found that the hysteresis loss of a magnetic material depends upon the following factors :

1. The maximum flux density that can be established in a magnetic material (B_{max}).
2. The magnetic quality of a material (η),
3. Frequency of reversal of magnetisation (f),
4. Volume of the magnetic material (V).

Combining all the above mentioned factors, the hysteresis loss is given by the relation.

$$W_h = \eta \cdot B_{max}^{1.6} \cdot f \cdot V \text{ watt}$$

The index 1.6 is empirical and holds good, if the value of B_{max} lies between 0.1 and 1.2 Wb/m^2. If B_{max} is less than 0.1 Wb/m^2 or greater than 1.2 Wb/m^2, the index value is greater than 1.6.

The eddy current loss of a magnetic material also depends upon the same factors as mentioned by hysteresis loss and is given by the relation,

$$W_e = \eta \cdot B_{max}^2 \cdot f^2 \cdot V \text{ watt}$$

It may be noted that here the index value of B_{max} is 2. Moreover, the eddy current loss is proportional to the square of the frequency of the reversal of magnetisation (*i.e.*, f^2). On the other hand, the hysteresis loss is proportional to 'f'.

The sum of eddy current losses and hysteresis loss are some times referred to as total iron losses or total core losses. Mathematically total iron loss,

$$W_i = W_h + W_e$$

Notes: 1. Hysteresis loss can also be obtained from the area of the hysteresis loop (or B/H loop) as shown below. We know that hysteresis loop represents the energy spent in taking the iron bar through one cycle of magnetisation. The value of this energy spent per cycle is given by the relation,

$$\text{Energy spent/cycle} = \text{(Area of B–H loop)} \times V \text{ joules}$$

where V is the volume of the iron bar and is equal to $A \times l$ where A is the area of cross-section and l is the length of the iron bar.

It the flux density is reversed at a rate of f cycle/sec (or Hz), then the net energy spent is called hysteresis loss. Mathematically the hysteresis loss,

$$W_h = \text{(Area of the B–H loop)} \times V \times f \text{ joules/sec or watt}$$

2. It may be noted that while calculating the area of the *B/H* loop, the scale factors of *B* and *H* must be taken into consideration. For example, if the scales are

1 cm = x AT/m for H and 1cm = y Wb/m^2 for B then hysteresis loss,

$$W_h = x\,y \text{ (Area of B–H loop)} \times V \times f \text{ watt}$$

In the above expression, the loop area is to be in cm^2.

Example 18.4 *Determine the power loss due to hysteresis in transformer core of 0.01 m^3 volume of 50 Hz frequency. Take area of the loop as 600 jm^{-3}.*

(A.M.I.E. Summer, 1993)

Solution, Given: $V = 0.01$ m^3 ; $f = 50$ Hz and ; Area of the loop = 600 jm^{-3}.

We know that the power loss due to hysteresis,

$$W_h = \text{(Area of the loop)} \times V \times f \text{ watt}$$
$$= 600 \times 0.01 \times 50 = 300 \text{ watts } \textbf{Ans.}$$

Example 18.5. *The hysteresis loop of a specimen of iron having mass of 10 kg is equivalent in area to 250 J/m^3 of iron. Find the loss of energy per hour at the rate of 50 cycles/sec. Assume the density of iron as 7500 kg/m^3.*

Solution. Given: Energy loss = 250 J/m^3 (per cycle).

Energy loss at the rate of 50 cycles/s

$$= 250 \times 50 = 12500 \text{ J/m}^3$$

Energy loss per hour,

$$= 12500 \times 3600 = 4.5 \times 10^7 \text{ J/m}^3$$

Density of iron = 7500 kg/m^3.

$$\text{Volume} = \frac{\text{mass}}{\text{density}} = \frac{10 \text{ kg}}{7500 \text{ kg/m}^3} = 1.33 \times 10^{-3} \text{ m}^3$$

or energy loss per hour

$$= \frac{4.5 \times 10^7}{1.33 \times 10^{-3}} = 3.41 \times 10^{10} \text{ J/hour } \textbf{Ans.}$$

18.23 FERRIMAGNETISM

We have already discussed in Art. 18.16 that if all the magnetic moments line up in the direction of the field, the substance is known as ferromagnetic. We have also discussed that in ferromagnetic materials, the magnetic moments are composed of two groups, which consist of electrons with opposite spin directions. It has been observed that in some materials, if both the groups are aligned in opposite directions and their magnetic moments are equal, the substance is known as antiferromagnetic. In these materials, the net magnetization is zero. However, the magnetic moments of the two groups may not be equal, if the substance consists of two different elements. As a result of this, complete cancellation of the magnetic moments do not take place and it results in some net magnetization. This phenomenon is known as ferrimagnetization. The materials, which exhibit this type of behaviour, are known as ferrimagnetic materials or simply ferrites.

18.24 FERRITES

These are complex compounds of various metals and oxygen which exhibit the phenomenon of ferromagnetism. They have a chemical formula of the form $MgAl_2O_4$. This indicates that the ferrite $MgAl_2O_4$ is a compound of magnesium and aluminium oxides. The other ferrites can be formed by replacing magnesium by other divalent metals (*i.e.,* metals having two valence elements). The commonly used divalent metals are iron, copper, nickel, cobalt, zinc, cadmium by ferric ions, which is trivalent metal (*i.e.,* metal having three valence electrons).

It will be interesting to know that the ferrite, so obtained, is non-magnetic, what when other metals, mentioned above, are used the ferrites obtained are magnetic and have a high permeability. The ferrites have a very high resistivity (greater than 10^5 ohm-cm) and extremely low dielectric loss. Thy have low eddy current loss, but large hysteresis loss. These days, the ferrites are widely used in high frequency transformers and inductors. They are also used in microwave applications and as computer memory core elements.

18.25 HIGH FREQUENCY APPLICATIONS OF FERRITES

It has been found that metals and alloys are not suitable for high frequency operation as required in communication equipments. This is due to the reason that eddy current losses at high frequencies become very large. However, ferrites are very useful for high frequency applications because their electrical resistivity is very high ($\approx 10^6$) than that of metals and alloys. Because of this, they behave as electrical insulators. This reduces the eddy current losses to a negligible value and hence makes ferrites very useful at high frequencies. However, the choice of ferrite material depends upon the application. Thus for audio and TV transformers, nickel-zinc ferrites are more suitable. For microwave isolators and gyrators, which operate in KHz and MHz range, magnesium manganese ferrites are highly suitable. For computer memory cores, magnesium-manganese ferrites with a higher magnesium to manganese ratio than that of microwave isolators are more suitable. For microwave isolators operating in GHz (gigahertz) range, garnets such as YIG (yttrium-iron garnet) are widely used.

18.26 CLASSIFICATION OF MAGNETIC MATERIALS

The magnetic materials may be classified into the following two types depending upon the ease with which they can be magnetised or demagnetised:

1. Hard magnetic materials.   2. Soft magnetic materials.

18.27 HARD MAGNETIC MATERIALS

These are the materials, which are difficult to magnetize or demagnetize. It has been observed that the hard magnetic materials retain a considerable amount of magnetic energy, even after the magnetic field is removed. These materials are also known as permanent magnets and are widely used in meters, motors, electron tubes, transducers etc.

(a) B – H curve of hard magnetic materials

(b) B – H curve of soft magnetic materials

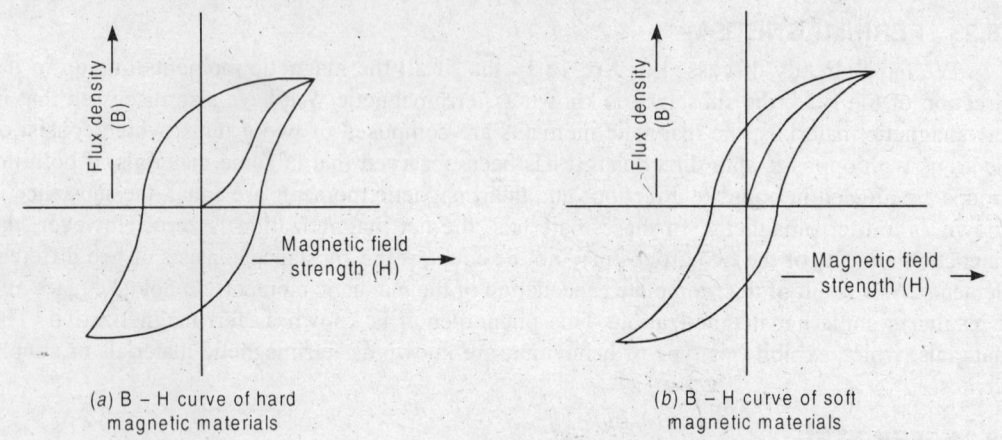

Fig. 18.10.

The permanent magnets must have a high remnant flux density (B_r) and a large coercive field (H_c). This means that area of the hysteresis loop must be as large as possible as shown in Fig. 18.10 (a) The area of the loop represents the energy required to demagnetize the permanent magnet. The maximum value of this area is ($B_r \cdot H_c$) and is called energy product. The energy must be as larger as possible for permanent magnets.

It may be noted that hysteresis losses are of no significance in permanent magnets. This is due to the reason that in permanent magnets, no reversal of magnetisation is required.

It will be interesting to know that to keep the permanent magnets magnetised for a longer time, the movement of domain walls must be prevented. This means that the resistance to domain wall motion must have a large value. It has been found that the same factors, which improve the mechanical hardness, impart better resistance to domain wall motion in permanent magnets.

Magnetic Materials

Table 18.3 Properties of hard magnetic materials

S. No.	Material	B_r Wb/m²	H_c kA/m	$B_r \cdot H_c$ kJ/m³
1.	High Carbon steel (0.9% C)	0.9	3.98	3.58
2.	Tungsten steel (5% W)	1.05	5.57	5.85
3.	Chromium steel (4% Cr)	0.95	5.17	4.91
4.	Cobalt steel (36% Co)	0.95	18.31	17.40
5.	Al-Ni-Co alloys 24% Co, 14% Ni, 8% Al)	0.8 – 1.2	60 – 120	48 – 144
6.	Barium ferrite (Ba 0.6 Fe₂O₃)	0.21	140	29.4
7.	Cobalt rare earths CO₅ (Sm, Pr)	1.0	200	200

Table 18.3 shows some of the important permanent magnet materials and their properties. Among the metallic alloys, the best known materials are the alloys of aluminium nickel and cobalt. These alloys are commonly known as alnico alloys. Among the non-metallic alloys, the barium ferrite is the most commonly used permanent magnet. This material has a coercive field of about 100 kA/m. The most recent materials developed as permanent magnets are the materials of cobalt alloyed with rare earth elements like samarium (*Sm*), praseodymium (*Pr*) etc. These materials have a very large value of energy products of the order of 200 kJ/m³.

18.28 SOFT MAGNETIC MATERIALS

These are the materials, which are easy to magnetize and demagnetize. It has been observed that soft magnetic material retains a small amount of magnetic energy, even, after the magnetic field is removed. These materials are also known as temporary magnets and are widely used in transformer cores, electric machinery cores, memory cores in computers etc.

The soft magnetic materials require a frequent reversal of the direction of magnetization. These materials must have a high remnant flux density (B_r) and low coercive field (H_c). This means that area of the hysteresis loop must be as small as possible as shown in Fig. 18.9 (*b*). This keeps the hysteresis losses to a minimum value.

The soft magnetic materials must have easily moving domain walls. This is achieved by annealing a cold worked material. A cold worked material has a high dislocation density. The annealing reduces the dislocation density and makes motion of domain walls much easier. Soft magnetic materials must be free from impurities and inclusions.

They hysteresis losses in soft magnetic materials are reduced by manufacturing materials with a preferred orientation of grains. The preferred orientation means a direction in which the material can be easily magnetized. For example, iron can be magnetized more easily along [100] direction than along [110] direction. Moreover, it is very difficult to magnetize the iron along [111] direction. The preferred orientation can be achieved by suitable rolling schedule and final recrystallization to produce what is called texture. The same can also be achieved in some cases by casting the liquid alloy in a metal mould.

The other source of energy loss, in soft magnetic materials, is the eddy current loss. These losses can be minimized by increasing the resistivity of the magnetic medium. This can be achieved by adding about 4% silicon into iron to give Fe-Si alloys. The resistivity can also be increased with the addition of nickel, molybdenum metals etc.

Table 18.4 Properties of soft magnetic materials.

S. No.	Material	Initial relative permeability	Electrical resistivity Ω-m	Hysteresis loss J/m^3	Saturation Flux density Wb/m^2
1.	Commercial iron	250	0.1×10^{-6}	500	2.2
2.	Fe-Si alloy (4% Si)	500	0.6×10^{-6}	100	2.0
3.	Fe-Si oriented	1500	0.6×10^{-6}	90	2.0
4.	Permalloy (45% Ni)	2700	0.55×10^{-6}	120	1.6
5.	Supermalloy (79% Ni, 5% Mo)	100,000	0.65×10^{-6}	21	0.8
6.	Ni-Zn Ferrite	200—1000	-10^{-6}	35	0.4
7.	Mn-Zn Ferrite	2000	-10^{-6}	40	0.3

Table 18.4 shows some of the important properties of soft magnetic materials. It may be noted from this table that the permeability of commercial iron is 250, the electrical resistivity is 0.1×10^{-6} Ω-m and the hysteresis loss is 500 J/m^3. However, the addition of 4% of silicon into iron, increases the permeability to 500, electrical resistivity to 0.6×10^{-6} (this reduces eddy current losses) and hysteresis losses reduces to 100 J/m^3. The Fe-Si alloys are suitable for operation at power frequencies of 50-60 Hz. But they are not suitable for high frequency operation as used in communication equipment. Fe-Ni alloys such as permalloy and supermalloy are very suitable for high frequency operation. It has been found that metals and their alloys are not suitable for frequencies exceeding a MHz range. For such high frequencies ferrites and garnets are very suitable. The commonly used ferrites and garnets such as YIG (yttrium-irongarnet).

A.M.I.E. (I) EXAMINATION QUESTIONS

1. Write a short note on ferromagnetism. *(Summer, 1993)*
2. Distinguish between paramagnetism and ferromagnetism. *(Summer, 1993)*
3. Describe briefly, paramagnetism, ferromagnetism and diamagnetism. *(Winter, 1993)*
4. Explain ferromagnetism. *(Summer, 1994)*
5. Discuss the paramagnetism, ferromagnetism and diamagnetism. *(Winter, 1994)*
6. Compare the properties of diamagnetic, paramagnetic and ferromagnetic materials.

   (Summer, 1995)
7. Describe briefly paramagnetism, ferromagnetism and diamagnetism. *(Summer, 1996)*
8. What are hard and soft magnetic materials Explain giving examples and applications.

   (Winter, 1996)
9. Discuss the behaviour of ferro, para and diamagnetic materials in detail. *(Winter, 1996)*
10. What are ferrites ? What are they used for : *(Winter, 1996)*
11. Explain the terms
    (*i*) Magnetisation,
    (*ii*) Demagnetisation,
    (*iii*) Paramagnetism and
    (*iv*) Ferromagnetism. *(Summer, 1997)*
12. What is the permeability of diamagnetic material ? *(Summer, 1997)*

Magnetic Materials

13. Define the following terms in brief:
    (a) Hysteresis loss
    (b) Eddy current loss
    (c) Soft and hard magnetic materials
    (d) Permalloys
    (e) Preferred orientation *(Winter, 1999)*
14. (a) What is diamagnetism and paramagnetism ? Explain in brief. Name one material of each type.
    (b) Cite the major similarities and differences between ferromagnetic and ferrimagnetic materials. *(Summer, 1999)*
15. Explain ferromagnetism with a B-H loop. *(Winter, 2000)*

PROBLEMS

1. Calculate the energy lost per hour in the specimen of iron subjected to a magnetisation at 50 cycles/sec. The mass of the specimen is 50 kg and the hysteresis loop is equivalent in area to 250 J/m^3. Density of iron is 7500 kg/m^3. (**Ans.** 1.5×10^7 J)

MULTIPLE CHOICE QUESTIONS

1. The magnetic properties of materials are a consequence of
   (a) magnetic moments associated with electrons
   (b) magnetic moments associated with protons
   (c) magnetic moments associated with neutrons
   (d) magnetic moments assoicated with the nucleus
2. The relative permeability of a magnetic materials is
   (a) 0   (b) 1   (c) 5   (d) 5000
3. The relative permeability of a diamagnetic material is
   (a) < 1   (b) = 1   (c) > 1   (d) > 5000
4. Hysteresis loss in a magnetic material is proportional to
   (a) $1/f$ (b) $1/f^2$ (c) f^2 (d) f
5. The cddy current loss is a magnetic material is proportional to
   (a) $1/f$ (b) $1/f^2$ (c) f^2 (d) f

ANSWERS

1. (a)   2. (d)   3. (a)   4. (d)   5. (c)

APPENDIX A

Values of Physical Constants

S. No	Constant	Symbol	Value
1.	Atomic mass unit	a.m.u.	1.660×10^{-27} kg
2.	Avogadro's number	N	6.023×10^{23} / mol
3.	Bohr magneton (Magnetic moment)	μ_B	9.273×10^{-24} $A.m^2$
4.	Boltzmann's constant	k	1.38×10^{-23} J/K = 8.620×10^{-5} eV/K
5.	Electronic charge	e	1.602×10^{-19} C
6.	Electronic mass	m	9.109×10^{-31} kg
7.	Gas constant	R	8.314 $J/mol.K$
8.	Permeability of free space	μ_0	$4\pi \times 10^{-7}$ H/m
9.	Permittivity of free space	ε_0	8.854×10^{-12} F/m
10.	Planck's constant	h	6.626×10^{-34} $J.s$

Conversion Factors

S.No.	Unit	Conversion factor
1.	1 angstrom (Å)	10^{-10} m
2.	1 atmosphere	0.1 MN/m^2
3.	1 kgf/cm^2	98.1 kN/m^2
4.	1 eV/entity	96.49 kJ/mol
5.	1 kcal	4.18 kJ

APPENDIX B

Physical Properties of Selected Metals

S.No.	Metal	Symbol	Atomic number	Melting point (°C)	Density g/cm^3	Crystal Structure (20°C)
1.	Aluminium	Al	13	660	2.70	FCC
2.	Antimony	Sb	51	630	6.70	RHOMBIC
3.	Arsenic	As	33	814	5.78	RHOMBIC
4.	Beryllium	Be	4	1280	1.85	HCP
5.	Bismuth	Bi	83	271	9.80	RHOMBIC
6.	Boron	B	5	2300	2.30	ORTHO-RHOMBIC
7.	Cadmium	Cd	48	321	8.65	HCP
8.	Chromium	Cr	24	1875	7.19	BCC
9.	Cobalt	Co	27	1495	8.83	HCP
10.	Copper	Cu	29	1083	8.90	FCC
11.	Gold	Au	79	1063	19.25	FCC
12.	Iron	Fe	26	1539	7.87	BCC
13.	Lead	Pb	82	327	11.38	FCC
14.	Magnesium	Mg	12	650	1.74	HCP
15.	Manganese	Mn	25	1245	7.47	—
16.	Molybdenum	Mo	42	2610	10.22	BCC
17.	Nickel	Ni	28	1453	8.90	FCC
18.	Platinum	Pt	78	1769	21.45	FCC
19.	Silver	Ag	47	961	10.49	FCC
20.	Tantalum	Ta	73	2996	16.60	—
21.	Tin	Sn	50	232	7.30	BCT
22.	Tungsten	W	74	3410	19.30	BCC
23.	Uranium	U	92	1132	19.07	ORTHO-RHOMBIC
24.	Vanadium	V	23	1900	6.10	—
25.	Zinc	Zn	30	420	7.13	HCP

INDEX

A

Addition polymerization, 278
—to polymers, 278
Advantages of cold working, 191
—hot working, 194
Agglomerated materials, 294
Allotropic forms of pure iron, 208
Alloy steels, 258
Aluminium, 262
—alloys, 263
Amorphous solids, 80
Angle between two planes or directions, 75
Annealing, 225
—of a cold worked metal, 191
Antimony, 268
Application of Drude-Lorentz theory, 107
—superconductors, 120
—voltage across a P-N junction, 318
Aromatic organic compounds, 276
Asbestos, 340
Atmospheric corrosion, 246
Atomic models, 7, 8, 15, 17
—number, 6
—packing factor, 69
—radius in cubic system, 59
—radius in hexagonal system, 59
—weight, 6

B

Bauschinger's effect, 184
Bearing metals, 268
Beginning and development of Material Science, 1
Bending strength, 138
B.H. curve for diamagnetic materials, 348
—ferromagnetic materials, 352
—paramagnetic materials, 349
Biomaterials, 269
Body centered cubic (BCC) structure, 45
Bohr's atomic model, 9
Bond energy, 102
—length, 103

Bonding, Causes of, 92
—of semiconductors, 310
Bonds, Classification of, 92
Bragg's law, 85
Brass, 264
Bravais lattices, 41
Brillouin, zone theory, 110, 111
Brinell's hardness test, 162
Brittle fracture, 198
Brittleness, 136
Bronze, 264
Burger's vector, 182

C

Cadmium, 268
Capacitor, 332
Carburising,
Case hardening, 230
Castability, 141
Cast iron, 254
Cathodic protection, 251
Causes of bonding, 92
—fracture, 197
Cementation, 249
Cement concrete, 297
Cements, 296
Ceramics, 301
Changes in steel structure below critical temperature, 216
—volume in agglomerated materials, 296
Charpy test, 159
Chemical dip coating, 250
Cladding, 249
Classification of bonds, 92
—corrosion, 237
—electron theory of metals, 106
—fractures, 197
—magnetic materials, 356
—magnetism, 347
—metal deformations, 170
—semiconductors, 311

362

Index

—solids based on band theory, 116
—solids based on zone theory, 114
Clod working, 191
Common planes in a simple cubic structure, 50
Comparison between cold and hot working, 194
—edge and screw dislocation, 180
—elastic and plastic deformations, 171
—hardness and hardenability, 227
—ionic, covalent and metallic bonds, 99
—recovery and recrystallisation, 193
—slip and twinning, 177
—thermoplastics and thermosetting plastics, 286
Composition cells, 242
Compressive strength, 138
—test, 158
Concentration cells, 243
Condensation polymerization, 280
Conductivity, 122
Conductors, 117
Control of corrosion, 247
Coordination number, 79
Copolymerization, 279
Copper, 263
Covalent bonds, 95
Creep, 137
—curve, 166
—fracture, 204
—test, 165
Crevice corrosion, 246
Critical points, 209
—resolved shear stress, 174
—temperature, 211
Crystal, 37
—defects, 81
—directions, 53
—lattice, 37
—structure, 85
Crystalline solids, 80
Crystallinity in polymers, 282
Cynading, 231

D

Decomposition of austenite, 214
Defects, Crystal, 81
Deficiencies in Bohr's atomic model, 15
Deformation of polycrystalline materials, 185

Degree of polymerization, 281
Density and porosity of materials, 295
Density of energy levels in Brillouin zone model, 114
Determination of crystal structure, 85
Diamagnetism, 347
Dielectric constant, 333
—loss, 334
—polarization, 331
—properties, 333
—strength, 336
Difficulties in Rutherford's atomic model, 8
Dimensional stability of refractory material, 304
Dimensions of a unit cell, 76
Dipole bonds, 100
Dipping, 249
Direct chemical corrosion, 237
Direction of thermo e.m.f., 130
Disadvantages of cold working, 191
—hot working, 194
Dislocation climb, 183
Dislocations, 178
Dispersion bonds, 99
Domain structure, 351
Drude-Lorentz theory, 106
Dry corrosion, 237
Ductile fracture, 201
—to brittle transition, 203
Ductility, 135

E

Eddy current loss, 354
Edge dislocation, 178
Effect of frequency on dielectric constant, 334
—impurities on cast iron, 254
Elastic after effect, 184
—deformation, 170
—strength, 137
Elasticity, 135
Elastomers, 287
Electrical behaviour of polymers, 284
Electric field, 329
—flux density, 330
Electrochemical corrosion, 239
Electrode potential, 236
Electrons, 25
—structure of elements, 25

Electronic configuration, 23
Electroplating, 248
Electrostriction, 339
Endurance, 137
Energy bands, 116
Energy of an electron in Bohr's atomic model, 10
Energy level diagram, 19, 20
Engineering materials, 4
—metallurgy, 3
Erosion corrosion, 246
Equation of motion of an electron, 121
Eutectoid steel, 211
Expression for conductivity of a semiconductor, 313
—current density in metal, 123
External voltage method, 252
Extrinsic semiconductors, 311

F

Face centered cubic (FCC) structure, 45
Factors affecting creep resistance, 167
—mechanical properties of a metal, 141
—machinability, 139
—impact resistance, 161
—resistivity of metals, 117
Fatigue corrosion, 246
—fracture, 205
—test, 164
Fermi-Dirac distribution, 109
Ferrimagnetism, 355
Ferrites, 355
Ferroelectric materials, 337
Ferromagnetism, 349
Ferrous alloys, 253
Flame hardening, 232
Flow of current in forward biased P-N Junction, 318
Formability, 140
Frank-Reed generator, 183
Free cutting steels, 261
Frequency of an electron, 13
Fretting corrosion, 247
Fundamental particles, 5

G

Galvanic cell, 241
—series, 237
Glass, 303
—fibre, reinforced plastic, 299
Grain growth, 192

Griffith's theory, 199
Gun metal, 265

H

Hardenability, 227
Hardening, 226
—and setting of cement, 297
Hard magnetic materials, 356
Hardness, 136
—test, 163
Heisenberg's uncertainity principle, 26
Hexagonal closed packed (HCP), structure, 46
High frequency applications of ferrities, 356
High resisting steel, 261
—speed steel, 260
Hook's law, 144
Hot working, 194
Hydrogen bond, 100
Hypereutectoid steels, 211
Hypoeutectoid steels, 211
Hysteresis curve, 337

I

Impact test, 159
Imperfections, 81
Important terms in agglomerated materials, 295
—feature of miller indices of crystal plane, 47
Improving machinability, 140
Induction hardening, 231
Inertness of noble gases, 91
Insulating materials, 328
Insulators, 115
Integrated circuits, 325
Intergranular corrosion, 244
Intrinsic semiconductors, 311
Ionic bonds, 93
Iron-carbon system, 209
Isobars, 6
Isotopes, 6
Izod test, 161

K

Knoop's hardness test, 164

L

Laminated materials, 297
Laser, 326
Lattice parameters of a unit cell, 38
Lead, 266
Line defects, 83

Index

M

Machinability, 139
Magnesium, 267
Magnetic field, 342
    —field strength, 344
    —flux density, 344
    —hysteresis, 353
    —moment, 343
    —particle test, 206
    —permeability, 344
    —susceptibility, 346
Magnetization, 345
Magnitude of thermo e.m.f., 130
Major defects in a metal due to faulty heat treatment, 232
Malleability, 139
Materials for hip and joint replacement, 269
Maximum number of electrons in an atom, 18
    —main shell, 18
    —sub-shell, 19
Mean free path, 126
Measurement of hardenability, 227
    —of temperature with thermocouple, 132
Mechanical behaviour of polymers, 281
Mechanism of addition polymerisation, 280
    —brittle fracture, 199
    —corrosion, 239
    —creep fracture, 204
    —ductile fracture, 202
    —electrochemical corrosion, 240
    —fatigue fracture, 205
    —slip, 174
    —twinning, 177
Metallic bonds, 97
    —coatings, 248
    —structure, 44
Mendeleev's periodic table, 32
Mica, 339
Micromechanical systems (MEMS), 306
Microstructure of mortensite, 220
Miller indices, 46
Mobility, 124
Modern atomic model, 17
    —periodic table, 32, 33
Modes of plastic deformation, 172

Modification, of corrosive environment, 248
Modified iron-iron carbide phase diagram, 212
Modulus of material, 144
Multiplications of dislocations, 183

N

Natural rubber, 287
Nickel, 267
    —nonferrous metals, 262
Nitriding, 231
Non-destructive tests, 205
    —metallic coatings, 248
N-P-N transistors, 323
Normalising, 224
N-type semiconductors, 312
Number of atoms in a cubic structure, 62
    —body centered cubic (BCC) structure, 65
    —face centered cubic (FCC), structure, 67
    —simple, cubic structure, 63
    —per square millimetre in cubic crystal planes, 63
Nylon reinforced rubber, 299

O

Objectives of heat treatment, 223
Optical fibres, 305
Orbital frequency of an electron, 13
Organic compounds, 273
Origin of magnetic,
Orgin of thermo e.m.f., 130
Oxide coatings, 250
Oxygen absorption corrosion reaction, 240

P

Packing factor, 295
Paints and lacquiring, 250
Paramagnetism, 348
Particle size, 295
Passivity, 241
Pauli's exclusion principle, 26
Periodic table, 32
Perpendicular distance between a plane and origin of the cube, 72
Permittivity, 330
Phase diagrams, 210, 212
Photoelectric cells, 325
Piezoelectricity, 339
Piezoelectric Ceramics, 307
Pitting corrosion, 244

Plain carbon steels, 257
Plastic coatings, 250
    —deformation, 171
    —strength, 137
Plasticity, 135
Plywood, 298
Point defects, 81
Poisson's ratio, 145
Polarisation, 331, 332
Polymers, 276
Polymer crystals, 283
Polymorphism, 80
P-N junctions, 317
P-N-P transistors, 323
P-type semiconductors, 312
Preferred orientation, 190
Prevention of corrosion, 247
Primary bonds, 92
    —solidification, 213
Primitive cell, 38
Procedure for finding miller indices of crystal planes, 47
    —crystal directions, 54
    —sketching the plane from the given miller indices, 49
Processes of heat treatment, 223
Properties of covalent solids, 97
    —ionic solids, 94
    —metallic solids, 98
    —refractory materials, 303

Q
Quantum number, 15

R
Radiographic test, 206
Recovery, 191
Recrystallisation, 192
Refractory materials, 303
Reinforced materials, 298
    —cement concrete, 298
Relation between energy and wave number, 112
Representation of crystal directions in a cubic unit cell, 54
    —crystal planes in a cubic unit cell, 47
Resilience, 136
Resins, 284, 340
Resistivity, 122

Retained austenite, 220
Rockwell's hardness test, 163
Rubber, 340
Rutherford's atomic model, 8
    —Difficulties in, 8

S
Sacrificial anode method, 251
Salient points of TTT diagram, 218
Saturated organic compounds, 274
Schrodinger theory, 25
Science of metals, 2
Screw dislocation, 179
Season corrosion, 245
Secondary bonds, 92
    —solidification, 213
Selective corrosion, 247
Semiconductor devices, 321
    —diodes, 321
    —materials, 320
Semiconductors, 116, 310
Setting and hardening of cement, 297
Shape of the polymer, 281
Shape memory alloys, 269
Shear strength, 138
Significance of atomic number in a periodic table, 33
Silicon controlled rectifier, 324
Single crystal, 37
Slip, 172
Soft magnetic materials, 357
Solid phases in iron-iron carbide phase diagram, 210
Sommerfeld's atomic model, 16
    —theory, 108
Space lattice, 37
Special steel, 260
Spheridising, 226
Spraying, 249
Spring steel, 262
Stainless steel, 260
Steel, 256
Stiffness, 136
Strain, 143
Strength, 137
Strengthening mechanism of polymers, 281
Stress, 143
    —cells, 242
    —corrosion, 245

Index

Stress-strain diagram for different materials, 157
  —curve for low carbon steel, 154
Sub-divisions of material science, 2
Suitable design and fabrication procedure, 247
Superconductivity, 120
Surface defects, 84
  —hardening, 230
Susceptibility of diamagnetic materials, 347
  —ferromagnetic materials, 352
  —paramagnetic materials, 349
Synthetic rubbers, 287

T

Table, Periodic, 32
Temperature, of polymer, 282
Tempering, 228
Tensile strength, 138
  —test, 146
Thermistors, 325
Thermoelectricity, 130
Thermoplastic resins, 284
Thermosetting plastics, 285
Thomson's atomic model, 7
Time-temperature transformation diagram, 217
Tin, 266
Tool and die steels, 259
Torsional strength, 138
Total polarization of a dielectric material, 332
Toughness, 136
Transformation of austenite, 215, 218, 219
Transistors, 322
Twinning, 176
Type of cast irons, 254
  —composite materials, 294
  —crystal systems, 39
  —corrosions, 243
  —dislocations, 178
  —galvanic cells, 241
  —heat treatment process, 223
  —mechanical properties of metals, 134
  —mechanical tests, 146

  —non-destructive tests, 205
  —organic compounds, 273
  —polymerization, 278
  —primary bonds, 92
  —rubbers, 286
  —secondary bonds, 99
  —strength, 137
  —stresses, 143
  —synthetic plastics, 284
  —technological properties, 139
  —transistors, 323

U

Ultrasonic test, 205
Underground corrosion, 246
Uniform corrosion, 244
Unit cell, 38
Unsaturated organic compounds, 275
Use of inhibitors, 248
  —protective coatings, 248

V

Vanadium, 268
Variation of mechanical properties in annealing, 193
Varistors, 325
V-I characteristics of P-N junction, 319, 320
Vicker's hardness test, 163
Vitrous coatings, 250
Vulcanization, 288

W

Wave mechanics, 25
Weldability, 140
Wet corrosion, 239
Whiskers, 37
Workability, 140
Work hardening, 189
Wood, 289

Z

Zener diodes, 322
Zinc, 266
Zinc base alloys, 266
Zirconium alloys, 269

IMPORTANT BOOKS ON MECHANICAL ENGINEERING

HEAT AND MASS TRANSFER (MULTICOLOUR)
(IN SI UNITS)
R.K. Rajput

CONTENTS: ●·BASIC CONCEPTS ● **PART I: HEAT TRANSFER BY CONDUCTION** ● CONDUCTION ÄSTEADY - STATE ONE DIMENSION ●·General Heat Conduction Equation in Spherical Coordinates, ●·Heat Conduction Through Plane and Composite Walls,·●·Heat Conduction Through Hollow and Composite Cylinders,·●·Heat Conduction Through Hollow and Composite Spheres, ●·Critical Thickness of Insulation ●·Heat conduction with Internal Heat Generation ●·Heat Transfer from Extended Surfaces (Fins)·●·CONDUCTIONÄSTEADY-STATE TWO DIMENSIONS AND THREE DIMENSIONS ●·CONDUCTION ÄUNSTEADY - STATE (TRANSIENT) **PART II: HEAT TRANSFER BY CONVECTION** ● INTRODUCTION TO HYDRODYNAMICS ●·DIMEN-SIONAL ANALYSIS ●·Characteristic Length or Equivalent Diameter,·●·FORCED CONVECTION A. LAMINAR FLOW ●·Laminar Flow over a Flat Plate·●·Laminar Tube Flow ●·Introduction ●Turbulent Tube Flow ●·Empirical Correlations·●·FREE CONVECTION ●·Simplified Free Convection Relations for Air, ●·Combined Free and Forced Convection, ● BOILING AND CONDENSATION ●·Introduction·●·Boiling Heat Transfer, ●·Condensation Heat Transfer ●·HEAT EXCHANGERS ●·Heat Exchanger Effectiveness and Number of Transfer Units (NTU) **PART III: HEAT TRANSFER BY RADIATION** ● THERMAL RADIATIONÄBASIC RELATIONS ● RADIATION EXCHANGE BETWEEN SURFACES **PART IV: MASS TRANSFER** ● MASS TRANSFER **PART V: OBJECTIVE TYPE QUESTION BANK** ● Index

Code : 10 202	4th Edn. 2006	ISBN : 81-219-1777-8

ELEMENTS OF MECHANICAL ENGINEERING
C.S. Chethan Kumar & B.P. Mahesh

CONTENTS : ● Sources of Energy ● Boilers ● Primemovers ● Refrigeration & Air Conditioning ● Machine Tools ● Soldering, Brazing and Welding ● Lubrication and Bearing ● Power Transmission ● Mechatronics ● Foundry Technology ● Metal Forming ● Simple Machines

Code : 10 280	1st Edn. 2004	ISBN : 81-219-2383-2

TOOLING DATA
P.H. Joshi

CONTENTS: ● Jigs and Fixtures ● Press Tools ● Cutting Tools ● Appendices

AHW 1720		ISBN : 81-7544-172-0

A TEXTBOOK OF FLUID MECHANICS AND HYDRAULIC MACHINES
R.K. Rajput

CONTENTS : Part I Fluid Mechanics ● Properties of Fluids ● Pressure Measurement ● Hydrostatic Forces on Surfaces ● Buoyancy and Floatation ● Fluid Kinematics ● Fluid Dynamics ● Dimensional and Model Analysis ● Flow Throuch Orifices and Mouthpieces ● Flow over Notches and Weirs ● Laminar Flow ● Turbulent Flow in Pipes ● Flow Through Pipesi Boundary Layer Theory ● Flow Around Submerged Bodiesadrag and Lift ● Compressible Flow ● Flow in Open Channels ● **Part II-Hydraulic Machines** ● Impact of Free Jets ● Hydraulic Turbines ● Centrifugal Pumps ● Reciprocating Pumps● Miscellaneous Hydraulic Machines ● Water Power Development ●·Index

Code : 10 185	3rd Edn. 2004	ISBN : 81-219-1666-6

ELEMENTS OF FRACTURE MECHANICS
Prashant Kumar

CONTENTS: ● Background ● Energy Release Rate ● Strees Intensity Factor ● SIF of More Compelx Cases ● An Elastic Deformation at the Crack Tip ● Elastic Plastic Analysis through J-Integral ● Crack Tip Opening Displacement ● Test Methods ● Fatigue Failure ● Questions ● Problems ● References

AHW 1569		ISBN : 81-7544-156-9

A TEXTBOOK OF FLUID MECHANICS
R.K. Rajput

10 192	2nd Edn. 2004	ISBN : 81-219-1667-4

A TEXTBOOK OF HYDRAULIC MACHINES
(Fluid Power Engineering)
R.K. Rajput

10 194   2nd Edn. 2004   ISBN : 81-219-1668-2

A TEXTBOOK OF HYDRAULICS
R.K. Rajput

CONTENTS: Properties of Fluids ●•PRESSURE MEASUREMENT Pressure Measurement Pressure ● Hydrostatic Forces on Surfaces ●•Bouyancy and Floatation ● Fluid Kinematics ● Fluid Dynamics ●•Dimensional and Model Analysis ● Flow Through Orifices and Mouthpieces ●•Flow over Notiches and Weirs ●   Laminar Flow ● Flow Through Pipes ●•Flow Around Submerged Bodiesadrag and Lift ● Flow in Open Channels ●•Impact of Free JETS ● Hydrauli Turbines ●•Centrifuga Pumps ●•Reciprocating Pumps ● Miscellaneous Hydraulic Machines ● Experiments

Code : 10 198   1st Edn. 1998   ISBN : 81-219-1731-X

ISO 9000 CONCEPTS, METHODS AND IMPLEMENTATION
Tapan P. Bagchi

CONTENTS: ● Quality Management Practices Worldwide ● Quality, Customers and ISO 9000 ● ISO 9000: A Management Overview ● Writing the Company's Quality Policy ● Interpretation of Key ISO 9000 Clauses ● Aspects of Manufacturing and ISO 9000 ● ISO 9000 Standards and Production ● Process Control by Control Charts ● Problem Solving with 7 Tools ● Data, Records and Traceability ● Procurement System Standards in ISO 9000 ● Inspection Test Standards and Calibration ● An Overview of Acceptance Sampling and Documenting the Quality System ● What is a Quality Audit? ● Design and Change Control Standards ● What are Taguchi Methods? ● Implementing ISO 9000 ● ISO 9000 Registration: Two Case Studies ● Total Quality Management: Concepts and Accessories ● A Prototype Quality Manual ● References

AHW 0112   ISBN : 81-7544-011-2

ENGINEERING MATERIALS
R.K. Rajput

CONTENTS: Introduction ● Building Stones ● Bricks and other Clay Products ● Lime ● Cement ● Mortar ● Concrete ● Timber and Wood-Based Products ● Metals and Alloys ● Paints, Varnishes, Distempers and Anti-Termite Treatment ● Asphalt, Bitumen and Tar ● Asbestos, Adhesives and Abrasives ● Plastics and Fibres ● Glass ● Insulating Materials ● Fly-Ash, Gypsum and Gypsum Plaster ● Rubber and Composite Materials ● Lubricating, Belting and Packing Materials   ● Cutting Tool Materials ● Electrical Engineering Materials ● Material Science of Metals

Code : 10 210   2nd Edn. 2004   ISBN : 81-219-1960-6

PRESS TOOLS: DESIGN AND CONSTRUCTION
P.H. Joshi

CONTENTS: ● Tool and Workpiece Materials ● Cutting Action ● Chip Formation ● Tool Wear ● Lubrication and Surface Finish ● Cutting Power ● Machining Economics ● Single Point Turning and Threading Tools ● Tools for Holes ● Milling ● Broaching ● Gear Cutting Tools ● Index

AHW 4465   ISBN : 81-85814-46-5

STRENGTH OF MATERIALS (MUTLI COLOUR)
(Machanics of Solids S.I. Units)
R.K. Rajput

CONTENTS: Simple Stresses and Strains ● Principal Stresses and Strains ● Centroid and Moment of Inertia ● Bending Stresses ● Combined Direct and Bending Stresses ● Shear Stresses in Beam ● Thin Shells ● Thick Shells ● Riveted and Welded Joints ● Torsion of Circular and Non Circular Shafts ● Springs ● Stain Energy and Deflection due to Shear & Bending ● Columns and Struts ● Analysis of Framed Structures ● Theories of Failure ● Rotating Discs and Cylinders ● Bending of Curved Bars ●•Unsymmetrical Bending ●•Material Testing Experiments ●•Index

Code : 10 174   5th Edn. 2006   ISBN : 81-219-1381-0

CUTTING TOOLS
P.H. Joshi

CONTENTS: ● Tool and Workpiece Materials ● Tool Geometry ● Chip Formation ● Tool Wear ● Lubrication and Surface Finish ● Cutting Power ● Machining Economics ● Single Point Turning and Threading Tools ● Tools for Holes ● Milling ● Broaching ● Gear Cutting Tools ● Standards for Cutting Tools ● Index

AHW 4538   ISBN : 81-85814-53-8

A TEXTBOOK OF ENGINEERING MECHANICS
(Applied Mechanics)
R.S. Khurmi

CONTENTS: Introduction ● Composition and Resolution of Forces ● Moments and their Applications ● Parallel Forces and Couples ●· Equilibrium of Forces ●·Centre of Gravity·●·Moment of Inertia ●·Friction ● Principles of Lifting Machines ● Simple Lifting Machines ● Support Reactions ●·Analysis of Perfect Frames (Analytical Method & Graphical Method) ● Equilibrium of Strings ● Virtual Work ● Plane Motion ●·Motion Under Variable Acceleration ●·Relative Velocity ● Projectiles ●·Motion of Rotation ●·Combined Motion of Rotation and Translation ●·Simple Harmonic Motion ●·Laws of Motion ●·Motion of Connected Bodies ●·Helical Springs and Pendulums ●·Collision of Elastic Bodies ●·Motion Along a Circular Path ●·Balancing of Rotating Masses ●·Work, Power and Energy ●·Kinetics of Motion of Rotation ●·Motion of Vehicles ●·Transmission of Power by Belts and Ropes ●·Gear Trains ●·Hydrostatics ●·Equilibrium of Floating Bodies ●·Index.

Code : 10 023   20th Edn. 2005   ISBN : 81-219-0651-2

APPLIED MECHANICS AND STRENGTH OF MATERIALS
R.S. Khurmi

CONTENTS: Introduction ● Composition and Resolution of Forces ● Moments and their Application ● Parallel Forces and Couples ● Equilibrium of Forces● Centre of Gravity ● Moment of Inertia ● Principles of Friction ● Applications of Friction ● Principles of Lifting Machines ●Simple Lifting Machines ● Linear Motion ●·Circular Motion ● Projectiels ●Laws of Motion ● Work, Power and Energy ● Simple Stresses and Strains ●Thermal Stresses and Strains ●·Elastic Constants ●·Strain Energy and Impact Loading ●·Bending Moment and Shear Force ●·Bending Stresses in Beams ●·Shearing Stresses in Beams ●·Deflection of Beams ●·Deflections of Cantilevers ●Torsion of Circular Shafts ●·Riveted Joits ●·Thin Cylindrical and Spherical Shells ● Analysis of Perfect Frames (Analytical Method & Graphical Method)

Code : 10 025   13th Edn. 2005   ISBN : 81-219-1077-3

A TEXTBOOK OF APPLIED MECHANICS
R.S. Khurmi

CONTENTS: Introduction ● Composition and Resolution of Forces ● Moments and their Applications ● Parallel Forces and Couples ● Equilibrium of Forces ● Centre of Gravity ● Moment of Inertia ● Principles of Friction ● Applications of Friction ● Principls of Lifting Machines ● Simple Lifting Machines ● Linear Motion ● Circular Motion ● Projectiles ● Laws of Motion ● Work, Power and Energy ● Analysis of Perfect Frames (Analytical Method) ● Analysis of Perfect Frames Graphical Method ● Index

Code : 10 191   13th Edn. 2004   ISBN : 81-219-1643-7

STRENGTH OF MATERIALS
R.S. Khurmi

CONTENTS: Introduction ●·Simple Stresses and Strains ● Stresses and Strains in Bars of Varying Sections ● Stresses and Strains in Statically Indeterminate Structures ● Thermal Stresses and Strains ● Elastic Constants ● Principal Stresses and Strains ● Strain Energy and Impact Loading ● Centre of Gravity ● Moment of Intertia ● Analysis of Perfect Frames (Analytical Method & Graphical Method) ● Analysis of Perfect Frames (Graphical Method) ● Bending Moment and Shear Force ● Bending Stresses in Simple Beams ● Bending Stresses in Composite Beams ●·Shearing Stresses in Beams ● Direct and Bending Stresses ● Dams and Retaining Walls ● Deflection of Beams ● Deflection of Cantilevers ● Deflection By Moment Area Method ●Deflection by Conjugate Beam Method ●Propped Cantilevers and Beams ●·Fixed Beams ●Theorem of Three Moments ●·Moment Distribution Method ●Torsion of Circular Shafts ●·Springs ●·Riveted Joints ● Welded Joints ●·Thin Cylindrical and Spherical Shells ●·Columns and Struts ●·Introduction to Reinforced Concrete ●·Index

Code : 10 024   23rd Edn. 2005   ISBN : 81-219-0533-8

A TEXTBOOK OF HYDRAULICS, FLUID MECHANICS AND HYDRAULIC MACHINES
R.S. Khurmi

CONTENTS: Introduction ● Fluid Pressure and its Measurement ● Hydrostatics ● Applications of Hydrostatics ● Equilibrium of Floating Bodies ● Hydrokinematics ● Bernoulli's Equation and its Applications ● Flow through Orifices (Measurement of Discharge & Time) ● Flow through Mouthpieces ● Flow over Notches ● Flow over

Weirs ● Flow through Simple Pipes ● Flow through Compound Pipes ● Flow through Nozzles ● Uniform Flow through Open Channels ● Non-uniform flow through Open Channels ● Viscous Flow ● Viscous Resistance Fluid Masses Subjected to Acceleration ● Vortex Flow ● Mechanics of Compressible Flow ● Compressible Flow of Fluids ● Flow Around Immersed Bodes ●·Dimensional Analysis ●·Model Analysis (Undistorted Models and Distorted Models ●·Non-Dimensional Constants ●·Impact of Jets ●·Jet Propulsion ●·Water Wheels ●·Impulse Turbines ●·Reaction Turbines ● Performance of Turbines ●·Centrifugal Pumps ●·Reciprocating Pumps ●·Performance of Pumps ●·Pumping Devices ●·Hydraulic Systems ●·Index

Code : 10 026　　　　　　　　　19th Edn. 2004　　　　　　　　ISBN : 81-219-0162-6

STEAM TABLES
(With Mollier Diagram in S.I. Units)
R.S. Khurmi

CONTENTS: Rules for S.I. Units ● Introduction to Steam Tables (Temperature) and Mollier Diagrams ● Saturated Water & Steam Tables ● Saturated Water & Steam Tables (Pressure) ● Specific Volume of Super-heated Steam ● Specific Enthalpy of Super-heated Steam ● Entropy of Super-heated Steam ● Specific Volume of Super-critical Steam ● Specific Enthalpy of Superecritical Steam ● Specific Entropy of Super Critical Steam

Code : 10 044　　　　　　　　　8th Edn. 2005　　　　　　　　ISBN : 81-219-0654-7

A TEXTBOOK OF PRODUCTION ENGINEERING
P.C. Sharma

CONTENTS: Jigs and Fixtures ●·Press Tool Design ●·Forging Die Design ●·Cost Estimation ●·Economics of Tooling ●·Process Planning ●·Tool Layout for Capstans and Turrets ●·Tool Layout for Automatics ●·Limits, Tolerances and Fits ●·Gauges and Gauge Design ●·Surface Finish ●·Measurement ●·Analysis of Metal Forming Process ●·Theory of Metal Cutting ● Design and Manufacture of Cutting Tools ● Gear manufacturing ● Thread manufacturing ● Design of Machine Tool elements and machine tool testing ● Machine Tool installation and maintenance ● Design of product for economical production ● Statistical quality control ● Kinematics of machine tools ● Production planning and control ● Manufacturing systems and automation ● Computer integrated manufacturing ● Plant Layout ● Production and Productivity ● Appendix ● Index

Code : 10 038　　　　　　　　　10th Edn. 2005　　　　　　　　ISBN : 81-219-0421-8

A TEXTBOOK OF THERMAL ENGINEERING
R.S. Khurmi & J.K. Gupta

CONTENTS: Introduction ● Properties of Perfect Gases ● Thermodynamic Processes of Perfect Gases ● Entropy of Perfect Gases ● Thermodynamic Air Cycles ● Formation and Properties of Steam ● Entropy of Steam ● Thermodynamic Processes of Vapour ● Thermodynamic Vapour Cycles ● Fuels ● Combustion of Fuels ● Steam Boilers ● Boiler Mountings and Accessories ● Performance of Steam Boilers ● Boiler Draught ● Simple Steam Engines ● Compound Steam Engines ● Performance of Steam Engines ● Steam Condensers ● Steam Nozzles ● Impulse Turbines ●·Reaction Turbines ●·Performance of Steam Turbines ●·Modern Steam Turbines ●·Internal Combustion Engines ●·Testing of Internal Combustion Engines ●·Reciprocating Air Compressors ●·Air Motors ●·Gas Turbines ●·Performance of Gas Turbines ●·Introduction to Heat Transfer ●·Air Refrigeration Systems ●·Vapour Compression Refrigeration ●·Psychrometry Air Conditioning Systems ●·General Thermodynanuc Relation ●·Variable Specific Heat ●·Index

Code : 10 172　　　　　　　　　15th Edn. 2004　　　　　　　　ISBN : 81-219-1337-3

A TEXTBOOK OF HYDRAULICS
R.S. Khurmi

CONTENTS: Introduction ● Fluid Pressure and its Measurement ● Hydrostatics ● Applications of Hydrostatical Equilibrium of Floating Bodiesl Hydrokinematics ● Bernoulli's Equation and its Applications ● Flow Through Orifices (Measurement of Discharge) ● Flow Through Orifices (Measurement of Time) ● Flow Through Mouthpieces ● Flow Over Notches ● Flow Over Weirs ● Flow Through Simple Pipes ● Flow Through Compound Pipes ● Flow Through Nozzles ● Uniform Flow Through Open Channels ● Non Uniform Flow Through Open Channels ●·Viscous Flow ●·Viscous Resistance ●·Impact of Sets ●·Hydraulic Turbines ●·Hydraulic Pumps ●·Pumping Devices ●·Hydraulic System ●·Index

Code : 10 027　　　　　　　　　19th Edn. 2004　　　　　　　　ISBN : 81-219-0135-9

THEORY OF MACHINES
R.S. Khurmi & J.K. Gupta

CONTENTS: Introduction • Kinematics of Motion • Kinetics of Motion • Simple Harmonic Motion • Simple Mechanisms • Velocity in Mechanisms (Instantaneous Centre Method) • Velocity in Mechanisms (Relative Velocity Method) • Acceleration in Mechanisms • Mechanisms with Lower Pairs • Friction • Belt, Rope and Chain Drives • Toothed Gearing • Gear Trains • Gyroscopic Couple and Precessional Motion • Inertia Forces in Reciprocating Parts • Turning Moment Diagrams and Flywheel • Steam Engine Valves and Reversing Gears • Governors • Brakes and Dynamometers • Cams • Balancing of Rotating Masses • Balancing of Reciprocating Masses • Longitudinal and Transverse Vibrations • Torsional Vibrations • Computer Aided Analysis and Synthesis of Mechanisms • Index

Code : 10 013　　　　　　13th Edn. 2005　　　　　　ISBN : 81-219-0132-4

MATERIAL SCIENCE
R. S. Khurmi & R.S. Sedha

CONTENTS: Part I: Science of Metals : • Introduction • Structure of Atom • Crystal Structure • Bonds in Solids • Electron Theory of Metals • **Part II: Mechanical Behaviour of Metals:** • Mechanical Properties of Metals • Mechanical Tests of Metals • Deformation of Metals Fracture of Metals • **Part III: Enginering Metallurgy:** Iron Carbon Alloy System • Heat Treatment • Corrosion of Metals • **Part IV: Engineering Materials** • Ferrous and Non-Ferrous Alloys • Organic Materials and Ceramics •　Composite Materials • Semiconductors •　Insulating Materials • Magnetic Materials • Appendices • Index

Code : 10 109　　　　　　4th Edn. 2004　　　　　　ISBN : 81-219-0146-4

A TEXTBOOK OF MATERIAL SCIENCE
K.G. Aswani

CONTENTS: Selection of Materials • Structure of Atoms, Crystal Structure and Bonds in Solids • Mechanical Properties of Materials • Deformation of Metals • Alloy Systems and Phase Diagrams • Cast Iron • Carbon and Alloy Steels • Heat Treatment of Carbon and Alloy Steels • Non-Ferrous Metals and Alloys • Polymers and Polymeric Products • Corrosion and its Prevention • Composites • Electrical and Magnetic Properties of Materials • Manufacturing Processes • Refractories • Cementitious Materials • Tables • Index

Code : 10 157　　　　　　2nd Edn. 2001　　　　　　ISBN : 81-219-1199-0

A TEXTBOOK OF PRODUCTION TECHNOLOGY
(MANUFACTURING PROCESSES)
P.C. Sharma

CONTENTS: • Introduction • Engineering Materials and Heat Treatment • The Casting Process • Mechanical Working of Metals • The Welding Process • The Machining Process • Cutting Tools materials and Cutting Fluids • Machine Tools • Unconventional Manufacturing Methods • Power Metallurgy • Procesing of Plastic • Special Processing Methods • Ceramic Materials and Their Processing • Composite Materials and Their Processing • Tracer Controlled Machine Tools • Numerically Controlled Machine Tools • Appendix • Index

Code : 10 160　　　　　　6th Edn. 2005　　　　　　ISBN : 81-219-1114-1

ENGINEERING FLUID MECHANICS
K.L. Kumar

CONTENTS: • Introduction • Fluid Statics • Fluid Kinematics • Fluid Dynamics • Flow Measurement • Ideal Fluid Flow • Laminar Flow • Boundary Layer Flow • Flow Around Immersed Bodies • Flow Through Pipes • Flow Through Open Channels • Compressible Flow • Dimensional Analysis and Similitude • Fluid Machines • Appendices

Code : 10 020　　　　　　7th Edn. 2004　　　　　　ISBN : 81-219-0100-6

INDUSTRIAL MAINTENANCE
H.P. Garg

CONTENTS: Part I: Restoration and Manufacture of Machine Parts • Fits, Tolerances and Surface Finish • Materials for Machine Parts and Heat Treatment of Steels • Guide Surface • Gear Transmission • Bush Bearings, their Shank and Housing • Ball and Roller Bearings their Shank and Housing • Key Fitting • Spline Fitting • Coupling • Clutches • Lead Screw and Nut • Machine Spindle • Vee Belt Drive • Chains and Sproket Wheels • Machine Hydraulics • Tailstock Repairs • Repair of Three Jaw Chucks • Repair of Cracks in C.I. Body • Restoration of Parts by Welding, Metalization, Chromium Plating • Threads and Threaded Joints • Seals and Packings • Special Features of the Repair of Cranes, Hammers, Power Presses / Shears and Furnaces • Fitters Common Tools, Appliances and Devices, Handling Facilities and Measuring Instruments

●·Typical Manufacture and Machining Norms ●·**Part II: Planned Maintenance** : Preventive Maintenance Planning ●·Repair Cycle ●·Repair Complexity Maintenance Stages ●·Machine Kinematics and Schedule of Complete Overhaul ●·Pert in Maintenance ●·Lubrication and Lubricants ●·Materials and Standard Spares Planning ●·Spare Parts Stock Pate Accuracy and Technological Test Charts ●·Organisation of Maintenance Department ●·Depreciation and Machine Life ●·Appendices ●·Index

Code : 10 021   3rd Edn. 2002   ISBN : 81-219-0168-5

A TEXTBOOK OF REFRIGERATION AND AIR-CONDITIONING
R.S. Khurmi & J.K. Gupta

CONTENTS: Introduction ●·Air Refrigeration Cycles ●·Air Refrigeration Systems ●·Simple Vapour Compression Systems ●·Compound Vapour Compression Systems ●·Multi-evaporator and Compressor Systems ●·Vapour Absorption Refrigeration Systems ●·Refrigerants ●·Refrigerant Compressors ●·Condensers ●·Evaporators ●·Expansion Devices ●·Food Preservation ●·Low Temperature Refrigeration (Cryogenics) ●·Steam-Jet Refrigeration system ●·Psychrometry ●·Comfort Conditions ●·Air Conditioning Systems ●·Cooling Load Estimation ●·Ducts ●·Fans ●·Refrigeration Tables and Charts ●·Applications of Refrigeration and Air-Conditioning ●·Index

Code : 10 097   3rd Edn. 2005   ISBN : 81-219-0268-1

A TEXTBOOK OF MACHINE DESIGN
R.S. Khurmi & J.K. Gupta

CONTENTS: Introduction ●·Engineering Materials and their Properties ●·Manufacturing Considerations in Machine Design ●·Simple Stresses in Machine Parts ●·Torsional and Bending Stresses in Machine Parts ●·Variable Stresses in Machine Parts ●·Pressue Vessels ●·Pipes and Pipe Joints ●·Riveted Joints ●·Welded Joints ●·Screwed Joints ●·Cotter and Knuckle Joints ●·Keys and Coupling ●·Screwed Joints ●·Levers ●·Columns and Struts ●·Power Screws ●·Flat belt Drives ●·Flat Belt Pulleys ●·V-belt and Rope Drives ●·Chain Drives ●·Flywheel ●·Springs ●·Clutches ●·Brakes ●·Sliding Contact Bearings ●·Rolling Contact Bearings ●·Spur Gears ●·Helical Gears ●·Bevel Gears ●·Worm Gears ●·Internal Combustion Engine Parts ●·Index

Code : 10 012   13th Edn. 2005   ISBN : 81-219-0501-X

MECHANICAL ENGINEERING
(Objective Type)
R.S. Khurmi & J.K. Gupta

CONTENTS: ● Engineering Mechanics ● Strength of Materials ● Hydraulics and Fluid Mechanics ● Hydraulic Machines ●Thermodynamics ● Steam Boilers and Engines ● Steam Nozzles and Turbines ● I.C. Engines and Nuclear Power Plants ● Compressors, Gas Dynamics and Gas Turbines ● Heat Transfer, Refrigeration and Air Conditioning ● Theory of Machines ● Machine Design ● Engineering Materials ● Workshop Technology ● Production Engineering ● Industrial Engineering and Production Management ● Automobile Engineering ● Index

Code : 10 042   5th Rev. Edn. 2004   ISBN : 81-219-0628-8

A TEXTBOOK OF ENGINEERING DRAWING
(Geometrical Drawing)
R.K. Dhawan

CONTENTS: Section I: Introduction & Drawing Instruments ●·Layout of Drawing Sheet ●·Conventions ●·Lettering ●·Dimensioning ●·Scales ●·Geometrical Constructions ●·**Section II:** Loci of Points ●·Conic Sections ●·Plane and Space Curves ●·**Section III:** Theory of Projection & Orthographic Projection ●·Orthographic Reading or Interpretation of Views ●·Identification of Surfaces ●·Missing Lines & Views ●·Sectional Views ●·Isometric Projections ●·Auxiliary Views ●·Freehand Sketching ●·**Section IV: Projections of Points** ●·Projections & Traces of Straight Lines ●·Projections of Planes ●·Projections of Solids ●·Sections of Solids ●·Intersection of Surfaces ●·Development of Surfaces ●·Perspective Projections.

Code : 10 147   2nd Edn. 2005   ISBN : 81-219-1431-0

INDUSTRIAL MAINTENANCE MANAGEMENT
S.K. Srivastava

CONTENTS: Introduction and Objectives ●·Quality, Reliability and Maintainability (QRM) ●·Maintenance Jobs and Technologies ●·Defect List Generation & Defect/Failure Analysis ●·Maintenance Types/ Systems ●·Condition Monitoring (Equipment Health Monitoring) ●·Maintenance Planning and Scheduling ●·Systematic Maintenance ●·Computer Managed Maintenance System (CMMS) ●·Total Productive Maintenance (TPM) ●·Other Concepts of Maintenance Types/Systems ●·Maintenance Organisation ●·Maintenance Effectiveness & Performance Evaluation/Audit ●·Maintenance Budgeting, Costing and Cost-Control ●·Training (HRD) of Maintenance Personnel

●·Bibliography ●·Index

Code : 07 300   3rd Edn. 2005   ISBN : 81-219-1663-1

A TEXTBOOK OF WORKSHOP TECHNOLOGY
(Manufacturing Processes)
R.S. Khurmi and J.K. Gupta

CONTENTS: Introduction ●·Industrial Safety ●·Fundamentals of Metals and Aloys ●·Properties, Testing and Inspection of Metals ●·Ferrous Metals and Alloys ●·Non-ferrous Metals and Alloys ●·Heat Treatment of Metals and Alloys ●·Mechanical Working of Metals ●·Carpentry and Joinery ●·Pattern Making ●·Foundry Tools and Equipments ●·Moulding and Core Making ●·Special Casting Processes ● Smithy and Forging ●·Welding ●·Bench Work and Fitting ●·Sheet Metal Work ● Rivets and Screws ●·Limit System and Surface Finish ●·Measuring Instruments and Gauges ●·Quality Control ●·Powder Metallurgy ●·Plastics ●·Metallic and Non-metallic Coatings ●·Pipes and Pipe Fittings ●·Machine Tools (Introduction) ●·Index

Code : 10 155   6th Edn. 2004   ISBN : 81-219-0868-X

THE AUTOMOBILE
Harbans Singh Reyat

CONTENTS: The Automobile ●·Construction and Working of Automobile ●·Suspension System ●·Engine I ● Engine II ●·Engine Lubrication System ●·Engine Cooling System ● Fuels and Fuel Systems ●·Fuel Injection System ● Electrical Systems I ● Electrical Systems II ●·Transmission System ●·Clutch ●·Transmission ●·Drive Shaft and Drive Axle ●·Steering System ●·Brakes ●·Air Conditioning in Automobiles ●·Maintenance of Automobiles ●·Trouble Shooting ●·Appendices ●·Glossary of Technical Terms

Code : 10 019   6th Edn. 2004   ISBN : 81-219-0214-2

INDUSTRIAL ENGINEERING AND PRODUCTION MANAGEMENT
Martand Telsang

CONTENTS: **Part I : Industrial Engineering** ●·Introduction to Industrial Engineering ●·Productivity ●·Work-Study ●·Method Study ●·Work Measurement ●·Value Engineering ●·Plant Location ●·Plant Layout ●·Material Handing ●·Job Evaluation and Merit Rating ●·Wages and Incentives ●·Ergonomics ●·**Part II: Production Management** ●·Introduction to Production/Operations Management ●·New Product Design ●·Demand Forecasting ●·Production Planning and Control ●·Capacity Planning ●·Material Requirement Planning (MRP) ● Process Planning ●·Project Scheduling with CPM and Pert ●·Production Control ●·Inventory Control ●·Production Cost Concepts and Break-even Analysis ●·Maintenance Management ●·Make or Buy Decisions ●·Planning and Control of Batch Production ● **Part III: Advanced Topics in Production Management** ●·Application of Linear Programming Technique in Production Management ● TQM - Concept and Philosophy ●·Business Process Reengineering ●·Group Technology ●·Just in Time (JIT) Manufacturing ●·Operations Strategy ●·Materials Management ●·Project Management ●·Service Management ●·Product and Service Reliability ●·Theory of Constraints (Toc) ●·Advanced Manufacturing Technologies and Systems ●·Supplements ●·Additional Solved Problems ●·Appendices

Code : 10 197   2nd Rev. Edn. 2004   ISBN : 81-219-1773-5

A TEXTBOOK OF MACHINE DRAWING
(In First Angle Projections)
R.K. Dhawan

CONTENTS: **Section I:** Introduction & Drawing Instruments ● Layout of Drawing Sheet ●Conventions ●Lettering ●Dimensioning ●Scales ●·**Section II:** Theory of Projection & Orthographic Projection ●·Orthographic Reading or Interpretation of Views ●·Indentification of Surfaces ●·Missing Lines & Views ●·Sectional Views ●·Isometric Projections ●·Auxiliary Views ●·Freehand Sketching ●·**Section III:** Detail & Assembly Drawings ●·Limits, Fits & Machining Symbols ●·Rivets and Riveted Joints ●·Welding ●·Screw Threads ●·Fastenings ●·Keys Cotters and Joints ●·Shaft Couplings ●·Bearings ●·Brackets ●·Pulleys ●·Pipe Joints ●·Steam Engine Parts ●·I.C. Engine Parts ●·Valves ●·Gears ●·Cams ●·Jigs & Fixtures ●·Miscellaneous Drawings

Code : 10 148   2nd Edn. 2005   ISBN : 81-219-0824-8

MECHANICS
(Dynamics & Statics)
P. Duraipandian, Laxmi Duraipandian
& Muthamizh Jayapragasam

CONTENTS : ● Introduction●Kinematics ●Force● Equilibrium of a Particle●Forces on a Rigid Body ● A Specific Reduction of Forces ● Centre of Mass ● Stability of Equilibrium ● Virtual Work ● Hanging Strings ● Rectilinear

Motion Under Constant Forces ● Work, Energy and Power ● Rectilinear Motion Under Varying Force ● Projectiles ● Impact ● Circular Motion ● Central Orbits ● Moment of Inertia ● Two Dimensional Motion of a Rigid Body ● Theory of Dimensions

Code : 14 199   6th Edn. 2005   ISBN : 81-219-0272-X

INDUSTRIAL AND BUSINESS MANAGEMENT
Martand T. Telsang

CONTENTS : **Part One : Business Environment** ● Introduction To Business ● Forms of Business Organisation ● Business Environment ●Growth, Development and Internationalisation Of Business ● Business Ethics and Social Responsibility **Part Two :** PRINCIPLES OF MANAGEMENT ● Functions of Management ● Evolution and Development of Management Thought ● Planning ● Decision Making ● Organising ● Staffing [Human Resource Management] ● Communication ● Motivation ● Leadership ● Controlling ● **Part Three:** Human Behaviour At Work ● Organisational Behaviour ● Individual Behaviour ● Group Behaviour ● Job Satisfaction●Organisational Change And Development **Part Four :** Functional Areas of Management ● Production Management ● Materials Management ● Marketing Management ● Financial Management ● Financial and Cost Accounting ● Quality Management ● Project Management **Part Five :** Entrepreneurship Development and Small Scale Industries ● Entrepreneurship Development ● Management of Small Scale Industries **Part Six :** Industrial Relations and Safety Management ● Industrial Relations ● Labour Legislations ● Industrial Safety Management **Part Seven :** Industrial Economics ● Managerial Economics ● Engineering Economics **Part Eight :** Special Topics in Management ● Management By Objectives (MBO) ● Management Information System ●·Office Management ● Management of Technology ●·Business Process Re-engineering ● Introduction to Enterprise Resource Planning (ERP) ● Kaizen—Continuous Improvement **Part Nine :** Quantitative Techniques for Managerial Decisions ● Introduction to Operations Research ● Pay-Off Matrix and Decision Trees ●·Games Theory ●·Linear programming Problems ● Assignment Model ● Transportation Model ●·Sequencing Models ● Queueing Models ●·Simulation ●·Replacement Models ●·Break Even Analysis ●·Forecasting Techniques **Part Ten :** Case Studies ●·Case Method ●·Case Problems ●·APPENDICES ●·Selected References

Code : 10 228   1st Edition: 2001   ISBN : 81-219-2056-6

TRIBOLOGY IN INDUSTRIES
Sushil Kumar Srivastava

CONTENTS: ● Tribology vis-A-vis Terotechnology ● Friction ● Wear ● Lubrication ●·Behaviour of Tribological Components ● Other Tribological Measures ● Tribo-Technical Systems ●Tribological Monitoring of Equipment's Condition ● Tribology in Metal Working Processes ● Tribology in Steel Industries ● Tribology in Mining Industries ● Tribology in Paper and Pulp Industries ● Tribology in Glass Fiber Industry ● Tribology in Transport Sector ● Bibliography ● Index

Code : 10 209   1st Edition : 2001   ISBN : 81-219-2045-0

PRODUCTION MANAGEMENT
Martand T. Telsang

CONTENTS: ● Introduction to Production and Operations Management ● Product Design and Process Selection ● Managing the Supply Chain ● Quality Management ● Design of Facilities and Jobs ● Supplement

Code : 07 389   1st Edition : 2005   ISBN : 81-219-2458-8

A TEXT BOOK OF MACHINE TOOLS AND TOOL DESIGN
P. C. Sharma

CONTENTS: ● Theory of Metal Cutting ● Cutting Tool Materials and Cutting Fluids ● Machine Tools ● Machining Variables and Related Relations ● Tracer Controlled Machine Tools ● Gear Manufacturing ● Tool Layout for Capstans and Turrets ● Tool Layout for Automatics ● Jigs and Fixtures ● Press Tool Design ● Appendix-I ● Appendix -II ● Index

Code : 10 285   1st Edition : 2004   ISBN : 81-219-2362-X